A TEXTBOOK OF ENVIRONMENTAL STUDIES

(FOR UNDERGRADUATE STUDENTS)

AS PER UGC MODEL CURRICULUM

Dr. D. K. ASTHANA
M.Sc., Ph.D.
Associate Professor,
Department of Botany,
D.A.V. College
Kanpur (U. P.)

Dr. MEERA ASTHANA
M.Sc., Ph.D.
Associate Professor,
Department of Zoology,
A.N.D. College
Kanpur (U.P.)

S Chand And Company Limited
(ISO 9001 Certified Company)
RAM NAGAR, NEW DELHI - 110 055

S Chand And Company Limited

(ISO 9001 Certified Company)

Head Office: 7361, RAM NAGAR, QUTAB ROAD, NEW DELHI - 110 055
Phone: 23672080-81-82, 66672000 Fax: 91-11-23677446
www.**schandpublishing.com**; e-mail: **info@schandpublishing.com**

Branches:

Ahmedabad	:	Ph: 27541965, 27542369, ahmedabad@schandpublishing.com
Bengaluru	:	Ph: 22268048, 22354008, bangalore@schandpublishing.com
Bhopal	:	Ph: 4209587, bhopal@schandpublishing.com
Chandigarh	:	Ph: 2625356, 2625546, 4025418, chandigarh@schandpublishing.com
Chennai	:	Ph: 28410027, 28410058, chennai@schandpublishing.com
Coimbatore	:	Ph: 2323620, 4217136, coimbatore@schandpublishing.com (Marketing Office)
Cuttack	:	Ph: 2332580, 2332581, cuttack@schandpublishing.com
Dehradun	:	Ph: 2711101, 2710861, dehradun@schandpublishing.com
Guwahati	:	Ph: 2738811, 2735640, guwahati@schandpublishing.com
Hyderabad	:	Ph: 27550194, 27550195, hyderabad@schandpublishing.com
Jaipur	:	Ph: 2219175, 2219176, jaipur@schandpublishing.com
Jalandhar	:	Ph: 2401630, jalandhar@schandpublishing.com
Kochi	:	Ph: 2809208, 2808207, cochin@schandpublishing.com
Kolkata	:	Ph: 23353914, 23357458, kolkata@schandpublishing.com
Lucknow	:	Ph: 4065646, lucknow@schandpublishing.com
Mumbai	:	Ph: 22690881, 22610885, 22610886, mumbai@schandpublishing.com
Nagpur	:	Ph: 2720523, 2777666, nagpur@schandpublishing.com
Patna	:	Ph: 2300489, 2260011, patna@schandpublishing.com
Pune	:	Ph: 64017298, pune@schandpublishing.com
Raipur	:	Ph: 2443142, raipur@schandpublishing.com (Marketing Office)
Ranchi	:	Ph: 2361178, ranchi@schandpublishing.com
Sahibabad	:	Ph: 2771235, 2771238, delhibr-sahibabad@schandpublishing.com

First Edition 2006
Reprint with corrections 2007, 2009, 2010, Second Edition 2012, Reprints 2013, 2014, 2018 (Twice)
Reprint 2019 (Twice)

ISBN : 978-81-219-2764-2 **Code** : 1005C 121

PRINTED IN INDIA

By Vikas Publishing House Pvt. Ltd., Plot 20/4, Site-IV, Industrial Area Sahibabad, Ghaziabad-201010 and Published by S Chand And Company Limited, 7361, Ram Nagar, New Delhi-110 055.

PREFACE TO THE SECOND EDITION

The second edition of the ***Textbook of Environmental Studies*** provides a comprehensive review of the state of natural resources, different aspect of the problem of environmental degradation and degeneration of biodiversity. Nearly thirty-five years following the Stockholm conference in 1972, have yielded little fruitful results. Depletion and degeneration of natural resources, environmental degradation and degeneration of biodiversity continue. The threat of global climatic change haunts us. As human number multiplies per capita availability of resources decline. With shrinking freshwater supplies, degraded cropland, and an eroded gene-pool, we may not be able to feed the ever growing human number in times to come.

We hope that this edition of the book shall present a comprehensive and complete resume of the state of our environment and natural resources and shall be useful for the students and anyone who is interested in diverse aspects of the problems associated with environment, wildlife and natural resources. A thorough understanding of problems facing the mankind among young impressionable minds, the students at undergraduate and post graduate levels will be of enormous help to redress the problems as the movements which start at grass-root level often acquire tremendous momentum and soon transform into effective action.

AUTHORS

PREFACE TO THE FIRST EDITION

Life started nearly three and a half billion years ago in a hot humid and reducing environment characterized by total lack of free oxygen. The transition from such an environment to the environment in which we live today has been very slow and gradual. It took nearly three billion years of persistent activity of countless generation of countless number of species which have come and gone to develop an environment in which this oxygen breathing terrestrial life exists.

Today life is again in the process of bringing about another change in the environment. This time, however, the transition is being brought about by a single species, the *Homo sapiens* and is characterized by a marked rapidity. We have raised significantly the concentration of green house gases in the atmosphere, introduced waste materials in the stratosphere and have altered natural systems all over the world to plant crops found cities and establish industries. Our planet is warming up, polar ice caps are melting, the protective ozone shield is thinning out, our soils are degenerating, freshwater is becoming a scarce commodity in many regions of the world and species after species are being pushed extinction every day. The threat of global warming looms large before us. The spectre of global hunger by the close of 21st century haunts. We may not be able to feed the teaming millions, about four billion of whom are expected to be added to the world population, nearly 90% of which shall live in Asia and Africa alone, by the close of twenty first century. This could have grave socio-economic and political consequences. We have done all this within a span of about two and half hundred years. We have learnt a lot, unraveled the mysteries of life, conquered diseases, epidemics and even death but have not learnt to live in harmony with nature. We are dismantling, bit by bit the vital life support system which supports the very existence of mankind. Humanity has placed a heavy burden on nature and its resources which grows as human number multiplies.

A fundamental change in our life styles, the way we think, value and behave is required. Life in a viable ecosystem should be the most desirable goal not wealth or consumption of resources. We shall have to move beyond compartmentalization and outmoded patterns to draw the very best of our intellectual and moral resources from every field of endeavour. We shall have to integrate economic development and environmental imperatives to enhance and strengthen the natural resource base so that the legitimate needs of all are fulfilled and capacity of our future generations to fulfill their needs is not undermined.

Many welcome signs are apparent. There has been a rising concern about environmental issues all over the world and protection of wildlife and environment have appeared on the policy agenda of many national and international agencies. World population growth has slowed down. A positive action has already been initiated at several places. But the distressing trends are also there. Reduction in biodiversity and environmental degradation continues. Although world population growth has slowed down nearly 70 million shall still be added to the world population every year until 2100 AD or so. About 90% of this growth shall occur in developing countries, crowding the already crowded regions of the world intensifying thirst, poverty, malnutrition and hunger. Sincere efforts and multipronged strategy are required to halt the damaging trends. Lest, the future could be grim, beset with shortages of all types and life difficult to live through and all this could happen within the life times of our children and grandchildren.

The authors have tried to present a simple, comprehensive, illustrated and well documented account of the state of environment, wildlife and natural resources today. This book covers all aspects of the subject which students of graduate and post graduate classes should be aware of not only for their own sake but for the sake of forging a pattern of right conduct toward the nature, natural resources and the environment.

The authors are grateful to all friends and colleagues for their valuable help, constructive criticism and constant encouragement in writing this book. We wish to express our grateful thanks to S. Chand and Company Ltd for publishing this book.

AUTHORS

CONTENTS

PART I

The Life Support System on our Planet

Part II

Natural Resources

Part – III
Biodiversity

Part V
Social Issues and the Environment

PART – I

The Life Support System on our Planet

1

Chapter

The Environment: Introduction

Everything which surrounds us may be referred to as the environment and the study which covers different aspects of environment, its structure, functions, its quality and the maintenance of its quality including conservation of its living and non-living constituents can be collectively referred to as "Environmental science" or "Environmental studies". The air, soil, water, all living and non-living things around us constitute the environment, which influences our lives. It is from the environment surrounding us that we get food to eat, water to drink, air to breath and all necessities of our daily lives. The environment around us constitutes a "**life support system**".

The concern for environment is an expression of a fundamental change in human perception of nature, natural resources and wildlife on our planet. The traditional concept, that natural resources are abundant for man to use or abuse, has been responsible for massive degeneration of nature, natural systems, environment and wildlife. Departing from the traditional perception of human dominance over nature, a more realistic view that man is just a species among millions of species and his well being is intimately linked to well being of all other species has now emerged. Man cannot survive alone and aloof from other living beings. The natural systems in which man exists along with all other species must be maintained in a healthy and functional state.

(I) MULTI-DISCIPLINARY NATURE OF ENVIRONMENTAL STUDIES

Existence and behaviour of living and non-living constituent of the environment are covered by laws of physical, chemical, geological and biological sciences. Human behaviour and management of human societies are covered by psychological, political and social sciences. Rules framed for maintenance of environment in a healthy state come under jurisdiction of national legislation while agreements reached regarding common problems faced by two or more than two countries come under purview of international law. Different aspects linked with sciences listed above have to be taken into consideration for any study involving environment. The science of environment is, therefore, a multidisciplinary science, which may require attention of experts from different branches of science when decisions regarding environmental matters have to be taken.

(II) IMPORTANCE OF ENVIRONMENTAL STUDIES AND THE NECESSITY OF PUBLIC AWARENESS

Man has long effected his local environment but it is in the 20th century only, particularly in the last fifty years that the scope of his influence has expanded to a global scale. Today we affect earth systems significantly extracting materials, using energy and emitting pollution in our quest to provide food, shelter and a host of other products for the world's growing population. Over-exploitation of natural resources and pollution of environment are corroding the vital life support systems on which all life depends for its subsistence. As natural systems degenerate, it will be difficult to maintain productivity of our agriculture and obtain necessities of our day-to-day life. Chemically altered environment shall make our lives more and more difficult. This is not a healthy sign. We have to reverse the damaging trends. The future of entire humanity is at stake.

Much improvement in environmental quality can be achieved by individual life style decisions or by the action of local bodies. For example, couples may decide to have only two children and thus help in population control. Individuals can use energy more efficiently. They can use a bicycle instead a car thus saving a little petrol. The few drops saved by everyone shall make a huge quantity – a little less Carbon dioxide shall go into the atmospheric air. The few grains saved by each of us shall add up to make surpluses, which will lower the prices enabling the poorest to afford it.

For issues concerning environment not only attention but also active co-operation of every one, at every level of social organization, scientists, educationists, social scientists, politicians, and administrators is needed. Individuals collectively make a society or a state. Movements, which begin at grass-root levels, affect the ideologies and policies of a country or the nation as a whole more effectively than the policies introduced from top downwards. Social and economic changes start with individuals. Changing public opinion can bring changes in government policies, which transform into action later. Many of the environmental problems are simply there because so many people around the world contribute little bits and pieces to it. Likewise a little effort on the part of each individual shall add up to produce significant improvement in conditions of the environment.

(III) COMPONENTS OF THE ENVIRONMENT

The Earth has an outer solid crust of silicate rocks, a highly viscous mantle, a liquid outer core that is much less viscous than the mantle, and a solid inner core. Many of the rocks now making up Earth's crust formed less than 100 million years ago, however, the oldest known mineral grains are 4.4 billion years old, indicating that the Earth has had a solid crust for at least 4.4 billion years. This crust ranges from 5 to 70 km in thickness. The crust is thinner under oceans where it is largely made up of dense silicates of iron and magnesium. The thicker parts of the crust constitute the continents of our planet and are made up of a little less dense rocks of sodium, potassium and aluminium silicates. The abundance of water on Earth is a unique feature that distinguishes our Planet from others in the solar system. Earth's solar orbit, volcanism, gravity, green house effect, magnetic field and oxygen-rich atmosphere make Earth a unique place for life to flourish. Earth and its environment have the following three basic components which are:

Table 1.1: *Composition of Earth's atmosphere*

	Gases	Concentration*	Percentage
1	Nitrogen	780840.00	78.08
2	Oxygen	209460.00	20.95
3	Argon	9340.00	00.93
4	Carbon dioxide	383.00	00.038
5	Neon	18.00	00.0018
6	Helium	5.20	00.00052
7	Methane	1.75	00.00017
8.	Krypton	1.14	00.00010
9	Hydrogen	0.3	00.00005
10	Nitrous oxide	0.50	00.00005
11	Xenon	0.09	00.000009
12	Ozone	0.07	00.000007

*As parts per million. *Source*: NASA Earth Fact sheets 2007.

1. The atmosphere or air.
2. The hydrosphere or water
3. The lithosphere or the rocks and the soil.

Approximately 70.8 % of Earth's surface is covered by water and constitutes what can be referred to as the ***Hydrosphere***. The rest (only 29.2%) is landmass which makes up the five continents of the planet, with its rocks and the soil and can be referred to as the ***Lithosphere***. The average depth of the oceans is 3,794 m (12,447 ft), more than five times the average height of the continents. The hydrosphere and the lithosphere are covered by an envelope of gases which is referred to as the ***Atmosphere.***

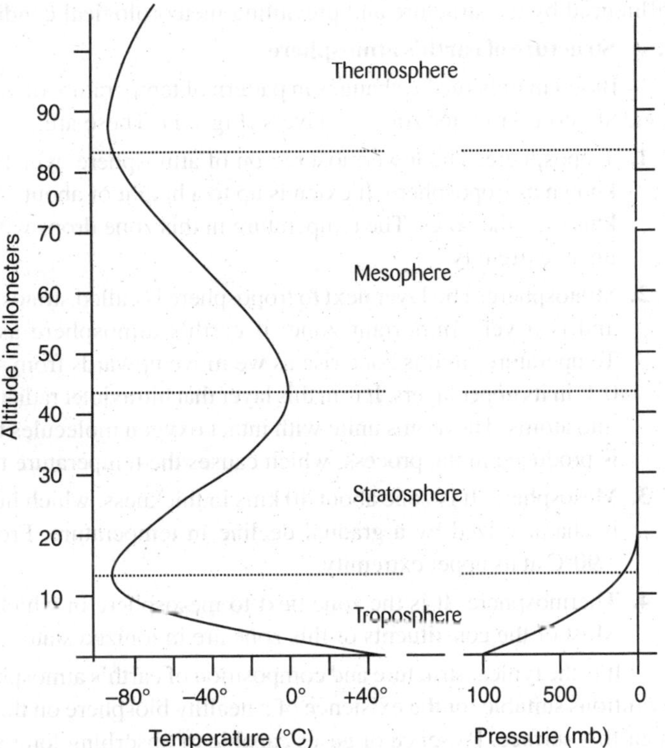

Fig 1.1 *The structure of arth's atmosphere*

(1) The atmosphere or air

The mantle of gases, which surrounds our planet, is referred to as "atmosphere". It is a complex mixture of gases, water vapours and a variety of fine particulate material (Table 1.1). It consists of about 5.15 x 10^{15} metric ton of gases, which exert a pressure of about 1 kg per sq cm. on earth' surface (1). Most of these gases are compressed in the lowermost layer of the atmosphere. Pressure decreases as we move upward. The envelope of gases, which surrounded our planet when it was originally formed, was notably different from the atmosphere of today. It consisted largely of Hydrogen and Helium. This atmosphere, often referred to as the "**First atmosphere**" was lost early when earth's surface was not solidified. About 4.4 billion years ago when our planet had cooled enough to develop a crust, earth's surface possessed numerous volcanoes, cracks and crevices which released steam, carbon dioxide, and ammonia. As these gases collected around our planet another atmosphere entirely different from the first atmosphere was developed. This "**Second atmosphere**" had nearly 100 times as much gas as the current atmosphere, but as it cooled much of the carbon dioxide dissolved in sea-water and precipitated out as carbonates. Finally, this atmosphere was composed largely of water vapours, nitrogen, carbon dioxide, methane, traces of ammonia and other reducing gases. There was no oxygen in the atmosphere. It is believed that the greenhouse effect which the high concentration of carbon dioxide and methane caused prevented our planet from freezing into a ball of ice. It was in such a hot humid and reducing atmosphere that the primitive life first appeared on our planet. The present day oxygen-rich, "**Third Atmosphere**" was developed by the photosynthetic activity of green plants the most primitive of which appeared about three billion years ago.

About 90% of man's total daily intake of materials – food, water and air – is contributed by air. An average human being breathes about 22,000 times a day and inhales about 16 kg of oxygen. The pollution of air, therefore, may have a profound influence on living organisms. Unfortunately, an alarming quantity of gases, particulate material, fumes, vapours and smoke is discharged daily into the atmosphere. Dilution, dissemination, transformation and clearance of the pollutants are strongly influenced by the structure and prevailing meteorological conditions.

1. Structure of earth's atmosphere

Based mainly on the changes in pattern of temperature of different layers, the atmosphere can be divided into four major zones or layers (Fig 1.1). These are:

1. Troposphere: The lowermost region of atmosphere, which is in contact with earth's surface, is known as troposphere. It extends up to a height of about 20 kms above the equator and about 8 kms over the poles. The temperature in this zone drops with height being as low as -80°C at its upper extremity.
2. Stratosphere: The layer next to troposphere is called stratosphere. It is about 30 kms in thickness and is a very important zone of earth's atmosphere as it contains the vital ozone layer. Temperatures in this zone rise as we move upwards from -80°C at its lower extremity to about 0°C in its upper layers. It is in this layer that ultraviolet radiations from sun split oxygen molecules into atoms. The atoms unite with intact oxygen molecules to give rise to ozone molecules. Heat is produced in the process, which causes the temperature to rise with height.
3. Mesosphere: It is zone about 40 kms in thickness, which lies next to the stratosphere. This zone is characterized by a gradual decline in temperature. From 0°C in its lower layers to about – 90°C at its upper extremity.
4. Thermosphere: It is the zone next to mesosphere in which temperature increases with height. Most of the constituents of this zone are in ionized state.

It is the typical structure and composition of earth's atmosphere which is responsible for creating conditions suitable for the existence of a healthy biosphere on this planet. It regulates the temperature of earth's surface. Presence of gases capable of absorbing long wavelength radiations is responsible for maintaining the temperatures under which life activities are possible. Harmful ultraviolet radiations, which could severely damage the terrestrial life, are absorbed high above earth's surface, in the stratosphere. A constant mixing of the contents of atmosphere occurs due to air currents and vertical temperature gradient which prevents accumulation of harmful gases and vapours at any particular spot. Atmosphere is therefore a quick and effective media for transport and dissemination of gaseous wastes. As troposphere derives its heat from earth's surface, warm air being lighter rises and cools down adiabatically. Cooling condenses water vapours. The entire load of pollutants is brought down with snow or dew or rains which cleanses the atmosphere (2).

2. Meteorological conditions and air circulation

Man introduces a large amount of gaseous wastes and fine particulate material into the atmosphere. Meteorological conditions play an important role in the dilution and dissemination these wastes in the atmosphere. Even in badly polluted localities there are times when the atmosphere appears quite clean. These fluctuations are not caused by gross changes in the emission of pollutant but are due to the variations in meteorological conditions of the locality such as temperature pressure humidity etc. (3;4).

1. Dilution of pollutants in the atmosphere: It is actual temperature gradient in the troposphere which determines the extent of mixing or dilution of gaseous material introduced into the atmosphere. All gases when transferred from high pressure to a zone of lower pressure expand and undergo cooling. Similarly contraction and warming occurs if they are transferred from a zone of lower pressure to higher pressures. As the atmospheric pressure decreases with height, movement of a parcel of the introduced gases upwards causes its expansion and thereby lowering of its temperature. Likewise displacement downwards results in its contraction and warming. This happens adiabatically, i.e., no heat is either withdrawn or provided to the system. The decline or lapse in temperature of a dry gas is 1°C for every 100 metres of height covered and is known as *Dry adiabatic lapse rate* (DALR). It is a theoretical value calculated for the pressure difference which exists between the altitudes 100 metres apart. Any gaseous pollutant tends to expand and cool down by 1°C if it is moved 100 metres upward from its original position (5).

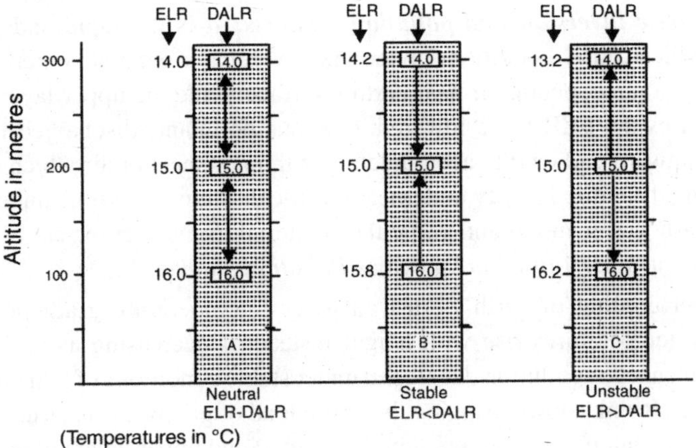

Fig 1.2 *Netural, stable and unstable*

Much of earth's atmosphere is compressed into a relatively thin shell of dense air around the globe. Higher up the air is thinner and rarer. Solar radiations pass through atmosphere and heat earth's surface. It is from earth's crust that air derives most of its heat. Lower layers, therefore are warmer that upper layers. As warm air rises, lower pressures above cause it to expand and cool down. This brings it back towards earth's crust to be heated again and an active churning of the contents of atmosphere is brought about. The actual decline in temperatures with increasing altitude or the temperature gradient in the atmosphere is largely determined by meteorological conditions of the locality. It is referred to as the *Environmental Lapse rate* (ELR). It is the actual decline in the temperature per 100 metres of height covered. Under dull and cloudy weather earth's surface may cool down rapidly cooling the air nearly. As upper layers are still at the same temperatures, the ELR is lowered than the DALR. Under conditions of bright sunshine the ELR may become significantly higher than the DALR. A parcel of introduced gas on rising in the atmosphere may find different temperatures at different heights while its own temperatures vary at a steady rate of 1°C for an upward or downward displacement of 100 metres. Following three situations may occur:

A. Neutral atmosphere: In cases where ELR is equal to DALR, vertical movement of mass of air, either upward or downward, shall cause no variations in its temperature and the density with respect to the surrounding air. The atmosphere, therefore, shall neither resist nor promote any mixing (Fig 1.2 A).

B. Stable atmosphere: The term stable atmosphere is used to denote the set of conditions under which vertical mixing of pollutants is actively resisted. This happens when ELR is smaller than DALR. Any displacement of a mass of air upward shall cool it faster than the surrounding air. Being cooler and therefore, heavier, it shall show a tendency to move back to its original position. Likewise any movement downward shall make it warmer and therefore, lighter than the surrounding air. It shall have the tendency to move up to regain its earlier position. Thus any vertical movement shall be actively resisted and no mixing shall occur (Fig 1.2 B).

C. Unstable atmosphere: It denotes a condition of the atmosphere which promotes rapid mixing of contents of the air. This happens when ELR is greater than DALR. Any vertical movement upward of a mass of air shall cool it but the cooling shall not as much as the cooling of the surrounding air. The mass of air shall be warmer and hence lighter than air around it. As a consequence it shall have a tendency to rise up on its own. Similarly a movement of the air mass downwards shall result in its being heavier than the surrounding medium. It shall show a tendency to sink down on its own. The atmosphere will promote active mixing of its contents or the materials released in it. The mixing shall occur in a thick layer of air and the pollutants shall be subjected to greater dilution (Fig 1.2 C).

*2. **Atmospheric inversion and pollution blankets:*** It is the rapid and prolonged cooling of earth's surface which causes conditions under which vertical mixing of contents of the atmosphere is resisted. Cooling of atmospheric air near earth's surface while the upper layers are still at the same temperatures causes the ELR to be lowered. Gaseous pollutants discharged in the atmosphere are unable to move upward and tend to accumulate in a thinner sheet of air which expands in transverse direction covering the entire locality underneath. Little dilution is possible under such conditions and the blanket of waste gases, fumes and particulate material forms over the earth's surface. The canopy of gaseous pollutants thus formed in known as *Pollution blanket*.

With further cooling of earth's crust an inverse temperature gradient is established in the atmosphere. The temperatures rise with height instead of decreasing as is the case under normal conditions. The blanket of pollution developed under stable conditions of atmosphere earlier becomes cooler and heavier. It sinks down bringing the entire load of gaseous wastes to earth's surface. Such a condition under which temperatures gradient is inverted and upper layers of atmosphere tend to sink back towards earth's surface is referred to as *atmospheric inversion.* In fact atmospheric inversion is the main cause of most of the air pollution episodes so far recorded in the world (5).

(2) THE HYDROSPHERE OR WATER

Water is a natural resource of fundamental importance. It is absolutely essential for life. In fact, it is the abundance and unique characteristics water which has enabled life to be developed on this planet. The atomic structure and properties of water seem to be especially designed for the biosphere.

1. Structure of water molecules

A water molecule consists of two atoms of hydrogen and an atom of oxygen. Two hydrogen atoms share their only electrons with those of the oxygen atom. The shared electrons are attracted more towards the oxygen nuclei, which have a greater positive charge. This result in their asymmetric distribution and oxygen atoms gets a slight negative charge while hydrogen atoms a little positive charge (6). The molecule acquires a polarity, which causes it to form aggregates, a structure which is referred to Lattice structure (Fig. 1.3). It is these aggregates, which are responsible for the cohesive nature and a number of unique properties of water.

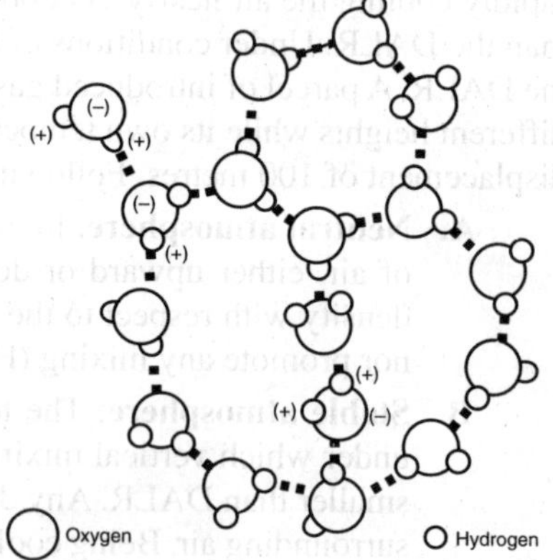

Fig 1.3 *The lattice structure of water moleclas*

2. Unique properties of water:

The unique properties of water can be summarized as follows:

1. Water has maximum density at 4°C. When it is cooled below 4°C it expands and freezes at 0°C, so that ice is lighter than water.
2. Water has an extraordinary high specific heat. Variations in its temperature require large amounts of heat energy or its withdrawal as compared to other liquids.
3. As water-molecules are bound together by hydrogen bonds much more energy is required to evaporate it as compared to other liquids. Hydrogen bonds have to be broken for evaporation, which requires additional energy.
4. Water is characterized by the highest surface tension among all other liquids. This causes an extraordinary amount of water to be retained in the soils due to capillary action.

Life evolved under water. Life cannot occur without it. Water is the very medium in which all biochemical reactions within a living organism and a large number of chemical reactions involving components of rocks, soils and pollutants of the environment occur. It is the availability of water, which determines the nature composition, and abundance of terrestrial life. Regions with very low rainfall become deserts. Lush green vegetation develops where water is in abundance. The specific heat of water is 1, which is higher than any other liquid. Due to which changes in its temperature need large quantities of heat energy or its withdrawal as compared to other objects. This causes temperature differences between land and sea when heated by solar radiations and an active air circulation is maintained. The air circulation determines precipitation pattern and climatic conditions of a locality (4;7;8).

A high latent heat of vapourization causes a large amount of solar energy to be used in evaporation of water. The solar heat could have, otherwise, raised the global temperatures significantly. Our globe receives more intense solar radiations at the equator than those at temperate and sub-polar regions. Differential heating of waters coupled with rotation of our planet generates major streams of cold and heated water in our oceans, which moderate the climate of a locality. Water vapours effectively absorb long wavelength radiations as CO_2 does. A large amount of heat is trapped by water vapours present in the atmosphere, which plays an important part in regulating the temperature of earth's crust. The unique property of expansion when cooled below 4°C causes water to freeze from top downwards as most of the aquatic bodies lose heat from their surface. Aquatic life stays safe under the ice-sheet (9).

It is because of extra-ordinary high surface tension that a considerable amount of water is retained in the soil due to capillary action. It is on this moisture that green plants thrive during drier periods. Water is an efficient means of transfer and transport of materials dissolved or suspended in it. Low-lying regions of the world, river basins and land along sea coasts are, therefore, much more productive than upland areas. Water transports dissolved materials silt debris, nutrients and pollutants etc. from upland areas to low lying regions. Nearly 90% of world population is concentrated in these areas. Water plays an important role in reducing atmospheres burden of gaseous pollutants. Water vapours condense around fine particles suspended in the air in which gaseous materials dissolve and the entire load of waste gases and vapours is brought down with rains or snow. This cleans the atmospheric air. Water plays an important role in weathering of all types of rocks and the formation of soils (10).

(3) LITHOSPHERE OR ROCKS AND THE SOIL

Earth's crust consists of rocks and the loose material derived from these rocks, most of which is modified into soil. Like a precious carpet of lose material spread over the bed rocks, it is our soils which sustain life by providing nutrients to the green plants which in turn support all life on earth's surface. Like a sponge it stores up and provides passage to ground water, shelter to microbes and burrowing animals and supporting foundations to buildings and other utility structures. There are about

Table 1.2 *Elementary composition of earth's crust*

	Element	%		Element	%
1	Oxygen	49.85	8	Magnesium	2.11
2	Silicon	26.03	9	Hydrogen	0.97
3	Aluminium	7.28	10	Titanium	0.41
4	Iron	4.12	11	Chlorine	0.20
5	Calcium	3.18	12	Carbon	0.19
6	Sodium	2.33	13	Others	1.00
7	Potassium	2.33			

a dozen elements, which constitute the bulk of earth's crust – nearly 99% by weight (Table 1.2). These elements go into the formation of minerals, which in turn form rocks from which all soils are derived.

1. Minerals and rocks in earth's crust:

Parent rocks in earth's crust are a complex aggregate of minerals whose properties determine the physical and chemical nature of the rocks, which they form. Though there are more than two thousand minerals known, almost 90% of earth's surface is made up of only 10-12 mineral. About 58% of the mass of terrestrial rocks is constituted by Feldspars, 16% Pyroxenes, 13% by Quartz and about 4% by Mica. These four mineral groups make up well over 90% of the rocks. The remaining 10% of the bulk is made up of other minerals (11).

Disintegration of rocks results into the formation of gravel, sand, silt and clays. Burial, compaction and hardening of the products of disintegration of rocks result in the formation sedimentary rocks. When sedimentary rocks are subjected to high temperatures and pressures, metamorphic rocks are produced. Igneous rocks also transform themselves into metamorphic rocks when subjected to intense heat and pressures. Melting, upward movement and subsequent cooling form igneous rocks again. The disintegration and formation of different types of rock is a cyclic process which operates on geological time scale. Clay, silt, sand and gravel are just intermediates in the cycle – the materials from which soil formation occurs (12). Rocks which occur in earth's crust are basically of three types:

1. **Igneous Rocks:** Igneous rocks are formed by cooling and solidification of molten rock material called Magma. Some of the common igneous rocks are Diorite and Basalts.
2. **Sedimentary Rocks:** Sedimentary rocks are those rocks which develop as a result of gradual accumulation, consolidation and hardening of products of weathering of mineral particles brought about wind, water or biological agencies. These rocks are characterized by presence of distinct sedimentary layers. Some of the important sedimentary rocks are: Shale, Sandstone, Limestone etc.
3. **Metamorphic Rocks:** Metamorphic rocks are rocks formed as a result of metamorphosis of igneous and sedimentary rocks. This transformation is brought about by intense heat and high pressures. Important metamorphic rocks are: Schist, Slate, Marble, Quartzite etc.

2. Disintegration of rocks and formation of soil

Earth's crust is subjected to continuous battering of wind, rains and changes of temperature, which decompose rocks into small fragments. The modification of earth's crust as a result of its interaction with atmosphere and the hydrosphere is called weathering, the final products of which are gravel sand, silt and clay. Three main type of weathering can be distinguished on the basis of the nature of agencies, which cause it.

1. Physical Weathering: Physical weathering of rocks is a mechanical process, which is brought about by changes in temperature, water and wind. Differential thermal expansion and contraction due to changes in temperature produces internal tensions in the body of rocks, which develop cracks, and splits apart. Freezing waters in crack and fissures expand and create intense pressure much more than the tensile strength of rocks, which tear rock fragments apart. Abrasions produced by wind, rapidly flowing waters and glaciers also cause slow disintegration of rocks.
2. Chemical weathering: Chemical weathering of rocks involves relatively slow and simple chemical reaction like dissolution, hydration, hydrolysis, carbonation, oxidation and reduction etc. Principal agents of this type of weathering are water, air oxygen, carbon dioxide and other materials carried by these agencies, which act and bring about gradual transformation of rock fragments into gravel sand silt and clay. Mineral constituents of the rocks may also be altered in the process.

3. Biological Weathering: A number of organisms play an important part in weathering of rocks. Lichens, bacteria, a number of fungi and algae etc. in presence of plenty of moisture, produce metabolites, which aid in the process of disintegration of rocks. Roots of plants growing in cracks and fissures create considerable pressures, which tear the rocks fragments apart.

As a result of weathering, rocks are broken down into smaller particles amidst which numerous bacteria, algae, fungi, lichen mosses and other small organisms grow. These slowly contribute organic matter when they die and decay. Blue green algae and bacteria fix atmospheric nitrogen and with other microbes solubilize nutrients from soil particles. With time, various other types of plant and animals appear which add more and more organic matter. Different types of bacteria, fungi, actinomycetes, etc. decompose various kinds of polysaccharides, proteins, fats, celluloses, lignins, pectins, waxes, resins and their derivatives. The residual, incompletely decomposed organic matter left after the microbial action is called **humus**. The process of organic enrichment, the decomposition of organic matter and the accumulation of humus are referred to as **humification**. The generation of plant nutrients from dead and decaying organic matter is called mineralization. Weathering, mineralization and humification the three processes taken collectively are referred to as **pedogenesis** or the soil formation.

Decaying plant roots leave a network of channels in the soil. Worms bore through the soil leaving their secretions and excreta to be decomposed by other microbes. This improves porosity of the soil and its water and nutrient retention capacity. It is the microbial life of the soil, which is responsible for development, as well as the natural capacity of nutrient regeneration of the soils on which much of world agriculture depends (13). Decomposition and mineralization of organic matter and regeneration of nutrients is another very important role of a fertile soil. Soils contain organotrophic bacteria, nitrifying bacteria, denitrifying bacteria, pigmented microbes, nitrogen fixing organisms and a number of fungi and protozoans. Activity of these microbes keeps the soil in a fertile state for plants to flourish on it. Waste materials or pollutants discarded on the soil are also effectively decomposed by these microbes. The soil atmosphere and gases dissolved in water percolating through the pores in the soil provide enough oxygen to the microbes to carry on the decomposition of organic matter and mineralization with efficiency.

Fertile soil is a natural resource of fundamental importance. Most of our necessities are fulfilled by the communities of plants and animals, which develop on it. Soil is a solid medium underneath which provides mechanical support to the plants. Plants and animals including man live on it. Soils serve as a huge sponge. Its porosity and the moisture-holding capacity enable it to soak up large quantities of rainwater or the water flowing over or through it. In drier periods, subsurface strata of soils serve as a reservoir of water on which a number of living organisms including man depend for their water requirement. Even when land surface is completely dry, plants can obtain water through their roots, which go deep into the soil. Ion-exchange capacity of the soil is responsible for retaining micro- and macronutrients in the soil for long durations. The colloidal component, which consists of clay micelles and humus particles (particles smaller than 0.002 mm in size) tightly, binds a number of ions to their surface. Water carrying useful ions through the soil distributes them evenly. Lower concentration of nutrient ions following uptake by plants causes these ions to be released into the soil solution. Therefore, ion-exchange capacity of the soil ensures even distribution and availability of nutrients as and when required by plants. It is also helpful in preventing excessive leaching of the nutrient ions while maintaining an appropriate pH, which directly influences nutrient availability and the growth of plant, animals and microbes. Major steps of biogeochemical cycles of elements, which constitute the biosphere, occur in soil. Through the microbial machinery which occurs in it, soils regulate composition of air, water and even the biosphere, adding or deleting materials of biological significance. Healthy soils are therefore very important for the well-being of the communities, which stand on it (14;15).

(IV) MODIFICATION OF EARTH'S ENVIRONMENT BY THE BIOSPHERE

The environment in which the first life appeared on this planet was altogether different from the present day environment. For the first 1.0 billion years, earth's surface was virtually barren – inhabitable. Temperatures were high, much of the radioactivity, which was associated with earth's formation, persisted. Ultraviolet radiations lethal to life penetrated right up to earth's surface. Earth's atmosphere was rich in ammonia, carbon dioxide, water vapours, methane and nitrogen. There was no free oxygen in the environment. In brief it was a reducing atmosphere as indicated by the presence of metals in their reduced state (such as ferrous iron), in the oldest rocks of the earth's crust – rocks which were formed during the first one or two billion years of earth's existence.

It was in this very hot, humid and reducing atmosphere that life was born under the water surface. Electrical discharges, high-energy ultraviolet rays, radiations emitted by radioactive isotopes etc. were probably helpful in generating a variety of organic molecules. Concentration of these compounds gradually rose in primitive seas of the time, as there was virtually nothing to decompose them. Reducing atmosphere was also helpful in formation of organic molecules as well as in preventing their disintegration. Scientists believe that it was this **'organic soup'**, which gave birth to the primitive life and provided nourishment to it in its early stages. The remains of the earliest life have been preserved in some sedimentary rocks deposited about 3.5 billion years ago. These rocks are referred to as **Stromatolites** (or layered rocks). They contain organically preserved micro-fossils resembling the present day bacteria and blue green algae (16). Many of these organisms were capable carrying out biological nitrogen fixation, which enriched the ancient waters with compounds of nitrogen. Some were capable of carrying out photosynthesis but were unable to use water as electron and hydrogen source (anoxygenic photosynthesis). They did not produce oxygen as oxygen comes from splitting of water molecules. They used H_2S, H_2 and organic compounds as hydrogen source.

It was about 3.0 billion years ago that some of the ancestors of these bacteria developed the capacity to split H_2O molecules to obtain hydrogen atoms and evolve oxygen. It was a very significant step in organic evolution, which gave birth to the first 'plants' of the time – organisms capable of carrying out oxygenic photosynthesis. These 'plants' were similar to present day Cyanobacteria or blue green algae. For about half a billion years these 'plants' gradually multiplied, diversified and proliferated all over the world. The vast track of time, starting from 2.9-3.0 billion to about 0.5 billion years ago, is known as the Age of Cyanobacteria or Cyanophytes. Being the first to carry out oxygenic photosynthesis, while sequestering carbon dioxide in organic molecules, cyanobacteria played a major role in oxygenating the environment. Other photosynthesizing plants would later evolve and continue releasing oxygen and sequestering carbon dioxide. Photolytic dissociation of water under ultraviolet rays might have produced some oxygen, however, it was mainly oxygenic photosynthesis, which has added most of this vital gas to the atmosphere. As the level of oxygen in the atmosphere determines the extent of ozone concentration, in oxygen deficient atmosphere there was no ozone layer to protect the terrestrial life till about half a billion years ago. It was about 450 million years ago that enough oxygen could accumulate in the atmosphere to sustain an effective concentration of ozone which could filter out ultraviolet radiations. This enabled life to colonize the land surface. The critical level of oxygen, in the presence of which ozone layer develops, is about one tenth of the present-day oxygen content of the atmospheric air (17). Over time, excess carbon became locked in biomass, fossil fuels, soils, sedimentary rocks (notably limestone), and animal shells. Today, the biomass accounts for about 3.7 kg carbon per square meter of the earth's surface averaged over land and sea, making a total of about 1900 gigatonnes of carbon

As oxygen was released, it reacted with ammonia to release nitrogen, in addition, bacteria would also convert ammonia into nitrogen. But most of the nitrogen currently present in the atmosphere resulted from sunlight-powered photolysis of ammonia released steadily over the aeons from volcanoes. As more plants appeared, the levels of oxygen increased significantly, while carbon dioxide levels dropped. At first the oxygen combined with various elements (such as iron), but eventually oxygen accumulated in the atmosphere, resulting in mass extinctions of form which could not live in presence of oxygen. With the appearance of an ozone layer, life-forms were better protected from ultraviolet radiation. This oxygen-nitrogen atmosphere is the "third atmosphere". About 200 – 250 million years ago, up to 35% of the atmosphere was oxygen (as found in bubbles of ancient atmosphere embedded in amber). Stromatolites formed during the period, contain metals in oxidized state, in shallow seas almost everywhere forming enormous carbonate-reefs on the margins of islands and other landmasses (18).

For synthesis of life in addition to carbon other elements like nitrogen, phosphorus, sulphur, and micronutrients etc. were also required. Earth's surface had enough of these nutrients but lacked sufficient nitrogen for luxuriant growth of plant and animal life. The process of nitrogen fixation, therefore, should have evolved simultaneously with anoxygenic photosynthesis or even earlier. Bacteria which were already there about 3 billion years ago and blue green algae which evolved as early as 2.9 billion years ago gradually enriched the ecosystems of the time with organic nitrogen. The organic enrichment of earth's surface resulted in formation of soils and dense stands of vegetation were developed. It was, therefore, the result of slow but persistent activity of about three billion years of the biosphere that conditions to suite the requirement of present day oxygen breathing life have evolved on our planet.

QUESTIONS

1. Discuss the structure of earth's atmosphere. Pollution of which part of the atmosphere is responsible for causing Ozone
2. depletion? Enumerate the consequences.
3. What do you understand by Atmospheric inversion and Pollution blankets? Under which conditions active mixing of waste
   gases discharged in the atmosphere is actively resisted?
4. Discuss the unique properties of water molecule and its role in shaping the biosphere.
5. Discribe the rock cycle, pedogenesis and the role soils in shaping the biosphere.
6. Discuss the modification brought about in the earth's environment by the biosphere.

2

Chapter

Ecosystem and Ecosystem Services

The existence of living organisms depends on a system of complex interactions within and in-between the components of the environment – the lithosphere, hydrosphere, atmosphere and the biosphere itself. It is these interactions, which satisfy the needs of all living organisms such as food, shelter, water, oxygen to respire, mates to reproduce etc – essential for sustained life on this planet. The complex system, in which interactions between the different components of the environment occur, is referred to as an ecosystem. To be precise *any spatial or organizational unit which includes the living and the non-living constituents interacting with each other and producing an exchange of materials between the two, is termed as an ecosystem* (1;2). A community of plants depends on sun light for energy. Water and mineral nutrients come from soil. Atmosphere provides carbon dioxide. It is on the organic matter produced by green plants that herbivores live, grow and built up animal communities. Carnivores depend for their food supply on the herbivores. Man an omnivorous animal depends both on plant and animals for its food requirements.

(I) STRUCTURE OF AN ECOSYSTEM

An ecosystem consists of living and non living components. The non-living or abiotic components of an ecosystem are mineral nutrients, temperature, light, water and air etc. They shape the living or biotic component. The living component of the ecosystem is made up of populations of different organisms which occur in the system. A large number of individuals belonging to different species which adjust, adopt, interact with each other, share the same general environment and use the same resources form the biotic community of the ecosystem.

Green plants alone are able to trap solar energy. It is stored as chemical energy within chemical bonds of large organic molecules. When organic matter is set on fire, these bonds are suddenly broken. The energy appears as heat energy. Heterotrophs or consumers get their energy by splitting up these bonds in a step-wise manner. It is rather efficiently used in growth, development and other life activities. The transfer of energy from green plants through a series of organisms which consume food-energy and in turn are themselves consumed, constitute the food chain. A nutritive series is formed in which plants are **primary producers** of organic substances which are consumed by herbivorous animals which in turn are preyed upon by carnivorous animals. There may be larger animals which prey upon these carnivores. Thus, nutrition is passed on in gradual steps to the ultimate consumers. The term **trophic level** refers to the parts of this nutritive series in which organisms obtain food in the same general way. Thus, all animals which obtain food or in other words energy by consuming green plants, such as grass-hoppers, rodents, cattle etc, shall be at the same trophic level which is termed as the level of **primary consumers**. All those animals or predators which live on primary consumers or herbivores are said to be at a higher trophic level which is termed as the level of **secondary consumers** and so on. An assemblage of **trophic levels** within an ecosystem is referred to as **trophic structure** or the **food chain** (2).

An ecosystem may have two to six trophic levels through which energy and nutrients flow. In a more simple language, food chains usually have two to six links or group of organisms which derive their nutrition in the same general way. An example of a short food chain would be:

Grasses ⟶ Cattle ⟶ Man

This chain consists of three links only. A four-link terrestrial food chain could be constituted by grasses, as primary producer, grass-hoppers as primary consumer, birds as secondary consumers and hawks as tertiary consumer as shown below.

Grasses ⟶ Grass-hopper ⟶ Birds ⟶ Hawks

In an aquatic system, algae, phytoplankton and other aquatic plants are at the level of primary producers and herbivorous animals, which include crustaceans, insect's larvae and fishes, are primary consumers. There may be present large fishes and other carnivorous animals, which feed on herbivores, constituting the level of secondary consumers. There may also be some large birds which feed on the fishes. Man may eat these birds. Such food chains possess five links or trophic levels.

Alage and other aquatic plants	→	Crustaceans, insect larvae, fishes	→	Larger fishes, other animals	→	Birds which feed on fishes	→	MAN
Primary producer		**Primary consumer**		**Secondary consuner**		**Tertiary consumer**		

Two basic types of food chains can be distinguished in a natural ecosystem. The first type of food chain, which starts from autotrophs or green plants constituting the first trophic level, the herbivores and predators forming the second and third trophic levels respectively, can be referred to as the main food chain or **Grazing food chain.** The second type of food chain which starts with debris and detritus produced by green plants is termed as **Detritus food chain**. Consumers which consume debris, detritus and often partially decomposed animal and plant material form the first link in this food chain. Predators which consume these organisms constitute the second trophic level and so on. A large proportion of biomass of green plant is often diverted to the detritus food chain which forms a very important route of energy flow and transfer of materials in an ecosystem. Even in the most heavily grazed pastures, more than 50% of the primary production is passed on to detritus food chain (2).

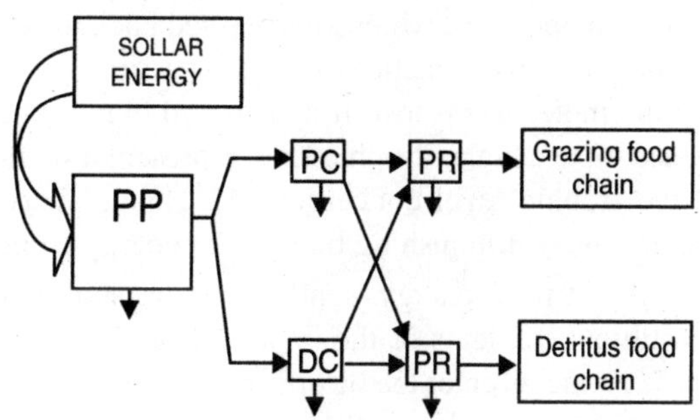

Fig 2.1 *Grazing and Detritus food chain*

(Pc = Primary producers DC = Dertritus consumers, PR = Priduators)

Parasites which derive nutrition from other plants and animals also constitute a link in yet another type of food chain which may be designated as the Parasitic food chain. For example helminthes parasitizing roots of higher plants or other animals may in turn be parasitized by other organisms. The parasitic food chain, thus, commences from any level in a trophic structure and may at times result in heavy losses of energy. It will be apparent from the nature of the three types of food chains that all of them are simply different pathways of flow of the energy and other materials. The entire complex can be integrated into a single trophic structure with several interlinked trophic routes characterizing an ecosystem. When there are several interlinked alternatives at each trophic level the trophic structure becomes very complicated and is referred to as a food-web (2;3).

(II) THE FLOW OF ENERGY AND MATERIALS THROUGH AN ECOSYSTEM

In an ecosystem, the total amount of organic material produced by autotrophs (green plants), gross primary production, meets the following three general fates:

1. It is oxidized by autotrophs themselves during respiration producing CO_2 and H_2O.
2. It is consumed by herbivores and is thus passed on to higher trophic levels.
3. With death of the plants (or its detached parts) their organic remains are passed on detritus food chain.

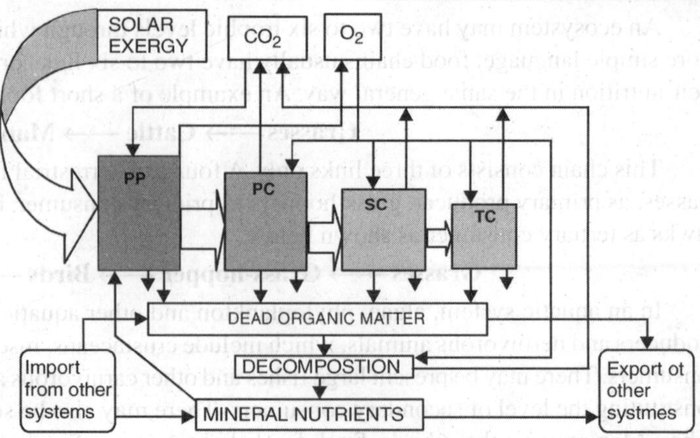

Fig 2.2 *A simplified diagram showing the flow of energy and mineral cycling in a natural ecosystem.*

(PP = Primary producers, PC = Primary consumers,SC = secondary consumers. TC = Tertiarv consumers)

At all trophic levels, organisms are subjected to a similar fate as plants are. Primary consumers and consumers at higher trophic levels also oxidize a part of the organic material they consume in respiration. When a carnivore eats them, the organic matter is passed to a higher trophic level. If they escape predation, the ultimate fate is their death followed by decay and decomposition of their bodies to simple components, which are recycled back into the system. Therefore, at each trophic level, a huge amount of organic matter or in other words energy trapped in various chemical bonds within the organic molecules is lost. It is scattered in highly diluted form in the system. The amount of energy passed on to a higher trophic level represent a small percentage of the total biomass or energy content of the trophic level concerned (4). Hence successive trophic levels in an ecosystem possess a progressively diminishing biomass or energy content as we move up a food chain (Fig 2.2).

Apart from energy, synthesis of organic matter requires mineral nutrients also which are the constituent elements of the organic world. These are taken up from the surrounding environment. These elements enter the biosphere at the initial step of trophic structure, through primary producers. From primary producers they are transferred from one organism to another along the food chain to reach the ultimate consumer. Much is lost in the process as exudates, excreta and faecal matter, which return to the surrounding medium. Finally, when the living organisms die, decomposition of their bodies releases rest of these elements in the environment from where they were drawn earlier (Fig 2.2). The entire process can be visualized as a series of reservoirs or pools through which an element moves in a cyclic fashion in an energy driven stream, finally coming to rest in its original pool or reservoir (4).

All elements, which form the basic structural and functional component of a living system, circulate in the environment in characteristic channels from one reservoir to another and constitute cycles, which are known as **bio-geo-chemical cycles** (2). For example, the carbon of atmospheric pool of carbon dioxide may form a part of green plants at one time, the limb of an herbivore at another and the delicate heart of a child at still another time. Thus the same atom may journey through a series of living organisms or pools but it comes to rest in the vast atmospheric reservoir as carbon dioxide again (Fig 2.2). Thus, the flow of materials in an ecosystem is cyclic whereas the flow of energy is unidirectional (3;4).

(III) SHORTER FOOD CHAINS HAVE A HIGHER PRODUCTION EFFICIENCY:

The shorter the food chain, the greater is the biomass which can be produced from a given amount of energy. A considerable amount of energy is lost during transfer from one trophic level to another. Thus, trophic chain composed of five or six trophic levels shall involve four or five such transfer steps and consequently, the energy loss shall be much greater. Only a small part of biomass or organic material produced by autotrophs is consumed by primary consumers or herbivores. This either is utilized to meet the metabolic needs of herbivores or is passed on to decomposers after the death of the organisms. Only a very small part is passed on to a higher trophic level. With involvement of many trophic levels wastage of energy shall multiply and this will lower the production efficiency (1;4).

Antarctic seas are the most productive systems of the world during summers where days are as long as six months at a stretch. They receive a twenty-four hour solar energy, which results into the production of a substantial biomass of phytoplanktons. Baleen Whales are the only consumers in most cases. Thus the food chain consists of two links only:

Phytoplanktons → Baleen Whales.

Only one energy transfer step, therefore, is involved. The sharp turn over of various water layers ensures a thorough mixing. Nutrients dissolved out from bottom strata of rocks, silt and detritus are in plenty. Lower temperatures also cause a reduced rate of respiration and therefore, relatively little amount of energy is lost in meeting the metabolic needs of component organisms. All these factors combine to make the Antarctic oceans the most productive seas of the world during the polar summers (3;4).

(IV) DIVERSITY AND STABILITY IN AN ECOSYSTEM:

All life has a common biochemical basis. There is a certain range of environmental conditions within which life processes proceed with greatest ease and efficiency. This does not mean that all plants and animal require similar conditions of environment. But it does imply that within the confines of this range a large number of plants and animal species can live thrive and multiply. This common biochemical basis of life processes ensures that more plants and animals are going to live and thrive between 27°C and 32°C than at freezing temperatures (0°C). Of course many plants and animals can be well adapted to life in freezing temperatures but such organisms shall be few number. Natural selection for this will be rigorous which shall eliminate a number of species. The severity and harshness of the conditions of environment act as a limiting factor for the variety in species composition. Thus we can expect Arctic and Antarctic regions, snow capped mountain ranges, deserts and ocean depths etc. to possess little diversity in composition of plant and animals because of harsh and severe conditions. In tropical zones a large number of species can thrive because of the milder conditions of environment (1;5).

An important consequence of increased complexity of trophic structure is the comparative stability of the ecosystems. Simpler trophic structures are more vulnerable to catastrophic changes as compared to more complex trophic structures (6). For example, in Antarctic seas, shrimp (*Euphasia* sp.) is the main food source for sea birds, marine fishes and squids – to these consumers there is no other alternative available. Baleen Whales are also partially dependent on it. If the shrimp is eliminated by some accident, there will be drastic reduction in population of marine fishes, whales, squids and sperm whales. If the production of lichen (*Cladonia* sp.) is impaired by some accident, the entire Arctic terrestrial system shall collapse as the entire life in the system depends on it (Please see Fig 2.3). Complicated trophic structures such as are found in tropical and sub-tropical climates, possess several alternative inter-linked food chains and a temporary or permanent elimination of one or a few species does not cause drastic changes in the system. There are several alternative sources available to the organisms at the next trophic level and the flow of energy and operations of biogeochemical cycles continue. At the most, species obligately dependent on the eliminated species shall suffer. Thus, the system stability is dependent upon the complexity and diversity in the ecosystem, which possesses several alternative inter-linked trophic channels. Simpler ecosystems are much less stable than the complicated ones (1;6;7).

(V) THE LAW OF DIMINISHING RETURNS AND THE CARRYING CAPACITY:

With an increase in diversity and complexity in an ecosystem, the energy requirement or the cost of maintenance of the system usually goes up at a greater rate. Doubling the size or complexity, therefore, requires more than double the amount of the energy to maintain the system in proper functional state. As an ecosystem becomes larger with many trophic levels and a greater number of species at each trophic level the portion of gross production which must be oxidized or respired to maintain the trophic structure, rises while the part which is added to bring about further growth diminishes. A stage is ultimately reached when the cost of maintenance of the system becomes so high that nothing is left to go into further increment and growth. The entire production is used up in the maintenance of the system. The energy input and its use balance themselves and a steady state is reached. The amount of biomass that can be sustained at each trophic level under such a steady state is termed as maximum or **optimum carrying capacity** of the system (1;3;4).

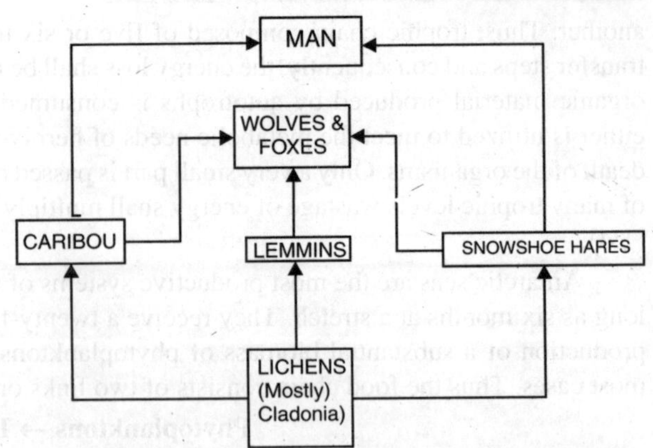

Fig 2.3 *Simplified trophic structurein Arctic terrrestrial system*

(VI) TROPHIC STRUCTURES IN POLAR, TEMPERATE AND TROPICAL REGIONS:

Drastically cold and rough conditions permit only those life forms to survive and thrive which are capable of withstanding the rough environment of the Polar Regions of our planet. The few life forms, which occur, are often found in abundance. The trophic structure, therefore, tends to be simple in these places. It becomes more and more complicated as we move towards equator since the diversity and complexity of trophic structure increases. A simplified diagram of terrestrial system in Arctic Region is given in Fig 2.3.

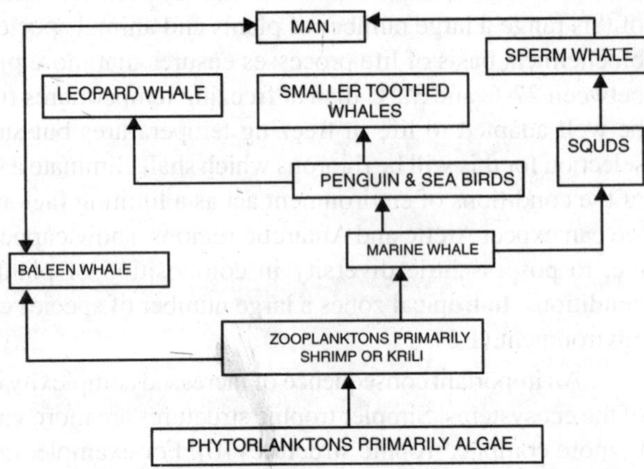

Fig 2.4 *A simplified diagram of Antarctic aquatic system*

In Arctic terrestrial ecosystem, primary producers are represented mainly by Lichens, the most common of which being *Cladonia.* This commensal combination of an alga and a fungus forms the main food for primary consumers, the herbivores such as caribou, lemmings and snowshoe hares. There are other plants and animals as well in the Polar Regions and Fig 2.3 does not provide a complete picture. But all of them, like the flowering plants, musk oxen, birds, fishes, polar bears etc. can be fitted in a relatively simple trophic structure (8).

Likewise in Antarctic region also a relatively simple trophic structure occurs. In an aquatic system in the Antarctic, the primary production is represented largely by phytoplanktons – mainly algae. These support a rich population of zooplankton, *Euphasia superba* also know as Shrimp or

Krill. Marine fishes, baleen whales, penguins and sea birds etc. depend mainly on these zooplanktons for their food. Baleen whale may also consume phytoplankton directly (Fig. 2.4). Man, toothed whales, leopard seal, squids, sperm whales etc. are the main predators (8).

The pattern of energy flow and trophic structure is not so simple in Temperate and Tropical regions of our globe (9). Relatively milder conditions of environment permit the growth of a variety of species of plants and animals, which have their own nutritional requirement, preferences and susceptibilities. Variety in floral composition supports a rich variety and number of herbivores and like-wise, the population of carnivores at all trophic levels is highly diversified. The tropical island of Barro, Colarado, situated in the Panama Canal Zone, has a community of over 1200 species of trees, over 30,000 species of insects, 310 species of birds, 32 species of amphibians, 68 species of reptile and over 70 species of mammals – all of these confined within a space of only 6 sq. kms, the total area of the island. It is almost impossible to make out the pattern of energy flow, food chains and trophic structure in such a complicated system which involves many thousand component species. In such an ecosystem resources are shared and organisms may not depend entirely on one another. For example in the chain consisting of:

Grasses → Mouse → Snakes → Predator birds →Hawk

Birds may be versatile omnivores, consuming mouse, snakes, and at times fruits produced by green plants. Apart from snakes and birds mouse may be eaten by a variety of predators. Snakes, apart from consuming mouse may also consume a variety of other insects. This type of inter-relationship within a system produces a complex interlinked trophic structure in which several alternative resources exist for an individual or a group of individuals. The trophic structure no longer consists of simple straight chains but becomes a complex web of interlinked trophic relations, which is often termed as a **Food web** (9;10).

(VII) ECOLOGICAL PYRAMIDS:

Energy losses at each trophic level and the relationship between size and rate of metabolism of individuals within a community result into a typical trophic structure, which is, usually, characteristic of a particular type of ecosystem. A graphical representation of trophic structure or the food chain results into a pyramid like figure and hence it is termed as **ecological pyramid**. Based on the parameters selected to depict the trophic relationship an ecological pyramid may be of the following types:

1. Pyramid of numbers wherein the trophic structure is depicted in terms of numbers of organisms present per unit area.
2. Pyramid of biomass wherein the trophic structure is depicted in terms of biomass per unit area.
3. Pyramid of energy wherein the trophic structure is depicted in terms of rate of flow of energy or productivities.

The usual shape of a pyramid, which is widest at the base and gradually tapers to a narrower tip may at times be inverted, i.e., the lower tier may be smaller than the upper ones. This may happen in case of pyramid of numbers or of biomass. This happens when the size of organisms constituting successive trophic levels differ tremendously. On the other hand, the pyramid of energy is always upright, *i.e.*, shaped truly like a pyramid (4).

A phytoplankton population consisting of millions of small floating organisms may support a few thousand zooplanktons on which only a few fishes subsist. The pyramid of numbers in this case shall be upright with largest lowermost tier and successive tiers being smaller and smaller (Fig 2.5:1). In an isolated patch of large trees of which a dozen or so may be present in the system, a large number of insects and insect-eating birds may live and thrive. If numbers are used to represent the trophic structure, the pyramid will become an inverted one as shown in Fig 2.5:2. Thus, the shape of pyramid of numbers varies widely depending upon the size of organisms constituting the different trophic levels. Hence, the true effect of the food chain and size factor are not properly represented. The number may, at times, be so large as to render the depiction of each of the trophic level on paper difficult or impossible (4).

The pyramid of biomass provides a little better picture. When the total mass of individuals at each trophic level is plotted successively, a gradually tapering pattern emerges as long as the sizes of the component organisms do not differ significantly. In trophic chain composed of four links, i.e., Grasses – Grass-hoppers – Birds – Hawke, there is no significant difference in sizes of the constituent organisms and a sloping or tapering pyramid is produced. However, in cases where there is a tremendous difference in size of individuals at successive trophic levels, such as in Antarctic aquatic systems during polar summers, the pyramid of biomass is usually of the inverted type. The food chain in Antarctic aquatic system usually consists of two links only: Phytoplankton – Baleen Whales. The weight of a single whale may be enormous as compared to the phytoplankton, which are microscopic organisms. In such cases, though more energy is canalized from producers to primary consumer, the total weight of the producers may be much lower than that of the consumers. The system is sustained by the rapid metabolism and quick turn over of small organisms, which results into a large output on which huge organisms like Baleen whales survive (1;2)

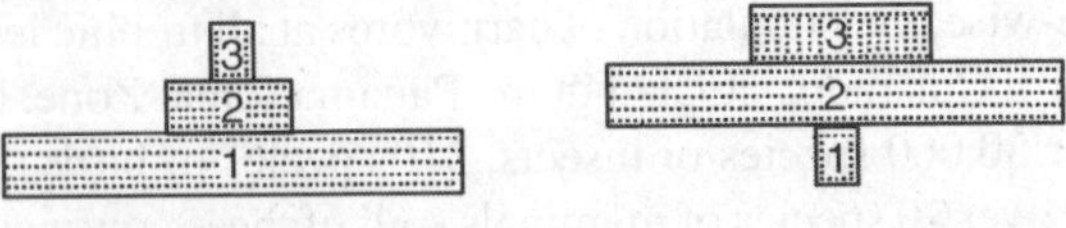

1.Phytoplanktons.
2. Zooplanktons & Insects,
3. Fishes

Fig 2.5:1 *Pyramid of numbers for an aquatic system.*

1.Trees 2. Insects 3. Birds

Fig 2.5:2 *Pyramid of numbers for a terrestrial*

It is actually the rate of production of food materials which is important rather the total standing crop or biomass. It is the quantum of energy flowing into the system from one trophic level to the other or rather more precisely the rate of the flow of energy from lower to higher trophic levels, which provides the best graphic representation of the functions of an ecosystem (3;4).

(VIII) THE DIVERSITY OF ECOSYSTEMS:

Three fourth of earth's surface is covered with water which thrives with life. In fact, it is these aquatic ecosystems only which gave birth to biosphere about 3.5 billion years ago and nurtured it for another three billion years till some organisms left their original abodes to colonize the land. It was a slow process of persistent activity of green plants, which transformed earth's atmosphere from a reducing, oxygen-less mass of gases to an oxygen-rich environment in which ozone could form and accumulate. By Silurian era of earth' geologic history an effective ozone layer was formed. Life could now, colonize the land surface and terrestrial ecosystems appeared. From then onwards, both aquatic and terrestrial ecosystems have existed side by side.

I. Terrestrial ecosystems:

The incidence of solar radiations, which determines the temperature of a locality, is dependent on latitude. Precipitation is strongly influenced by wind patterns, which are also associated with latitudes. Therefore, a remarkable feature of the distribution of the terrestrial ecosystems of the world is that many division lines between different regions tend to run parallel to the lines of latitudes. The same type of biome occurs within the same general latitude. However, on mountain ranges, the pattern is changed. The biotic regions change with changes in altitude and therefore, lines demarcating various regions tend to run parallel to contours depicting heights from mean sea level (11).

1. FOREST ECOSYSTEMS:

a. Arctic and Alpine Forests (Tundra):

The northern zone of earth, which extends from permanently frozen ice caps on Northern Pole up to 45° N latitude, is the most difficult region for life to exist. It is also referred to as the Tundra biome. The region receives little solar radiations, as a result of which it is the coldest region of our planet. The ground is covered with ice for eight to nine months a year, which melts for a brief period of three or four months only on the surface. Water in the lower layers is in a permanently frozen state (the permafrost) which restricts the downward movement of water and solutes when ice on the surface melts. The precipitation is low mostly in the form of snowfall, whereas there is little evaporation, which usually results in water logged conditions. Due to low temperatures, decomposition, mineralization and recycling of nutrients are very slow and hence the soil is poor in plant nutrients. Strong cold winds, known as blizzards, blow for most of the year (12).

Severe cold conditions in the region make life extremely difficult. There are no trees. Plants are dwarfed, which protects them from blizzards. As temperatures at the ground surface are a little higher there is a thick mat of mosses and lichens on the frozen soils. During summers when ice melts for a brief period of two to three months some grasses and few flowering plants come up. These plants have to grow, form flowers, fruits and seeds before the winters return. The herbivorous animals of Tundra include reindeer, caribou, musk ox, arctic hare, voles and lemmings that depend largely on mosses and lichens and the few species of flowering plants found in the region. Arctic fox, wolves, polar bear are the carnivores found in this frozen wilderness. Birds include waterfowl, ducks, swans and gees etc. Tundra represents a fragile ecosystem. A few species of mosses, lichens and higher plants form the basis of entire life in these regions. Severe cold conditions limit the number of species. Fortunately, harsh living conditions have restricted human influence to its minimum and have preserved much of these systems in their original state (12).

b.Coniferous Forest (Taiga):

Temperate coniferous forests, also called Boreal coniferous forests or Taiga biome, occupy sub-arctic regions of North America, Europe and Asia forming a broad belt in the south of Tundra biome (Between latitude 45Ú to 57Ú N). This biome is absent in Southern Hemisphere as the corresponding areas are occupied by sea. The climate is characterized by bitterly cold and long winters and a brief cool summer season. The ground is frozen during winters, whereas in summers snow melts to create water logged conditions. Precipitation is rather uniform throughout the year, somewhat more concentrated during the summers. Taiga is slightly warmer and wetter than Tundra and this has caused a little more diversity in flora and fauna to develop in this region (11;13).

As the name suggests, this biome is dominated by coniferous forests which are a huge store house of soft wood in the world. Most of the trees are of evergreen types which assume a conical shape so that snow may not accumulate on branches and leaves. These forests have very little undergrowth – herbs and shrubs are scarce. Where the trees are spaced apart a rich ground cover of mosses, lichens, orchids and ericaceous shrubs may be present. Animals of Taiga include caribou, moose, bear, dear, wolverine, martens, wolf, snowshoe hare, voles, shrews and bats. A considerable variety of insects and birds inhabit coniferous forests. Amphibians are relatively common in Southern parts of Taiga forests. Temperate coniferous forest has been exploited by man for their supply of soft wood since times immemorial. However, this region is rather sparsely populated as the soil is poor and agriculture is not feasible. Most of the inhabitants of the region depend largely on forests and wildlife for their livelihood. Humans with his modern technology and equipments have inflicted considerable damage to these ecosystems (11;14)

c. Deciduous Forests:

These forests occur in regions characterized by moderate climate and are primarily composed of broad-leaved deciduous trees, which shed their leaves during the winter season. They occupy most of the eastern part of United States, much of Central Europe and Eastern Russia, parts of China, Japan, Australia and Chile. The average rainfall is about 75-200 cms per year whereas temperatures range between 10°C to 25° C.

Both flora and fauna is richer and more diverse than those of Taiga biome. Animals common in deciduous forests are deer, bears, squirrels, voles, mice foxes, bobcats, raccoon, mountain lions etc.

Smaller animals include amphibians, like frogs, toad, salamanders, lizard, snakes etc. Invertebrates are earthworms, snails, millipedes, coleopteran and orthopterans etc. A wide variety of birds inhabit deciduous forests. The range of animal size and adaptations are wide and a profound seasonality in behaviour is observed in a large number of animals as well as plants of deciduous forests. The region occupied by deciduous forests has been extensive affected by human activity. A large number of huge cities stand in this region and human influence has caused extensive destruction of natural forests. Industries abound and much of the land is used for agricultural purposes. Natural forests and wildlife have become confined to small isolated patches spared by man (11;14;15).

d. Tropical Rain forest Ecosystems:

Tropical rainforests occur within a broad belt delimited by the Tropic of Cancer and the Capricorn on either side of the equator and are among one of the richest and most diverse biotic regions of the world. They receive more that 200-250 cms of rainfall per year. The rains are usually distributed evenly throughout the year. The temperatures are also among the highest of all biomes, ranging between 20°C to 30°C. There is little seasonal fluctuation in temperatures, rainfall and photoperiods (10;11).

The most remarkable feature of tropical biomes is the relative stability of its climatic conditions. Another outstanding feature of biotic community in tropical forests is its richness and diversity of species which are organized in vertical tiers one over the other. A profusion of lianas or woody climbers and epiphytes, which grow across the different strata of the vegetation, characterize a tropical community. Epiphytes and lianas add considerable variety to the three dimensional configuration of the tropical forests. They provide convenient travel channels for ants, bees and other insects, tree frogs, snakes, monkeys and ocelots etc. There is a regular growth of plants throughout the year, which results in constant supply of food material often in abundance. However the soils are poor as most of the available nutrients are lodged in plant and animal biomass. Animals do not have to run or migrate to other places for food. Another characteristic feature of a tropical forest is that it is full of life activity all round the clock. The animals of different species live in various vertical strata some of which are active during the day and some active during the night. There is always hue and cry of some animal or the other throughout twenty-four hours of a day. Unlike other biomes, tropical forests never sleep (10;14;15).

The area of the world where tropical forests stand is largely occupied by poor developing countries. Poor nutrient status of the soil restricts agricultural activity and people have to depend largely on the resources of the forest, wildlife and fisheries for their livelihood. A rapid rise in population of man has enormously increased the pressure of demand on tropical ecosystems. Deforestation, soil degradation and impoverishment of biological components has been taking place rather at a fast pace in these systems (16).

2. GRASSLAND ECOSYSTEMS:

Grassland biomes occur in regions, which receive, and annual rainfall between 25-85 cms. Grassland biomes are found both in tropical as well as temperate regions of the world. The prairies, pampas, steppes, puszta etc. are more or less synonymous terms for these biomes, which differ from each other in species composition and some conditions of the environment. However, a common feature of all grasslands is intermittent and erratic rainfall, which is not enough to support a forest but is more than that of a desert (11;17).

As the name indicates, grasses are the dominant plant species of this biome. However, there are a number of herbaceous plants as well in the system. A number of small shrubs and trees also occur along with grasses. The annual production of organic matter in grassland is high which accumulates rapidly and the soils develop a thick layer of humus. Roots of grasses and other herbs may penetrate as deep as 2 meters in the soil. That is why grassland soils are among the thickest and richest soils and have provided to humanity some of the richest agricultural areas of the world (11).

Animal populations of grassland are also rich and diverse. Many mammals, ungulates, rodents and hares flourish with large populations of bison, antelopes, zebra, horses, donkeys, buffalo, giraffe, elephants, hippopotamus, rhinoceros and a number of marsupials. Lion, tiger, foxes wild dogs, coyotes etc. form the majority of carnivore populations. It is these grasslands, which have provided to humanity many of the draft and milch animals, which he uses today. Insects abound in grasslands. Typical birds of grassland ecosystems are prairie chicken, grouse, larke, bastards, game birds, ostrich, non-flying birds like emu etc (17; 18).

Grassland ecosystems have been extensively affected by human activity. Much of the natural grasslands have been converted into agricultural land while rapid increase in the number of domesticated animals and their grazing has put severe strain on these systems as a consequence of which much of the grasslands have shrunk to isolated patches. There has been an appreciable reduction in the number of species of plants and animals and insects in these systems (4;14;18).

3. Desert ecosystems:

Desserts are arid biomes, which usually receive less than 25 cms of rains annually. They occupy approximately 18% of the total land surface of our planet. There are some deserts, which may not receive any rains for periods as long as several years. Such deserts are practically devoid of any form of life. In the Sahara desert, South Libya, there is no vegetation or virtually no sign of life for several hundred kilometers at a stretch. However, most of the deserts are not so dry. They do have some water in the form of occasional rains or groundwater on which a variety of life forms depend (1;4).

The predominant vegetation of desert consists of succulent species with a waxy coating on the surface, which is helpful in conserving water for long durations, such as cacti. Along with adaptation to conserve water these plants also have an efficient system to absorb the water available in the arid environs. There root system is profuse with adventitious branches spreading out to cover as large surface area as possible. They also penetrate deep into the soil. There may also be present deciduous shrubs with thick waxy leaves. The growing season is very short. Flowering is strikingly abrupt occurring suddenly after a brief flash of rains. Cacti, creosote bush (*Larrea divericata*), sagebrush (*Artemisia tridentata*), a few genera of palms etc. constitute major vegetation of desert biomes. The biodiversity is very low. Animal life is similarly well adapted to low levels of water. Development of impervious integuments, excretion of uric acid instead of urea, or highly concentrated urine and reduction in direct water intake by conserving water produced in the metabolism are some of the adaptation of animals to extreme xerophytic conditions. The animal community of deserts is confined to the zone where some plant life exists and is dominated by burrowing and nocturnal rodent, reptiles, insects and arachnids (11;13;19).

Due to lower productivity, the leaf litter layer is thin and limited and as a consequence organic matter content of the soil is very low. Evaporation tends to concentrate soluble material near the surface, which makes the soil highly alkaline or saline. The nutrient status of the soils is very poor.

Water is apparently the primary limiting factor in desert biomes. The United States spent millions of dollar in development of irrigation facility in the mid 1950s in Afghanistan to make the deserts bloom. The deserts did bloom but quickly the poor nutrient status, salinity and alkalinity inhibited plant growth and the project turned into a dismal failure. The Afghanistan experience demonstrates that the desert agriculture requires much more than simple addition of water. It requires proper application of scientific skill, hard labour and management of nutrients to put the 18% of our land area under deserts to some productive use (20;21).

II. Aquatic biomes

Covering more than three fourth of earth' surface aquatic ecosystems are not only a dominant feature of earth but are also very rich and diverse in their species composition and complexity. These may be divided into three major categories:

1. Freshwater ecosystems.
2. Marine ecosystems.
3. Estuarine systems.

1. Freshwater ecosystems:

Freshwater ecosystems provide an enormous variety of habitats for aquatic organisms to grow. They may have running water or stagnant waters. The spectrum of inorganic salts, dissolved gases and organic matter present in dissolved state in freshwater system is highly variable. Presence or absence of water movements, the depth and shape of the depression, the gradient or slope of the bottom in case of streams and rivers, conditions of temperature, photoperiods etc. combine with the physicochemical parameters of water quality to produce in each water body a unique system.

Primary producers of aquatic systems are freshwater algae usually dominated by members of Chlorophyceae, Cyanophyceae and Bacillariophyceae. In shallow regions rooted plants occur. The zone of deeper waters is occupied by phytoplanktons – organisms which have adopted themselves to floating mode of existence. In running waters those organisms predominate which have some device to attach themselves to some substrate. Frogs, snakes, clams, and a considerable variety of adult and larval insects populate the margins of the water body. In deeper zones (limnetic zone) a variety of zooplanktons, several protozoans, arthropods etc. share the water with larger swimming organisms like fishes, amphibians and insects. The zone near the bottom of the water body is occupied by benthic organisms. Light is often a limiting factor in deeper waters. The primary source of energy in this zone is the debris and detritus which sinks down from above. The living community here consists mainly of decomposers which live on dead and decaying organic matter (2;4)

2. Marine ecosystems:

Marine ecosystems are the largest reservoir of water, plant nutrients and living things in the world. Total biomass in these systems far exceeds those of all freshwater ecosystems put together. Average depth of oceans is about 3750 meters with the maximum depth exceeding ten thousand meters (in Marianas Trench in the Pacific Ocean). Ocean waters are characterized by high salinity (about 3.5%) with Sodium chloride constituting about 27% of the total salt content. Unlike freshwater systems, which occur as isolated patches, sea is a continuous entity – all oceans of the world being interconnected and there is a regular circulation of water within caused by important water currents. In spite of a large salt content ocean waters may lack nitrogen and phosphorus, which are important plant nutrients. Most of these nutrients are derived from land through streams and rivers, which drain their water in sea or from shoreline by upwelling of water near the shore caused by air currents (1;2;4).

The shoreline, between the land and oceans and open sea or the zone between the high tide and low tide is referred to as the **Littoral zone.** It is subjected to extremes of physical turbulence, waves, and fluctuations of temperatures and light intensities etc. This zone possesses plenty of plant nutrients, which dissolve out from the shore and bottom, and the turbulence of the waters keeps the nutrients evenly distributed throughout the system. Along a rocky coast occur sessile organisms while on a sandy coast organism adapted to burrowing habit and those, which can adhere to the sands occur. Coral reefs formed at the edges of islands or littoral of mainland under water by colonial coelenterates may be regarded as marine equivalents of tropical rainforests. Plenty of light and nutrients give rise to enormous diversity of life forms in these systems. Today, however, much of littoral systems are being adversely affected by an increasing load of silt and pollution washed down from the land surface. **Neritic zone** is the zone of sea next to littoral region. It is a zone of shallow water with continental shelf running underneath. The depth ranges between few meters to 200 meters. This zone constitutes about 7.5% of the total ocean area and is a zone of high productivity. There are plenty of nutrients, which are derived from the shore, and there is plenty of light because of the sea is relatively shallow.

Extensive algal communities of giant kelps and smaller unicellular and multicellular forms develop in this zone. It is this region of the sea, which provides to humanity some of the best fishing waters of the world (4;22).

Pelagic zone is the zone of open sea. The depth is usually more than 200 meters. This zone constitutes about 90% of the total surface area of the world and is relatively poor in nutrient content. Photosynthesis is largely carried on by phytoplanktons consisting of diatoms and dinoflagellates. Copepods and arrow worms are usually the major zooplanktons along with shrimp, jellyfishes and ctenophores. These waters, however, are able to support animals as large as blue whales in region near Antarctica due to regular summer blooms of phytoplanktons. Organisms in the pelagic zone below the level of light penetration live on organic matter that sinks down from the upper layers. Crustaceans and copepods are abundant in the upper reaches of this non-photosynthetic region. **Benthic zone** extends from the edge of the continental shelf to the deepest trenches and crevices of the ocean floor. It receives practically no light though it has plenty of nutrients. The pressure of water mass above is often enormous which influences life in this zone considerably. Dominant organisms of this zone are heterotrophs often anchored in the muddy stratum made up largely of siliceous and calcareous deposits of foraminifera, snails, radiolarians and diatoms. Sea-lilies sea fans, sponges and brachiopods form rooted animal community while starfish, sea cucumbers, sea urchins are mobile forms of this zone (2;11).

3. Estuarine ecosystems:

The region where sea water meats with freshwater from rivers and streams by the action of tides are referred to as estuaries. These include gradually sloping riverbeds, u-shaped basins, river delta, shallow inland extension of sea etc. There are wild fluctuations in salinity in these waters. Estuaries are highly productive ecosystems, which are more productive than either fresh water or marine waters. They act as nutrient trap. Producers in estuarine ecosystems include seaweeds, marsh grasses and phytoplanktons, which are capable of carrying on rapid photosynthesis throughout the year. Oysters, crabs, shrimps various fishes etc. form major component of animal lives in these systems. Marsh grasses and other vegetation protects the shoreline from erosion while trap much of the silt, debris and nutrients from flowing directly into the sea (2;13;22)

(IX) ECOSYSTEM SERVICES:

Mankind benefits from a variety of resources and processes which are provided by natural ecosystems. These benefits are collectively referred to as ***Ecosystem Services*** which are broadly defined as the *benefits provided by ecosystems to humanity.* Nature works incessantly to provide these free benefits to mankind through the enormous collection of living organisms which it maintains. We are often not even aware of the benefits we derive from a healthy ecosystem. Replacing these benefits with human technology would be exorbitantly costly and for many almost impossible. The value of these benefits has been estimated to be about 2928 billion US dollars.

Recognition of how ecosystems provide these complex services to humankind dates back to the time of Plato (c. 400 BC) who wrote that deforestation could lead to soil erosion and the drying of springs. During latter half the 20th century authors like Osborn (25), Vogt (27) and Leopold (26) lamented on the plundered state of natural ecosystems while pointing out the dependence of humanity on the 'natural capital' for a number of benefits or services. In 1956, Sears (28) drew attention to the critical role of the ecosystem in processing wastes and recycling nutrients. Ehrlich, P.R. and A. Ehrlich in 1970 called attention to "the most subtle and dangerous threat to man's existence...... the potential destruction, of those ecological systems upon which the very existence of the human species depends" Finally the term 'environmental services' was, placed in a report of the Study of Critical Environmental Problems (30) which listed services including insect pollination, fisheries, climate regulation and flood control etc. In following years, many variations of the term were used, but eventually the term 'ecosystem services' became the standard in scientific concept (31). Modern expansions of the concept of ecosystem services include socio-economic and conservation objectives as well (30).

Some ecosystem processes confer direct benefits to humanity, but many of them provide benefits indirectly. These services include:

1. ***Provision of ecosystem goods*:** Ecosystems provide to man a number of useful things such as foods (including seafood and game) spices, medicines, pharmaceutical and industrial products as well as energy (hydropower, biomass fuels). Healthy soils with rich vegetation act as a sponge which soaks up large quantities of water to serve us during dry seasons.
2. ***Soil formation and its maintenance in state fit for plant growth:*** Soils is precious commodity. Our food production system shall collapse if the process of soil formation and its maintenance in healthy state is disrupted, both of which are carried out by innumerable microbes in the soil. Under natural conditions 25 mm of top soils forms in 1000 years (31). The cumulative activity of the soil biota redistribute nutrients, aerate the soil, increase the rate of water infiltration, facilitate top soil formation thereby enhancing plant productivity. One hectare of high quality soil may contain about 1300 kg of earthworms, 1000 kg of arthropods, 3000 kg of bacteria, 4000 kg of fungi and many other plants and animals (32).
3. ***Decomposition of organic wastes and recycling of nutrients*:** Decay, decomposition and mineralization of wastes and dead organic matter recycle a huge amount of nutrients naturally. Each year about 88.06 billion metric tons of organic wastes are produce worldwide disposal of which has been estimated to cost more than 760 billion US dollar. This cost does not include benefits derived from decreased level of environmental pollution, recycling of nutrients, significant reduction in human diseases and elimination of the need for landfills and disposal sites (31).

   Advanced science and technology has caused the production of a wide range of chemicals many of which contribute to pollution of ecosystems. Worldwide about 100,000 chemicals are used. Accumulation of toxic wastes has resulted in an estimated 500,000 - 600,000 hazardous waste sites in the United States alone which must be cleaned up. Biological degradation and cleanup by a variety of microbes is an effective way of decontaminating these sites. It has been estimated that the cost of cleaning up of these hazardous sites over next 30 years could be as high as 750 billion US dollars (33). Use of biological methods could reduce this cost to mere 75 billion US dollars (34). Maintaining biodiversity in soil and water is essential for improved effectiveness of the process of biodegradation and detoxification.
4. ***Biological nitrogen fixation:*** Nitrogen is essential for plant growth. More than 77 million metric tons of nitrogen is applied to our fields as commercial fertilizers which costs about 38.5 billion US dollar. Worldwide biological nitrogen fixation provides about 150 -180 metric tons of nitrogen to our soils which is worth about 90 billion dollar (35;36). With the drying of our oil wells, natural gas and oil which serve as an important raw material for synthesis of nitrogenous fertilizers are going to be increasing scarce and expensive. Biological nitrogen fixation shall serve us as an important alternative to synthetic fertilizers. Moreover it could cut down global energy use by about 30% which has to be spent for the manufacture of nitrogenous fertilizers (37).
5. ***Regulation of local and global Climate:*** Healthy ecosystems with rich and diverse biota play a critical role in regulating earth's physical, chemical, and geological properties. It was the activity of cyanobacteria, the first green plant to appear on our plant about 3.5 billion years ago which started producing oxygen to create the oxygen-rich atmosphere for

oxygen breathing organisms to appear (38;39). A huge quantity of CO_2 was buried as carbonate rocks and fossil fuels by green plants which reduced its atmospheric concentration. Man is reversing the clock. He has been adding an increasingly larger and larger amount of CO_2 in the atmosphere which is warming our planet. It is the precise composition of green house gases in the atmosphere which maintains the mean global temperatures at about 14°-15°C enabling life to flourish on our planet (40).

Albedo which is the proportion of solar radiation reflected back to the atmosphere has been shown to have a critical role in warming of earth's surface. Surfaces with low albedo reflect a small amount of sunlight, those with high albedo reflect a large amount. This causes earth's surfaces with low albedo to get heated more than places with higher albedo. Different types of vegetation have different albedos; forests typically have a low albedo, whereas deserts have high albedo. In Western Australia, the replacement of native heath vegetation by wheatlands increased regional albedo. As a result, air tended to rise over the dark heathland, drawing moist air from the wheatlands to the heathlands. The net effect was a 10% increase in precipitation over heathlands and a 30% decrease in precipitation over croplands (41)

Vegetation absorbs water from the soil and releases it back into the atmosphere through evapotranspiration, which is the major pathway by which water moves from the soil to the atmosphere. In the Amazon region, vegetation and climate is tightly coupled. Evapotranspiration of plants is believed to contribute an estimated fifty percent of the annual rainfall (42). Deforestation in this region leads to a complex feedback mechanism, reducing evapotranspiration rates, which leads to decreased rainfall and increased vulnerability to fire (43).

6. ***Resistance to invasion by alien species:*** Invasions of species beyond their native range constitute a major concern for the conservation of natural and managed areas all over the world. Invasive species threaten biodiversity (44), change ecosystem functioning (45) and cause economic losses (46). The hot spots for biodiversity are particularly at risk of invasion by introduced species which could result in loss of a number of rare species found nowhere else on our planet. The available evidence indicates that higher species richness in a healthy ecosystem can increase the biotic resistance against invasion by exotic species. Biotic resistance is the ability of resident species to inhibit the establishment, growth, survival and reproduction of invasive species which may vary from habitat to habitat due to changes in the composition, and diversity of the species (44).

7. ***Control of insects, pests and diseases:*** Nearly 99% of insects and pests are controlled by natural enemy species and host plant resistance . Without pesticides, natural enemies, host plant resistance and other non-chemical control methods about 70% of crops could be lost (48). In natural ecosystems a process of checks and balances (prey-predator interactions) is operative which automatically controls the populations of diverse insects, pests and pathogens. Bird predation on insects has been estimated to provide an annual benefit of about 180 US dollars per hectare (50). Huge quantities of pesticides have to be applied to our crop fields because most of our agro-ecosystems are monocultures. Mixed cropping, landscape diversity (such as the intermixing of crop and non-crop patches) and crop rotation nearly always decreases yield losses caused by pathogens, insects and pests (49). Rice blast disease, caused by a fungal pathogen of rice, has been controlled in a large region of China by planting alternating rows of two rice varieties (52). The use of mixtures of different crop varieties has been shown to effectively retard the spread or evolution of fungal pathogens of grains-crops (53;54). More than 60% of human pathogens are naturally transmitted from animals to humans by insects and pests. Human health, particularly the risk of exposure to

many infectious diseases, is appreciably reduced if the populations of insect and pest vectors of infectious disease are kept under control. Greater wildlife species richness in natural ecosystems keeps the population of insect vectors or carrier of pathogens to humans under control, thereby reducing the probability of human infections and outbreak of infectious disease epidemics (55).

8. ***Pollination:*** Pollinators, such as bees, butterflies, birds and bats provide substantial benefits to the maintenance, diversity and productivity of both agricultural and natural ecosystems (56). At least one third of world's food production relies heavily, either directly or indirectly on insect pollination. Poor reproduction observed in several rare plants has been linked to the loss of specialized pollinators (57;58). Earlier, low fruit production in plants was usually attributed to nutrient limitation, but a number of studies have shown that failure of fertilization due to pollen limitation could be the cause of fruiting failure and thereby low fruit yield (59). The economic value of pollination service rendered by diverse bees, insects and others has been estimated to be about 200 billion US dollars.

    Pollination limitation also leads to increased inbreeding, reduced genetic fitness, and increased susceptibility to environmental stress (60). In order to persist in agro-ecosystems pollinators need local floral diversity and nesting sites. Large monocultures, which we set up, fail to provide these. Cultivated orchards surrounded by other orchards have significantly fewer bees than orchards surround by uncultivated land (61). Only farms near natural habitats sustain communities of pollinators large enough to provide the needed levels of pollination (62). About 20% more coffee is obtained from plantations in Costa Rica surrounded by natural forests as the stingless bees which provides the pollination-service to coffee plant nests only in the adjacent forests. It is not available in farms away from natural vegetation resulting in decline of coffee production by at least 20% (63).

9. ***Seed dispersal***: Seed dispersal is an important process in natural ecosystems and population dynamics. Absence of an effective population of animal agents of seed dispersal causes the plant population to be confined to localized areas which results in crowding, wastage of seeds, inbreeding depression and loss of genetic diversity in the species concerned. Seed dispersal is carried out in number of ways. Most of the plants, however, including economically useful ones to man, depend on seed dispersal by animals. The seeds of a large number of woody plants are dispersed by animals – about 80-95% in the tropics and about 30-60% in temperate forests (64). Seeds can be dispersed by animals that eat the fruit and discard the seeds (frugivores) or by seed eaters. Many seeds do not survive, but those few which survive serve the purpose.

    Seeds of a number of plants could be dispersed by a single species of animals, i.e., most of the seed dispersal systems are of generalized type (65). However, each animal species could disperse seeds in distinct manner which affects the pattern of distribution of the plant concerned (66). Reduction in population size of fruit eating dispersal agents could cause a shift in preferred items or fruits and could result in adverse consequences for the least preferred ones. Seed of least preferred species are not dispersed or are carried away at a highly reduced rates resulting in crowding and other such consequences. Thus even reduction in population density (rather than total absence) may be sufficient to dramatically change the structure and functioning of an ecosystem (67;68). The value of seed dispersal systems is hard to estimate. Many tree crops of high economic importance depend on the seed dispersal services of animals. On the other hand the persistence of populations of wild vertebrates depends on the availability of fruits of such crops. The overharvesting of Brazil nuts (*Bertholletia excelsa*), acai palm (*Euterpe oleracea*), and Araucaria pine seeds (*Araucaria angustifolia*) in many areas — including protected areas as well — is threatening not only the plant populations but also the animals that depend on their seeds, such as peccaries, toucans, and other large-bodied frugivores (69; 70; 71).

QUESTIONS

1. What do you understand by 'Ecosystem'? Discuss the structure of a complete self sustaining ecosystem.
2. What do you understand by carrying capacity of an ecosystem? Discuss the effect of diminishing diversity on productivity and stability of an ecosystem.
3. How does trophic structure in Polar Regions differ from those of tropics?
4. What are ecological pyramids? Why the pyramid of energy provides the best graphical representation of structure of an ecosystem?
5. Briefly discuss different types of ecosystems, which occur, on our planet.
6. What do you understand by Ecosystem services?
7. Enumerate the services provided by a healthy ecosystem.

3

Chapter

The Impact of Human Activity on the Environment

Man has become a powerful force now. He has tremendous power at his command. His actions have a global impact. He has altered natural flora and fauna over a large surface area of the world. Human establishments and agriculture are fast expanding and encroaching upon natural systems. Forests are cleared, grasslands invaded, hill tops leveled, marshes drained and even land under water is reclaimed to provide space for human establishments and his agriculture. The soil is bared, flora and fauna exterminated, natural ecosystems are destroyed and are replaced by such artificial systems as agriculture, horticulture, animal farms etc or else a jungle of steel and concrete structures comes up in place of lush green vegetation.

(I) APPEARANCE OF MAN AND GROWTH OF HUMAN POPULATION

It was in congenial atmosphere created by biosphere's relentless activity of more than three billion years that man appeared on our planet barely a million years ago. During the first 40-50

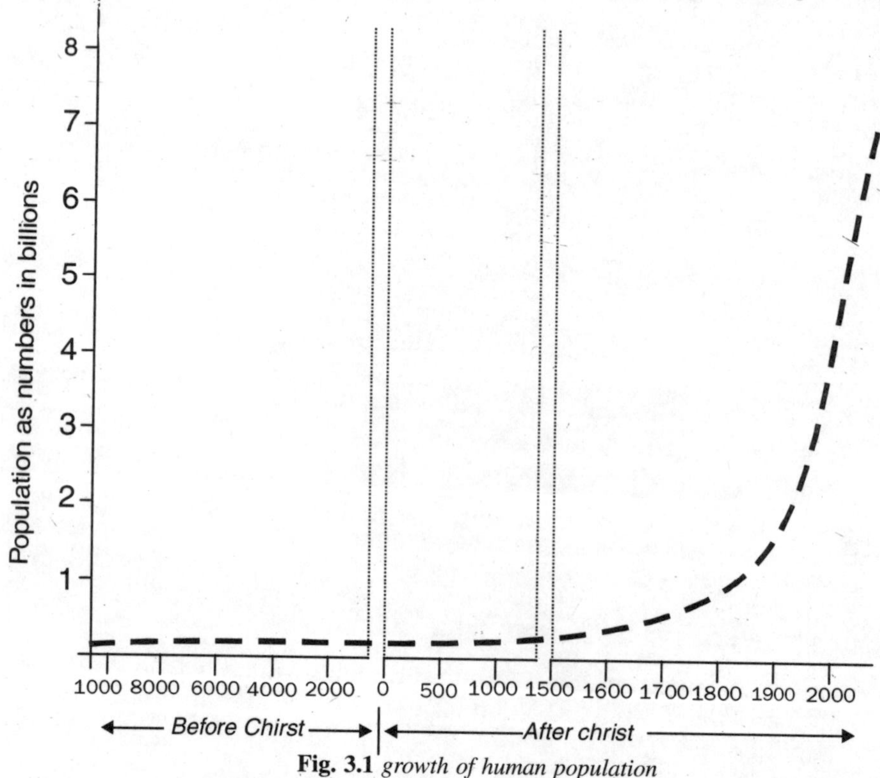

Fig. 3.1 *growth of human population*

thousand years of his existence man lived as a scavenger, hunter and gatherer of food material. Under the circumstances, the global ecosystem could support only about 10 million people. About 50,000 years ago, the total world population has been estimated to be about 10 million only.

The rapid growth of human population is mainly a result of rising food supply. The beginning of agriculture and domestication animals marked the first step of humanity towards a civilized society. As a result, the food production now rose to match the requirement of about 100 million people by 10,000 BC. It was about this time that human settlements sprang up in fertile regions of the world. For several thousand years following the development of settled life, human population remained well below the 250 million mark, which was crossed only about the time of Jesus the Christ. Within a span of 1650 years following the birth of Christ, the global population grew to about 500 million people and by 1800 AD, it was around 1 billion. It was only in the twentieth century that growth of world population entered its exponential phase. This was made possible by spectacular rise in food supplies and advancement in science technology. In 1930 AD, it was two billion, in 1960 AD, it was about three billion, in 1975 AD, it was four billion and in 1987 AD, it was five billion. By the close of the second millennium, the world population crossed the 6 billion mark. It grew at a rate of about 90×10^6 per year during the following years with a slight decline in growth rate till the end of the century. By a conservative estimate, the world population is expected to be above 7.0 billion by the year 2010 AD and 8.25 billion by the year 2025 AD (1;2;3).

Modern man has built up an extraordinary accumulation of 'human biomass' in which energy and materials flow in enormous quantities from all other systems. The wastes discarded by 'him' pollute the environment and threaten to disturb the global biogeochemical cycles. He lives a life of comfort and luxury, which requires additional resources. The consequences of his actions have a global impact (4). He successfully diverts a huge share of global production, 3.78×10^{17} to 4.59×10^{17} kcals for his own use out of the total world production of about 7.7×10^{17} kcals (5; 6). Though these estimates are crude, yet it is almost certain that more than half of the total world production is now consumed by a single species, the *Homo sapiens*. Naturally, the consequences of over-exploitation of natural resources and pollution of environment have resulted into steady impoverishment of biological systems and changes in global biogeochemical cycles. In general the impact of human activity results into:

A. Reduction of ecosystem complexity and diversity.

B. Changes in Biogeochemical cycles.

C. Adverse effects on Ecosystem services.

a. Reduction in ecosystem complexity and diversity

Man has been over-simplifying the structural complexities and diversity, which occur in undisturbed natural ecosystems. Agriculture, for example, is just an applied management of trophic chain. The complicated many-tiered trophic structure of a natural system is reduced to two links: primary producer and man. Man sets an artificial agricultural system, protects the biomass, which develops in his fields from pathogens, pests, birds and other herbivores. All this increases the efficiency of production as the energy needed for maintenance of a simple structure is much lower than the complicated ones (7). However, simple trophic structures are more vulnerable to catastrophic changes. An unforeseen pathogen could reduce or eliminate the single species, which happens to constitute the entire population of producers. In a complicated ecosystem with each trophic level composed of a number of species, several alternatives are available for energy to flow, for materials to circulate in the system, which is thus maintained in operative state (8).

Man also tends to oversimplify microbial community present in soils or water bodies. Use of chemical fertilizers and intensive agriculture lead to depletion of organic material on which microbial populations depend for nutrition while insecticides washed down into water and the soil kill susceptible organisms directly. The diversity of animal communities is also adversely affected by human activity. Man has taken a few species, domesticated and protected them to form huge populations. Those species, which are left out, are finding it increasingly difficult to cope up with the stresses imposed by human activity around them. The overall effect of this is the replacement of the diverse community with a community consisting of few species only which man chose to domesticate. This has led to a severely damaging effect on grasslands, forests and has caused outbreak of animal diseases, which have often acquired dimensions of an epidemic (9;10;11).

The process of reduction in genetic diversity has also been taking place along with over-simplification of natural ecosystems. A number of plant and animals have become extinct or are on verge of extinction. Evolution of new taxa and extinction of ill adopted ones have been taking place in a balanced way ever since life came into existence. However, the strain placed by human activity has speeded up the process of elimination of weaker species. This is a sad state of affairs. Each extinct species takes away with it a combination of traits or gene pool, which took millions of years to evolve (12;13). The loss is irreparable. To date about 150 species of plants are cultivated on large scale. Of these 150 species, only 29% provide the total food stock available to the humanity. Man has slowly and consciously been reducing the number of plants and animals, which he breeds. Selection of varieties for better output, resistance to the environmental conditions, better tastes and flavours etc. has led to a drastic reduction in genetic diversity of plants and animals. The varieties or species in which he is no longer interested are vanishing one after the other. In 1970 A D, only two varieties provided 70% of American wheat harvest. Only four varieties provided 72% of entire potato harvest in United States in 1980 AD. About 75% of the total apple production in France consisted of North American varieties in 1980 AD of which 71% was the famous Golden variety alone (14).

The reduction in genetic diversity of plants cultivated on large-scale places restrictions on the possibilities of creating new varieties through breeding programs. A more immediate consequence of growing genetically uniform plants or livestock is their increased vulnerability to unkind environment or unforeseen pathogens. For example, the Bezoska winter variety of wheat, which was grown over an area of 15 million hectares during successively mild winters in Russia, was destroyed by a single spell of severe winter in the year 1972 A.D. In 1970 A D, American farmers lost millions of dollars in maize harvest as their genetically uniform crop was attacked by a virulent fungus. For many Russians as well as American this brought famine like conditions (14).

b. Changes in biogeochemical cycles

Ruthless exploitation and pollution of the environment has disturbed the operations of all-important biogeochemical cycles. The magnitude of waste materials discarded by man has been growing persistently. These wastes or their decomposition products are regularly added to the various components of the environment in significant quantities to disturb even the global cycles. Man has been extracting substantial quantities of materials (such as fossil fuels) which represent biological output of past ages, at a rate much faster than the rate of their formation. So are the deposits of various minerals elements and metals formed by persistent biogeochemical activity of millions of years, which are being depleted at a very fast rate. These resources belong to our children and grand children as much as they do to us. The policy of reckless exploitation of natural resources is one, which our future generations will rightly deplore.

1. The Carbon Cycle:

Carbon occurs as carbon dioxide in the atmosphere, as organic compounds in plants and animal bodies, in coal and petroleum deposits and as inorganic carbonate in water, rocks, shells and testes etc. Photosynthesis brings carbon of atmosphere into the biotic pool while respiration and decomposition of organic matter adds most of this element to the atmosphere. A part of carbon of biotic reservoir may be fixed in coal petroleum and carbonate deposits, which constitute the biological output of the carbon cycle. Some carbon dioxide may dissolve into water from the atmosphere and enter the hydrosphere while inorganic carbonates in aquatic system may be precipitated and turned into limestone (15). A highly simplified carbon cycle is given in Fig 3.2.

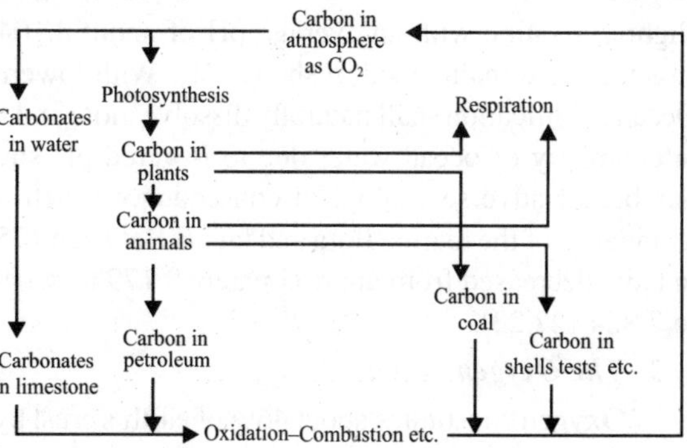

Fig. 3.2 *The Carbon Cycle*

Human activity has led to an enhanced rate of input of carbon into the atmosphere, which has caused a measurable rise in the concentration atmospheric carbon dioxide. This is due to increased use of organic matter, coal, petroleum and natural gas as fuel and the combustion of carbonate rocks for the manufacture of cement and lime. Rapidly growing human population is also modifying the natural ecosystems over an increasingly larger area of earth's surface by extensive deforestation, faulty agricultural practices, intensive grazing etc. The universally distributed poisons like lead and DDT tend to depress the photosynthesis efficiency of green plants on global scale. So while there is more input of carbon dioxide in the atmosphere its output is somewhat reduced.

The carbonate rocks in earth's crust is largest reservoir of carbon on our planet which contains about 66,000,000 to 100,000,000 x 10^9 metric tons of carbon. The atmosphere contains about 750 x 10^9 metric tons of carbon as carbon dioxide. We have a little more than 3500 x 10^9 metric tons of carbon locked in fossil fuels, which represent photosynthesis of past ages. Today, about 1900 x 10^9 metric tons of carbon is locked in terrestrial biosphere and soil organic matter. In oceans of the world about 42000 x 10^9 metric tons of carbon is present as dissolved CO_2, HCO_3, CO_3, organic compounds and in ocean biota. In pre-industrialized world (1750 AD) the concentration of CO_2 in the atmosphere was 278 ppm. Today it is about 380 ppm (16;17;18). The total mass of carbon in earth's atmosphere, during the period, rose by 12-14% of which about 150 x 10^9 metric tons has been introduced as a result of burning of fossil fuels alone. Rising concentration of CO_2 in the atmosphere threatens the mankind with:

a. Global warming and Climatic change: A small rise in the concentration of carbon dioxide in the atmosphere shall have no effect on plants and animals. In fact it could be beneficial to plants as photosynthesis should go up. However, carbon dioxide is a green house gas and absorbs infrared and heat waves effectively. A high concentration of carbon dioxide in the atmosphere causes it to acts like a big blanket around the globe which obstructs loss of heat from earth's surface. Earth's surface and oceans warm up. This effect is referred to as ***Green house effect*** or simply as ***Global warming***. Mean global surface temperatures have increased by 0.74 ± 0.18 °C during the last 100 years ending in 2005. The Intergovermental Panel on Climate Change (IPCC) concludes that most of the temperature increase since 1950 is "very likely" due to the rise in anthropogenic greenhouse gas concentrations. Climate model projections indicate that there could be further rise of 1.1 to 6.4 °C in global surface temperature by end of twenty-first century. A rise of a few degrees may seem small but this could have disastrous consequence for the entire world (19).

b. The danger of ocean acidification: Nearly 48% of the Carbon dioxide added to the atmosphere is absorbed by oceans which cover three fourth of earth's surface (20). Carbon dioxide when dissolved in water produces an acid which lowers the pH of water concerned. With rising concentration in the atmosphere more and more CO_2 shall enter oceans causing acidification of its waters. Our oceans are slightly alkaline with an average pH of about 8.104, due to which calcium encrusted live or dead structures are maintained in shape (21). With lowered pH ocean waters shall grow more corrosive. Ocean acidification shall naturally dissolve more and more calcified structures and the resultant change in chemistry of ocean water due to lowered pH shall trip the ecological balance. It could cause a number of adverse ecological consequences such as changes in nature, composition and relative abundance of the marine flora and fauna. Between 1751 and 1994 surface ocean pH has been estimated to have decreased from approximately 8.179 to 8.104. By the close of 21st century it could go down to 7.824 (22;23).

2. *The Oxygen cycle:*

Oxygen constitutes about 46% of earth's crust by weight and its cycle is complicated one because of the innumerable chemical combinations in which it occurs on this planet. By far the largest reserve of oxygen (about 99.5%) on our planet is earth's crust where it occurs in silicate, carbonate and oxide minerals. Atmospheric oxygen is a major reserve of this gas, in combination with hydrogen it occurs in water, with carbon it occurs as carbon dioxide, in carbonates and bicarbonates etc. It forms an important constituent of organic matter. During photosynthesis, water molecules dissociate to form oxygen, which is fed into the atmosphere from where it is drawn in aerobic respiratory activity of living organisms to yield water molecules again. A number of other oxidation reactions require oxygen (24; 25). A highly simplified oxygen cycle is given in Fig 3.3.

Apart from being an essential requirement for all aerobic organisms, oxygen has played a very important part in evolution of life on this planet. It has given rise to ozone umbrella, which protects terrestrial habitats from harmful solar radiations. Early life occurred in water only as there was no effective ozone concentration in the atmosphere, which could prevent biocidal ultraviolet radiations from reaching earth's surface. Since the level of oxygen in atmosphere determines the extent of ozone concentration it was only when enough oxygen could accumulate in the atmosphere so as to form an effective ozone shield that life could emerge from water and colonize the land. This happened about 440 million years ago sometime in the Silurian era when the oxygen concentration in the atmosphere rose to about 1/10 of the present day oxygen level (26;27). Disturbed Oxygen cycle poses the following problem to the humanity:

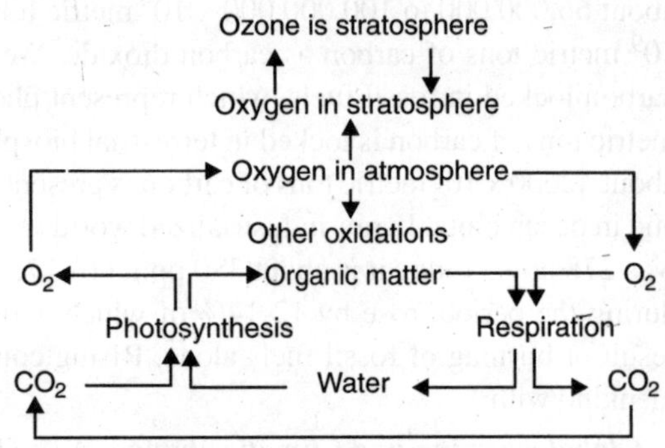

Fig. 3.3 *The Oxygen Cycle*

a. Our diminishing oxygen reserves: Though photochemical dissociations of water in the atmosphere has been estimated to produce about 1.5 gms of oxygen per sq. meter per year, the main process responsible for production of oxygen is photosynthesis which has been estimated to produce about 260 gms of oxygen per sq meter per year. Almost equal amount of oxygen is consumed in respiration of animals and other organisms. Oxygen is also consumed in reactions involving oxidation of large number of substances such as CO, Fe, H_2, CH_4 etc. but the precise quantitative significance of these reactions is not well understood. The quantity of oxygen required for burning coal, petroleum and natural gas has been estimated to be 10-11 gms per sq meter per year which come from atmospheric pool and this has no detectable effect on the global concentration of this gas. However, we are reducing our oxygen reserves by about 3-5 ppm every year (28).

b. The threat of stratospheric Ozone depletion: Of more immediate concern is the concentration of ozone in the atmosphere. In lower atmosphere, ozone is harmful to both plants and animals and causes photochemical or oxidizing type of air pollution. But high up in the stratosphere its presence is essential as it absorbs harmful ultraviolet radiations and provides a protective umbrella to the biosphere below. Ultraviolet rays split up oxygen molecules to produce oxygen atoms, which combine with molecular oxygen to form ozone. Photolytic dissociation of ozone also occurs naturally and its formation and dissociation balance each other. However, we have been introducing a variety of such materials in the upper atmosphere, which accelerate the dissociation of ozone catalytically. This has the undesirable consequences of thinning out the protective ozone shield which could in the long run turn the geologic clock back to the times when earth wore a very thin mantle of this gas and harmful ultraviolet radiations of sun could reach right up to earth's surface. Terrestrial life shall be drastically affected (24).

3. The Nitrogen cycle:

Nitrogen an essential element for all living organisms has a more complicated cycle. The main reserve of nitrogen is atmospheric air, rock deposits and the living and dead organic matter. Biological nitrogen fixation carried on by some bacteria and blue green algae brings the nitrogen of the atmospheric pool into the soil or aquatic bodies from where it is taken up in the biosphere. Abiological nitrogen fixation also provides some nitrogen to the living organisms. Decay and decomposition of plants and animal wastes yields ammonia, which is converted to nitrite by bacteria, *Nitrosomonas*, and then nitrate by the activity of bacteria like *Nitrobacter*. The process of dinitrification brought about by the activity of bacteria such as *Pseudomonas, Thiobacillus, Micrococcus dinitrificans* etc. converts nitrates to elemental nitrogen which is finally added to the atmospheric pool. Fig 3.4 shows a simplified nitrogen cycle. It should be obvious that the cyclic flow of nitrogen in an ecosystem involves a precise balance of activity of a few species of bacteria so that adequate levels of nutrients are maintained without excessive accumulation of inorganic and organic compounds of nitrogen, which are beyond a certain concentration harmful to a living system (9;11;29).

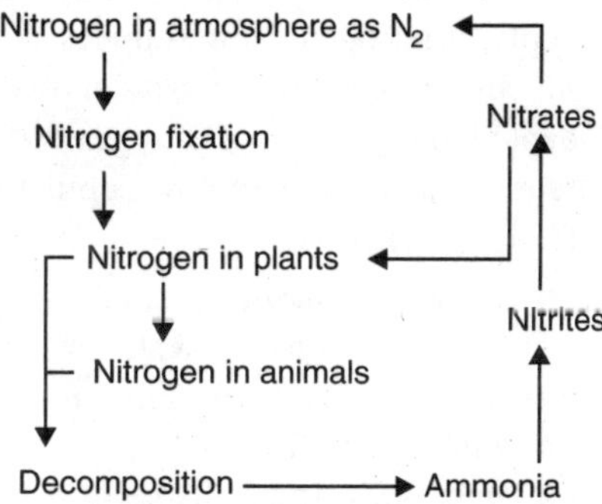

Fig. 3.4 *The nitorgen cycle*

Human activity has greatly altered the nitrogen cycle. It is having serious impacts on ecosystems around the world because nitrogen is essential to living organisms and its availability plays a critical role in the organization and functioning of the world's ecosystems (30).

a. *The case of the Missing Carbon Sink:* All over the world the plant growth and standing crop biomass have been limited by scanty nitrogen supplies – this is particularly so in temperate and boreal regions. Raised levels of nitrogen deposition in natural ecosystems must have caused additional plant growth, thereby sequestering some atmospheric carbon into the plant biomass. Important question which are yet to be answered are: How much plant growth has been caused by man-generated nitrogen addition? How much extra carbon has been stored in plant biomass rather than contributing to the atmospheric CO_2 content? Answers to these questions could help explain the imbalance in the carbon cycle that has come to be known as the '**missing carbon sink**' which is the only the positive anthropogenic effect on nitrogen cycle (30;31).

b. *Raised concentrations of Oxides of Nitrogen in the atmosphere:* Increased global concentrations of nitrous oxide (N_2O) and nitric oxide, (NO) have been causing great concern these days. Nitrous oxide is a potent greenhouse gas and with nitric oxide it drives the formation of photochemical smog in the atmosphere. Atmospheric reactions may also convert oxides of nitrogen into an acid which could contribute to acid rains. In the stratosphere oxides of nitrogen contribute to ozone depletion catalytically (32).

c. ***Acidification of soils and of the waters of streams and lakes:*** As organic matter is decomposed, ammonia accumulates in the soil. It is converted to nitrates by bacterial action, a process which releases hydrogen ions. Acidity develops in the soil. Higher concentrations of nitrates enhance emissions of nitrous oxides from the soil.

d. ***Loss of soil nutrients such as calcium and potassium essential for long-term soil fertility:*** Highly water-soluble nitrates are leached into streams or groundwater. As these negatively charged nitrate ions seep away, they carry with them positively charged alkaline minerals such as calcium, magnesium, and potassium. Thus human modifications to the nitrogen cycle decrease soil fertility by greatly accelerating the loss of calcium and other nutrients that are vital for plant growth (30).

e. ***Accelerated loss of biological diversity:*** Most of the natural systems are adapted to a limited supply of nitrogen. New supplies of nitrogen showered upon these ecosystems can cause a marked reduction is species diversity and changes in mix and relative dominance of species. Species well adapted to high level of nitrogen in the soil out-compete the ill adapted ones which are either lost or are represented by very low populations. With changes at lower levels higher trophic levels are also affected (33).

f. ***Detrimental effects on estuarine and near-shore ecosystems:*** Greatly increased transport of nitrogen by rivers into estuaries and coastal waters results in changes in the plant and animal life and ecological processes of estuarine and near shore ecosystems, and contributes to long-term decline in coastal marine fisheries.

4. ***The Phosphorus cycle :***

Phosphorus plays an essential role in almost every step of organic synthesis. This element is more abundant in living organisms than in abiotic system. The main reservoirs of phosphorus on earth are living organisms and relatively insoluble calcium phosphate deposits in rocks and sediments. Phosphorus is solubilized from its deposits under conditions of low pH and is taken up by plants, which pass it on to the food chain. Plant and animal secretion and decomposition of dead organic material bring it back to the surrounding medium. A part of this phosphorus is leached away to oceans through rivers where it forms marine deposits being available to fishes and marine birds which convert it into phosphate rocks, guano-deposits and bone deposits as relatively insoluble tricalcium phosphate. Therefore, only a little amount of phosphorus returns to land from oceans through fishes and guano birds while much of it is lost to relatively deep-sea deposits (10;34). Fig 3.6 depicts a highly simplified phosphorus cycle.

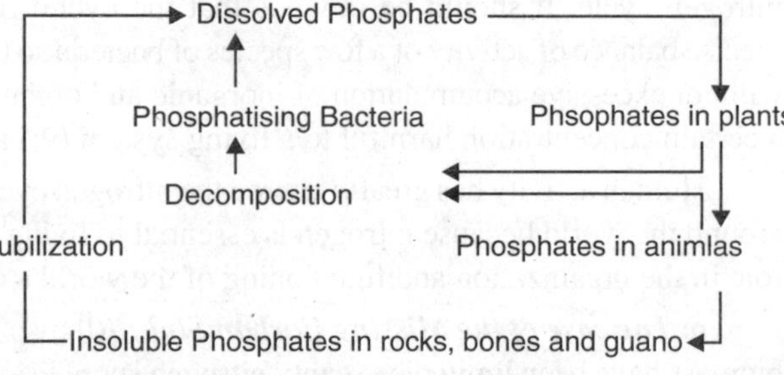

Fig. 3.5 *The Phosphours Cycle*

Our understanding of the rate of transfer of phosphorus between various components of the cycle is very poor. It is likely that only up to 1900 A.D. phosphorus cycle was in a balanced state in which the rate of its solubilization was almost equal to the rate of its transfer through rivers to oceans. However, rapid soil erosion has greatly enhanced the rate of loss of phosphorus from soils to the sea. Hence, phosphorus has to be added to agricultural fields to sustain productivity. The industrial production of phosphorus overtook its natural rate of cycling around 1910 A.D. and about 70-80% of the industrially produced phosphorus is ultimately applied to agricultural fields. The remaining phosphorus is usually used in soaps, detergents and water softening plants, which is discharged directly into fresh water systems, or oceans wherein it causes nutrient enrichment, excessive algal growths and other problems (35).

The practice of burning dead plant material and cow dung cakes instead of using them as natural manures cuts short the natural cycling and releases the primary nutrient rather in one stroke without providing the much needed infra-structure – the humus – to the soil which is helpful in retention of mineral nutrients. The released phosphorus is quickly leached away by running water. Poor yields and subsequent degradation of agricultural land are the major consequences of the practice, which has been prevalent in many Asian countries including India (36).

5. *The Sulphur cycle:*

Major reserves of sulphur on earth's crust are deposits of sulphide and sulphates while as organic sulphur it occurs in bodies of all living being. Plant and animal secretions and decomposition of dead organic matter yields hydrogen sulphide gas, which is converted to sulphates by bacterial action. Plants take up sulphur as sulphates. Bacterial action may turn some sulphates to hydrogen sulphide which is changed to sulphates again. A simplified sulphur cycle is presented in Fig 3.6. Sulphur in coal and petroleum yields sulphur dioxide gas upon combustion. A number of metals occur as sulphide or sulphate deposits which during their mining and processing operations give rise mostly to sulphur dioxide gas. Sulphur dioxide is changed to corresponding acid which, finally, yields sulphates which are ultimately added to the sulphate content of the environment. Some sulphur is also added to the soil or to aquatic system as a result of volcanic activity.

There are two major factors contributed by human being, which disturb the natural sulphur cycle. Combustion of fossil fuels yields an enormous amount of sulphur dioxide worldwide. It is the injection of sulphur into the atmosphere, which has been causing great concern. A lot of sulphur is added to atmosphere as a result of mining and processing operations during extraction of metals. Volcanic emissions, which are the only known natural source of sulphur dioxide, have been estimated to contribute 2×10^9 to 5×10^9 kg of sulphur per year, while the annual amount of industrial sulphur injected into the atmosphere is thought to be about 83×10^9 kg (35). Sulphur dioxide is a serious local pollutant with a wide range of harmful effects. This gas is mainly responsible for causing acid rains. Rapid exploitation of sulphur deposits is another feature, which causes anxiety to us. These deposits are believed to have been produced by the activity of anaerobic bacteria (*Baggiota* sp) from hydrogen sulphide gas. We are consuming these deposits faster than they are formed naturally. A day might come when there will be no good deposits of sulphur left for human use (36).

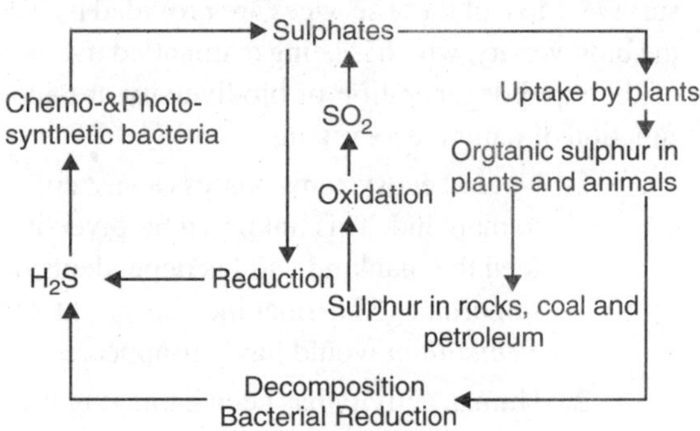

Fig. 3.6 *The Sulphur Cycle*

6. Biogeochemical cycles of trace elements:

In addition to major elements which compose the bulk of living organisms, there are a number of trace elements, which also circulate in the biosphere. Some of these are essential ones, some may enter a living being as a substitute for essential ones, while there are some elements, which are simply drawn in and circulated in the biosphere along with the general flow of materials. They are released in more soluble state in the soil as a consequence of weathering of parent rocks from where they move to aquatic bodies as well. Primary producers take them up and pass them on higher trophic level in the food chain. Exudates of living organisms and products of decomposition of dead plant and animals bodies bring them back to the soil A part of these elements are leached away with rapidly flowing waters to rivers and then to the seas where they are finally laid down in sedimentary deposits. Some of these elements may also combine with organic molecules to form volatile organometallic compounds and contaminate the atmosphere as well.

It is mainly the human activity, which has disturbed the natural cycles of these elements. Important ones which are causing environmental problems these days are lead, mercury, cadmium, chromium, arsenic, zinc and nickel etc. Industrial production of most of these metals has already exceeded their natural rate of circulation in global ecosystem. Coal and petroleum crude is a complex mixture of a number of such elements. Combustion of coal, petroleum and their mishandling during processing, transportation and storage etc. release these elements in a much higher concentration than those attained during their natural cycling.

We are exploiting the deposits of these elements at a much faster rate than they are being laid down. Obviously, there may come a time when there is no good quality, economically viable deposits left for our use. Unlike consumptions of world' stock of fossil fuels, metals are not irreversibly consumed – instead they tend to be scattered and broadcast in diluted state by human activity. In the process, these elements may get accumulated at places in sufficient concentration to be harmful to a biological system as has been the case of mercury, cadmium and lead (36).

C. DISRUPTION OF ECOSYSTEM SERVICES.

Services rendered by natural ecosystems to humanity are invaluable and irreplaceable. Apart from providing a number of commodities, like edibles, spices, medicines, pharmaceutical and industrial products, healthy ecosystems carry out a number of essential jobs without which mankind may not survive. Most of these services are provided by a complex collection of microbes, plants and animals, the biodiversity, which is being dismantled in bits and pieces by the human activity. The consequences of the rapid degeneration of biodiversity are apparent if we take a closer look of the structure and function of natural ecosystems:

1. Natural ecosystems possess a vast collection of biodiversity which is not even fully known to mankind. This unknown biodiversity may possess many species which could be used to feed the mankind and his dependents or could be of potential pharmaceutical or industrial important (37). Imagine the plight of mankind if plants like *Cinchona* or fungus like *Penicillium* would have disappeared before their curative properties were discovered.
2. Human activity has been damaging the biodiversity in our soils which causes its formation and maintains its fertility by faulty agricultural practices and excessive use of fertilizers and pesticides. This has resulted in nearly 70% of the global area of arable land being affected up to varying degrees by soil degradation and has lost nearly one third of its productive capacity – with about 19.85% being in a critical condition (38).
3. About 88 billion metric tons of organic wastes are generated by man and his dependents which have to be decomposed to harmless constituents and the mineral constituents recycled to be used again by the primary producers. Man tends to concentrate these wastes in isolated pockets on the outskirts of human establishments on land surface or dispose them of in river channels where they create problems of waste disposal, breed insects pests and pathogens which endanger human health and those of his dependents. The products of decomposition of these wastes are washed off with rains. Natural ecosystems, which are used to limited supplies of nitrogen are saturated with salts of nitrogen. Decomposition of organic matter under oxygen deficient conditions produces methane a potent green house gas while combustion yields CO_2 which adds to the concentration of green house gases in the atmosphere. All this affects the structure and functioning of natural ecosystems which disrupts the 'services' they provide (37).

4. Worldwide biological nitrogen fixation provides about two third of the nitrogen needed for world agriculture. Faulty agricultural practices, excessive use of nitrogenous fertilizers, insecticides and pesticides have a detrimental effect on the natural nitrogen fixing machinery necessitating the addition of more and more inorganic nitrogen after each harvest (39).
5. Photosynthesis by green plants and the phytoplanktons ubiquitously present in oceans removes a substantial quantity of CO_2 emitted in the atmosphere. Today, while adding more and more of this gas to the atmosphere we are disturbing this vital cycle by land use modifications and excessive use of pesticides which tend to depress the efficiency of photosynthesis. This is expected to enhance the rate of global warming and climatic change. Through land use modification we are altering the albedo over a large area of earth's surface which results in changes in local climate and pattern of precipitation (40).
6. Invasions of species beyond their native range constitute a major concern for the conservation of natural and managed areas all over the world. Invasive species threaten biodiversity, change ecosystem structure and function and adversely affect ecosystem services. In healthy ecosystem the rich diversity of life forms provides biotic resistance to invading species. They are simply unable to establish, grow and multiply. But in systems disturbed by human activity this resistance is lost (41).
7. Nearly 99% of insects and pests are controlled by natural enemy species and host plant resistance. In natural ecosystems a process of checks and balances (prey-predator interactions) is operative which automatically controls the populations of diverse insects, pests and pathogens. Human activity disturbs this system of checks and balances and reduces richness and biodiversity. Insects, pests and pathogens multiply causing greater losses and higher incidences of plant and animal diseases including those of humans (42).
8. Nearly one third of world's food production relies heavily, either directly or indirectly on insect pollination. Poor reproduction observed in several rare plants has been linked to the loss of specialized pollinators. Low fruit production in plants was usually attributed to nutrient limitation, but a number of studies have shown that failure of fertilization due to pollen limitation is the cause of fruiting failure and thereby low fruit yield. In order to persist in agro-ecosystems pollinators need local floral diversity and nesting sites. Large monocultures, which we set up, fail to provide these. Pollination limitation also leads to increased inbreeding, reduced genetic fitness, and increased susceptibility to environmental stress (43).
9. Seed dispersal is an important process in natural ecosystems and population dynamics. Absence of an effective population of animal agents of seed dispersal causes the plant population to be confined to localized areas which results in crowding, wastage of seeds, inbreeding depression and loss of genetic diversity in the species concerned which makes the species susceptible to stresses of environment (44).

QUESTIONS

1. Describe the structure of earth's atmosphere. Which of the different layers of atmosphere have been chemically altered by anthropogenic activity? Enumerate the consequences?
2. Discuss the unique properties of water molecule and its role in shaping the biosphere.
3. What are biogeochemical cycles? Enumerate the consequences of changes which man has brought about in global biogeochemical cycles.
4. Discuss the modification brought about in the earth's environment by the biosphere.

PART – II

Natural Resources

4

Chapter

Natural Resources

With a phenomenal rise in human population, natural resources are being heavily taxed all over the world. No doubt at present, total global production is nearly enough to match the human demands for energy and materials if we judiciously distribute the resources available to us. However, looking forward to future scenario, the situation appears pretty grim. Nearly all-agricultural production, produce to animal husbandry, the world fisheries etc. have to be diverted to support the human society. While the number of mouths to be fed is regularly rising, degradation of natural ecosystems, deterioration of fertile soils and pollution of environment threatens world food production. In near future, it could be difficult for the world food production to keep pace with the growing demands.

(I) WHAT IS A NATURAL RESOURCE?

Anything, which is useful to man, or can be transformed into a useful product or can be used to produce a useful thing, can be referred to as a *Resource.* A natural resource is the resource obtained from nature. It is these natural resources, which form the very basis of entire life on this planet. A natural resource can be of the following two types:

1. Biotic Resource
2. Abiotic Resource

A biotic resource is the resource, which is directly or indirectly derived from photosynthetic activity of green plants. Food, fruits, wood, fibre, milk and milk products, fish meat, leather etc. are termed as biotic resource. Coal, oil and natural gas are also biotic resources as they were produced by photosynthetic activity of plant, which occurred millions of years ago. Mineral material, fresh water, rocks salts and chemicals etc. are termed as abiotic resources as biological activity is not involved in their formation.

(II) RENEWABLE AND NON-RENEWABLE RESOURCES

A resource can be renewable or non-renewable. Renewable resources are those resources, which can be regenerated, whereas non-renewable resources are those, which cannot be regenerated once they are exhausted. Our high-grade mineral deposits and deposits of fossil fuels are non-renewable resources as a finite quantity of mineral elements, coal, oil and natural gas is present on our planet, which may be consumed completely. There formation requires millions of years, which cannot occur within the human scale time. Unlike fossil fuels, mineral elements are inexhaustible, i.e., we cannot consume them irrecoverably, yet the concentrated deposits, which occur today, can disappear at some point of time in future.

Wood, fibre, fodder fruits vegetables, meat milk and milk products etc. – resources that are developed directly or indirectly by recent photosynthetic activity are renewable resources. They can be generated again and again as long as photosynthesis continues on this planet. However, it does not mean that these resources are unlimited. Consumption faster than their regeneration not only causes their depletion but also tend to damage the very system, which is responsible for their production.

(III) BASIC HUMAN REQUIREMENT:

The biosphere is largely composed of about a dozen of lighter elements (.... with molecular weights lower than 40) which are linked together by chemical bonds during the formation of which energy is used. It is primarily the green plants which are the starting point for the entry of both material and energy into the biosphere. The requirements of greens plants are few and simple. They need an inorganic carbon source, water mineral nutrients and light. All animals including man depend directly or indirectly for their food requirement on green plants. As animals do man can also satisfy almost all of his basic requirements from biosphere around him, e.g., from plants, animals and microbes.

However, in addition to food and water the modern man also requires a large number of other things. Shelter, clothing, transportation, entertainment, defense, medicines etc. are some of the necessities of human life for which man needs both material and energy resources of different kinds often in huge quantities. Therefore, resources importance to man of today can be grouped as:

1. Food Resources
2. Water Resources
3. Energy Resources
4. Mineral Resources
5. Forests and Wildlife

These resources are not equally distributed throughout the world. We tend to recognize the value of a resource only when it is rather scarce. With plenty of rains and fresh water, all around us we do not realized its significance. But in dry, semi-desert regions no one can ignore the importance of fresh water – simply because it is in short supply. When a resource is in abundance, its division, distribution and conservation is often meaningless. It is only when per capita consumption approaches per capita availability of any resource that competition develops and the resource acquires its true importance. However, over exploitation threatens most of our natural resources now. It will be wise policy to practice economy, equitable distribution and start conservation efforts with future demands in mind so as to make our natural resources last longer.

(IV) FINITE NATURE OF NATURAL RESOURCES:

Neither energy nor matter can be created or destroyed. No one, howsoever advanced technologically, can create something out of nothing nor can anything be so discarded as to become nothing. A çonstant flow of materials is needed to maintain living beings, a society or an economy, which must come from somewhere, whereas a continuous of stream of discarded wastes has to go somewhere. The total amount of matter present on our planet is fixed – except for some cosmic particles entering and some gases leaving from outer atmosphere. This has been so through the known history of our planet. The elemental composition of earth's atmosphere, hydrosphere and the lithosphere is fixed, stable and known.

The enormous changes, which take place every day, involve changes in the state, mix and distribution of materials on earth. Plant communities grow, die, or burn away. Large quantities of water evaporate, condense and rain back to earth's surface. Volcanoes erupt, emit lava, ash and gases, create new islands or bury cities. Each year man extracts billions of tons of earth, transform them chemically creating new molecular combinations, which never existed before. All these activities require energy, which comes only from just three sources – the incoming solar radiations, energy from nuclear fission and the residual heat of earth's core. Fusion energy is the only source of energy which man has not been able to harness till date.

In other words, as far as the matter is concerned, our planet is nearly a closed system – nothing enters or leaves it. It is this finite quantity which circulates again and again in repeating channels to sustain life on this planet. However, for energy, earth is an open system. It receives large amounts of energy from sun, much of which has to be radiated back to space in order to maintain a controlled temperature. Mankind's immediate environment, therefore, the planet earth, is limited in size and space as well as in its material resources.

1. Limitations of non-renewable resources:

Most of our mineral deposits and deposits of fossil fuels are non-renewable in nature. Long periods of geochemical or biological activity have collected these materials to form concentrated deposits. The slow, insidious, almost imperceptible processes, which have created these deposits, are still operating and most of the deposits are being developed even today. But consumption faster than their regeneration shall exhaust them at some point of time in future. We may not irreversibly consume these materials but their exploitation does scatter them in a highly dispersed state in the environment. With a sophisticated technology, we may recover these materials from their highly dispersed state. However, the cost shall be enormous and effort could be economically non-viable. Unlike mineral deposits, we can use the deposits of fossil fuels only once. Once they are depleted, there is no way to re-form them. Thus continued over-exploitation shall exhaust many of our valuable deposits which took millions of years to form .They cannot be duplicated within the human scale of time. They require time on geological scale to form (1)

2. Limitations of renewable resources:

Most of the biotic resources, which are developed as a result of only recent photosynthetic activity of green plants can be referred to as renewable resources . The biosphere (minus man) constitutes an excellent life support system, which can fulfill all human needs. Nevertheless, its size and productivity is limited by availability of water, nutrients and environmental conditions.

Though an enormous quantity of water is present on our planet, for fresh water life depends largely on precipitation, which too is available only in a finite quantity annually. Its uneven distribution over earth's surface has caused large surface areas to become infertile deserts. A large part of earth's surface is too cold or rocky for any productive use, whereas a large area has to be devoted to forests and wild life, which due to their obvious importance cannot be curtailed beyond certain limits. This leaves only a limited area at our disposal for agriculture and pastures. No wonder during the past thirty years the area used for food grain production has not increased at all – in spite of earnest human efforts to do so.

Growing more and more from the same crop fields involves expensive use of fertilizers, energy inputs, irrigation and high yielding varieties. Global livestock and fisheries resources can also not be expanded beyond certain limits. Products of forestry and wild life too cannot be extended indefinitely. It is only up to a limited extent that resources of biosphere can be safely exploited. Over-exploitation tends to damage the biotic systems which produce them and thereby reducing the overall productivity. It could cripple the very resource base which is so important to our existence. Thus like non-renewable resources, renewable resources are also finite. They cannot be stretched beyond a certain limits. With careful management, this limit can be enlarged. However, there is an ultimate line within which humanity shall have to confine itself (2).

(VI) LACK OF SUSTAINABILITY OF MODERN SOCIETY:

Consumption of natural resources whether slow or at a fast rate shall finally result in exhaustion of non-renewable resources at some point of time in future. However, we have little to worry about mineral wealth, as we cannot consume metals and minerals irreversibly. Recycling and reclamation of lower and lower grade of ores, shall provide something to us, though the product concerned may become more and more expensive. The impending threat of exhaustion of our fossil fuels presents to us a greater worry. While we may exhaust our oil within 50-65 years and natural gas in a hundred years or so, it is coal which is going to last the longest. As we learn to adopt ourselves to the use of alternative sources of energy, there shall be drastic changes in technologies, economies and our very life-styles. Man with his tremendous skill and adaptability should be able to withstand the *fuel shock* rather comfortably (3). Indeed humanity has lived without minerals or fossil fuels for thousands of years. Widespread use of metals started about three thousand years ago while extensive use of fossil fuels is only about three hundred years old. Man can survive without these resources. However, our greatest worry is the gradual deterioration in the quality of environment, which has been damaging the resources of biological origin. These are renewable resources and it is these resources on which all living beings on this planet depend for their food supplies. For a growing human population we need an ever-rising quantity of food, fresh water and other materials. This has placed enormous strain on our renewable resources, which have so far been mercilessly over-exploited. The deterioration in the quality of our environment has slowly been damaging the very system, which gives rise to these resources. Even under healthy conditions of environment, most of the biotic resources can be exploited only up to a limited extent. Over-exploitation leads to biological impoverishment causing a drastic reduction in its productivity. The system starts behaving like a non-renewable resource, which may become completely defunct due to continued over-exploitation. The danger to its productivity becomes acute if the system has to work under adverse conditions of environment (4;5).

Clearly, such a state of rapid consumption of our natural resources, a steady impoverishment of our biological systems and deterioration in the quality of our environment cannot continue for long. It is unsustainable. At any point of time in future, the system could collapse, leaving the humanity with acute shortages of food to eat cloths to wear and scarcities of all types. Aggravating the situation are enormous inequalities of distribution, appalling injustices in allocation of resources, vested interests of people trying to corner for themselves a larger share through unscrupulous means. There are countries whose entire economy rests on exploitation of natural resources only – there is no other way for survival. About 20% of the global population now enjoys a life of wasteful luxuriance whereas much of the remaining 80% live under conditions of utter poverty, scarcity and helplessness. Naturally, such a world is not going to last for long (6).

(VII) TOWARDS SUSTAINABLE DEVELOPMENT:

Both the affluence of developed countries and the desperate poverty of under-developed countries are injurious to the life support system on our planet. Human life in developed countries of the world requires large amounts of energy and material inputs while a ceaseless stream of wastes is generated which damages the environment and results in rapid depletion of resources of our planet. Life in under-developed countries strives to survive on a meager share, clamouring for the basic necessities and in ignorance or desperation often damages the very resource base on which rests the entire life support machinery of this planet.

We have to build a sustainable world – a world which should last forever. There should be a fair sharing of global resources among the living beings. Everyone in this world should get at least the basic amenities of everyday life – food clothing, drinking water, shelter etc. in such a way that there should be no damage to other life forms and the environment. The resources of the world if properly managed, distributed and utilized economically are sufficient for all living beings – as the biosphere stands today. In future, however, we may require sharp decline in growth rate of human population, which we are capable of bringing about with a little effort (7).

To build up a sustainable world, a world of permanence in which all living beings live in perfect harmony with each other and with the environment we shall have to adopt certain basic practices which can be enumerated as under:

- Protecting and augmenting regenerability of the life support system on this planet which can be achieved by avoiding wasteful use of natural resources, rationalized husbanding of all renewable resources, conserving all non-renewable resources and prolonging their life by recycling and re-use.
- Fair sharing of resources, means and products of development between and within the nations of the world. This should lead to a significant reduction in disparity in resource use, in its economy and shall curtail the associated environmental damages all the world over.
- Educating people regarding the concealed economic and environmental costs of over consumption of resources.
- Adopting willingly sustainability as a way of life by encouraging frugality (to be content with less) and fraternity (sharing things with others in a fair way).
- Meeting the genuine needs and legitimate aspirations of all by blending economic development with environmental imperatives to remove poverty.

Today the environment is no longer a concern about a locality or wildlife or deforestation or pollution. It is a crisis about the developmental pattern, which we have followed so far. It is a global issue, which forces us to think as to where we are going? What shall happen if we do not stop, reconsider and make necessary modifications in our means, methods and objectives. It is high time that we should re-think and take proper steps to build up a world of permanence – a sustainable society that lasts forever.

QUESTIONS

1. What do you understand by a Natural Resource?
2. Discuss the limitations of Renewable and Non-renewable resources.
3. Why our modern society lacks the elements of "Sustainability"?
4. What can be done to turn our society into a Sustainable system?

5

Chapter

Land: A Precious Resource

Soil is an extremely precious resource for mankind. It is a solid stratum underneath. Like a precious carpet of lose material spread over the bed rocks, it is our soils which sustain life by providing nutrients to the green plants which in turn support all life on earth's surface. Like a sponge it stores up and provides passage to ground water, shelter to microbes and burrowing animals and supports foundations of buildings and other utility structures.

(I) MAJOR COMPONENTS OF THE SOIL:

Important constituents of a soil are

1. Particles of sand silt, clay or gravel.
2. Organic material.
3. Soil water and air.

1. Particles of sand, silt and clay:

The principal component of the soil is inorganic material called sand, silt, clay or gravel which are derived from the decay and decomposition of parent rock material from which the soil is derived. The distinction between the particulate constituents of the soil is somewhat arbitrary and is based on size of particle. Sand particles are 0.02-2.0 mm in diameter and are mainly composed of quartz or silica. Silt is finer. It consists of particles between 0.002-0.02 mm in diameter being made up of quart along other silicate minerals. Particles less than 0.002 mm in diameter consists of wide variety of silicates and aluminosilicates. They are colloidal in nature and are collectively referred to as clay. Gravel consists of particles larger than 2.00 mm in diameter. Depending on the proportions of sand silt and clay a soil may be grouped into a number of classes such as sandy, sandy loam, clayey loam, clay etc.

2. Soil organic matter:

The organic matter content of soils usually ranges between 2-5% or at times it may be less than 2%. Partially decomposed organic matter present in a soil is referred to humus, in the formation of which most of cellulosic material disappears, lignins are modified while proteins are retained. Though present in rather a small quantity, organic matter or humus plays a very important role in maintenance of porosity, water and nutrient retaining capacity and the fertility of a soil. It binds soil particles in aggregates of smaller or larger size (the crumb) which are very helpful in restricting soil erosion. The top layers of soil consists of dead organic substances such as plant litter, decaying excreta and remains of a variety of living organisms which include bacteria, fungi, algae protozoa worms, arthropods, mollusks etc. The amount of organic material in soil depends on climatic conditions, the type of inorganic soil components and the topography. In temperate regions of the world where decomposition is slow, plenty of organic debris collects in the soil whereas in the tropical and subtropical regions warm and humid conditions cause quick decomposition and mineralization of organic matter.

3. Soil water and atmosphere:

Soil water and atmosphere or air fills up the spaces in between the soil particles. The quantity and the movement of water and gases in the soil are determined by the size of particles. Soils with a higher percentage of smaller particles, like clayey soils, usually have more pore space though the pores are smaller. Soils with larger particles in abundance, like sandy soils have little interstitial space but pores are larger. The water holding capacity of sandy soils is low although most of the water present is available to plants. However, in clayey soils the water holding capacity is higher but most of it is held tightly by mineral in the microscopic pores and is unavailable to plants. The water in soil is a solution of dissolved materials from which plants obtain nutrients as well as water. To a large extent the amount of dissolved nutrients depends upon the pH of soil solution. Pollution which affects the pH of the soil water, therefore, also influences the availability of nutrient ions to the plants.

Some space between the soil particles is also occupied by the soil atmosphere or air which is vitally important for oxidation process involved in the decay of organic material in the soil. Roots of higher plants and a number of soil microbes require oxygen to grow and carry on their activity. Absence of oxygen, as is the case of water logged soils, causes reducing conditions to occur and anaerobic activity is stepped up. The soil atmosphere is similar to but not identical to the air above as there is a higher percentage of carbon dioxide present in it. This is due to decay and decomposition of organic material.

(II) BASIS OF SUSTAINED FERTILITY OF SOILS

Plant nutrients in the soil are regularly regenerated by decay, decomposition and mineralization of organic matter deposited by living organisms, which also solubilize nutrients from rock fragments and soil particles. Atmospheric nitrogen is fixed by a number of bacteria and blue-green algae. A number of microbes are known to produce growth-promoting substances. Insects, helminthes, rodents etc. turnover and mix a large quantity of soil. Root decay leaves in the soil a network of channels. Decaying plant roots and burrowing animals create innumerable passages in the soil, which helps in exchange of gases. Mucilages produced by blue green algae, bacteria and a number of other micro-organisms along with organic matter bind the soil particles in crumbs or aggregates which prevents excessive leaching and erosion. Fungal hyphae weave a network, which helps in keeping these aggregates together (1;2).

We have been cultivating our soils ever since agriculture started thousands of years ago. Had soil been a lifeless material its store of plant nutrients, howsoever large, might have disappeared at some point of time. Living communities in the soil are the basis of its fertility and other properties, which promote growth of a healthy plant life on it years after years in succession. Living organisms in the soil require nutrients, water, air and like all other living beings their capacity to provide nutrients to the plants which grow on it is limited. Anything detrimental to the living community within the soils tends to cause its degeneration, loss of fertility and an increased rate of soil erosion. In absence of living communities and organic matter soil turns into a useless heap of sand silt and clay.

At a given point of time, only a definite amount of plant biomass can be sustainably harvested from the system – withdrawal over and above this limit damages the system. In man-made agro-ecosystems, once we have harvested a crop, much the available nutrients in the soil are also removed with it and subsequent crops are poorer in yield. The soil needs time to regenerate its store of nutrients. We can enhance the nutrient regeneration capacity of a soil by adding additional organic matter which nurtures soil biota, improves water holding capacity and soil texture. Of course, we can obtain higher yields by adding chemical fertilizers. However, the use of chemical fertilizers is detrimental to the soils as it is deprived of organic matter on which the soil microbes thrive and in time the soil loses is nutrient regeneration capacity.

(III) ARABLE LAND IS A SCARCE COMMODITY

Out of nearly 167.5 million sq kms of total available land surface distributed among five continents and more than 180 countries, only 18.25 million sq kms of land may be used for growing food. Large tracts of land have to be set aside as grasslands and pastures which have to be maintained to grow feed for animals. Some land is needed to grow fruits and vegetables and other cash crops. On about 44.1 million sq km stand forests with 40% or more of canopy cover. About 105.2 million sq.kms of land consists of barren wasteland or land under other uses. Large tract of land are occupied either by hilly terrains, barren deserts or else are covered with ice where food cannot be grown (3). Bringing barren wasteland under cultivation is virtually impossible or extremely difficult. Forests cannot be reduced due to their obvious important in maintaining global ecological balance and as storehouses of biodiversity. It is on the best and the most productive land that most of our cities and industries stand – we cannot convert it into cropland.

Global area used for cereal cultivation was 590 million hectares in the year 1950. It rose to about 720 million hectares by the year 1975-1976. This unprecedented rise of about 22% in the area devoted to cereal cultivation has altogether no parallel in the known history of mankind. However during the last thirty years, in spite of our efforts the crop fields have not extended at all, rather there has been a decline of about eight million hectares. We have probably reached the end. If we closely examine the data we shall find that much of this increment in global area devoted to cultivation of cereals has come from extension of our crop field over marginal soils – soils on hill slopes or water stressed areas which are too fragile for intensive cultivation. We have cleared hill slopes, carved steps to cultivate cereals and other edibles. We have dammed rivers, dug canals to carry water to our water starved field. Cultivation on marginal soils requires more skillful management. Carrying capacity of these soils is often poor and has to be supplemented with an ever rising fertilizer and energy inputs to obtain higher yields which is detrimental to the health of the soils.

This makes arable land a scarce commodity and leaves rather a small fraction of land area for growing food, which is not likely to expand any more in near future. To feed the ever-rising global population we shall have to grow more and more food from the existing croplands only (4). There are certain regions of the world only which are gifted with large tracts of fertile land. The largest area of arable land occurs in South America followed by North America and South Asia. In terms of percentage of total area, it is the South Asian countries, the Indian subcontinent to which goes the credit of having the highest proportion of richest and the most fertile land in the world (Fig 5.1).

(IV) LAND DEGRADATION

One of the basic reasons for degradation of land and fertile soils is the extensive clearing of the cover of vegetation which protects and nurses the soil underneath. Faulty agricultural practices, extensive modification of land surface, construction activities, mining and processing operation etc. also contribute substantially to land degradation. It is the soil life, the microbes, animals and plants which transform rock fragments – sand, silt and clay – into fertile and productive soils. Anything that is detrimental to soil life tends to degrade the soil fertility. The plant cover over the soil surface exercises a strong control on the microclimate within the soil. Once the plant cover is removed the soil is exposed to solar radiations and the battering action of wind and rains. As it heats up the organic matter present in the soil disintegrates and the soil biota disappears. The soil is turned into a dead heap of sand, silt and clay. With nothing to hold soil particles together and sustain soil crumb structure the top surface is loosened. In deeper layers of soil compaction occurs resulting into the development of hard impervious layer underneath. The loss of soil as silt and sediments from forested land is negligible, from grassland it is 0.1-0.2 tons per hectare, from agricultural land it is about 32 tons per hectare, whereas from bared, denuded soil and deforested wasteland it is as high as 59 tons per hectare (5). The loosened top soil is subjected accelerated erosion while the hard impervious layer developed underneath impedes sub-surface drainage resulting in water logging or accumul·tion of salts and alkalis. The process of land degradation is hastened by pollution of the soil which adversely affects the soil-life.

Nearly 70% of the global area of arable land has now been affected up to some degree by degradation and have lost nearly one third of its productive capacity – with about 19.85% being in a critical condition. Every year we have to use more and more energy and add an ever rising quantity of fertilizers to maintain productivity. The cost of food production is rising every year. Land degradation may be categorized into follows type:

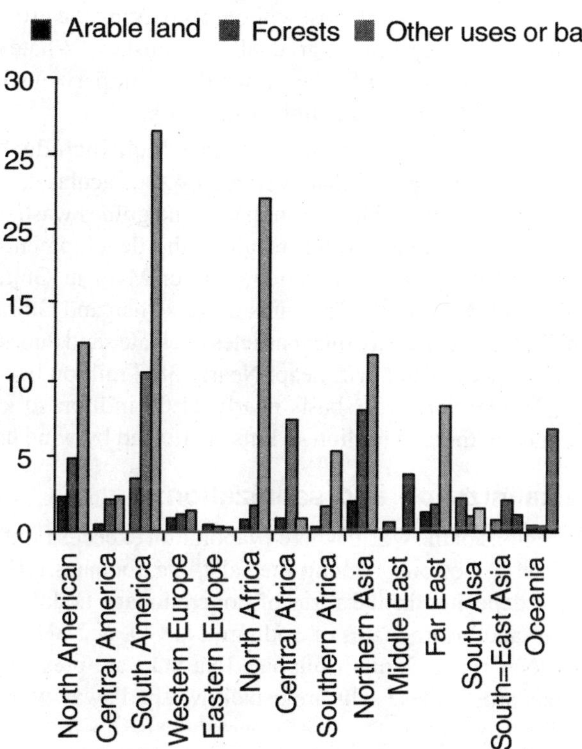

Fig. 5.1 *The pattern of global land use*

1. Soil erosion:

Erosion is displacement of soil, sediments, rock and other particulate material caused by air currents, flowing waters, movement of ice or subsidence of land under the influence of gravity. Erosion is a natural process but in many places it is greatly enhanced by the pattern of human land use. Poor land use practices include removal of the cover of vegetation, overgrazing, and unmanaged construction activity etc. With the loss of cover of vegetation, the soil loses its ability to soak up rain water and shear strength. The loss of soil moisture is amply testified by the appearance of drought tolerant plants or xerophytes. The roots of trees, shrubs and grasses bind the soil particles together and provide it strength while organic matter binds the soil particles in soil crumb structure. Their disappearance loosens the top soil and makes it vulnerable to erosion, denudation. Air or water currents carry away the dust and detritus in an ever rising quantity. Rain drops form millions of tiny craters on bare soil surface which connect to form rills. Rapidly flowing waters turn rill into gullies. Gullies eventually turn in ravines (Gully erosion). In water stressed regions, winds sweep away the sheet of finer surface particles and deposit them in the form of sand dunes elsewhere (Sheet erosion). The coarser material is left behind as skeletal drift heaps. Land that is used for the production of agricultural crops generally experiences a significantly greater rate of erosion than that of land under natural vegetation.

Today the rivers of the world carry about 18100 million metric tons of soil (silt and sediments) to the sea every year. Yellow river and Yangtze rivers of China transport, annually, about 1600 million metric tons of silt and sediment to the sea. Ganga alone carries nearly 411 million tons of precious soil every year from Uttar Pradesh and Bihar to the Bay of Bengal. To this is added the sediment load of river Brahmaputra which ranges between 402-710 million metric tons per year. An enormous quantity of soil, each centimeter thickness of which takes nearly a thousand years to develop, is lost. And also gone are the precious nutrients which would have otherwise nurtured a rich plant cover over the soil surface. As the organic matter and soil moisture disappears compaction occurs which prevents rainwater infiltration through it. With no plants, grasses herbs or shrubs to impede the flow of water much of rainwater is quickly lost without recharging the groundwater table. Underground water table a precious freshwater resource gradually diminishes. Whatever water that infiltrates the upper layers is prevented from moving further down due to impervious layer underneath – causing problems of water logging, salinity and alkalinity (6).

River Yamuna and its tributaries which include Betwa and Ken in Uttarpradesh and Madhyapradesh, India, are characterized by spectacular development of ravinous land or badlands. Torrents of rainwater gushing through rills and gullies wash the soil away while parts with calcareous deposits are left in place which result in the development of deep impenetrable ravines. Similar ravinous formations occur in valley of river Mahi in Gujrat and river Damodar in West Bengal. Deprived of plant cover the dry lands of Rajasthan and Gujrat in India are severely affected by wind erosion. Winds sweep away finer particles to create sand dunes at faraway places while coarser material is left behind as skeletal drift heap. Nearly 38.7 million hectares of Indian land is afflicted by wind erosion (7). On worldwide basis nearly 11.00 million sq kms of arable land is affected by water erosion and another 5.5 million sq kms is affected by wind erosion.

2. Alkalinization and salinization:

In regions of the world where evaporation exceeds rainfall and sub-surface drainage is poor salts accumulate in the soil. Among the salts, bicarbonates (HCO_3) and carbonates (CO_3) predominate which are deposited in the form of concretionary nodules, referred to as Kankar. In some places carbonates and bicarbonates of Sodium (Na_2CO_3 & $NaHCO_3$) along with a little amount of Sodium chloride ($NaCl$) and Sodium sulphate (Na_2SO_4) are present in higher quantities. Where sub-surface drainage is proper these salts are usually washed away with movement of water in the underground water table. However in localities where there is sub-surface pan or layer of hard impervious material the dissolved salts accumulate. As the land dries up, capillary action brings up salts and colloidal material to the surface from deeper layers of the soil. The soil turns saline or alkaline and its fertility is severely hampered. As little as 0.2% Sodium carbonate is sufficient to render the soil useless for cultivation. Taking water by canals to irrigate the cropland, or irrigation by tube well or river water in water stressed regions, a practice which is being followed all over the world is all right in monsoon fed localities where sub-surface drainage is proper. These waters are in fact dilute salt solutions. Plants take up what they require. Due to inadequate drainage the residual salt content accumulates in the soil and in the long run causes problems of salinity and alkalinity.

Because all Na^+ and K^+ and many Ca^{2+} and Mg^{2+} salts of chloride, sulfide, and carbonate are readily soluble, it is this set of ions that contributes most to soil salinity. At sufficiently high concentrations, the salts pose a toxicity hazard. Crop utilization of water and fertilizers has the effect of concentrating salts in the soil. In absence of careful management, irrigated soils can become saline or develop toxicity. There are numerous reports of reduced fertility and salinity caused by use of river water stored in huge dams and reservoirs in drier regions of the world. In drainage basins of rivers Colorado and Columbia, the Nile, Euphrates and Tigris, Indus basin, Irtysh, Yangtze etc salt accumulation has damaged soil fertility significantly.

A widespread example of irrigation-induced toxicity hazard is nitrate accumulation in groundwater caused by the excess leaching of nitrogenous fertilizer through agricultural soil. Human infants using groundwater containing higher concentration of nitrates as drinking water can contract methemoglobinemia ("blue baby syndrome") because of the transformation of NO_3^- to toxic NO_2^- (nitrite) in the digestive tract. Costly groundwater treatment is currently the only remedy possible when this problem arises. In India, even in the fertile Gangetic plains, there have been many set back. Yields have declined by 40-70% in the command area of the famous Sharda Sahayak Project (U.P.). Sriram Sagar (Andhra Pradesh) has laid waste close to 60,000 hectares of fertile land. Projects on Chambal (M.P. & Rajasthan) and Gandak (Bihar & U.P.) have destroyed 98,700 hectares and 2,11,010 hectares of fertile land. In 1980 about 7.0 million hectares of our soils were salt infested which represents 7% of the total irrigated area (7).

Roughly about 770,000 sq kms of global arable land has already been affected by excessive accumulation of salts. On worldwide basis we are losing about 16,000 sq kms of world's irrigated farmland every year due to alkalinization and salinization (8;9). Generally it is the over irrigated but poorly drained crop fields which have mostly suffered from the scourge of alkalinization and salinization. About 1.12 million hectares and 0.52 million hectares of rich and fertile farmland in Uttarpradesh and Punjab respectively, suffers from a heavy accumulation of alkali and salts and have lost their productivity. Needless to say, these states are situated in the most fertile and productive region in the world. In India a total of about 8.2 million hectares of fertile crop land has been damaged by alkalinization or salinization or both.

3. Water logging:

Water logging is a problem caused by rise of the underground water table very close to the ground surface. The water rises up to the zone of roots of the plants pushing out the air present in the interstitial spaces between the soil-particles. This disturbs all processes within plant-roots and the soil which need oxygen. The productivity of land declines considerably. The basements of buildings are plagued with persistent dampness. Water logged land loses its sheer strength – the capacity to bear load of buildings and other utility structures. As the foundations sink cracks appear in the buildings while salts dissolved in water may damage the concrete and stone structures chemically. Excessive infiltration of surface water due to prolonged rain, seepage from canal systems and impoundments and influx of water through over-irrigation is usually responsible for development of water logged conditions.

The upper surface of the ground water is usually referred to as water-table. It usually conforms roughly to the topography of the land and its depth depends on a number of factors. In localities having a gentle slope and heavy to moderate rains the water table may be within a few meters of ground surface. In area with sparse or nominal rains it could be as deep as 50 meters or more. The water table fluctuates regularly from season to season. In drier months ground water the table sinks to deeper levels. The rate of fluctuation depends on the rate of movement of ground water or permeability of the medium varying from a few centimeters to several meters per day. In impermeable soils, such as soils rich in fine textured clays, there is practically no movement of ground water. Such soils are more likely to develop water logged conditions as the water collects over the impermeable surface filling the pores and crevices in the soil and driving the air out.

Water logged soils have a hard impermeable pan or strata underneath which prevents downward or lateral infiltration of water. In absence of proper underground drainage, water accumulates in the surface layers and saturates the soil. Many of the dry and arid regions of the world are characterized by a poor underground drainage system. Most of these areas have traditionally been under rain-fed irrigation. The soils can support cultivation of course cereals, many of which can grow under dry and rough conditions. An attempt to irrigate such soils by constructing artificial reservoirs and canals tends to provide more water than these soils are used to absorb. Due to poor drainage, the excess water is unable to escape. It raises the underground water table, which comes to lie within a meter or so under the surface of the soil. The soils are unable to support the usual crops and the productivity declines drastically. Even houses and buildings over such soils may collapse due to waterlogged foundation. An unfortunate outcome water logging in such localities is concentration of the dissolved salts in the upper layers of the soil resulting into raised salinity and alkalinity. As these soils dry up under the scorching heat of the day during summers, the dissolved salts are brought up and deposited in the upper layers while water evaporates. High temperatures and wind velocities may evaporate as much as 10-12 mm of water per sq.meter per day.

All over the world, particularly in the regions where large scale irrigation projects have been taken up, water logging has wasted large areas of productive land. The canal system developed to transfer waters of Amu darya and Syr darya away from Aral sea and Irtysh-Karaganda canal (Or Satpoev's canal) water- logged nearly 0.92 million hectares of productive land. These canals were a part of massive irrigation projects taken up during Soviet era in Kazakistan which added 2.9 million hectares to the total irrigated area in the locality. Both the projects have failed now. The localities which these projects sought to irrigate, suffered from acute salinization and water logging. Aral Sea which had the reputation of being world's largest freshwater lake has diminished to one third of its original volume. Its waters are muddy and salt infested. The thriving commercial fishery has collapsed and the locality is now one of the worst poverty hit regions of the world. The project has been dubbed as "*A symbol of planned destruction*" by a World Bank team (10).

In India about 12.7 million hectares of precious agricultural land was reported to be waterlogged by 1992 (7). The command area of Sharda sahayak project (UP) has waterlogged almost 78 thousand hectares of productive land. The Sri Ram Sagar Irrigation project of Andhrapradesh has water logged close to about 60 thousand hectares. Irrigation projects on Chambal (Madhyapradesh and Rajasthana) and Gandak (Bihar and Uttarpradesh) have wasted about 98.7 and 211.0 thousand hectares of productive land to water logging. Indira Gandhi canal in the northwestern Rajasthan has brought about extensive water logging of dry land. In parts of southwestern Punjab intensive irrigation caused a rise of water table by 20-50 cms per year during 1975 to 1995 which has resulted in the expansion of waterlogged area from 60.1 thousand hectares in 1978 to 104.0 thousand hectares in 1997(11).

There are numerous reports of water logging associated with construction of large dams, reservoirs and the network of canals built to carry water to dry and arid regions of the world. A typical case of extensive water logging and salinity is the famous Indira Gandhi Nahar Pariyojna or the Rajasthan Canal. As initiated in 1960, the project proposed to irrigate 1.87 million hectares of land in Thar Desert, India, using 9.36 billion cubic meters of water annually through a network of canals spread over 8,840 kms. Nearly 60% of the project was completed by 2000. No doubt, the project was boon to the people of Ganganagar, Jaisalmer, Bikaner and Jodhpur district. However, water logging and raised salinity had already begun corroding the productivity with the commencement of the project. Nearly 6.5% of the area brought under irrigation by the canal water was reported to be water logged and about 30% of the total are intended to be served by the canal system was reported to be vulnerable to water logging by 2000AD.

4. Soil pollution:

Soil pollution can be defined as the build-up in soils of persistent toxic compounds, chemicals, salts, radioactive materials, or disease causing agents, which have adverse effects on plant growth and animal health. This type of contamination typically arises from application of pesticides, percolation of contaminated surface water to subsurface strata, leaching of wastes from landfills and waste dumps or direct discharge of industrial wastes to the soil.

Most of the soil contaminations are the result of pollutants adhering to the soil particle surface, or lodging in the spaces in between the particles of the soil. The accumulation of pollutants in the soil depends upon a number of factors. Some toxic materials may literally drain through soils such as sand and gravel and move under the influence of gravity to deeper layers. Most of the polar or organic chemicals discharged into a clayey soil will stay in place as they are strongly adsorbed to the soil particles. Of course the equilibrium reached is a dynamic one, where new pollutants may lodge on to soil particles releasing the older ones and the action of groundwater movement may over time transport some of the soil contaminants to other locations or depths. Soil contaminants may have significantly deleterious effects on the soil systems. There may be drastic changes in soil chemistry which can arise from the presence of many hazardous chemicals even at low concentration of the toxic agents. These changes bring about changes in the metabolism of micro biota within the soil. Many microbes useful to soil may disappear altogether. In most of the cases contaminants typically alter plant metabolism, which shows up as reduced crop yields. The adverse effects on soil life disturb the soil texture making it prone to soil erosion. Organisms living in the soil may absorb, accumulate and biomagnify the contaminants. Edibles produced on the contaminated land often possess higher concentration of the toxic materials and could be unfit for human consumptions. Soil pollution, therefore, degrades a significant proportion of land area every year.

In some areas of China soil pollution presents a genuine threat to area's food security with soils already suffering from various degrees of pollution. According to a scientific sampling, 100,000 sq kms of China's cultivated land have been polluted. Contaminated water is being used to irrigate further 21,670 sq kms of cropland. About 1,300 sq kms has already been covered or destroyed by solid waste. In total, the area accounts for one-tenth of China's cultivatable land which is mostly in economically developed areas. An estimated 12 million metric tons of grain are contaminated by heavy metals every year, causing direct losses of 2.57 billion US dollars. Harmful substances accumulate in crops and via the food chain, find their way into our bodies, where they can cause a variety of illnesses. Soil pollution also damages ecosystems and ultimately threatens their safety (12).

Reckless growth of industrialization and urbanization in third world countries ignores the norms which should be followed to treat and dispose of liquid and solid wastes. In most of the large cities in Asia and Africa huge dumps of solid wastes can be seen skirting the city. Leachates from these dumps, effluents discharged by industries, wastes from mining and processing operations, pesticides used for crop protection or storage pollute an ever increasing area of fertile soils. Effluents from the industries of Kanpur and Unnao in Uttarpradesh have raised heavy metal concentration in the silt and sediments transported by Ganga. As much 23.6 ppm of chromium, 14.7 ppm of cadmium, 12.2 ppm of tin, 3.2 ppm of zinc, 2.6 ppm of copper and 1.6 ppm of nickel – concentrations well above the permissible level – has been detected (13).

5. Desertification:

Desertification is the degradation of land in dry arid, semi arid or sub-humid areas of the world due primarily to human activities and climatic variations. It involves the depletion of vegetation and soils and usually stems from the demands of increased populations that settle on the land in order to grow crops and graze animals. Today, an increasing area of productive land is being damaged by desertification at a rate much faster than ever recorded in the known history of mankind.

The new deserts which are being created are not necessarily in hot and dry places. Instead, they could be developed at any place where the soil has been so mistreated by humans that it has become useless for growing crops. Soils can be ruined easily in areas where seasonal rainfall is unreliable. Cutting down forests and trees, over-cultivation of the soil and over-grazing can all contribute to desertification. But it is in Africa and India where the problems of desertification are worst. The soils are often poorly-managed, which means that they store less water and produce less grain even when there is enough rainfall. Today nearly 75% of the world's drier lands – 45 thousand sq kms – are affected by desertification. The Sahara desert is expanding southwards at the alarming rate of about 5-10 kms per year. Every year 6 thousand hectares of agricultural land are being lost and turning into virtual desert. It has been estimated that nearly 35% of the earth's land surface is threatened by prospects of desertification which could affect the livelihoods of about 850 million people.

(VI) LAND DEGRADATION – A SERIOUS THREAT TO GLOBAL FOOD SECURITY

Today land degradation and desertification is seen as a serious threat to global food security. An analysis of rise and fall of a number of ancient civilizations suggests that cropland soils have had a major role in determining how long past civilizations survived as progressive entities in one place before they collapsed. Historical records of the past six thousand years show that civilized man, with a few exceptions, was never able to continue as a flourishing civilization in one locality for more than 30-80 generations or 8-20 centuries. The three noteworthy exceptions are the Nile valley (200 human life-times), Mesopotamia (160 human life-times), and the Indus Valley (175 human life-times) (14). These three civilizations (Mesopotamian, Egypt, and Indus) are distinguished by the fact that all had large river valleys that replenished the soil and its fertility annually. They lasted over 150 generations. Civilizations that lacked this soil-renewal feature lasted only 50 or so generations. It was only the failure of the soil replenishment system that terminated these civilizations (14). Note that the bulk of the civilizations described above had temperate climates. Most soils in tropical climates have much lower organic matter contents (1% carbon vs. 3% carbon in temperate climates) and therefore tend to be far less productive than soils in temperate climates. This may explain why many tropical regions have never hosted progressive civilizations. Note also that the bulk of the developing world is in tropical climates (14).

Although the reasons for the collapse of the ancient civilizations are varied most of them bear an enduring legacy of massive soil erosion, deforestation, overgrazing, irrigation systems that have never recovered from the buildup of salts in the soil, and desertification reflecting attempts to convert semi-arid (grazing) lands into croplands and the formation, growth and migration of deserts. Deforestation and over-grazing in the Armenian highlands (eastern Turkey) are believed to be the source of the silt loads that the Tigris-Euphrates River carried into Mesopotamia. From there the silt would have clogged the huge system of irrigation canals to such a degree that the task of irrigation system maintenance overwhelmed the Mesopotamians. Piles of dredged silt can still be seen in these localities.

Most other irrigation-dependent civilizations probably collapsed due to salinization and/or water logging. The people of that period might not have understood the chemistry or, if they did, they lacked the knowledge or the technology required to prevent such damage. Salinization and water-logging have an element of treachery associated with them. They act slowly over time, giving many civilizations an opportunity to develop to significant sizes. When irrigators first recognized the problem they changed to increasingly salt-resistant crops, prolonging the civilization's lifetime by some decades or centuries. But eventually they ran out of edible crops that could tolerate the salt concentrations that had built up. Restoration of these areas would require huge reallocations of water that is currently allocated to food production. This is why one rarely hears of restoration of abandoned irrigation systems.

Wind erosion created a "dust bowl" in the 1930s in an area covering Oklahoma and a number of surrounding states in the South-central US, creating huge numbers of wretched "environmental refugees". During 1954-62 the former Soviet Union embarked on a massive "Virgin Lands" program, converting 300,000 sq kms of grazing land into semi-arid croplands. The results were predictable - windstorms wiped out increase in soil productivity. The "Virgin Lands" program was declared a disaster and abandoned (15). Wind erosion also has an element of treachery associated with it. Almost nothing bad happens for a while - until the topsoil depth diminishes to the depth of the root zone of the crops being grown. It has to be abandoned. A number of other modern-day sub-civilizations are seemingly bent upon following in the footsteps of the environmental refugees of South-central US and the Soviets. Globally, 4.6 million of the world's 12.3 million sq kms of rain-fed croplands are classified as dry lands and hence at risk of eventual abandonment due to wind erosion.

People in the developing world are primarily concerned about their next meal, not their future meals a few decades from now. Thus they ignore future food productivities, and this leads inherently to natural resource management practices that are far away from being sustainable. As estimated by the world's leading soil scientists, an area of about 1.2 billion hectares - about the size of China and

India combined - has experienced moderate to extreme soil deterioration since World War II as a result of human activities. Over three-fourths of that deterioration has occurred in the developing regions from such causes as overgrazing, deforestation, land clearing, un-wise agricultural practices, and increased soil salinity and water logging, largely from irrigation. Other environmental threats to the agricultural resource base include loss of water through contamination, loss of genetic resources, habitats, and species, adverse impacts of pesticides, and greater resistance of plant diseases, weeds and insects; and climate change, both local and global. During the years 1985-86 when the global food production peaked the net growth rate was about 1.86% per year. The required growth rate of 2% per year maintained through a period of fifty years in the global food production appears to be tall order. In the face of growing world population, with shrinking resource base it will be difficult to feed the teeming millions of the world (16).

QUESTIONS:

1. What are major components of our soils?
2. "Fertile land is a scarce commodity" Discuss.
3. What is basis of sustained fertility of our soils? How can it be augment?
4. Enumerate major reasons of degradation of our soils. How can it be prevented?
5. Discuss the role of modern agricultural practices in degradation of our soils.
6. "Land degradation poses a serious threat to global food security in near future" Discuss.

6

Chapter

Freshwater

Freshwater is a natural resource of fundamental importance. Without water, life is impossible. In many respects, the properties of water are unique among all other liquids. As a solvent for organic and inorganic materials water has no parallel. When cooled below 4° C the density of water decreases till at 0°C ice forms which is lighter that water. It forms a solid crust over water surface. The aquatic life continues safely under the ice sheet. The extraordinarily high specific heat of water requires large amounts of energy or its withdrawal to raise or reduce its temperatures. The high latent heat of evaporation needs large amounts of heat energy to evaporate a unit volume of water. Among other liquids water is characterized by a highest surface tension due to which large quantities of water are retained in the soil as soil moisture on which plant life thrives. The properties of water, therefore, seem to be especially designed for the living organisms. No other liquid can replace it.

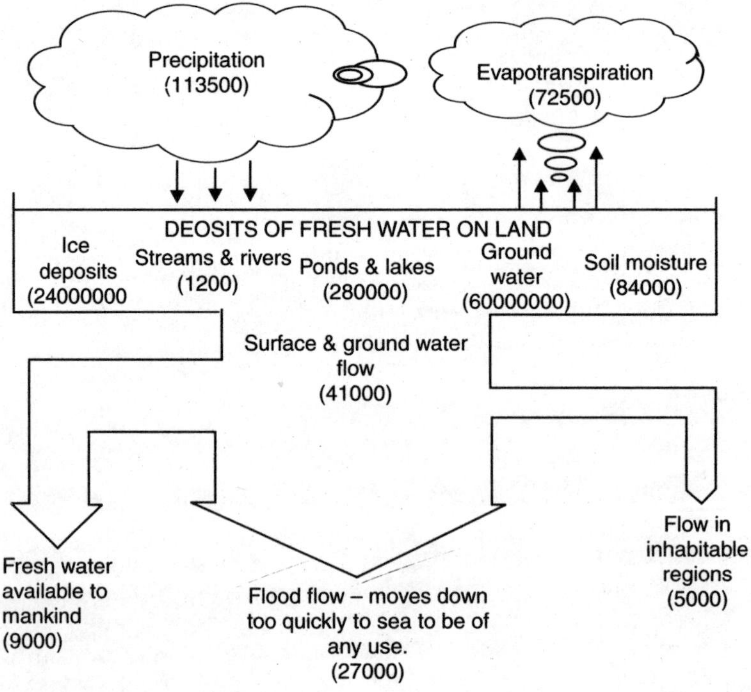

Fig 6.1 *Freshwater available to manking. (All quantities in cubic kms).*

(I) SOURCE OF FRESH WATER FOR TERRESTRIAL LIFE

We have a huge quantity of water on our planet, almost 1500 million cubic kms (1) but most of it is useless for us as it contains too much of salt. Total amount of available fresh water on our planet is only about 84.4 million cubic kms. Much of the freshwater on earth's surface and ground water represent deposits, which have accumulated over a long period of time. If input via precipitation exceeds the output, small amounts of fresh water accumulate as left over stock in ice deposits, lakes and reservoirs or underground surface. About 60,000,000 cubic kms of fresh water occur in deposits under the land surface. 24,000,000 cubic kms of water are locked as ice in snowcaps over mountains, ice-sheets and glaciers. About 85,000 cubic kms of fresh water are present as soil moisture on which plants survive. Nearly 1,200 cubic kms of freshwater is present in our streams and river at any point of time (Fig 6.1).

Solar radiations vaporize large quantities of water from both land and oceans. Plants transpire. A large quantity of water about 5,25,100 cubic kms is introduced into the atmosphere as water vapours. Air rises and cools down. The moisture condenses in the atmosphere and forms rain, dew or snow which are brought back to earth's surface. Every year withdrawals from surface and subsurface deposits are replaced by precipitation while the excess water is drained off (Table 6.1). It will be noticed from Table 6.1 that land surface contributes about 72,500 cubic kms of water to the atmosphere while it receives 1,13,500 cubic kms as precipitation. The excess water received by land surface, about 41,000 cubic kms has to flow back to sea in streams and rivers under the influence of gravity (2; 3).

Table 6.1 *Annual water budget of our planet*

	Water in cubic km
Total evaporation from sea surface	4,52,600
Total evaporation from land surface	72,500
Total precipitation on ocean surface	4,11,600
Total precipitation on land surface	1,13,500
Total Surface and ground water flow	41,000
Total evaporation from land and sea surface	5,25,100
Total precipitation on land and sea surface	5,25,100

These deposits of freshwater on land surface have their own significance. Ice sheets and glaciers regularly feed rivers and underground deposits at lower altitudes when ice melts and the water trickles down. Water in high altitude lakes and underground strata seeps down and trickles out under the influence of gravity. It is added to the flow of the streams and rivers or to the underground water table at lower altitudes. In sub-tropical and temperate regions of the world, where the precipitation is concentrated in a few rainy months, this maintains a regular flow in rivers and streams throughout the year. Streams and rivers flow over long distances catering to the need of people living all along its course. They transport large quantities of sediments rich in plant nutrients, which are deposited in low-lying areas in a broad zone all along its course. Water seeping down the beds of river replenishes ground water table on either side even in dry months making ground water available at convenient depths on either side of river channel. Drainage basins of streams and rivers of the world are underlain by huge stores of fresh water present in porous subsurface strata at variable depths. This makes underground water table a ubiquitous source of freshwater. Almost all of the ancient civilizations of the world developed in river valleys where plenty of freshwater was available which contained rich and fertile soils. It was easy, at these places, to obtain water all round the year from nearby rivers or underground deposits, which were often a few meters below the surface. Near sea-coasts underground water holding strata has to be saturated with fresh water to prevent salt-water infiltration from sea nearby.

The total annual precipitation of 1,13,500 cubic kms is not evenly distributed over earth's surface. At a given point of time, the amount of precipitable moisture present in the atmosphere is maximum at equator being equivalent to 44 mm of rains. At latitude of 45°-50° north and south, the available precipitable moisture would be about 25 mm during summers and 10 mm during winters as rain equivalent. At poles, this yield ranges from 2 mm in winters and 8 mm in summers as rain equivalent. The amount of precipitable moisture in atmosphere is subject to large variations which depend on a number of factors such as the pattern of temperature and pressure, flow of air currents and the topography of the region. However, it does make the equatorial belt the wettest zone. Rainfall decreases as we move out on either side of equator acquiring a seasonal character (2).

(II) THE LIMITS OF GLOBAL FRESH WATER RESOURCE

Like all other natural resources, global freshwater resource, also has its own limitations. There is a final limit up to which man-kind can draw water available in various deposits on earth's surface and sub-surface layers, without damaging the natural resource base or without causing any adverse changes in the environment. What is this limit? Up to what extent the withdrawal of fresh water by humanity is ecologically sustainable?

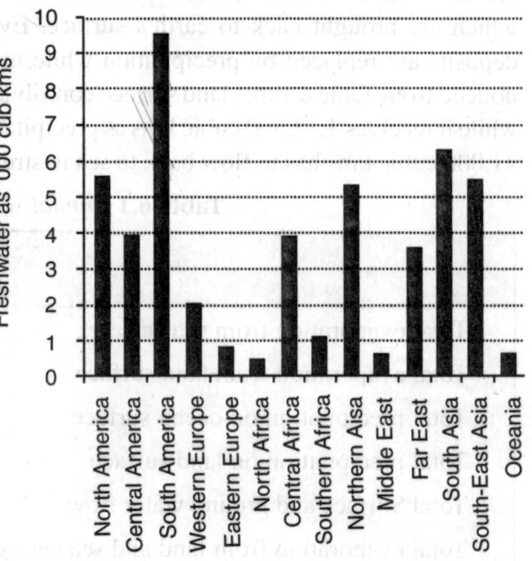

Fig 6.2 *Renewable fresh water available to mankind*

Precipitation is the chief source of fresh water for terrestrial life, which replenishes surface and ground water deposits, before trickling down and flowing back to the sea. In fact the amount, which can be safely used each year by humanity without causing any detrimental effect on ecology or environment, is the amount of fresh water which can be replenished by the net annual gain (41000 cubic kms) of water on land surface. Annual withdrawals up to the extent replaceable each year by precipitation shall not reduce the existing deposits on earth's surface or the underground water table. Any over draft beyond this quantity either from surface deposits or from ground water shall diminish the natural resource base, which in turn could bring about adverse changes in environment and ecology. Greenery shall disappear, the flora and fauna shall undergo changes and desertification will follow. Out of the 41,000 cubic kms of surplus water about 27,000 cubic kms consists of flood flows, which rushes down to sea, too quickly to be of any use to mankind. Of the remaining 14,000 cubic kms about 5,000 cubic kms, happen to be placed in inhabitable or very sparsely populated area and are of little use to the mankind. It is only about 9,000 cubic kms, which can serve the humanity (3; 1).

(III) MAN'S WATER REQUIREMENT

Water is needed in almost all spheres of human activity. It is required for direct consumption or indirectly for growing human food, washing, cleaning, cooling, transportation or even waste disposal.

Important sectors of human activity, which require water, can be grouped as follows:

1. Food production.
2. Power generation.
3. Industries
4. Domestic requirement
5. Miscellaneous.

According to an estimate, about 3,500 cubic kms of water were drawn for human use in the year 1970 AD (1). Human food production is the singular area which accounts for the bulk of water consumed by mankind. Almost 76% of the total water used by man has to be diverted to world agriculture, horticulture and livestock management. To produce 20 tons of organic matter in terms of fresh weight, 2000 tons of water has to be provided to the roots. Most of it is lost in transpiration. Water fixed is actually 3 tons for every 5 tons of dry organic matter produced. After agriculture, power generation (6.2%) and industries (5.7%) are the biggest consumer of fresh water. Domestic requirement consume only 3.5% of the total water drawn. Navigation fisheries, hydroelectric power generation, recreational activities etc (Placed under miscellaneous category here) also require huge quantities much of which flows down to the sea. The amount of water drawn for human use is never used up completely. A large fraction is returned to the surface deposits or stream-flows mostly in a polluted state, which can be used again as such or after treatment for many purposes. Out of the total quantity of water drawn (35,500 cubic kms) the amount of water irrecoverably consumed was estimated to be about 2,200 cubic kms in the year 1974. (2). Future estimates of fresh water consumption provide a grim picture According to an estimate, by the close of the twentieth century the world agriculture should require 30% more water than it did in 1970 and by 2025 AD this quantity is likely rise beyond 50-60% over the 1970 level of freshwater consumption. With a rapidly rising human population the freshwater requirement for domestic sector could rise almost 8-10 times than that of 1970 level. What is more troubling is that industries and power generation sector could draw as much as 15-16 times more water than they did in 1970. The demand of water for navigation, fisheries, hydroelectric power generation, and recreational activities shall also double itself.

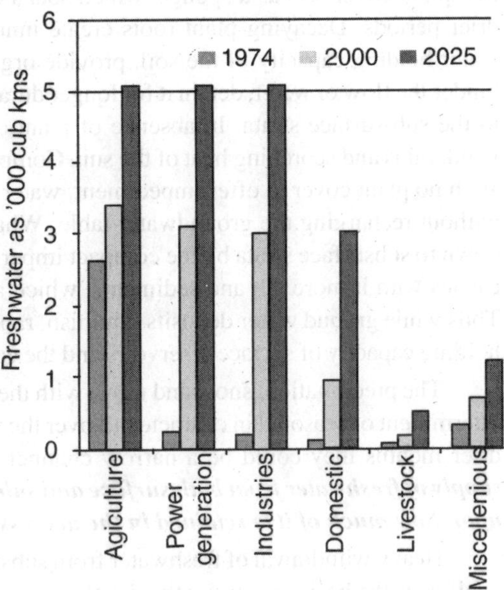

Fig. 5.1 *The pattern of global land use*

(IV) THE GENESIS OF FRESHWATER SCARCITY

The water, which trickles down and moves back as surface and subsurface flows, from higher altitudes to the sea, constitutes the primary source of freshwater for mankind. Early man settled at places where plenty of freshwater and fertile land was available on banks of streams and rivers which provided an all the year round supply of freshwater. Ancient human settlements sprang up along the course of almost all major rivers of the old world. Subsurface deposits of water, however, constitute a ***ubiquitous source of freshwater*** all over the world. All you have to do is to dig a well or a tube well to obtain freshwater. This has caused human settlements to spring up at places far away from the surface water source.

Probably the greatest singular cause of degeneration of our fresh water resources is extensive deforestation and land clearance which has bared a large area of earth's surface. Earth's crust with a rich plant cover acts as a sponge, which soaks up large volumes of water to cater to our need during drier periods. Decaying plant roots create innumerable channels in soil and improve porosity and water holding capacity of the soil, provide organic matter and help in retention of moisture. Plants hinder the flow of water, detain it for longer duration on land surface so that more of it percolates down to the sub-surface strata. In absence of plant cover, the soil is subjected to the battering action of wind, rains and scorching heat of the sun. Compaction of top layers of soil occurs. It loses its porosity. With no plant cover to offer impediment, water deposited by rains quickly flows down as flood flow without recharging the groundwater table. Whatever little of it is left is prevented from percolating down to subsurface strata by the compact impervious layer near the soil surface. The rapid flood flow carries with it more silt and sediments, which are deposits in riverbeds and reservoirs downstream. Thus while ground water deposits diminish, rapid silting and sedimentation tends to reduce the water holding capacity of surface reservoirs and the carrying capacity of channels of streams and rivers (4).

The precipitation, snow and rains, with the exception of tropical regions of the world, are highly intermittent or seasonal in character all over the world. During rainy seasons rivers are in spate, during drier months they could be a narrow channel of flowing water. ***A dependable all-the-year-round supply of freshwater from both surface and subsurface flows from areas at higher altitudes depends upon how much of it is retained in the depressions on land surface or in subsurface water table.***

Heavy withdrawal of freshwater from subsurface water table with a highly reduced input in these aquifers is the basic reason for the development of freshwater scarcity. In the catchments of the Gaula rivulet in Kumaun where the nearly 14% of natural forests were cut down in about 22 years the flow of water declined by about 40% (5). In the cultivated and moderately grazed land in Dugar valley in Pauri Garhwal the discharge during drier months declined by 36.5% during the period between 1994 and 1998 (6). The reduced discharge from these mountainous springs has grave consequences for the streams and rivers which emerge from the locality. The ratio of outflow between the rainy season and that of drier season of some streams in the Himalayas, such as Karnali has been estimated to be 120:1, for Sapta Kosi 150:1, for Kali Gandaki 180:1 (7). For most of the rivers which originate from greatly deforested stretches of Lesser Himalayas of Nepal the discharge during wet seasons has been estimated to be about 30,000 to 60,000 time higher than the discharge during drier season (8). This is indicative of the onset of desert like conditions as indicated by the appearance of drought tolerant xerophytes on the slopes of deforested Lesser Himalayas (9; 10). Apparently the nature has already started to prepare itself to cope up with the problem of aridity in the locality. With the dwindling flow of these rivulet and streams from higher altitudes, the dependable all-the year-round supply of freshwater as well as availability of groundwater at convenient depths is also affected adversely.

The ill effect of withdrawal of more water than the total annual input may be drastic. In United States, six states namely Colorado, Kansas, Nebraska, New Mexico and Okalahoma, relied heavily on the underground fresh water aquifer called Ogallala for the supply of fresh water. Its depletion due to huge over-drafts in these states caused the total agricultural area to decline by more than 15% (11). In Tamilnadu, India, heavy pumping has dropped the water table by 25 meters or so (12). In Beijing-Tienjing area of China, in Phoenix and Tucson, Arizona, a combination of agricultural and urban needs has been lowering the ground water table by several meters a year (13). In absence of rains, it is the water table below which sustains natural plant growth and keeps earth's surface moist and humid. With water table receding below the reach of plant roots natural vegetation disappeared, agriculture suffers and the process of gradual desertification starts.

(V) FLOODS

Generally, people think of floods as an outcome of accumulation of large volumes of water flowing through river channels spilling over the banks and causing extensive damage to human lives and property. The drainage basins and flood plains of rivers of the world river are in fact granaries of the world which account for more than 85% of the global grain production. The rich and fertile soils which develop in the flood plains regularly receive plenty of plant nutrients brought in by waters overflowing their banks and gushing down to the sea. However, devastating floods are common in drainage basin of most of the rivers (9). An enormous money and labour goes into flood-relief measures and efforts to control the fury of our rivers.

1. Major causes of recurrent floods:

Both natural and anthropogenic factors are responsible for causing floods in river basins although the relative importance of these factors may vary from place to place. In general extensive removal of plant cover steps up soil erosion causing the flood flow to carry more silt and sediments which are deposited in river channels downstream making them increasingly shallow and reducing their carrying capacity (14). As large volume of water has to flow down in shallow channels, it spills over the banks flooding the plains on either side. Important natural and anthropogenic factors include:

*1. **High intensity rains in the catchment area of the river:***

Heavy rains in the catchment area of the river causes more water to flow through the river channel. Rains downstream may also cause flooding and slow down the discharge of large volumes of water flowing down the channel in upper regions of the basin. The overall result is accumulation of large volumes of water in the channel, which spills over the banks and inundates large area of useful land.

*2. **Shallow channels and extensive flood plains:***

Flood flow passing through shallow channels flanked by extensive plains on either side shall spill over and submerge large tracts of land with a little rise in the volume of flow or accumulation of more water due to flooding of the channel lower down the course of the river.

*3. **Sudden changes in channel-gradient or blockage of the flow:***

Blockage of the flow of a river due to landslides in the upper reaches of drainage basin tends to flood the region upstream whereas abrupt removal of the debris may result in rapid discharge of large volume of water causing floods downstream.

*4. **Curves, bends and meandering course of the river channels:***

Sharp curves, turns and meandering of river tend to slow down the flow of water through the channel. With a lowered rate of discharge of water through the channel, large quantities of water accumulate upstream and cause extensive floods.

*5. **Extensive deforestation and removal of plant cover:***

For agriculture and other uses, man has cleared natural vegetation over large areas of most of the river basins. Trees have disappeared and much of the herbs, shrubs and grasses have dried out. On bared denuded land surface water flows down quickly. Large volumes of water, therefore, has to be discharged in brief period of time which is often too much for the capacity of the shallow river channels through which it has to pass. It often spills over the banks causing extensive floods.

*6. **Impact of urbanization and construction activity:***

Extensive pucca rooftops, land surface and asphalted roads considerably reduce infiltration of water to sub-surface strata. Large volumes of water accumulate in low-lying areas and have to be disposed off as quickly as possible through the drainage system of the city. Construction of embankments, barrages, reservoirs and channel manipulation etc have the over-all effect of slowing down the flood flow causing the water to accumulate above. Bridges also obstruct the flood flow lowering the amount of discharge of flood water causing floods up streams.

2. Flood control measures:

Most of the measures adopted today to control the flood fury and tame the river involve steps which delay the return of the flood flow deposited by heavy rains to river channel, hasten its discharge by clearing, straightening or deepening the channels and reduce the volume of flood water by storing it in large artificial reservoirs or diverting it to other places so that the flood flow remains well within the carrying capacity of the river channel. As the water deposited by rains at high altitudes has to flow downwards under the influence of gravity steps are also taken to construct dykes, embankments flood wall etc so that the flood flow does not enter human establishments.

1. Reforestation and development of a dense plant cover on hill slopes and land surface in the catchment area of the river: A thick cover of plants, grasses, herbs, shrubs and trees intercept rains, delay accumulation and the return of flood flow to the river channels. Decaying plant roots leave numerous channels making the soil act like a sponge and soak up large quantities of water. Plant cover protects the soil surface, which significantly reduces soil erosion. This lowers the sediment load of streams and rivers considerably. The carrying capacity of the river channels is maintained for much longer duration. Development of dense plant cover, therefore, reduces the devastating impact of floods.

2. Straightening the course of river channels to hasten the discharge of the flood flow: A highly sinuous, meandering course of the streams and rivers retards quick clearance of the flood flow causing floods in the shallow channels. The highly sinuous, meandering loops may be avoided by digging straight cuts joining the ends of the loops. This will straighten the course of river through which flood flow shall be cleared quickly. Probably the earliest effort of straighten the course of a river was taken up on Mississippi river in its lower reaches near Greenville (USA) in the year 1933. By 1936 the river was straightened and shortened to nearly 185 kms dissecting out loops measuring nearly 350 kms in length. This hastened the flood flow and moderated the intensity of floods in the upper reaches of the Mississippi river. Similarly, Missouri river an important tributary of Mississippi has been straightened and shortened by 52 kms before it discharges its waters in to the Mississippi river.

3. Reduction in the volume of flood flow by constructing huge storage reservoirs: The volume of flood flow may also be reduced by construction of dams across river channels, which impound large quantities of water. The huge amount of flood water thus accumulated is released slowly so that the flow is well within the carrying capacity of the river channels. The flood control reservoir on Miami River in Ohio (USA) built in 1913 and a series of reservoirs completed by 1921 effectively regulated the flood flow of these rivers. Three large reservoirs were built on Tennessee River, in 1933, now control the flood flow, provide water for irrigation and has tamed the ferocious river.

4. Diversion of floodwater to other channels to reduce its volume: The volume of flood flow may be kept within the limits of the carrying capacity of river channel by diverting a part of it to some low-lying area. For example Ghaggar Diversion scheme in India diverts more than 350 cubic meters per second of flood water into depressions and areas between sand dunes during floods before the river enters Rajasthan which keeps the flood flow of the river within the carrying capacity of the river channel.

5. Reduction in the impact of floods: In general the first response of people to overflowing rivers has been to protect themselves by erecting bunds and barriers along the river channels. Today several thousand kilometers of embankments constructed the world over try to confine the flood flow of various rivers to a narrow channel.

(VII) DROUGHTS

Droughts are more dangerous environmental setbacks as these are directly linked to the three basic requirements of life – air, food and water. Water is essential for growing food. No one can live without water. Drought is the cumulative effect of scarcity of water for prolonged periods.

It will not be out of place to point out here that increased dryness for prolonged periods causing drought conditions is linked to the amount of rainfall, its departure from normal annual average value and the local demand of water for various purposes. It is not the amount of total annual rainfall, which matters for drought or wet conditions, rather it is the regularity or irregularity of rainfall which is more important. For example, a regular annual precipitation amounting to 195 mm for years together may not cause any concern as the biosphere will adopt itself to the meagre supply. However, localities receiving a regular annual water supply amounting to 800 mm of rains, if provided supplies equivalent to only 400 mm of precipitation shall be disastrous, as the crops shall fail, ground water aquifers shall recede deeper down and greenery shall disappear. In other words, drought is a condition related to the failure of the usual pattern of rainfall at a particular time as most of the activities using water are adjusted to the supplies, which are normally available in the locality.

The economic impact of drought include economic losses due to highly reduced agricultural production, the diminished production of edibles from animal sources and even the industrial production suffers because of lack of water supply. The demographic affect of prolonged periods of drought includes depopulation of the locality and temporary migration of the people along with their necessities and belongings from affected areas to places where the resources are available. The area between the hot and dry deserts of Sahara in North Africa and the Tropical Savannah in the South running from East to west coast of Africa is termed as Sahel region, which is characterized by typical grassland climate. The most recent spell of severe drought in this locality began in the year 1968 and continued till 1975. The prolonged drought depleted subsurface water reserves. Water holes dried. In spite of relief measures from almost all over the world thousands of people died of hunger and starvation, thirst and diseases – only in Ethiopia the drought claimed about 50,000 lives. Nearly 5 million cattle died of thirst and starvation. In Ethiopia the drought conditions persisted a little longer till about 2001 and millions of children are still suffering from malnutrition and diseases in this country.

(VIII) CONFLICTS OVER WATER

The surface reservoirs, underground deposits and water flowing in streams and rivers provide most of the fresh water for human needs. Some rivers flow through many countries catering to the needs of people belonging to different ethnic groups, religious back grounds, ideologies and nationalities. Human manipulation of flow of rivers passing through many states or countries is a reason important enough to give rise to conflicts and controversies. Nowhere this is more evident than in the arid Middle East, where the scarcity of water has played a central role in defining the political relationships of the region for thousands of years.

Jordan River drains an area of slightly less than 20,000 square kilometers and flows 360 kilometers from its headwaters to the Dead Sea. Shared by Jordan, Syria, Israel, and Lebanon, the river is one of the most important sources of water in the Middle East and has been an object of intense international competition. Since the establishment of Israel in 1948, this basin has been the center of intense international conflict. In the 1950s, when Syria tried to stop Israel from building its National Water Carrier, a system to provide water to Southern Israel, fighting broke out across the demilitarized zone. When Syria tried to divert the headwaters of the Jordan away from Israel in the mid-1960s, Israel used force, including air strikes against the diversion facilities. Tensions also exist in the Jordan basin between Syria and Jordan over the construction and operation of a number of Syrian dams on the Yarmuk River, a tributary of Jordon.

Tigris and Euphrates Rivers are among the largest rivers in the Middle East. Both rivers originate in the mountains of Turkey, flow south through Syria and Iraq, and drain through the Shatt Al-Arab waterway into the Persian Gulf. Several tributaries of the Tigris drain the Zagros Mountains between

Iran and Iraq, and 15 percent of the Euphrates basin is in Saudi Arabia, though essentially none of its flow is generated there. Water-related disputes arose in the basin in the 1960s after both Turkey and Syria began to draw up plans for large-scale irrigation withdrawals.

The Nile River, the longest in the world, flows more than 6,800 kilometers from the highlands of central Africa and Ethiopia through nine nations to the Mediterranean. The nations that share the Nile are Egypt, Sudan, Ethiopia, Kenya, Tanzania, Zaire, Uganda, Rwanda, and Burundi, and the watershed covers nearly 10 percent of the African continent. Two major tributaries form the Nile: the White Nile, which starts in central Africa's Lake Plateau region, and the Blue Nile, which originates in the highlands of Ethiopia. More than 80 percent of the Nile's flow comes from the torrential seasonal flows of the Blue Nile. Ninety-seven percent of Egypt's water comes from the Nile River, and more than 95 percent of the Nile's runoff originates outside of Egypt in other countries. A treaty was signed in 1959 allocating the water of the Nile between Egypt and Sudan. Although this treaty has effectively reduced the risk of conflict between the two countries over water, none of the other seven nations of the basin are party to it, and several have expressed a desire to increase their use of Nile River water.

The drainage basin of River **Niger** (Africa) lies in Guinea, Mali, Burkina Faso, Benin, Niger and Nigeria. The drainage basin of **Zambezi** (Africa) lies in Congo, Angola, Zambia, Zimbabwe, Malawi and Mozambique. River **Paranas** of South Africa flows through Brazil, Paraguay and Argentina. The **Mississippi** basin, which is spread over about 3270000 sq kms, lies in several states of North America. The Multipurpose Irrigation project over **Mekong** River in South East Asia proposes to construct several dams, some with hydroelectric power generation units and a net-work of canals. It shall involve four countries namely Cambodia, Thailand, Laos and Vietnam. **Colorado River**, which flows through several states of America, has been hemmed with several dams and embankments and its flow is entirely under human control. Controversies over the use of its waters have developed many times which could only be resolved by federal intervention or adjucation by the Supreme Court of USA. A well-known example of such a dispute happens to be the waters of **River Cauvery** in India. The river originates in the hills of Western Ghats, runs through the rain shadow region of Karnataka in South India and flows eastward to Thanjavur delta region, the rice bowl of Chennai. It is an important river providing water to crops field in both Karnataka and Chennai. The rapids and the falls of the river are also used to generate hydroelectric power in Karnataka. Mettur dam control the flood flow of the river. During lean periods scarcity of water often develops. Karnataka people feel that they have stronger rights over the Cauvery waters. They refuse to release more water under their control and the Chennai farmers suffer.

(IX) FRESHWATER MANAGEMENT PRACTICES TODAY: SHORT COMINGS AND CONTROVERSIES

Freshwater management practices during 20th century largely involved construction of large and small dams and a network of canals to store and distribute freshwater wherever needed. All over the world huge reservoirs and dam have been constructed to store large quantities of fresh water. A net work of canals now carries water to water deficient places. However, human manipulation of the natural flood flow of rivers has caused many ecological setbacks and has given rise to a number of conflicts and controversies. Large volumes of water deposited by rains can be collected in these reservoirs and released slowly as and when desired. These dams and reservoirs serve the following purposes:

1. Water collected during rainy months serves as a reserve to be discharged during periods when there are little rains or no rains.

2. The water can be used for irrigation and other purposes through a network of canals developed to carry the stored water to needy places.
3. Regulated discharge of water deposited by rains prevents floods downstream.
4. The reservoirs can be used to breed fishes and other useful products.
5. The potential energy of the stored water can be used for hydroelectric power generation – the cheapest and cleanest source of electricity.

However, the artificial reservoirs, which accumulate huge quantities of water, are a potential threat to people of the locality. A breach in the dam or sabotage can cause floods of enormous magnitude. Entire localities downstream could be washed out of existence.

One cubic meter of pure water weighs one metric ton (or 1000 kg). Millions of cubic meter of water stacked at one place exerts enormous pressure on the strata underneath. The pressure may cause disequilibrium of isostatically adjust rocks which is further magnified by faults and fractures in rock strata under the reservoir bed. Many well-known tremours and earthquakes around the world have been associated with construction of large dams. In most of the cases, the intensity of earthquakes has been found to be positively co-related with the height of standing water table in the reservoir. The earthquake of 1931 in Greece followed the construction of Marathon dam in 1929. Tremours rocked localitics around the Hoover dam (Boulder dam) in USA constructed during the years 1929-1935, which created a huge artificial lake (Lake Mead). The Koyna earthquake of 1967 in India has been co-related with the construction of Koyna reservoir in 1962.

Artificial reservoirs submerge large tracts of fertile land much of which has been under cultivation since times immemorial and over which a number of human settlements and at times even architectural treasures lay. The temple of Abu Simbel, built by pharaoh Ramses II in 1200 BC was threatened by the proposed Aswan High dam. The temple was carefully cut into pieces and moved to a safer place in 1960. The dam was completed in 1972. However, Aswan High dam did submerge many of the architectural heritages of ancient Egypt. Aswan High dam is about 1 km long and 111 meters in height while the lake, which it created, extends 480 kms length and about 16 kms in average width. It displaced 90,000 people, which had to be re-located elsewhere.

The gigantic Multipurpose project on Narmada River in India envisages 30 major, 135 medium and 3000 minor dams on the river and its tributaries along with a network of canals in Madhyapradesh, Maharashtra and Gujrat. Of the 30 major dam only Sardar Sarovar in Gujrat shall submerge nearly 39,150 hectares of forests and agricultural land displacing about 66,675 people majority of whom are tribals. In 1950 the world had about 5000 large dams. By the close of the 20th century, this number has shot up to about 45000. The number of people affected or displaced by large dams and reservoirs during the period now exceeds 80 million people.

Submergence of large tracts of forests and agricultural land also wipes out the habitats of enumerable number of plants, animals and microbial species and brings about an ecological catastrophe. In fact in most of the cases we do not know much about the magnitude of the reduction in biodiversity we are likely suffer when the dams come up and impounded to its full water holding capacity. A thorough ecological impact assessment has been done only half-heartedly for most of these projects.

Huge dams, irrigation project and transport of water through network of canals involve large losses of precious water due to evaporation and seepage. Water logging of surrounding areas is quite common which makes large strips of land useless for cultivation. Aswan High dam loses nearly 15% of its precious water due to evaporation alone. Repeated use of water from these reservoirs to irrigate the crops fields having poor drainage results in problems of salinity and alkalinity. These waters are in fact salt solutions. Crop utilization of water and fertilizers has the effect of concentrating salts in the soil. In absence of careful management, irrigated soils can become saline or develop toxicity. There are numerous reports of reduced fertility and salinity caused by use of river water stored in huge dams and reservoirs in drier regions of the world. In drainage basins of rivers Colorado and Columbia, the Nile, Euphrates and Tigris, Indus basin Irtysh, Yangtze etc salt accumulation has damaged soil fertility significantly. Plants take up only those salts, which they require while large amount of water is lost in evapo-transpiration, which leaves behind much of the salt dissolved in water. Due to poor drainage, it accumulates in the soil. Raised sub-surface drainage may also bring about changes in chemistry of sub-surface aquifers. The salts dissolved from one layer may contaminate other sub-surface strata. Rapid silting and sedimentation reduces the life of these multimillion-dollar artificial reservoirs by about 35-45% of the originally intended span.

The World Bank conducted a review of nearly 50 large dams it had funded and a meeting of experts from all over the world with divergent views was convened at the initiative of International Union for Conservation of Nature (IUCN) in April 1997 at Gland, Switzerland which concluded that it high time to take a fresh look at large dams. In May 1998 a World Commission on Dams (WCD) was constituted under Prof. Kader Asmal, the then minister of Water and Forest, South Africa and 12 commissioners representing different interests and ideologies. The World Commission on Dams completed its 390-page report by November 2000, which was released by the President of South Africa, Mr Nelson Mandela in London. The report entitled ***'Dams and Development: A new Framework for decision making'*** has been published by Earthscan Publications Ltd. London 2000. The World commission concluded that large dams have had a negative impact on ecosystems, most of which cannot be mitigated through corrective measures. The record of large dams in flood control has also been a mixed one helping some communities making other more vulnerable. Although they have played an important role in irrigation and raised productivity in many localities, they have also resulted in widespread water logging and salinity. The overall picture that emerges at the end of WCD report is that while dams should not be ruled out altogether, they should be undertaken with the consent of communities concerned after other alternatives for the sustainable development of the locality have been assessed and found inadequate.

(X) FRESH WATER RESOURCES OF INDIA

India has plenty of freshwater. The sub-continent receives most of its fresh water during monsoon months (almost 75%). Remaining months are drier which necessitates the use of ground water or stored water during the dry spells. The uneven distribution of rains in different months of the year is matched by its equally uneven distribution over different regions of the country. Parts of Rajasthan receive very little rains while there are places like Cherapunji, which had the reputation of being the wettest place in the world.

(1) The Resource Base

India receives about 4,000 cubic kms of fresh water as precipitation every year. About 700 cubic kms of water thus received evaporate immediately and are lost to the atmosphere. About 2,150 cubic kms go to the soil whereas about 1,650 cubic kms are retained as soil moisture while about 500 cubic kms permeate through the soil surface to underground water deposits. Only 1,150 cubic kms of fresh water received annually are retained on land surface. Water resources in our country can be grouped as under:

1. Surface waters:

To 1,150 cubic kms of fresh water, which appear as surface water, may be added about 200 cubic kms of surface flow, which comes from outside India. The surface flow is further enlarged by addition of about 450 cubic kms of fresh water from ground water flow while about 50 cubic kms are added as runoff from irrigated areas. The surface loses almost 50 cubic kms of its water, which percolates down to the ground water deposits. The total surface flow per year is about 1,780 cubic kms, which are distributed among a number of river basins.

2. Ground Water:

The major portion of fresh water, which goes to earth's crust, is retained by its upper layers as soil moisture (about 1,650 cubic kms). Only 500 cubic kms percolate down to the ground water deposits. A large amount of fresh water applied to agricultural fields (about 120 cubic kms) moves down to ground water table while about 50 cubic kms of surface flow also end up as ground water. Therefore, a total of about 670 cubic kms of fresh water enters the ground water annually. It is upto this amount that we can withdraw fresh water from our sub-surface deposits. Any withdrawal above this limit shall be detrimental to the resource base.

(2) REQUIREMENT OF FRESH WATER IN INDIA

Table 6.2 *Estimated freshwater water requirement in India*

	Water needed for	1974 AD	2000 AD	2025 AD
1	Irrigation	350.0	630.0	770.0
2	Thermal power generation	11.0	60.0	160.0
3	Industries	5.5	30.0	120.0
4	Domestic requirements	8.8	26.6	39.0
5	Live-stock management	4.7	7.4	11.0
	Total	380	754.0	1100.0
All quantities in cubic kms. Source: Kathpalia 1985				

Like rest of the world, agriculture sector is the major consumer of fresh water in India. Thermal power generation sector is the next biggest consumer of fresh water, which is followed closely by industries. It is followed by domestic needs and requirement for livestock management which taken together useabout 13.5 cubic kms of fresh water. Table 6.2 summarizes the approximate requirement of fresh water in India as estimated for the years 1974 A.D., 2000 A.D. and 2025 A.D.

By 2000 AD, the total water requirement is expected to double itself while about 2025 A.D. we shall be requiring almost thrice as much water as we did in 1974 A.D. This extraordinary rise of demand for fresh water shall not be uniform in various sectors of our economy. In 2025 AD, the requirement for irrigation water will be doubled whereas our domestic needs as well as water requirement for livestock management shall be about four times than those of 1974 A.D. Our requirement shall exceed the availability by 2000 A.D. We shall have to bank more and more upon our ground water resources (15).

(3) FRESH WATER CONSERVATION IN INDIA

India has a long history of water resource development and conservation. Due to concentration of major portion of precipitation during monsoon months (75%) much of the fresh water is lost as flood flows and during the remaining months scarcity of water develops. Ponds, tanks, embankments have been made since time immemorial to resolve the water crisis during the drier months. The Delhi Sultanate and the late Moghuls engaged in large-scale irrigation works in Northern India during 1350 to 1780 A.D. By 1900 A.D., irrigation facilities already covered an area of about 40 thousand sq. kms. After independence as food problems plagued the Nation, huge irrigation projects were undertaken to raise agricultural productivity under a series of Five Year Plans. These projects involved an enormous amount of public money which was largely diverted to raise surface storage capacity of our country and dig huge canals for irrigation purposes. The Ministry of Water Resources, Govt. of India classifies various projects undertaken from time to time under the following two major headings:

1. Minor surface irrigaton sector.
2. Major and Medium irrigation sector

Minor surface irrigation sector (MSI) comprises of small projects, which irrigate only upto, 2,000 hectares each. Ground water irrigation sector is also placed under this category. This sector employs wells, tube wells, tanks, bunds, traditional lift irrigation etc. to provide water for agricultural purposes. As a matter of fact, Minor surface irrigation sector seems to be in a very poor shape, however, because of its grouping with ground water irrigation sector, which is almost doubly efficient and productive, the actual performance of MSI is masked from general observation.

Table 6.3 *Some information about Indian irrigation*

		Area in hectares
1	Total Land area	3,287,585,000
2	Net cultivated land	142,220,000
3	Irrigated crop-land	46,200,000
4	Net area served by surface waters	22,700,000
5	Net area served by ground water	23,500,000
6	Net area served by Major and Medium Irrigation sector	16,000,000
7	Net area served by Minor Surface Irrigation sector	6,700,000
8	Reported Irrigation potential of our country	78,100,000
9	Total surface water storage capacity	180 cubic kms
	Source: Diverse	

Major and medium irrigation sector involves huge projects under which huge dams and bunds are constructed with a capacity to hold billions of cubic metres of water and to irrigate millions of hectares of land surface. These projects have enhanced country's surface storage capacity to about 180 cubic kms and a net potential to irrigate about 78.1 million hectares of land. Some of the major projects completed till date are: Damodar valley project, Bhakranangal dam, Kosi project, Hirakud dam, Tungbhadra project, Nagarjuna sagar dam, Rehand dam etc. Along with these dams, a network of canals has been dug to cater to the needs of farmers in drier regions. The energy of flowing water from these dams is used to generate hydroelectric power, which is a completely pollution-free source of energy. By 1986-87 A.D. alone, India had developed an installed capacity to generate about 19,20,000 KW of hydro-electricity. These waters are also used to breed fishes.

QUESTIONS

1. Write an essay on Global fresh water resources. What are its limitations?
2. How much fresh water is available to mankind? Discuss the pattern of global use of fresh water. What are the future prospects?
3. Discuss major causes of wastage and degeneration of our fresh-water resources. What can be done to conserve them?
4. Give a brief account of availability and pattern of use of fresh-water resources in our country. Discuss the future scenario.

7 Chapter

Mineral Resources

Human welfare and availability of mineral resources have been linked together so closely and for so long that historians mark major periods of human history by reference to minerals, such as stone, bronze, copper and iron ages etc. Cheap and plentiful supply of minerals provided the physical foundation for our present-day Industrial Civilization. Even today, society's overall prosperity depends largely on the supply of mineral products. The term mineral resources refer to a wide variety of materials obtained from earth. In general they can be divided into following two categories:

1. Metallic minerals: Minerals which when processed provide metals such as Iron, Aluminium, Copper, Zinc etc.
2. Non-metallic minerals: Minerals, which yield products other than metals such as phosphate rocks, potash, soda ash, various salts, clay, sand, and stones etc.

   To these may be added materials, which provide energy such as coal, oil, natural gas etc. In view of their tremendous importance for the mankind, they have been discussed separately.

(I) Formation Of Mineral Deposits

Being a part of the inorganic matter, which constitutes this planet, most of the minerals or their decomposition products are widely distributed in earth's crust. It is the concentration of these minerals

Table 7.1 *Global reserve of metals and their major source*

	Metal	Reserves*	Chemical source	Important Producers
1	Iron	109,000	Haematite, Magnetite, Siderite	Russia, USA, Canada
2	Aluminium	2950	Bauxite	Australia. Guinea
3	Titanium	440	Ilmenite	Canada
4	Copper	310	Chalcopyrite	USA, Chile, Canada, Russia
5	Zinc	120	Sphalerite	Canada, USA, Russia
6	Lead	85	Galena	USA, Canada, Australia
7	Tin	5	Casseterite	Thailand, Malaysia
8	Manganese	800	Pyrolusite	South Africa
9	Chromium	780	Chromite	South Africa
10	Nickel	70	Pentlandite	Cuba, New Caledonia
11	Molybdenum	5	Molybdenite	USA, Russia
12	Cobalt	3	Co sulphide	Congo, Zambia
13	Tungston	2	Scheelite, Wolframite	China
14	Mercury	1	Cinnabar	Span, Italy
15	Silver	7000	Silver Sulphide	
16	Gold	400	Gold, Gold Telluride	South Africa, Russia

* As million metric tons. Gold and silver in metric tons. Source: Diverse

at a particular spot, in quantities sufficient to be exploited economically, which gives rise to a mineral deposit. The concentration of mineral material may be brought about by intense heat and pressure, simple evaporation, heated waters or molten rock material percolating through rock strata. Some deposits may be developed during the process of weathering transport and deposition. Microbial activity may be responsible for the formation of some deposits. The formation of these deposits is essentially a very slow geo-chemical or biological process of concentration of mineral material, which may take millions of years to develop an economically viable mineral deposit.

(II) The Mineral Wealth of our Planet

The slow but persistent action of a number of geo-chemical processes and biological activity has given rise to large deposits of different minerals in different parts of the world. In terms of total amount of metal content as known until the end of the 2nd millennium, Iron minerals like Haematite, Magnetite, and Siderite etc. are the most abundant minerals. In the form of these minerals a huge quantity of iron, about 109,000 million metric tons is available to us. Iron is followed by Aluminium, which occurs as Bauxite, Manganese that occurs as Pyrolusite, Chromium as Chromite, Copper as Chalcopyrite, Zinc as Sphalerite, and Lead as Galena in order of decreasing abundance. In Table 7.1 global reserves of metals, their major mineral source and important producer countries are summarized. The quantities of mineral reserves available to mankind change very quickly as new explorations and latest techniques of prospecting of mineral deposits bring to light new reserves almost every year. In practice, in-spite of the rapid extraction of enormous quantities of minerals during the century, we have more deposits as reserves at our disposal at the end of the 20th century than what we had at its beginning (1).

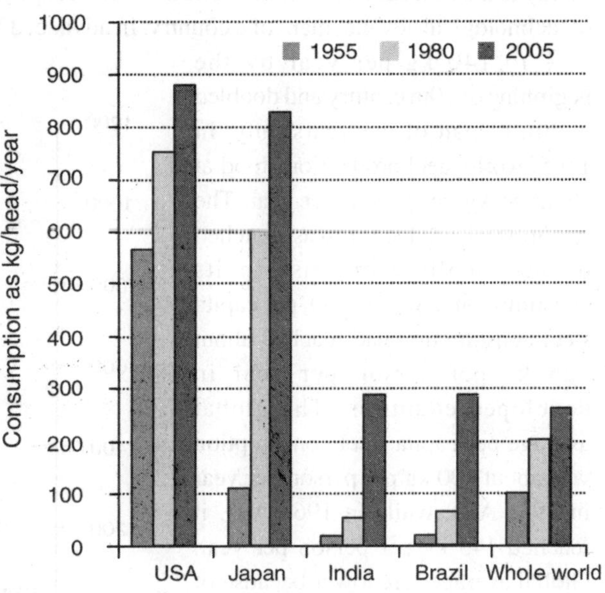

Fig 7.1 *Per capita steel consumption in selected ocuntries*

(III) Human use of Mineral Resources

Copper was the first metal to come into a widespread use. Gold and silver were probably known to humanity much earlier as they were often found in their metallic state. However, they are rare elements. They were never available in quantities large enough for a general use except perhaps for ornamental purposes. Though available in quantities larger than gold or silver, copper is also not very abundant in lithosphere. However, the metal can be easily extracted from its ores. The reduction temperature is low so that its smelting can be done in a simple furnace. Once the technology for copper-extraction was perfected by man its demand rose rapidly and the Age of Metals in human history commenced early by 4000 B.C.

It was about 1100 B.C. that furnaces capable of attaining temperatures high enough to reduce Iron ores were developed. The widespread availability of Iron minerals made it possible for the metal to be used on an unprecedented scale. A little amount of carbon had to be mixed to produce harder steel. Initially trees were used to produce this charcoal (carbon). As steel production expanded, much of the forests had to be felled, to be turned into coal required for the steel industry of the time.

It was in England in 17th century that the technique to convert mineral coal into a satisfactory form of carbon to be mixed with iron to produce steel was developed. This development was of tremendous importance to the humankind. It paved the way to development of steam engine and finally to the Industrial Revolution. The level of steel production and consumption are useful indicators of technological advancement of a country. In advanced Western countries, per capita steel production rose to 140 kg per year by the beginning of 20th century and doubled within a span of ten years only. In 1965, world steel production stood at about 55 kg per person per year. The rise in steel production was matched by an equally rapid rise in its consumption. By 1965 AD, per capita steel consumption had reached about 625 kg per person per year in developed countries. The global average per capita steel consumption was about 100 kg per person per year in 1955 A.D. while in 1965 A.D. it reached 145 kg per person per year. Global averages are lower because of the fact that little steel is used in developing countries of the world. Fig. 7.1 shows per capita steel consumption in five major countries of the world as estimated for the year 1965 AD. Consumption of metals other than iron, such as copper, zinc, lead etc. has remained remarkably constant over the last fifty years while those of certain other metals like tin, has decreased in relation to total steel consumption. Consumption level of light metals like Aluminium has rapidly increased (2).

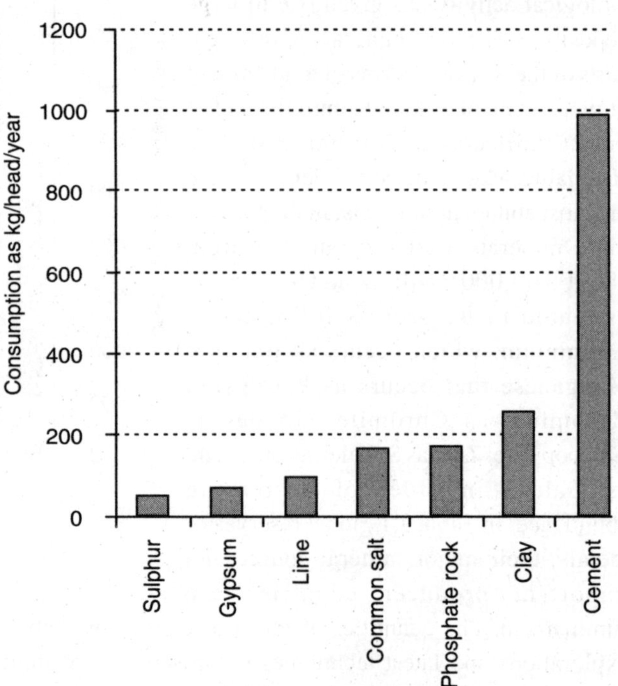

Fig. 7.2 *Global consumption of non-metallic minerals*

Although the quantities of metallic mineral deposits dug up and processed to support the human society are enormous, the quantity of non-metallic minerals used by man is growing even more rapidly. Approximate per capita consumption of some of the important non-metallic minerals is summarized in Fig 7.2. The amount of rocks and earth man moves each year is already gigantic. Today the total quantity of materials of all kinds obtained from earth amounts to an enormous 150 billion tons. This means that in order to support one person in our technologically advanced society having a population of 5 billion people, we have to dig up about 30 tons of materials of all types from earth's crust. Obviously, man has become an important geologic force (3).

(III) Non-Renewable Nature of Mineral Deposits and In-Exhaustible Nature of Mineral Elements

Table 7.2 *Crustal abundance, production and ore grade generally used and global production of selected metals*

	Element	Crustal abundance*	Ore-grade**	Global production***
1	Iron	58,000.0	20.00	552.00
2	Aluminium	83,000.0	18.50	18.10
3	Titanium	6,400.0	10.00	0.11
4	Copper	63.0	0.35	8.92
5	Zinc	94.0	3.50	7.30
6	Lead	12.0	4.00	3.35
7	Tin	1.7	0.35	0.21
8	Manganese	1300.0	25.00	8.60
9	Chromium	110.0	23.00	3.37
10	Nickel	89.0	0.9	0.95
11	Cobalt	25.0	0.2	0.002
12	Mercury	0.1	0.10	0.006

*As ppm. **Ore grade as percent metal content. *** As million metric tons. (*Source:Diverse*)

Mineral deposits are renewable on Geological time scale only. They are developed as a result of an enormously slow and gradual process of concentration which takes millions of years to form an economically viable mineral deposit. On human scale of time, therefore, these deposits are non-renewable. Extraction and use of mineral deposits at a rate faster than their renewal shall deplete even the largest of our mineral deposits in times to come. With the depletion of our high-grade ores, which provide more with lesser efforts, we shall have to look for other lower grade ores.

Though mineral deposits are exhaustible, the elements or the metal, which these deposits yield, cannot be irrecoverably consumed, as matter is indestructible. They can be changed from one state to another, from one chemical combination to another. Extraction, use, misuse and finally disposal of mineral elements tend to scatter them far and wide in the environment. We may exhaust high-grade deposits but the element concerned shall always be there, to be obtained from lower grade deposits, to be recovered from waste dumps or from any of its form or the chemical combination in which it happens to exist. As good quality deposits of mineral elements get depleted, progressively lower and lower grades of ores are used for their extraction. The element concerned becomes more and more expensive. Side by side goes an improvement in extraction technology, which often offsets the cost involved. Table 7.2 summarizes the global production of some metals and the approximate grade of the mineral deposit as percentage of the metal content from which the bulk of metal is obtained (1).

(IV) CAN WE EXHAUST OUR MINERAL WEALTH?

Early in 1970s it became fashionable to assert that the resources of the earth are finite. As these resources are being consumed at an increasingly faster rate to support our western life-style and to cater for the increasing demands of developing nations, the world is in danger of actually running out of many mineral resources. Of course, the resources of earth are indeed finite but the limits of supply of resources are so far away that the phobia of exhaustion of our mineral wealth has no rational basis.

Many of the resources concerned are either renewable or recyclable (1). Available reserves of 'non-renewable' resources are constantly being renewed, mostly faster than they are used (Fossil fuels are an exception). In fact, economically viable mineral resources are *created* by the interaction of three factors, which are geological knowledge, technology and economic forces.

1. Geologic knowledge:

The minerals in earth's crust cannot be considered usable resources unless they are known. Therefore, there must be a constant input of time, money and effort to find out what is there. We have to increase our geologic knowledge. A comprehensive study of ore-bodies has to be made so that the deposits can reliably be defined in terms of location, quantity and grade. They have to be technically and economically quantified as mineral reserves. This is being done regularly today. The measured resources of many minerals are increasing much faster than they are being used. Explorations bring to light new reserves almost every year. Simply on geological grounds, there is no reason to suppose that this trend will not continue. Today, proven mineral resources worldwide are more than what we had fifty years ago.

2. Technological knowledge:

It is meaningless to talk of a resource until the technology to use it is not there. Human ingenuity can create useful things out of useless wastes. It is the ingenuity and technology, which creates new resources by making particular minerals usable in new ways, which substitute, to some degree for others, which are becoming scarcer. More particularly, if a known mineral deposit cannot be mined, processed and marketed economically, it does not constitute a resource in any practical sense. Many factors determine the value of a particular mineral deposit as a usable resource. These are the location of the deposit in relation to processing set up and the markets, the ease of mining and processing operations, the cost of technology, labour costs and so on. The application of human ingenuity, through technology, alters the significance of all these factors and thus, in a way creates a resource.

For instance, Pilbara region of Western Australia is an excellent example of application of technology to create a resource. Until the 1960s, huge iron ore deposits of this area were simply geological curiosities, despite their excellent quality and grade. Australia has been a country noted for its lack of iron deposits. With the application of modern mining technology, heavy-duty railways and bulk shipping through the ports of Dampier and Hedland it was possible to obtain the ore from inland mines and carry it to Japan economically. For the last 35 years, Hamersley Iron (Rio Tinto), Mount Newman (BHP-Billiton) and others have been at the forefront of Australia's mineral exporters, drawing upon these 'new' ore bodies, and have creating much wealth for all Australians.

Only about a hundred years ago Aluminium, which is only next to iron in crustal abundance, was a precious metal because it was very difficult and expensive to reduce the oxide (Al_2O_3) to aluminium metal. With the discovery of the electrolytic process in 1886, the cost of producing metallic aluminium fell to about one twentieth of what it had been earlier. The metal now competes with iron, copper and others in many applications and has its own widespread uses in every aspect of our lives. Not only was a virtually new material provided for our use by this technological breakthrough, but the enormous quantities of bauxite worldwide progressively became a valuable resource. Without the technological breakthrough, they would have remained a geological curiosity (3).

3. Economic forces:

Whether a particular mineral deposit is sensibly available as a resource depends on the market price of the mineral concerned. If it costs more to get it out of the ground than its value requires, it can hardly be classified as a resource. Therefore, the value of a resource depends on the market price, which in turn depends on world demand for the particular mineral and the costs of supplying that demand. The dynamic equilibrium between supply and demand also gives rise to substitution of other materials when scarcity looms (or the price is artificially elevated). This is the third aspect of creating resources.

When in 1972 the Organization of Petroleum Exporting Countries (OPEC) suddenly increased the price of oil fourfold, many things happened both at producer and at consumer levels. The producers dramatically increased their exploration effort, and applied ways to boost oil recovery from previously exhausted or uneconomic wells. At the consumer level, increased prices resulted in large substitution of other fuels and greatly increased capital input in technology to develop more efficient plants. As a result of these activities, oil resources increased dramatically and the use of oil fell slightly. Globally oil use did not increase from 1973 to 1986 as fast as expected. The general forecast of 1972, of the doubling of the oil consumption in ten years proved to be quite wrong. Oil will certainly become scarce one day, probably before most other minerals. This will drive its price up. But as in the 1970s, this will also cause increased substitution and bring about greater efficiencies in its use as equilibrium between supply and demand is maintained by the market mechanism. Needless to add, oil will never disappear completely, it will become more and more expensive for us to use as our oils wells dry out one after the other.

(VI) CONSEQUENCES OF OVER EXPLOITATION OF MINERAL RESOURCES

In the entire history of human civilization such an unusually high demand has never been placed on natural resources of our planet. Although there is little danger of exhaustion of our mineral wealth the consequences of the rapid exploitation of mineral wealth have to be serious, drastic and enormously damaging to the entire biosphere. These can be summed up as follows:

1. Rapid depletion of high-grade mineral deposits:

Exploitation of mineral wealth at a rapid rate shall naturally deplete our good quality deposits. The ever-rising demands shall compel miners to carry on the extraction from increasingly lower and lower grade of deposits, which possess a poorer percentage of the metal. For example, copper was extracted from ores containing 8-10% of metal content before the industrial age commenced nearly three hundred years ago. Now we are using deposits, which contain only 0.35% of copper. To produce one ton of copper metal we have to dig out 285 tons of ore. This shall naturally involve a large amount of energy expenditure as well as a large quantity of waste material production. We may never reach an end as matter is indestructible. Most of the metals we require are present in highly dispersed state in the soil, the rocks and the trash or wastes we discard. With a sophisticated technology, we can fulfill most of our requirements from these sources, but the overall cost could be heavy, causing the metals to become more and more costly (4).

2. Wastage and dissemination of mineral wealth:

Most of our mineral deposits occur as a complex mixture of a number of mineral elements. After removal of top soil and rocks we dig out the desired mineral leaving behind others which are often left in the open as waste materials. Extraction of one element usually scatters and wastes a number of other elements, many of which are in short supply. This wastage rises as more and more ores are extracted and processed. Worldwide smelting of minerals for extraction of metals introduces an enormous quantity of sulphur, heavy metals such as mercury, cadmium, nickel, arsenic, zinc etc. into the environment which are separately mined elsewhere. We are technologically competent enough to extract these metals from the wastes produced from one mining industry rather than excavating fresh deposits. The cost could be heavier indeed but the practice shall pay in the long run. It will conserve our resources and also reduce the burden of pollutants which we have to introduce in the environment (5).

3. Pollution of environment from mining and processing wastes:

Mining is a dirty industry. It has created some of the largest 'Environmental disaster' zones in the world. The mining and processing of minerals generally involves following steps:

1. The soil and rock overlying the mineral deposits, called the 'over-burden' in miner's language, has to be removed before actual mining operations commence.
2. The ore is then mined and crushed.
3. After being converted to fine powdered state it is run through concentrators which remove unwanted material and impurities which are called 'tailings'.
4. The concentrated ore is then reduced to crude metal often at a high temperature by various methods depending upon the chemical nature of the ore.
5. Crude metal is then refined or purified in refineries.

Each step in mining and processing operations produces large quantities of waste materials. As most of today's mines are simple surface excavations, the first task of a miner is to remove whatever lies over the mineral deposit, be it a mountain, a forest or an agricultural field. Under-ground mining with a system of shaft and tunnels does not produce as much waste as open cast mining does. In 1988, over-burden, the material overlying the mineral deposits in U.S.A., amounted to about 3.3 billion tons of matter moved. This material even if chemically inert, clogs streams, gets deposited in lakes and clouds the air over large areas. If it contains sulphur and other reactive elements apart from wastage of our precious resource a number of other problems are caused (3;10).

Almost similar problems arise from the disposal of waste material produced after concentration of an ore. This material is called 'tailings' in miner's language. As most of the ores contain a large amount of sulphur, its oxidation and leaching results in formation of acidic leachates (Water containing dilute sulphuric acid). The finely grounded state of ores makes metal contaminants, which were earlier bound in solid rocks, available to acidic waters. Thus, these leachates contain appreciable amounts of heavy metals and toxic trace elements. Tailings may contain residue of organic chemicals such as toluene etc., which cause another type of problems. Ponds full of acidic leachates covering thousands of hectares of land surface now surround copper mines in U.S.A. These waters cause serious problems of water pollution if they happen to contaminate our surface or underground acquifers (3;8).

The grade of ore is important in determining the overall impact of mining activity. An ore containing 20% of metal content shall produce only four tons of tailings or waste material per ton of metal extracted but a low-grade ore containing 1% of metal shall produce 99 tons of tailings per ton of metal obtained. Gold mining is particularly damaging in this respect as the metal content of gold deposits is at best expressed as parts per million. Miners at Gold Strike mine in Navada – the largest in USA move about 325000 tons of ore to produce about 0.05 tons or 50 kg of gold per year (6;9).

In Amazon basin, Brazil, miners use a technique called hydraulic mining which involves blasting the gold bearing hillside with high pressure stream of water following by guiding the sediments through ducts where the gold being heavier settles down from tons of non-valuable material. This silt and sediments are finally washed down into some local stream. The practice has silted local rivers and lakes while the use of mercury to trap gold from sediments has contaminated large areas. Miners release an estimated 100 tons of mercury into the waters of Amazon River annually (6).

In North America, miners use 'Heap Leaching' a technique that allows gold extraction from a very low-grade ore. The technique involves sprinkling of cyanide solution over a heap of low-grade ore. While trickling down the solution dissolves gold. It is collected and later gold is recovered from it. Both cyanide solution reservoirs and contaminated tailings are left behind after the gold extraction. These pose hazards to wild life and threaten surface waters as well as under-ground acquifers. In October 1990, about 45 million litres of cyanide solution from a reservoir at Brewer Gold Mine, South Carolina, spilled over into a tributary of local Lynch River, killing more than 10,000 fishes. Thousands of birds die each year when they mistakenly consume cyanide solution from these impoundments (3).

4. Pollution Caused by Heavy Energy Requirement of Mining Industry:

Moving huge amounts of sand silt and clay etc. requires energy. Concentration of ore requires energy. Smelting and refining operations require energy. Electrolytic processes used for refining of some metals, like Aluminium, require energy. Transportation and disposal of wastes or tailings requires energy. Transportation of finished products requires energy. The overall worldwide requirement of energy in mining industry adds up to an enormous amount. This energy comes from diverse sources which mostly include fire-wood, coal, petroleum, natural gas and electricity. In order to provide energy to mining industry a huge quantity of these materials are burned which causes a variety of pollution problems (7).

(VII) Conservation of Mineral Resources

We have attained such an enormous level of consumption of our mineral wealth that good quality deposits of many elements have already disappeared while many others are in the process of being depleted. These resources belong to our children and grand children as much as they do to us. It is high time now that we should seriously think about conserving them. Important steps, which may be taken to make these deposits, last longer are:

1. Economy in use of mineral resources:

The simplest way to conserve mineral resources is to practice economy in the use of metals and minerals. Careful use or use only where it is necessary can reduce much of consumption of metals and minerals. Most of the metals and minerals are cheap because of the importance of mining industry in the development of a country, governments provide plenty of subsidies in the form of land at a throw away prices, cheap power, subsidized fossil fuels, large loans and generous tax exemptions. Many industrialized nations also try to ensure access to cheap mineral supplies through their international trade and policies. The lure of large revenues, generous financial assistance, finished products, arms and other necessities force poor countries to sell their mineral wealth at a cheap price. It is mainly due to their low prices that minerals and mineral products are freely used, even at places where they are hardly required. A little restriction and taxation could cause the minerals to become more expensive which in turn shall curtail unnecessary over-consumption.

2. Making finished products long lasting:

The 'use and throw away practice' of Western society after a product has lost its utility is a wasteful practice. The metal components in the product are also discarded and wasted. A study by the United States Office of Technology (Washington D.C.) in 1979 pointed out that repairs and re-use is a very promising method of conservation of metals and mineral products. Metal containing products, for example automobiles, should be so manufactured as to last longer, and be repairable to be used again or their components if in working condition may be used again and again.

Some metals are irrecoverably lost to the environment due to mishandling, corrosion, wear and tear. Of about 600 kg of per capita steel used in USA, for example, more than one-third is lost to be never recovered. The average life span of all steel in use varies between 25-30 years while that of other metals is somewhat shorter. To make a mineral product or an object made of metal last longer we shall have to prevent its corrosion, wear and tear and the irrecoverable losses.

3. Re-use and re-cycling of metals:

As rapid consumption of virgin minerals depletes our resources – why not use mineral products more efficiently or again and again. A machine having brass components can be pulled apart, its components used to make another object of utility. In India most of the copper, brass, bronze and aluminium objects are regularly recycled.

4. Use of cheaper substitutes:

There are a number of finished products in which cheaper material other than metals may be used. Why use a metal bucket when cheaper plastic buckets are available. Wherever possible cheaper materials may be substituted for mineral products and metals. Synthetic plastics offer a promising opportunity which are becoming cheaper and cheaper every year.

5. More efficient recovery of materials from minerals:

A number of minerals occur as a complex mixture of a number of elements. An ore is never pure. Even ores with a metal content as high as 20% may contain other elements, which are discarded as tailings either during concentration or during smelting. We are technologically competent enough to devise methods to recover or separate out most of the useful elements present in an ore. The cost may be heavy but the practice shall pay in the long run.

6. Search for new deposits:

We have entered an era of an advanced state of science and technology. Still we do not know much about our planet. Much of the rock formations and earth's treasures are still unknown to science. Greater part of South America, Africa, and Asia have not yet been thoroughly explored. An intensive search is expected to reveal a number of deposits, which shall naturally enlarge our mineral resources.

7. Protection of existing mineral Deposits:

Most of the mineral deposits are left as such on the mercy of agencies of weathering, decay and dissemination after extraction of good quality materials. As we deplete our better grade ores, it is likely that low quality ores which were once discarded as waste material could be in demand again. The reclamation of mining site, which involves restoration of deposit's original plant and soil cover, should be made a statutary responsibility of every miner. Apart from preventing wastage of resources, the practice shall also curb a number of pollution problems.

(VIII) The Ultimate Question

The ultimate question regarding our mineral deposits which faces us today is: **Are we running out? The answer is: not yet.**

Mineral material is a part and parcel of the inorganic constituents of our planet, which cannot be destroyed or consumed. Man tends to scatter them far and wide in the environment. He can extract them, in case of absolute necessity, from rocks, from the soil, from ocean waters, with a little more efforts and of course cost. However, there is another question, which worries the scientific community more than anything else does. It is, **"Shall we be able to afford the Environmental cost of such a rapid and enormous mineral extraction?"** No one is able to answer this question with precision. We are already facing a number of environmental problems associated with mining industry, which are enlarging every day. It is high time we should do something about it (2).

(IX) INDIAN MINERAL RESOURCES

India, if not very rich, is fairly rich in mineral resources. We possess good deposits of most of the mineral elements which are needed in large quantities by a technologically advanced society. It is often our inability to explore, excavate and extract the required metal, which is responsible for the shortage, not the lack of deposits.

Iron minerals, which are the most important ingredient of today's economy, are found in sufficient quantity in our country. We are not only producing sufficient amounts of iron ore for our own use but are presently exporting it to other countries. Starting from a meagre 3.0 million tons of ore production in 1950 A.D., we are now producing about 55 million tons of iron ore (1993-94). However, Indian steel industry has been reared in highly sheltered state, free from competition. Though it has provided for the developmental needs of our country so far, it will not be unfair to say that Indian Steel industry has not developed satisfactorily. We still have to import finished high quality steel from outside.

Table 7.3 *Recoverable reserves of mineral deposits in India*

	Minerals	Reserves in million metric tons	Life span in years
1	Iron ores	13460.00	257
2	Aluminium ores (Bauxite)	2462.40	299
3	Copper ores	416.80	34
4	Zinc ores	10.10	5
5	Tin ores	29.00	4
6	Tngston ore	54.00	3
7	Nickel	281.00	12
8	Gold metal	57.00	97
9	Gold ore	67.90	8
10	Lead ores	179.10	67
11	Molybdenum ores	2.50	5
12	Chrome ore	146.00	93
13	Manganese ores	167.30	65
14	Magnesite	245.10	267
15	Mica	60.00	300
16	Rock Phosphate	145.00	130
17	Asbestos	9382.80	–
18	Potash	404.00	–
19	Gypsum	237.60	–
20	Rock salt	3.50	–
21	Fire Clays	518.10	–
	Based on (11;12)		

Aluminium reserves in India have been estimated to be about 2462.0 million tons, out of which about 1890 million tons are of rich metallurgical grade. Our country ranks fifth among the Aluminium rich countries of the world. Production of Aluminium rose from about 4 thousand tons in 1950 to 511 thousand tons in 1991-92. At present, we have sufficient aluminium stock for the domestic market. We are exporting it to other countries as well.

Copper ores are in short supply in India. We have already run out of our good quality deposits of copper. Hindustan Coppers Ltd. is the sole producer of the primary metal in India. Malanjkhand Copper mines have been estimated to contain about 275 million tons of copper ore. Average grade of ore worked out in India contains about 1.23% of the metal whereas ores worked out in other countries such as Zambia, Chile or Peru have about 3.2% of metal content. Imported copper is, therefore, cheaper than the copper produced in our country. Removal of import tax on copper, which has so far enjoyed Government protection, shall affect the viability of Indian copper Industry adversely.

As far as identified until date, Zinc-Lead ore reserves in India are estimated to be about 390 million tons. However, mineable reserves, the bulk of which is localized in Rajasthan, are about 167 million tons only, with 8.16% of zinc and about 2.17% of lead content. Hindustan Zinc Ltd. a public sector metal producer has a dominant role in the development of zinc-lead industry in India. Zinc production is about 149 thousand tons and lead production is about 65 thousand tons in India (1993). Hindustan Zinc Ltd. has emerged as one of the major Zinc and Lead producers in the world. However, good quality ores of these elements are being depleted at a fast rate and this has caused the Indian products to be costly as compared to the imported zinc and lead which are cheaper because of a slump in world market these days.

Metallurgical grade ores of number of other metals are available in India. These usually occur along with deposits of other metals as complex mixtures. Important among these metal reserves are those of silver, cobalt, cadmium, chromium, manganese, tin, titanium zirconium etc. However, for many of these elements we have to look to other countries as their mining and processing is in a poor state in our country. Surveys & feasibility studies for extraction of these metals are in progress and for some metals we have already set up the extraction process. Though we may not be self-sufficient at present, it is certain that in near future we shall be able to produce large quantities of these elements in our country (10).

QUESTIONS

1. Give a brief account of formation of mineral deposits.
2. Metals cannot be irrecoverably consumed but we tend to disperse them in diluted state all over our planet. Discuss the statement.
3. What are the major consequences of over-exploitation of mineral resources? How can they be conserved?
4. Write a brief account of non-fuel mineral resources of India.

8

Chapter

Energy Resources

Up to 1700 AD most of the power available to human society was limited to solar energy trapped by green plants which produced organic matter. It was the biological oxidation of this organic matter, which fuelled the muscle power while combustion of organic matter provided energy for other purposes such as lighting, cooking, heating etc. The formation of fossil fuels (coal, oil and natural gas) is also due to photosynthesis carried on by plants, which occurred millions of years ago. These were, however, not in general use before the beginning of Industrial Revolution. Energy requirements of man were modest and could be fulfilled by solar energy 'recently' fixed by green plants. The situation has now changed drastically.

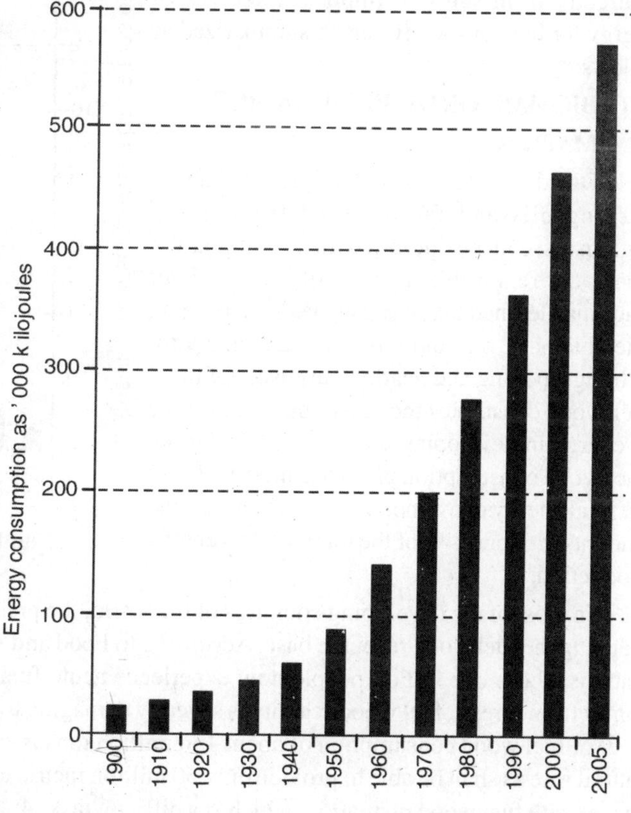

(I) GLOBAL ENERGY CONSUMPTION

Minimum per capita energy requirement of man is about 2000 kcals which is the quantity required to keep him alive and is obtained from food he eats. In a primitive society, apart from cooking, lighting, heating etc. there was little need of more energy. Industrial Revolution has, however, brought in an era of concentrated use of large amounts of energy. Per capita energy consumption, which was a little more than 2000 kcals has now shot up to about 100 times in technologically advanced countries of the world. There has been a rapid and steady rise in global energy consumption ever since the fossil fuels came into wide spread use (Fig. 8.1). We have entered an 'Age' of rapid consumption of fossil fuels, which represents the photosynthesis of millions of years ago. Per capita consumption of energy is not the same all over the world. It is highest in advanced Western countries. Only 20% of the world's people consume about two-third of the total energy produced by man while the rest of the population has to live with only one-third of the energy supply. To millions of people living in developing countries of the world electricity is still a dream, fossil fuels are difficult and costly to obtain and biomass constitutes the only source of energy (1;2).

(II) CONVENTIONAL SOURCES OF ENERGY FOR MANKIND

Conventional sources of energy for the human society are those, which have been in use since a long time and have become a convention. Since times immemorial, man has been using diverse sources of energy. However, most of energy which humankind has been using or which humanity still uses, is derived directly or indirectly from sun (3). Important sources of energy for human society can be summarized as follows:

(1) BIOMASS OR DRIED ORGANIC MATTER

Dried twigs, wood, leaves, cow-dung, burning oils and fats derived from living organisms can be included in this category. These are renewable sources of energy. Ever since man learned to use fire he has been burning dried biomass, oils and fats to obtain energy for lighting, cooking and heating purposes. Of these fuel-wood constitutes the most important source of energy in developing countries of the world. Fuel wood consumption provides almost 43% of the total energy consumed in these countries and amounts to about 14% of the total world's energy production.

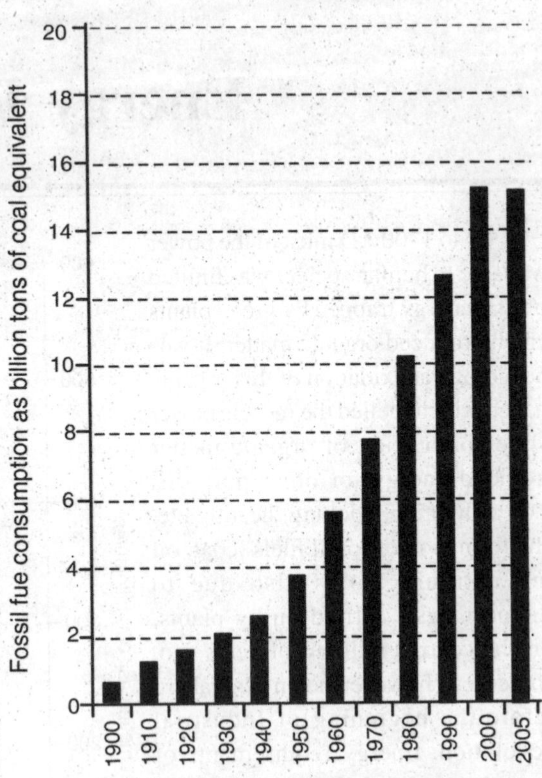

Fig. 8.2 *Global fossil fuel consumption*

In most of the developing countries of the world, the pressure of heavy demand has slowly been reducing the fuel wood resource base. According to Food and Agricultural Organization of the United Nations, about one billion people now experience acute fuel wood shortage. As fuel wood largely comes from forests, fuel wood scarcity is severely damaging the forests and wild life. In India the 1985 level of fuel wood consumption of about 150 million tons is expected to be doubled by 2005 A.D. As natural forests shall be able to provide only 50 million metric tons, the rest of the demand shall have to be met with fuel wood plantation, which is a difficult task (4;2).

(2) FOSSIL FUELS

India is one of the regions of the world which is rich in deposits of coal, oil and natural gas. However, there has been a rapid growth in demand for energy which has resulted in shortage of oil and natural gas. We have not been able to explore our fossil fuel deposits as thoroughly as was required which have naturally resulted in the widening gap between the indigenous supplies and the demand.

1 Coal

Since Independence, coal production has steadily been rising in India. From a meagre 32.3 million tons in 1950, annual coal production jumped up to about 162.3 million tons in 1985. The upward trend continues even today. In the year 1994 we produced about 262.7 million tons of coal of all types. Coal reserves in India have been estimated to be about 200 billion metric tons. About 26.2 billion metric tons of this quantity occurs below a depth of about 600 meters. As mining is carried out only up to about 600 meters in Indian collieries, for our practical purpose the total reserves available are approximately 173.8 billion metric tons. Future coal demand is expected to rise to about 340 million tons by 2000 A.D., 430 million tons by 2005, and to about 520 million tons by 2010 AD (5). So if we compare the projected demands with our available reserves, it is apparent that India has plenty of coal to last for at least four to five centuries. Most of the coal produced in India, about 65% is used to produce electricity.

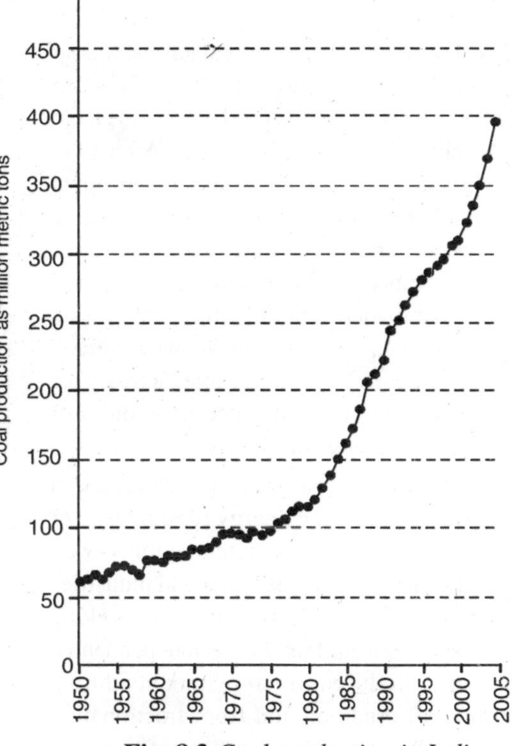

Fig. 8.3 *Coal production in India*

2 Oil:

We produced only 2.3 lakh tons of petroleum crude at the time of independence. Within a span of about 25 years, we were able to raise our crude oil production by about a hundred times. In 1985 the production was 302 lakh tons. The rising trend was maintained till 1990 in which year we extracted about 341 lakh tons of crude oil, after which there was a slow decline which continued for a period of about four years. This downward trend has now stopped and a slow and steady increase is expected in years to come (Fig. 8.4).

India has a large proportion of tertiary rocks and alluvial deposits in which oil usually occurs. The potential oil bearing area in India has been estimated to be about a million square kilometres which implies that there should be much more oil on the sub-continent than is presently known. In India oil was first found at Makum in Assam. Later it was discovered at Digboi, plains of Gujarat and at Bombay high. Recently oil has been located in the off-shore areas of deltaic coasts of Godavari, Krishna, Kaveri and Mahanadi. Oil exploration is an expensive business. It is the prohibitive cost involved, which has so far obstructed the exploratory activity. It is, however, certain that intensive search shall reveal plenty of oil in our country.

Indian oil reserves, as known till date are about 4.45 billion tons, out of which about 1257 million tons are recoverable. We have nearly extracted 483 million tons. Therefore, the recoverable balance is now about 774 million tons, which shall last for about 25 years if we produce crude oil at a rate of about 30 million tons per year. However, there has been rapid rise in demand for petroleum products. In our country the consumption of petroleum products has always been higher than the production. In the year 1990-1991, we imported 29.36 million tons of petroleum crude and other products. The following year we had to import 33.80 million tons and in the year 1992-93 we had to import about 50.66 million tons of crude oil and petroleum products. So we have to spend large amount of money to import crude oil and other petroleum products. In the year 1996-97 the demand for petroleum products is likely to rise to 78 million tons, by 2001-2 it shall be 101 million tons and by 2010-11 it shall be 164 million tons. At present, the import of petroleum crude and products is expected to rise higher and higher every year unless we are lucky enough to strike some new and large deposits of oil in our country. In view of the cost involved in exploration of oil, Indian Government has decided to invite foreign companies to assist in the task which, however, appear to be reluctant to join in the venture.

3. NATURAL GAS

The production of natural gas was about 2.2 billion cubic metres in 1980-81. Between the years 1980-1990, its production rose from 2.2 to about 17 billion cubic metres. After 1990 the production has somewhat slowed down fluctuating between 17 to 20 billion cubic metres. As known till date, gas reserves in India have been estimated to be about 1549 billion cubic metres out of which 889 billion cubic metres are recoverable. We have already produced about 175 billion cubic metres. This leaves a recoverable balance of about 714 billion cubic metres only.

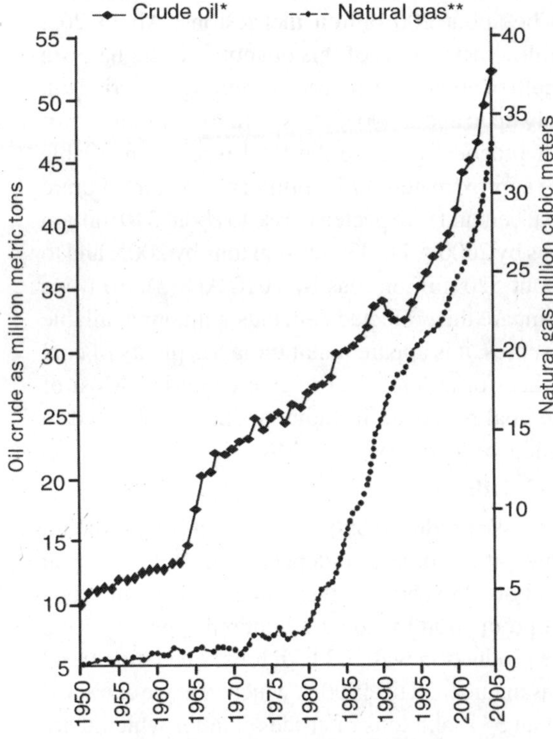

Fig 8.4 *Production of Crude oil and natural gas in India*

Natural gas is a gift from nature. It can be used as a source of energy for domestic or industrial use or for power generation as well as a raw material for petrochemical industries and fertilizer plants. Natural gas occurs along with petroleum crude as well as independently. It is also produced in large amounts during refining of crude oil and from fractionation plants. Due to lack of adequate storage, compression and transportation facilities, a large amount of natural gas has to be flared, simply burned. About 17 million cubic metres a day or about 6200 billion cubic meters per year of the precious commodity is simply wasted. It was only late in the seventies that efforts to check this wastage and use the gas were started. Through a 1730 km pipeline the gas from Bombay High and Gujarat gas fields is now taken to Rajasthan, Madhya Pradesh and Uttar Pradesh. This pipeline is called Hazira-Bijaipur-Jagdispur pipeline or simply HBJ pipeline. A similar pipeline is proposed for South India, which shall feed the natural gas from western offshore deposits and gas imported from West Asia to southern states. A gas grid is proposed for Assam as well. Construction, of a number of smaller pipelines is underway to feed a number of power generation units, fertilizer plants and LPG bottling centres. A number of big and small projects are coming up along the route of these pipelines.

The demand for natural gas in India is higher than its indigenous availability. In 1994, we produced only 2.8 million tons of natural gas against a demand of 3.4 million tons. The deficit is expected to rise to 1.18 in the year 1996-97, 3.30 in 2001-2 and about 7.8 million tons in the year 2010-11. To meet this ever-rising demand our hopes are pinned on supply of huge amounts of gas from Middle East countries. An inter-country pipeline under the sea has been proposed to be laid between Middle East and India outside the territorial waters of Pakistan. A memorandum of understanding has already been signed with Oman and Iran.

Like oil, it is expected that natural gas deposits in our country are also much larger than what we have been able to find out till date. An intensive search should reveal many new deposits as tertiary rocks and alluvial deposits are abundant in India with which natural gas is usually found associated. Moreover, a large quantity of natural gas is simply wasted in our country, which if used effectively should curtail much of the quantity of natural gas, which we import. A thorough exploration, reduction of wastage, proper distribution and management shall definitely satisfy much of our demand for natural gas.

(3) HYDRO-POWER

The energy of flowing waters can be used to drive turbines to convert it into electricity. In India the generation of hydro-electricity has been emphasized right from the beginning of our First Five Year Plan. By the end of IV Plan in the year 1974 we were able to develop an installed hydro-electricity generation capacity of about 6.9 thousand megawatts which constituted about 42% of the total installed power generation capacity. However, as the demand for electric power grew hydro-electricity could not keep pace with the power generation from other sources and by 1995, the last year of VIII Five Year Plan, its share in total installed power generation capacity fell down to about 25%.

The clean, renewable energy resource has some drawbacks. The initial cost of hydro-electricity generation is high and there is a long period after which the expenditure incurred pays dividends. The huge volumes of water which have to be stored for power generation sub-merge large portions of land, displace natural communities and could be hazardous for people living down-streams in case of earth quacks, accidents or sabotage as a breach in the dam may wash away and destroy the entire area below. This has caused a lot of controversy about a number of important projects in India in the face of which some very important projects proposed for power generation and irrigation have been slowed down.

(4) NUCLEAR ENERGY

Unit weight of fissionable Uranium ($_{92}U^{235}$) provides 3,000,000 times more energy and for every unit weight of Helium produced as a result of fusion reaction 24,000,000 times more energy is produced as compared to the energy obtained from burning a unit weight of coal. However, nuclear energy is costly and risky to generate because of the strict construction requirements of a nuclear power plant and expenses involved in processing and disposal of spent fuel. Nuclear power is a viable source of energy for mankind after we have consumed our fossil fuels in near future.

The nuclei of an atom have a density of about 1.6×10^{14} gms per ml. This means that one drop (0.05 ml) of nuclear matter shall weigh about 5.8 million metric tons. To split or to build (fission and fusion) such a super dense material tremendous amount of energy would be involved. In either case the energy released is actually due to the loss in mass of the matter or the 'mass-defect' which occurs during the process and can be calculated by the famous Einstein's equation:

$$E = m \times C^2$$

Where E is the energy released in ergs, m is the mass in grams and C is the velocity of electromagnetic radiations (3×10^{10} cms per second). We have so far been able to utilize only the fission reaction for generation of electricity. Use of fusion reaction is still in an experimental stage.

1. Nuclear power generation:

Nuclear power plants are essentially thermal electricity generating plants, which use steam to drive turbines. In thermal plants, this steam is generated by combustion of coal while in a nuclear power plant the energy obtained from fission of Uranium-235 ($_{92}U^{235}$) is used to produce steam. The fission reaction is carried on in Nuclear Reactors. These are nuclear furnaces covered with several feet thick concrete mixed with material like cadmium, boron etc. alloyed with steel. These elements have a larger neutron-cross-section and are more effective in trapping radiations, thereby protecting persons working near the Reactor. Two main types of nuclear reactors are being used these days:

a. Thermal Reactors or Slow Reactors: These reactors use a mixture of $_{92}U^{238}$ and $_{92}U^{235}$ in ratio of 19:1 as fuel. The fusion reaction is initiated by slow neutrons. As the neutrons strike $_{92}U^{235}$ nuclei it splits up to produce smaller nuclei, free neutrons and a large amount of energy, about 82×10^9Joules per gm of $_{92}U^{235}$, is released. The reaction can be represented as follows:

$$_{92}U^{235} + {}_0n^1 \rightarrow {}_{51}Sb^{133} + {}_{41}Nb^{99} + 4\ {}_0n^1 + \text{Energy}$$

$$_{92}U^{235} + {}_0n^1 \rightarrow {}_{51}Sb^{133} + {}_{41}Nb^{99} + 4\ {}_0n^1 + \text{Energy}$$

About 30 pairs of smaller nuclei are formed when $_{92}U^{235}$ undergoes fission reaction. In order to maintain the chain reaction neutrons released from the fission reaction have to be slowed down or moderated, as slow neutrons trigger the fission reaction. This is done by capturing excess neutrons with the help of moderators. Water, heavy water (deuterium oxide) or graphite is used as moderators. Adequate amount of $_{92}U^{235}$ must be present in the fuel so that the neutrons have chances to strike with the nuclei and continue the chain reaction. Natural Uranium occurs mostly as $_{92}U^{238}$ with only 0.71% of $_{92}U^{235}$. It has to be enriched so as to possess at least 5% $_{92}U^{235}$ before it can be used as a fuel for the reactor. As the amount of $_{92}U^{235}$ available on our planet is quite small, about 6800 metric tons only, its supplies are expected to get exhausted in near future unless more deposits are discovered.

b. Breeder reactors or Fast reactors: These reactors produce more fuel than the actual fuel input (fissionable) and hence the name breeder reactor. Naturally, occurring Uranium deposits contain about 99.29% $_{92}U^{238}$, the normal isotope, which is used as fuel in a breeder reactor. Fast, unmoderated neutrons are used to trigger the reaction, which can be depicted as follows:

$$_{92}U^{238} \xrightarrow{+_0n^1} {}_{92}U^{239} \xrightarrow{-\beta} {}_{93}Np^{239} \xrightarrow{-\beta} {}_{92}Pu^{239}$$

$_{92}U^{238}$ captures a neutron and forms $_{92}U^{239}$ which decays to Neptunium ($_{93}Np^{239}$) and then to Polonium ($_{94}Pu^{239}$). It is Polonium, which finally undergoes fission reaction. It captures fast neutrons and splits in the same way as $_{92}U^{235}$. Thus starting from a normal isotope of Uranium the breeder reactor provides fissionable material. The initial quantity of fissionable material used is small while the fissionable matter produced is quite large. One gram of $_{92}U^{238}$ through $_{94}Pu^{239}$ fission provides 21 × 10^9 Joules of energy. Thorium ($_{90}Th^{232}$) can be similarly used in a breeder reactor to provide energy:

$$_{90}Th^{232} \xrightarrow{+_0n^1} {}_{90}U^{233} \xrightarrow{-\beta} {}_{91}Pa^{233} \xrightarrow{-\beta} {}_{92}U^{233}$$

$_{92}U^{233}$ thus produced also undergoes fission reaction, which is initiated by fast neutrons.

2. The Potential of Fusion Reaction:

Harnessing the energy of a fusion reaction is like confining a sun to a power plant. Nearly all of the earth's energy stems from fusion reaction occurring in the sun and carrying out these reactions on earth promises an enormous supply of energy. Fusion reaction may be summed up as follows:

$_1H^2 + {}_1H^2 \rightarrow {}_2H^3 + {}_0n^1$ (79.5 × 10^9 Joules per gm)

$_1H^2 + {}_1H^2 \rightarrow {}_1H^3 + {}_1n^1$ (96.0 × 10^9 Joules per gm)

$_1H^2 + {}_1H^3 \rightarrow {}_1H^4 + {}_0n^1$ (338.5 × 109 Joules per gom)

Deuterium ($_1H^2$) is plentiful. For every 6700 atoms of hydrogen, one atom of deuterium is present on our planet in seawater. Tritium ($_1H^3$) for the last reaction may be produced by bombardment of Lithium with slow neutrons.

However, these reactions need an ignition temperature of about 40,000,000° C to 100,000,000°C followed by temperatures of about 25,000,000°C to continue the reaction. Everything on earth including the material of which it is made of shall turn into vapours at these temperatures. How to build a furnace or a reactor to carry out the fusion reaction? At these temperatures, matter turns into its ionized state and can be contained within strong magnetic fields while such high temperatures may be attained by concentrating strong laser beams. These are some of the possibilities. We are still not able to develop and perfect the technology to use fusion reaction as a source of energy for mankind.

3. Uncertain future of nuclear energy:

Needless to say, nuclear energy can provide unlimited power to mankind. The future of nuclear energy is, however, uncertain because of the problems caused by dissemination of radioactivity. Radiations disturb the structure and function of the genetic material (DNA & RNA) and threatens the very existence of a healthy biosphere.

Long half-lives of some of the fission products of a nuclear reactor raise the problem of disposal of radio-active wastes. The spent fuel contains a number of radio-active isotopes such as Sr^{90}, Cs^{137}, Cm^{244} etc. which decay during the first three or four hundred years. However, some isotopes, which have a half-life ranging between 1000 to 1000000 years, necessitate safe disposal or storage for thousands of years to come. Bombs placed at Hiroshima and Nagasaki released an estimated 1 million curies, whereas spent fuel from a Light water reactor has 177 million curies per ton of radioactivity at the time of its discharge. Even after 1000 years, it possesses about 1752 curies per ton of radioactivity (6). By the year 1990, about 84,000 metric tons of irradiated fuel had already accumulated (7). Even the safest places, amidst the toughest rock formations on either the surface or underground, cannot be relied upon to hold the radio-active wastes for such a long time as required for the radio-activity to disappear. Some of the burial sites, which represent the best of human efforts, have already been suspected to be vulnerable to leakages.

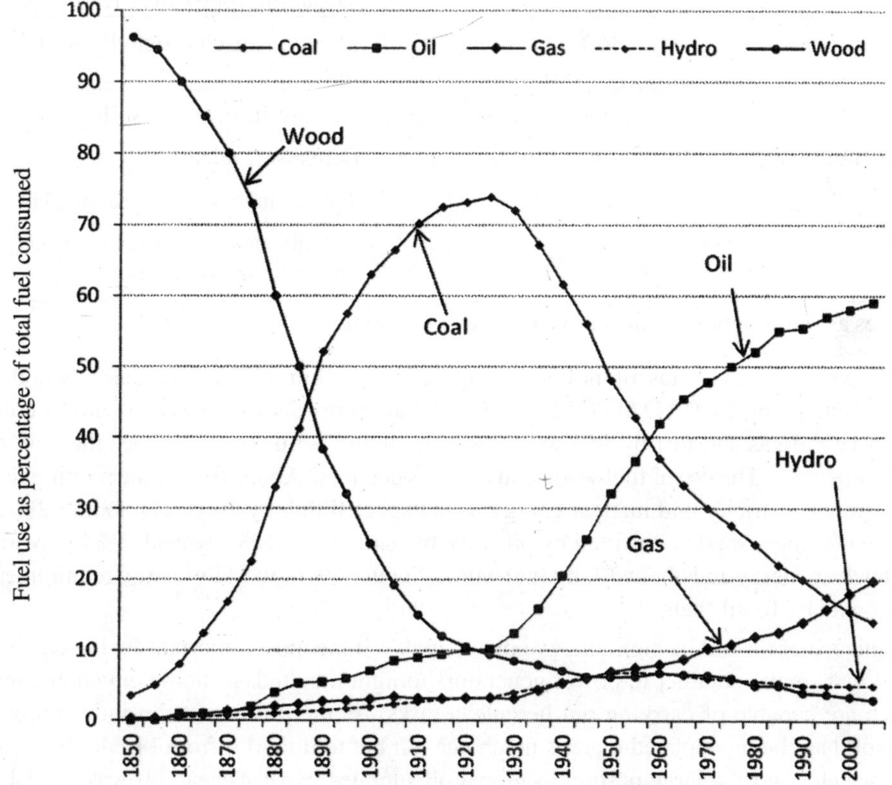

Fig. 8.5 *Pattern of global fuel use*

Accidents like those, which occurred at Three Miles Island, America in 1979 and at Chernobyl in former USSR in 1986, have added a fearful dimension to the problem and introduced further uncertainties to the use of nuclear power. Chernobyl was followed by violent demonstrations all round the world and many Governments were forced to reconsider the entire issue. Many nuclear power plants have already been closed down (8).

(III) THE DEPENDENCE OF HUMAN SOCIETY ON FOSSIL FUELS

Fossil fuels have become essential for human society. Their consumption in terms of coal equivalent was about 1 billion metric tons in 1900 A.D. and about 3 billion metric tons in 1950 AD. It jumped to about 12 billion metric tons by 1985-86 and was about 18 billion metric ton in the year 2005 AD in terms of coal equivalent (Fig. 8.2). Electricity, which we use, largely comes from thermal power plants, which consume more than half of the world's total coal production. Planes fly and trains, trucks, buses, cars etc. run on oil. Natural gas lights our kitchen stoves. We cannot imagine a life without fossil fuels. Hydro-electricity and nuclear power contributes only about 10-14% of the total global energy consumption. About 10-11% of energy comes from fuel wood and biomass (9). The rest, about 75-80% is contributed by fossil fuels (Fig. 8.5).

Table 8.1 *Global release of pollutants into the atmosphere from natural sources*

		Discharge	Causes of the discharge
1	Carbon dioxide	78000	Forest destruction, humus oxidation, forest fires respiration
2	Carbon monoxide	360	Metabolism, chlorophyll loss, forest fires.
3	Sulphur (total)	3600	Release from forest soils.
4	Nitrogen (total)	170	Forest soil release.
5	Hydrocarbons	175	Tree foliage, soil release of CH4
6	Particulates	191	Pollens, gas particle conversion, photochemical, forest fires.

* All quantities as million metric tons per year. *Source: Diverse*

In 1850 AD, wood provided most of the energy required by mankind. Coal had replaced wood as the main source of energy by 1900 AD as 70-75% of the global energy demand was met with by coal, 2-3% by oil and natural gas and rest by fuel wood. Today, oil and natural gas provides the bulk of global energy requirement. The use of fuel wood and coal has declined. About 10% of the requirement is met with by hydro-electricity and nuclear energy. It is expected that by 2005 AD about 62% of world's energy requirement shall be fulfilled by oil, 14% by natural gas, 12% by coal, 5% by hydro-electricity and nuclear energy and about 3% by fuel wood. About 80% of world's energy requirement shall stili be met with by fossil fuels (10).

The convenience of oil and its high-energy content makes it the most important fuel today. It is easy to carry oil in a small container to power generators to light hundreds of home or run gigantic machines, which are capable of carrying out herculean tasks. As the oil reserves in industrialized nations of the world are being depleted there is no choice left but to import it from Middle East and Gulf countries, which control a large portion of global oil supplies. Earlier cheap oil was available from these countries. Things were fine till early seventies. However, the vulnerability of oil-based economy of Western world became apparent in 1973, when oil-exporting countries of the world imposed an embargo, demanding more prices for their product. This created a fuel shock. Another shock came in 1980 when OPEC (Organization of Petroleum Exporting Countries) doubled the price of oil. This has led the world to think in terms of alternative sources of energy.

(IV) THE CONSEQUENCES OF RAPID CONSUMPTION OF FOSSIL FUELS

Table 8.2 *Global release of pollutants into the atmosphere due to human activity*

		Discharge	Causes of discharge
1	Carbon dioxide	3500	Fossil fuel combustion, cement production.
2	Caron monoxide	600	Motor vehicle exhausts,
3	Sulphur (total)	80	Coal combustion, smelting, petroleum refining industrial.
4	Nitrogen (total)	23	Coal and other combustions, fertilizers, waste treatment.
5	Hydrocarbons	97	Motor vehicle exhausts, oil combustion.
6	Particulates	296	Combustion, photochemical, gas particle conversion, industries
* All quantities as million metric tons per year. *Source: Diverse*			

Fossil fuels are 'non-renewable' resources. Over-consumption of these resources shall naturally cause their depletion, which shall result in a chain of reactions culminating into drastic changes in economies, pattern of agriculture and even our daily lives. As estimated till date our oil supplies shall be the first to run out in about 45-50 years. This will be followed by natural gas in about 65-70 years.

Coal will be the last to disappear, serving the mankind for about 350-400 years to come. Crude oil not only yields petrol, diesel, kerosene etc. but provides a number of products which are exceedingly important for the mankind. Oil crude and its products find wide usage in fertilizer industry. Ethylene, propylene, Benzene, 1-3 Butadiene etc. derived from crude oil which also yields polyethylene, PVC, monoethylene glycol, polypropylene, acrylonitrile, styrene, synthetic rubber and fibres for textile industry. With drying of oil wells acute crisis of these products is likely to develop which will affect our lives directly. Shortage of fertilizers could show up as reduced agricultural productivity. Feeding the world population will become even more difficult. Plastics, polymers, synthetic fibres shall become expensive. Some alternatives shall have to be developed to supplement the products of Petrochemical industry (11).

Combustion of fossil fuels, which yields energy, also produces a number of pollutants. Major pollutants produced are Carbon dioxide, carbon monoxide, Sulphur dioxide, Oxides of nitrogen, Hydrocarbons and Particulates. Total annual release of these materials from diverse sources has been summarized in Table 8.1 and 8.2. In addition to these, a number of toxic heavy metals and trace elements are expelled into the environment as a result of combustion of fossil fuels (11;12;13)

Coal is the most environmentally damaging source of energy. It is followed by oil. Combustion of natural gas causes the least pollution as compared to coal and oil. A shift from coal and oil to natural gas shall drastically curtail overall discharge of pollutants into the atmosphere. As we consume our oil, natural gas shall be an important fossil fuel, which in turn may reduce global emission of oxides of carbon, sulphur and nitrogen, hydrocarbons and particulates. After natural gas coal and biomass combustion shall be the main source of energy, which is almost as damaging a source of energy as coal. We shall have to develop more efficient, clean and environment friendly technology for their use in future.

(V) ALTERNATIVE SOURCES OF ENERGY FOR MANKIND

What are the alternative sources of energy after the depletion of fossil fuels? The fuel for future should be convenient, clean, less wasteful of energy and should be renewable so that there is no threat of its depletion (14). Important energy resources, which may be tapped as an alternative to fossil fuels, are:

(1) Wind power:

Approximately 11 × 10^{21} Joules of solar energy incident on our planet surface are used up in generating air and water currents. In a number of countries where persistent strong winds blow, wind power has been in use since ancient times. Across the passing air current large fans are placed whose revolving motion is carried down, through a shaft to drive water pumps, wind mills, turbines etc. Wind power provides cheap, clean and inexhaustible energy to small villages, domestic establishments, small-scale industries etc. The main drawback of wind power is its erratic and irregular supply. However, there are places where strong winds blow for most of the day or nights. At these places this inexpensive, inexhaustible energy resource can be used to save power obtained from other sources by concentrating work during the periods when wind power is available.

(2) Energy from Oceans:

Oceans are vast reservoirs of water covering about three-fourth of earth's surface. They can also be used for generation of power in following ways:

1. Tidal energy:

Gravitational pull exercised by sun and moon causes tides to develop. Sea level rises and falls depending upon the position of sun and moon. As the sea level rises water may be diverted through suitable channels to inshore reservoirs, driving the turbines during its entry. The stored water may gradually be released driving the turbines again during the periods of low tide. In this way an inexhaustible, clean and cheap power shall be available to mankind. Total tidal energy potential has been estimated to be about 2 × 10^{18} Joules per year.

2. The energy of waves:

Air currents rubbing the ocean surface produce waves, which are pushed to the shores where its energy is dissipated as waves break when they strike the shoreline. Floating propellers placed in shallow waters along the shores may be kept in a state of continuous motion by these waves. Their kinetic energy can be used to drive turbines maintained on floats or on platforms erected for the purpose in shallow waters. This will provide a cheap, clean and inexhaustible source of energy to mankind.

3. Thermal energy of oceans:

There is often a large temperature difference between the upper and the lower layers of sea water. In tropical regions surface waters (at 28-30°C) are warmer by 5-12°C than the layers about 1000 meters below. This temperature difference can be utilized to generate electricity with the help of some low boiling point working fluid (like liquid ammonia or propane). In reservoirs maintained on sea surface, the fluid picks up heat, evaporates and is guided through turbines. As the turbines rotate, the vapours are pushed down to lower temperature zones where they cool, condense and are brought back to warm water zone to be recycled again. The concept of Ocean Thermal Energy Conversion (OTEC) was first put forth by a French Scientist as early as 1881 A.D. and has already been tested. It has a huge potential to provide clean, cheap, inexhaustible energy to the mankind.

(3) Geo-thermal energy:

Temperature of earth increases at a rate of 20-75°C per km as we move down from earth's surface. The heat could be used by circulating waters through pipes to raise steam and generate electricity. For satisfactory results, it would be useful to drill and locate the turbines near already known hot springs or thermal springs, volcanoes etc. For this reason, geothermal energy shall be of a local interest only. The total geothermal energy potential of the world, if properly utilized could provide about 2 × 10^{18} to 20 × 10^{18} Joules of energy per year.

(4) Direct use of solar energy:

Earth receives about 75,000 × 10^{11} KW of energy from sun every day. Just 0.1% of this energy is sufficient to meet the energy requirement of the entire world. At noon, the solar energy striking an area of 12,550 sq. kms if converted to electricity shall be equal to the peak power generation capacity of all power plants in the world. Only a part of the roof of an average house in India if covered with solar panels can provide sufficient energy to meet entire energy requirement of the house.

In parts of the world, which receive abundant sunshine, solar heat can be used directly for heating and cooking purposes (As in solar cookers). Intense sunlight focused with reflectors, which track the sun as it moves across the sky, can save a large amount of energy, which we use to keep space warm, heat waters, and for cooking purposes. Solar energy can be converted to mechanical, chemical or electrical energy also. Silicon solar cells can convert solar radiations directly into electricity and can be used for domestic lighting, run television sets, radio-instruments and for community lighting. These systems are costly to install, however, once installed there is little expenditure involved and the device works for years and years together.

(5) Biomass based energy:

Biomass has been an important source of energy for mankind since ancient times. For today's needs, biomass is often not a convenient fuel. It has to be converted to suitable and convenient state before it can be used. Some of the important energy resources which originate basically from photosynthetic activity of green plants are:

1. Bio-gas:

Biogas consists mainly of methane, which is produced when organic matter decays under anaerobic conditions. Cow-dung, faecal matter, and other biodegradable wastes are allowed to decay under anaerobic conditions in digesters equipped with device to collect methane thus formed (The Gobar-gas plants). The residue is exceedingly rich in plant nutrients, which can be used as fertilizers. Methane as a fuel is pollution free, clean and cheap source of energy since it is obtained from the wastes, which we have to dispose of.

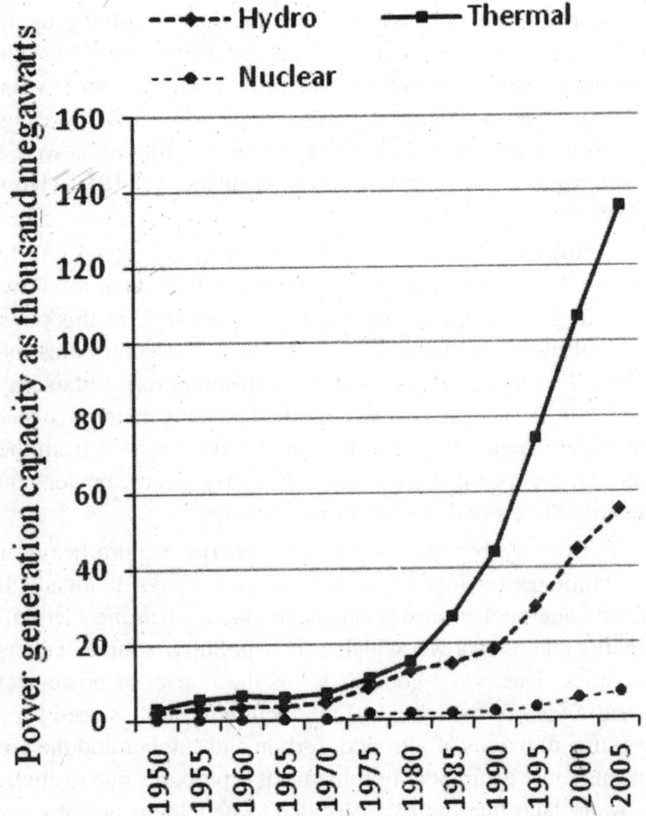

Fig. 8.6 *Growth of installed power generation capacity in India*

2. Petro-plants:

A number of plants belonging to families Euphorbiaceae, Asclepiadaceae, Apocynaceae, Convolvulaceae etc. possess hydrocarbons in their saps and latex. These can be used to produce liquid and gaseous fuels and used just as petroleum products are used. About 15 species of plants have yielded encouraging results.

3. Dendro-thermal energy:

Denuded wastelands can be used to produce fast growing shrubs and trees with high calorific value.

These can be used to provide fuel-wood, charcoal, fodder, and through gasification system gases to be used where fuel-wood and charcoal is not convenient. Similarly, baggasse, the pulp and waste discarded after expulsion of juice from sugarcane during the manufacture of sugar can be used to generate energy for local use.

(6) Hydrogen as the future fuel for humanity:

Like natural gas and oil, hydrogen can also provide the concentrated energy needed in domestic establishments, factories and motor vehicles. Hydrogen when burned produces 284 kilo-joules per mole of energy (or 142 kilo-joules per gm) and the product of combustion are water vapours only.

On weight-basis hydrogen is a better fuel than methane, the common constituent of natural gas. Methane produces only 55.6 kilo-joules per gm of energy as compared to hydrogen, which yields 142 kilo-joules per gm of gas burnt. Hydrogen is a cleaner fuel than any other gas or oil as it produces water only upon combustion. However, the storage of hydrogen poses problems because of its low density. Its storage in pressurized tanks in liquid state makes the containers too heavy to carry. Its storage as metal hydrides from which the gas can be recovered by heating is also expensive. Though abundant supply of this gas occurs in the form of water (H_2O), at present hydrogen is about four to five times costly to obtain than methane. Active research is underway to produce cheaper hydrogen and develop adequate and less costly means for its storage and use. The promise of inexhaustible pollution free energy resource makes hydrogen an important energy carrier of the future.

(VII)GENERATION OF ELECTRIC POWER IN INDIA

The importance of electric power cannot be denied. The convenience with which it can be transported from one place to other over long distances through a thin wire, the convenience with which it can be converted to light, heat, sound or mechanical energy with the help of simple appliances, the convenience with which it can be used simply by operating switches without any detrimental effect on the environment are some of the advantages of electrical energy. That is why a large part of energy used by mankind is in the form of electrical energy for the generation of which other forms of energy are used, a practice which often involves much wastage. In India also the use of electric power has been rising ever since electricity generating units were first introduced in the country. From a meagre production of 5.1 billion KWH in the year 1950 it has shot up to 446.5 billion KWH in the year 1998.

Total installed electric power generation capacity in India is about 188 thousand megawatts. About 73% of this power comes from combustion of fossil fuels or thermal power, which mostly uses coal to generate electricity. Oil contributes little of this power-generating capacity while the use of natural gas is only beginning. Hydro-electricity contributes about 25% while nuclear power contributes about 2% only. The relative share of different forms of energy has not been the same over the past 45 years. In 1974 we had a total installed power generation capacity of about 16.66 thousand megawatts of which about 42% came from hydro-power, 54% from thermal power plants and about 4% from nuclear energy. Fig 8.6 shows the growth of electric power generating capacity in India since 1950 with relative shares of different forms of energy.

Both hydropower and nuclear energy require heavy initial investment while there is a long gestation period after which the investment pays dividends. It is due to this reason that the growth of hydro and nuclear-power has been slower than the thermal power generation in our country. The share of thermal power which is most pollutive source of energy has been rising since last two or three decades. This is not good as far as the impact of power generation on environment is concerned. Nearly 65% of the total coal production in India is used for generation of electricity. The enormous volume of oxides of nitrogen, carbon and sulphur and the amount of fly-ash discharged from power plants cause a number of problems. It is probably due to environmental problems caused by coal fired power plants that India has decided to step up the use of nuclear energy, hydro-electricity and power obtained from gas-based plants. While nuclear energy and hydropower does not cause such environmental problems as are caused by coal, natural gas is also the cleanest and the least pollutive source of energy amongst all other fossil fuels.

(VIII) DEVELOPMENT OF NON-CONVENTIONAL ENERGY RESOURCES IN INDIA

It was the energy crisis of 1973 and 1978, which forced people to recognize the vulnerability of oil-based economy all over the world. Sincere efforts were undertaken to develop non-conventional sources of energy. In India, also, efforts to utilize non-conventional, renewable sources of energy were started only during seventies and a separate Department of Non-conventional Energy Sources (DNES) was established.

Conventional sources of energy, coal, oil, natural gas and nuclear energy – are non-renewable and their use is invariably associated with problems of environmental pollution. Hydroelectric power generation has its own drawbacks. Large-scale use of wood, which has an important source of energy in Indian villages, leads to deforestation. Moreover, the centralized system of power generation, which we have developed, with conventional sources of energy involves huge distribution networks. These are wasteful and expensive to maintain. Non-conventional sources provide energy in decentralized manner to small areas and can reach places where it is difficult to carry fossil fuels or power lines. Large-scale use of non-conventional energy resources tends to reduce the burden from conventional energy systems and therefore is helpful in enlarging their life span. The Indian efforts in the direction of utilizing non-conventional energy resources have been centered on utilization of solar thermal energy, conversion of solar energy into electricity, wind energy and biomass based energy resources

Questions

1. What are the conventional sources of energy for the humanity? Which of these is the cleanest, cheapest and renewable source of energy? Discuss.
2. Discuss humanity's dependence on fossil fuels as a source of energy. What are the consequences of rapid over-exploitation of fossil fuels?
3. What is nuclear energy? Discuss its advantages and disadvantages.
4. Discuss the advantages and disadvantages of natural gas over other forms of fossil fuels. What is being done to promote its use in India?
5. Discuss the alternative sources of energy for humanity. Which of these is the cleanest, cheapest, continuous and feasible source?
6. Write an essay on energy resources in India.
7. Discuss the efforts being made to use non-conventional sources of energy in our country.

9

Chapter

Forest Wealth

It was with the beginning of agriculture, animal husbandry and settled life that the destruction of natural ecosystems started. Expanding human establishments needed space, which was obtained at the expense of natural vegetation, which in turn was modified or destroyed. A number of products, such as food, timber, firewood, fruits and nuts come from forests. Human need, therefore, started a systematic exploitation of forest wealth, which intensified as the size of human population grew. Exploitation turned into over-exploitation. Even the technological advancement of modern times could not reduce society's dependence on forest products. As Nations develop, wood remains the basic raw material for construction, furniture, railway ties, paper, cellophane, rayon, plastics and so many things.

(I) Global Forest Wealth

Our understanding about the quality and extent of forests of the world is very poor even today. In most of the countries statistics on forests are either unreliable or absent altogether. Where some data exists, it is plagued by definition problems. How many trees and how close together they should be to be referred to as a forest? At what point does selective logging becomes deforestation? Does planting of one or few species, often on degraded land or once established cropland or on land cleared of standing forests mean reforestation? These are some of the basic aspects, which have to be resolved properly before an exact assessment of global forests can be made.

However, some idea of the density and the extent of world's forests can be had from the estimate of net primary production and total standing crop (1). Though much uncertainty exists as far as area under each forest type is concerned, the data do provide a crude idea of global forest wealth. The existing forests of the world can be distinguished into the following three main types,

1. Boreal Coniferous forests.
2. Forests of the temperate belt.
3. Tropical rain and seasonal forests.

(1) Boreal Coniferous Forests:

Around the Arctic sea in the North, between latitudes 55° to 65° occurs a zone of coniferous forests which covers greater parts of Alaska, Canada, Scandinavia and Siberia with extensions further south at many places. Due to cold climatic conditions, the decomposition of organic matter is slow and forest floor tends to accumulate a thick layer of organic material, which has an excellent water holding capacity. Decomposition of conifer needles renders the soil acidic. As far as nutrient status is concerned, the soil is poor to very poor. The flora and fauna is characterized by low species diversity.

Coniferous forests are huge storehouse of soft wood to the humankind. These forests roughly stand over an area of about 1.2 billion hectares with a total standing biomass of about 240 billion metric tons on dry weight basis. Net primary production ranges between 0.4 - 2.0 kg per meter sq. per year with a total of about 9.6 billion metric tons per year on dry weight basis.

(2) The Forests of the Temperate Belt:

Temperate region extends from 30° to 55° latitudes on either side of the equator. This zone is characterized by marked seasonal changes in the pattern of temperature and precipitation, which shape the vegetation. The forests of the temperate zone are spread over an area of about 1.2 billion hectares with a mean productivity of about 1.242 kg per sq. meter per year amounting to a total annual production of 14.9 billion metric tons on dry weight basis. The total standing biomass of these forests in terms of dry weight amounts to about 385 billion metric tons.

In regions, which receive scanty rains, extensive grasslands and shrub lands develop. Regions with a climate characterized by alternating dry and wet seasons give rise to deciduous forest in which the leaf fall is almost synchronous. Trees become leafless for periods between the fall and the appearance of new leaves. At places where sufficient moisture is available throughout the year forests of evergreen type develop. In these forests, leaf-fall is almost continuous and the canopy is never naked.

Grassland and shrub land constitute a very important natural resource for the mankind. It is these small plants – grasses, herbs, shrubs etc – which support a large crop of wild, and domestic animals such as horses, mule, ass, cows, buffaloes, sheep, goats, camel, deer, elephants, rabbit, zebra, antelopes etc, which in turn provide food, milk, wool, hide and transport to man. Under the grasses and shrubs, a thick layer of organic material accumulates on the soil, decomposition of which yields large amount of humus. This renders the soil exceedingly rich and fertile. Much of the grasslands and shrub lands of temperate and sub-tropical zone of the world have, therefore, been brought under cultivation. Much of the rich, productive and intensively cultivated soils of the world lie in the temperate and subtropical belt. Extensive deforestation has been brought about in these regions of the world.

(3) Tropical Rain and Seasonal Forests:

World's densest forests are found in the warm and humid belt on either side of equator, between the latitude 30°N to 30.5°S, across Southern America, Central Africa, parts of India and East Asia. An outstanding feature of these forests is the richness and diversity of the biotic community. The vegetation is dense, stratified into many layers with tall trees usually covered with vines, creepers, lianas, epiphytes etc. Under the tall trees there is a continuous ever-green canopy. The lowest layer consists of an under story of small trees, shrubs and herbs which become denser where there is a discontinuity in the overlying canopy. These forests are spread over an area of about 2.45 billion hectares with a mean annual production of about 2.017 kg per sq. meter per year amounting to a total annual production of 49.4 billion metric tons per year on dry weight basis. The overall standing biomass of these forests has been estimated to be 1025 billion metric tons, which is about 55.5% or more of the entire weight of all living things on earth's surface.

In tropical regions, warm climatic conditions promote rapid decomposition and mineralization of organic debris. The soil is subjected to heavy leaching while nutrient uptake by plants is also rapid. Cycling of mineral nutrients is, therefore, very quick. At all times a large proportion of mineral nutrients occurs in biomass in a tropical forest whereas in a coniferous forest much of the nutrients and organic matter lie on forest floor. A large portion of plant nutrients are removed when we clear a tropical forest for agriculture. The soil is made poorer and is unable to support cultivation for long durations.

(II) Deforestation

According to a global survey conducted in 1970 AD about one-fifth of earth's land was covered by closed forests with a canopy cover of over 20% or more while another 12% was under the open woodland with 5-19% of canopy cover (2). This forest cover is already considered a meagre one and even this too is shrinking at a fast rate.

The area under the Coniferous forests in the North has undergone little change since the beginning of this century. This forest belt lies in Alaska, Canada, Northern Europe and the Russian states. The pressure of human activity and demands has never been heavy in these woodlands as compared to the tropical and temperate regions of the world. Due to early technological advancement and industrialization, the bulk of population has shifted to urban settlements from these areas. The evolution of intensive agriculture has enabled the small rural population left behind to feed the large numbers concentrated in towns and cities. In United States of America, barely 2% of the population is involved in agriculture, animal husbandry, and forestry etc. The small numbers, not only produce enough to feed the entire nation but also export food grains to needy countries. The reduced pressure of demands and application of better technology in utilization of forest produce, have preserved much of the Northern belt of Coniferous forests across the globe from wasteful destruction.

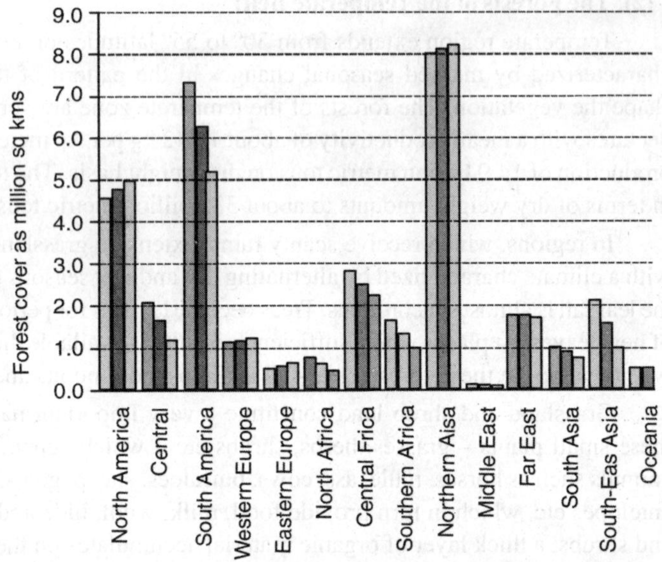

Fig. 9.1 *Gloabl forest cover as estimated for for the years 2000 AD, 2025 AD and 2025 AD.*

However, elsewhere on this globe conditions are different. Extensive deforestation has been taking place in developing countries, which lie in tropical and temperate regions of the world (Fig 9.1). We began 20th Century with about 7.0 billion hectares of forest cover over the land surface of our planet. By 1950 AD, it was reduced to about 4.8 billion hectares. If the present trend continues, we will be left with only 2.35 billion hectares in 2000 AD. In a FAO/UNEP study, it was concluded that an estimated 7.3 million hectares of rich tropical forests are lost every year. We are losing about 14 hectares of closed forest every minute (3).

Such regional or global estimates of deforestation, however, do not yield a clear picture of the deteriorating situation. The extent of wasteful deforestation is usually found to be much higher than the estimated or reported figures. In Africa, the decline in woodlands is almost 2.3 million hectares per year. Large tracts of forests are cleared which results in ecological setbacks causing further reduction in forested area. Dense forests of Western Africa and those of Madagascar, Rwanda, and Burundi etc are diminishing at a faster rate. Almost 75% of woodlands of Ivory Coast have been lost since 1960 A.D. Ghana has lost almost 80% of its forest wealth. In Nigeria, as well, the rate of deforestation was about 10% per year during 1975-1980 (4). In countries like Peru, Columbia, Ecuador, Venezuela etc. woodland clearance for grazing and farming is so rapid that within next twenty years the only forested area left shall be those of national parks and sanctuaries (5).

In India according to national policy laid down in 1952, a forest cover of about 33% was considered desirable. However, surveys conducted in the early seventies found a forest cover of about 22.7% only. The reduction in forest cover has been rapid during the preceding decade. We are losing our forest wealth at a rate of about 1.3 million hectares per year.

(III) Major Causes Of Deforestation

Deforestation is a consequence of over-exploitation of our natural ecosystems for space, energy and materials. The basic reasons for such extensive deforestation are:

(1) Expansion of Agriculture: Expanding agriculture is one of the most important causes of deforestation. As demands on agricultural products rise more and more land is brought under cultivation for which forests are cleared, grass-lands ploughed, uneven grounds leveled, marshes drained and even land under water is reclaimed. The Brazilian Government began the construction of Trans-Amazon Highway in 1970 in hope to colonize its largely empty Amazon basin, which was covered with lush green tropical forests. But it was soon realized that apparent fertility of lush green jungle soil was an illusion. Crops after crops failed. Roads were washed away. Newly settled communities disintegrated and high hopes that the Amazon basin shall feed the hungry millions of the world were dashed (6).

(2) Extension of Cultivation on Hill Slopes: Outside the humid tropical zone, in most of the third world countries, major forests often occur on hilltops and slopes. Though agriculture has nearly always been concentrated on plains and floors of valleys, farming on narrow flat steps cut one after another across the slope or terrace farming is an age-old practice. It has never been extensive because of the grueling labour and low productivity. However, the ever-rising human numbers and their necessities have forced many to go up to mountain slopes for cultivation. More and more slopes are cleared of plants, steps carved out and against many odds cultivation is attempted. After a few crops, the productivity declines and torrential sub-tropical rains carry down massive quantities of precious top soils to streams and rivers. While denuding hill slopes, the silt and sediments settle further down raising stream bottoms and riverbeds aggravating the flood situation.

(3) Shifting Cultivation: Shifting cultivation is often blamed for destruction of forests. In fact, it is poor fertility of soil, which has given rise to such a pattern of farming. A small patch of tropical forests is cleared, vegetation slashed, destroyed and burned. Crops are grown as long as the soil is productive, after which the cultivation is abandoned and cultivators move on to fresh patch of land. The abandoned land was allowed to lay fallow for long periods during which re-growth of vegetation took place and natural ecosystem was restored. Shifting cultivators, therefore, worked in harmony with nature. However, the demands of growing population have shortened the fallow periods drastically. The soil is unable to regain its fertility before it is put to use again. The loss of plant cover is nearly permanent. The influx of shifting cultivators in water-shed around Panama Canal has caused extensive soil degradation resulting in large-scale erosion of the soil. Future utility of Panama Canal and the Panama City's water supply system are threatened by massive deposits of silt and sediments. The Government has launched massive programme of reforestation of the watershed around Panama Canal (7).

(4) Cattle Ranching: Large areas of tropical forests in Central and South America have been cleared for use as grazing land to raise cattle and cash in on the lucrative beef exports to USA. Nevertheless, in these cases also, the problem of poor productivity of tropical soils makes the venture non-viable. The soil degenerates within a short span of time due to over-grazing and massive soil erosion occurs. Cattle ranching have done much damage to the tropical forest cover in South and Central America (8; 9).

(5) Firewood Collection: Firewood collection contributes much to the depletion of tree cover, especially in localities, which are lightly wooded. Denser forests usually produce a lot of combustible material in the form of dead twigs, leaves etc. In Madhya Pradesh, India, a recent observation revealed that felling of small trees for use as firewood and timber exceeds fresh plant growth. In some places in the state, the Government allows people to collect head loads of dead wood from forests for personal use. However, dead wood is actually manufactured, trees are axed, and their barks girdled and live trees become personal head loads to find their way to local markets. If the present trend continues, within twenty years, it is feared that half of the State, which has the largest area under forests in India, will become treeless (10). Outright felling of live trees to meet firewood and charcoal requirement is common in lightly wooded areas in many countries. In Upper Volta, Sudan, Nigeria etc well-organized gangs exist which cut live trees in widening circles around towns and cities, illegally convert them into charcoal for sale in cities. In Sudan, authorities have to use armed guards to protect live trees.

(6) Timber Harvesting: Timber resource is an important asset for a country's prosperity. Commercial wood finds ready national as well as international markets. As a consequence of which natural forests are being mercilessly exploited. Logging or felling of forest trees for obtaining timber is an important cause of deforestation in third world countries. Live trees with thick and straight trunks are felled and transported to commercial establishments elsewhere, to consumers who are ready to pay. In the process, large stretches of forests are damaged and the system, which could have provided resources worth much more to the local people, is disrupted. Ironically, local people get a tiny share in the benefits while axing their own resource base.

Commercial logging in tropical countries usually involves felling of trees of only selected species, which fetch better prices. This process of creaming or removing a few selected trees amidst dense vegetation on rather a delicate soil causes much more destruction than the actual number of trees or the volume of timber taken out would suggest. In a study in Indonesia, it was found that the logging operations destroyed about 40% of the trees left behind. In many third world countries, logging operations have been observed to lead to a permanent loss of forest cover. Loggers after removing a select group of trees move on to other areas. They are usually followed by others who move into the cut over area hoping to start farming and put down roots. The remaining vegetation is slashed and burned and agriculture is attempted. When cultivation fails, it is replaced by cattle ranching or by useless tenacious grasses (11).

(IV) Consequences of Deforestation

An undisturbed terrestrial ecosystem naturally develops into a sparse or dense forest. Factors like, humidity, temperature, rainfall and soil types etc. determine the nature and composition of the biotic community within a forest. Those very factors of abiotic environment, which influence and shape a forest, are also in turn modified by the population of living organisms within the system. Deforestation involves removal of plant biomass, which cripples the system. Various useful products such as firewood, timber, honey, fruits, nuts, resins and medicinal plants etc. are no longer available. A chain of events is set into motion the consequences of which can be summed up as follows:

1. Soil degradation and erosion.
2. Changes in climatic conditions.
3. Destruction of natural habitats.
4. Destruction of a valuable sink for environmental pollutants.

(1) Soil Degradation and Erosion:

Plants check rapid movement of air and water. Flowing waters stay in the area for a longer duration during which time nutrients are re-absorbed and as water percolates down, ground water table is recharged. Plant cover keeps the ground surface humid. Trees with the help of deep root systems are able to draw water from sub-surface water table. Humidity

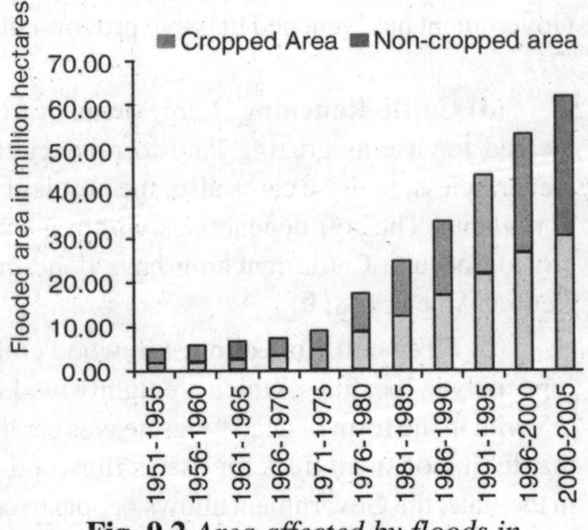

Fig. 9.2 *Area affected by floods in India*

prevents excessive water loss and rapid desiccation. Plants contribute organic matter, which upon decomposition adds humus to the soil. Porosity, water-holding capacity and productivity of the soil improve. Organic matter binds the soil particles in soil crumbs, which makes it more stable against forces of erosion.

Deforestation leaves the ground surface bare. In humid tropics, a large portion of available mineral nutrients is taken away when the biomass is removed. Herbaceous plants and grasses are exposed to the action of sun, wind and rapidly flowing waters. There is further loss of mineral nutrients. Grazing may remove much of the organic matter with which there is further loss of nutrients. Where remaining vegetation is burned to clear the land and agriculture attempted, loss of nutrients is even more rapid. Already poor tropical soil is made poorer. All this further reduces the cover of small plants and grasses as well.

With plant cover gone the battering action of wind and rains loosen the top soil which is thus carried along with water or air currents and deposited elsewhere. The top soil which is thus lost is irreplaceable. It has been estimated that prior to man's influence on earth's crust, oceans received about 9-10 billion tons of sediments annually. Today about 25 billion tons of precious top soil flows into oceans as silt and sediments every year. India loses about 5300 million tons of soil every year. Of this enormous amount nearly 2000 million tons get deposited in stream and riverbeds, about 480 million tons get lodged in dams and reservoirs and the rest is flushed into the sea (12). Most of the multipurpose reservoirs in India are silting up at a faster rate – silting being about 146-874% faster than what was assumed earlier. This has curtailed the life span of our multipurpose, multibillion-rupee reservoirs drastically. The life of Ramganga reservoir in the Gangetic watershed has been curtailed to about one-fourth of its originally intended span of existence (12).

Massive soil erosion aggravates flood situation in two ways. Firstly, the deposition of silt and sediments in riverbeds makes them shallow. Secondly, land devoid of forest cover loses its water holding capacity. About 10 million hectares of forested land can hold enough water to fill completely a reservoir as large as that of Bhakhra-Nangal dam. In absence of plant cover, this water flows down in rapid torrents. In streams and rivers, it has to flow through shallow channels where it spills over its banks inundating low-lying areas (13). Due to deforestation and extensive soil-erosion in water sheds of almost all major rivers in India total area of land affected by floods has been rising steadily (Fig 9.2)

(2) Changes in Climatic Conditions:

Forests shape our natural environment and local climatic conditions. They maintain humidity, regulate temperatures, break wind velocities and influence precipitation. The extent up to which forests influence our natural environment is a controversial subject. However, it is almost certain that dense growth of green plants has a moderating influence on local climatic conditions and the global environment in a number of ways:

1. Maintenance of Humidity: Active transpiration by green plants keeps the environment humid. Roots of trees penetrate deep down to the sub-surface water table. They are able to draw water even when the surface is dry.

2. Regulation of Atmospheric Temperatures: Average air temperatures under tree cover have been found to be appreciably lower than those measured in open fields. The type of plants also influences the temperatures in a forest. Evapo-transpiration exercises profound cooling effect–air cools down by 2-5°C when it comes in contact with cool vegetation. A record of maximum temperature of Mussoorie (U.P.) bears ample testimony to the phenomenon. All around Mussoorie there were lush green forests earlier. Rapid deforestation has been taking place around the city since 1950 AD, which was checked only by 1975 following demands of agitated inhabitants of the area. The mean maximum temperatures of the city rose by 6.5°C within a span of thirty years. A significant depression in the rising trend can be observed from 1975 onwards as an extensive reforestation drive was undertaken to remedy the situation.

3. Moderation of Wind Velocity: Forests check wind velocity by obstructing its passage physically. The velocity of wind through a forest is profoundly affected by the density of vegetation. Higher wind velocities accelerate transpiration and evaporation, which in turn speed up desiccation. Under drier conditions, soil particles are loosened and transported by air currents resulting in a higher rate of soil erosion.

4. The Role of Forests in Enhancing Precipitation: The effect of forests on annual precipitation has been one of the most debated topics among scientists. Experts have come up with varying assessments starting from nominal rise to an enhancement of about 12% in planes and almost twice as much in hilly areas. While studying the forest-climate link it is important to realize that not every shower is due to forests. Rains depend on a number of factors such as location, topography, sea-surface temperatures etc. Of these forests could be one of the factors, which promote rainfall.

An interesting example of deforestation affecting rains is provided by rain fall data from Chhotanagpur Plateau of Bihar which had a considerable area under forest cover in the beginning of 20th century. There were plenty of afternoon showers during summer months. Though there is no apparent reduction in monsoon rains over the plateau, these afternoon showers of summer months have declined. This decline in pre-monsoon showers (April-May-June) is attributed to the extensive degeneration of forests in the area. Excessive summer heat caused the humid air under dense forest cover to be heated up during the earlier part of the day. It rose to considerable heights to cool down, condense into water droplets and by the end of the day fall back as pleasant showers. Such rains are known as Convectional rains. These pre-monsoon showers were a boon for tea-plantations, which came at a crucial time, just when after a long spell of dry season water was badly needed. The number of rainy days excluding those of monsoon months, *e.g.*, June, July and August, were 374 during the four-year period between 1870 AD to 1874 AD. The number had gone down to 271 days during 1978-1982. The decline in the convectional rains has caused the tea gardens of Chhotanagpur plateau to disappear (14).

However, it is not easy to correlate precisely the large variations in rainfall with deforestation as a number of factors act cumulatively to produce a rain. Monsoon rains and the rains of cyclonic origin caused by differential heating of land and sea surface are hardly affected by a forest cover. However, it can be said with some certainty that forests do influence convectional rains or thunderstorms.

5. Forest Cover and Global Warming: An enormously large quantity of Carbon is trapped in forests and forest soils of the world. Coniferous forests possess about 117.6 billion metric tons, forests and woodlands of temperate zone about 188.6 billion metric tons while the tropical seasonal and rain forests possess about 502.25 billion metric tons of carbon. Needless to say, the oxidation of this carbon shall yield a huge quantity of carbon dioxide, which will be added to the atmosphere. Apart from being the basic raw material for photosynthesis Carbon dioxide is an important green house gas. Rising concentrations of this gas in atmosphere could cause excess warming up of our planet just as a thicker blanket makes us too much warm and uncomfortable. A rise in global temperatures, howsoever mild, could cause serious problems for the mankind.

About 81.50 billion metric tons of carbon is added into the atmosphere annually out of which about 3.5 billion metric tons are contributed by combustion of fossil fuels, organic matter, forest fires, deforestation and other human activities. A major part of this Carbon dioxide, almost 70 billion metric tons out of 81.5 billion metric tons are absorbed by ocean surface green plants and is converted to carbonates or organic matter. Of the remaining 11.5 billion metric tons, some carbon dioxide dissolves in rain water and is brought down to earth's surface while a continuous exchange of this gas occurs between water and air above at the air-water interface. Deforestation substitutes lush green forests with agriculture, grassland or herbs and shrubs with low productivity and little biomass. Enormous quantities of carbon dioxide are set free while loss of plant cover reduces the overall photosynthetic efficiency of the system. Thus while the input of carbon dioxide to the atmosphere is increased its output decreases. Many scientists believe that deforestation has been contributing significant amounts of carbon dioxide to the global atmosphere and thereby to the greenhouse effect or global warming.

(3) Destruction of Natural Habitats and Reduction in Biodiversity:

No doubt, the existence and well-being of man is intimately linked with other life forms which provide the basic raw materials for his existence. However, deforestation caused by man himself tends to disturb or eliminate the very habitats of millions of species. The bio-diversity is collapsing at an alarming rate. The greatest threat comes from the tropical regions where we are losing forests at an equally alarming rate of about 1.7 million sq. kms per year. The home of almost half of the living species on this planet is being destroyed and almost one species of mammal, birds or plants is condemned to extinction per day.

(4) Destruction of an Important Sink for Pollutants of the Environment:

Table 9.1 *Estimated reduction in pollutants by forests and soils*

	Pollutants	Quantities eliminated annually
1	Carbon dioxide	70×10^{12}
2	Carbon monoxide	55×10^{10}
3	Sulphur dioxide	201×10^{12}
4	Oxides of nitrogen	36×10^{11}
5	Reactive hydrocarbons, (Mainly C_2H_4).	40×10^{9}
6	Ozone	45×10^{13}
7	Partculates (lead only)	100×10^{6}

Source: Smith 1990

Forest soils and vegetation have a large capacity to absorb, transform and accumulate various pollutants of environment. Vegetation acts as an effective sink for a number of undesirable constituents of the environment. Deforestation not only destroys this sink but also reduces soil's capacity to eliminate pollutants. Removal of plant cover leaves the soil bare. Its organic matter content is quickly mineralized. Without organic matter, microbes fail to survive and the soil is turned into a lifeless mass of sand silt and clay. In absence of microbial machinery, the soil is unable to perform the bio-chemical activity involved in absorption, accumulation or transformation of pollutants.

On global scale an enormous quantity of carbon dioxide, carbon monoxide, sulphur dioxide, oxides of nitrogen, reactive hydrocarbons, ozone and other particulates are cleared by vegetation and the soil (Table no. 9.1). These pollutants may serve as fertilizer to the plant community. As much as 1-13 kg per hectare per year of Nitrogen, 6.1 kg per hectare per year of sulphates and about 20-40 kg per hectare per year of calcium and potassium may be provided to natural ecosystem through atmosphere in the shape of dry or wet deposition (15;16).

No doubt, many of the contaminants eliminated by green cover and the soil originate naturally from plants and the soil microorganisms. However, there is an efficient natural sink to eliminate them. The natural system is also capable of making adjustments for limited inputs of pollutants added by human activity. If we destroy this natural sink while introducing more and more pollution to the system, it is bound to fail with grave consequences for the human race.

(V) Conservation of Forests

To maintain a healthy environment and obtain a sustainable supply of a number of forest products, natural forests should be carefully managed and conserved. Conservation of forests should involve the following two aspects:

1. Prevention of deforestation.
2. Extension of our forest wealth.

(1) Prevention of Deforestation:

We have, so far, relied upon the natural regeneration capacity of the forests to obtain various products such as fuel-wood, timber, fodder etc. However, at many places and in most of the third world countries we have been extracting materials much more than the average productivity of the system. Consequently, our forests are shrinking and are gradually degenerating into useless wastelands. In order to prevent further degeneration of our forest wealth we should adopt rather strict measures, which involve:

1. Controlling unregulated expansion of agriculture and cattle ranching at the expense of our natural forests.
2. Controlling unregulated grazing and destruction of green cover.
3. Controlling unregulated fuel wood collection and timber harvesting.

Table 9.2 *Forest cover as percentage of total land area in India.*

	1980-81	1985-86	1990-91	1995-00
Dense forests (Crown density above 40%)	10.99	11.51	11.70	12.0
Open forests (Crown density 10-40%)	8.41	7.83	7.59	7.42
Scrub-land (Crown density less than 10%)	2.34	2.01	1.82	1.64
Mangroves.	0.12	0.13	0.13	0.12
Other land use	78.14	78.52	78.76	78.82
Source: Government of India, State of Forest Report 1991 and other estimates.				

Most of our flat and fertile land, where there are some means of irrigation – rains, rivers, lakes or ground water resource – have already been brought under cultivation or are in use as pastures. Any further extension of agriculture now shall involve cultivation over marginal land – degraded soils or land in arid, semi-arid and hilly regions of the world, which shall require grueling labour and large investments. While the productivity shall be low, this will involve further destruction of our forests. We can comfortably manage our existing cropland to produce sufficient food grains and edibles to feed the world, even in near future, as full production potential of our agriculture has not been tapped so far. We should concentrate on proper management and intensification of existing agriculture rather than expanding it on virgin lands.

In lightly wooded localities, uncontrolled grazing causes extensive destruction. Young plants, saplings, are nibbled before they have any chance to grow up into a tree. Grazing should be permitted periodically so that young plants could get a chance of growing stronger while herbs, shrubs and grasses could regenerate enough biomass for the cattle to graze upon without damaging the system. Indiscriminate felling of trees for fuel wood and timber harvesting should be completely abandoned while selective harvesting should be done in such a manner that all species should get adequate time to regenerate themselves. In no case, extraction of fuel wood and timber should be allowed to exceed the natural regeneration capacity of the system.

This could involve drastic reduction in advantages, which local people derive from the forests and without which their life could become miserable. If they are not provided fuel wood from local forests, what shall they do to cook their meals or to keep their homes warm. If their cattle are not allowed to graze on herbs and grasses of the local forest wherefrom they will feed themselves. These are some of the vital issues, which have to be resolved for the success of our forest conservation efforts. An alternative has to be provided to the local people if we have to carry on our forest conservation drive successfully.

(2) Extension of Our Forest Wealth:

There is plenty of space around us, large stretches of degraded barren wasteland, hill slopes, arid and semi-arid regions on which we can extend our forests. Even amidst our agricultural fields, on outskirts and fences, trees may be planted which shall be of enormous benefit to the people, the environment and agricultural productivity. To be successful, a reforestation drive must involve proper management of soil environment, as most of the plantation has to be done on degraded wasteland and hill slopes. To improve soil environment we shall have to reduce soil erosion, provide water and improve soil fertility by adding fertilizers wherever possible. A combination of soil conservation measures and techniques of growing and maintaining plant life are needed to carry out the reforestation drive. Active co-operation of local people in reforestation drive should be sought as without it all our efforts could end in utter failure.

As forests help in maintaining a healthy environment and provide a number of valuable products and benefits, they should also be treated like our agriculture, which receives the top priority today. In order to promote better plant growth, our forests should be looked after properly. Wherever necessary adequate, soil conservation measures and proper inputs of water, fertilizers and other chemicals should be used to maintain and augment healthy growth of forests.

(VI) The Forest Wealth of India

The practice of scientific forestry in India began with Sir Dietrich Brandis, who took over as Inspector General of Forests in 1864 AD and under whose guidance State forest departments were created in various British ruled provinces. It was in 1894 A.D. that Government of India formulated a Forest Policy, the first of its kind, which has been a trendsetter for forest management throughout the British Empire. It was this policy, which formed the basis of forest management in India until Independence. Recognizing the importance of forests in maintaining a healthy soil-environment and agriculture this policy sought to preserve the forest cover over hill slopes and marginal lands, restrict shifting cultivation and manage densely wooded forests on commercial lines while forests, which yielded inferior timber, fuel wood, fodder etc., were set aside for use of local people.

Needless to say, the policy though good for the times in which it was formulated and implemented half-heartedly. Wanton destruction of forests was permitted in interest of agriculture, commercial activity and military requirements of the rulers. However, the demarcation of protected or reserved forests following the implementation of the policy did help to preserve a large part of our forest wealth until independence.

Following independence a Central Forestry Board was set up in 1950 and in 1952, a National policy was developed on the lines of which Indian forests were managed until about 1988 AD. As there was more emphasis on development, industrialization and rapid growth in all spheres of our national activity, immediately following the independence, conservation of forests received little attention. Enormous destruction of our natural forests continued for about two or three decades and we have lost about 3.9 million hectares of precious forest wealth since 1950 AD. About 2.50 million hectare forests have been destroyed for expansion of agriculture, 0.41 million hectares have been lost to river valley projects, about 1.5 million hectares for industries, 0.56 million hectares for road construction and about 0.40 million hectares for miscellaneous purposes.

Table 9.3 *Different types of forests in India.*

	Forest Type	Area	Percentage
1.	Tropical moist deciduous	23.686	37
2.	Tropical dry deciduous	18.304	28.6
3.	Tropical wet evergreen	5.120	8
4.	Tropical semi-evergreen	2.624	4.1
5.	Tropical dry evergreen	0.128	0.2
6.	Tropical thorn forests	1.664	2.6
7.	Littoral and swamps	0.384	0.6
8.	Sub-tropical broad leaved	0.256	0.4
9.	Sub-tropical pine	4.224	6.6
10.	Sub-tropical dry evergreen	1.600	2.5
11.	Montane wet temperate	2.304	3.6
12.	Himalayan moist temperate	2.176	3.4
13.	Himalayan dry temperate	0.192	0.3
14.	Moist alpine scrub	1.344	2.1
15.	Sub-alpine	–	–
16.	Dry alpine scrub	–	–
	Total	64.000	100

Area as million hectares. Source *:* Indian Forester July 1971, p.436.

We do not have precise information about the extent of deforestation, as earlier data about our forests are unreliable. According to a report on State of Forests, Govt, of India, 1991, which is based on satellite images and is, therefore, more reliable, actual vegetation cover is about 64.00 million hectares out of the total land area of about 328.72 million hectares. This means that about 19.4% of our land is covered by forests, which is rather inadequate. The National Forest Policy of 1952 considered a forest cover of about 33% as desirable for our country. It is, however, satisfactory to note that the total forest cover has undergone little change since 1980 AD. During the previous decade (*i.e.*, from 1980 to 1990 A.D.) we have been able to control deforestation effectively. It will be apparent from the figures in Table 9.2 that though open forests, the scrubland have diminished area under dense forest, and mangrove vegetation has actually gone up.

Indian forests have been grouped under 16 different categories, which are listed in Table 9.3. However, the most common forests of India are tropical forests of different kinds, which together constitute about 81% of our forest wealth. A number of products are obtained from our forests, which are far from being adequate for our needs. This is naturally due to depletion of our forest wealth while the demand for forest products has gone up considerably due to rapid rise in population. Much of the gap is covered by plantations, imports, other alternative sources or substitutes. The National Forests Policy 1988, which replaces the Forest Policy of 1952, lays emphasis on conservation of our existing forests, extension of our forest wealth through soil conservation, reforestation and maintenance of environmental stability through restoration of necessary ecological balances. Earnest efforts are being made to implement its directives.

Questions

1. Write a brief note on Global Forest Wealth and enumerate the importance of forests to human society.
2. What are the major causes of deforestation? Discuss its consequences.
3. Is there any connection between forests and rains? Discuss how forests are helpful in enhancing precipitation.
4. What is shifting cultivation? Discuss its impact on deforestation.
5. Give a brief account of the strategy you would adopt to conserve and extend our forest wealth.
6. Write an essay on forest wealth of our country. What is being done to conserve and extend our forest wealth?

10

Chapter

Food Resources

Every day the world has to find food for more than six billion people and this number is steadily growing. Each of us requires sufficient food to provide energy to carry out our day-to-day activities and materials from which our bodies can be built and maintained in healthy state. To keep our bodies in proper functional state and ensure good health our diets must also include foods, which have vitamins and minerals. Energy comes largely from carbohydrates and fats, materials to build our bodies are provided by proteins whereas minerals and vitamins usually come from fresh fruits and vegetables. Global food supplies largely come from plants, which provide cereals, legumes and pulses, fruits and vegetables. On worldwide basis, produce from animals constitute rather a small part of our daily diets.

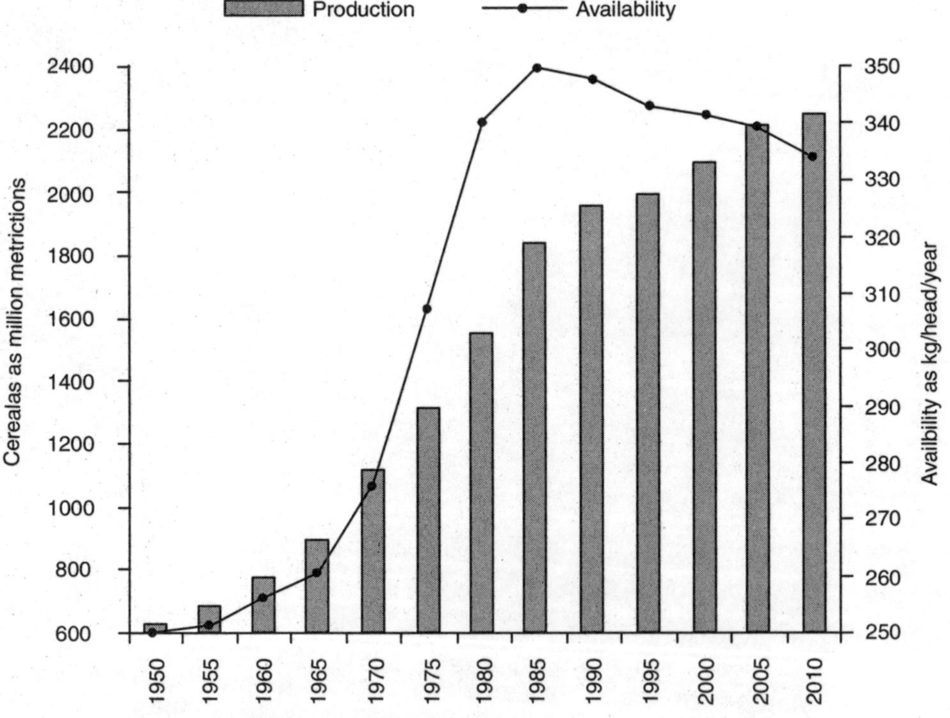

Fig. 10.1 *Global cereal production and per capita availability*

For calculation of food values, our food consumption is usually measured in terms of calorie, which is the unit of heat energy. A calorie is equivalent to the amount of heat, which raises the temperature of one gram of pure water by one degree centigrade. Food value of eatables is estimated by the amount of heat energy or calories produced when one gram of the material is burnt or completely oxidized. The average daily energy requirement of an individual varies from person to person according to their body weight, sex and the physical work they do. Pregnant female and lactating mothers require more calories to nourish the baby. On an average a daily per capita food intake which provides 2,500 to 3000 kilocalories (one kilocalorie = 1000 calories) is usually considered sufficient for a common man. In terms of grains, a per capita intake of about 200 kg/year is considered sufficient to provide the required energy. Our body is a versatile machine, which can utilize carbohydrates, fats, as well as proteins as a source of energy. However, to built up and maintain our bodies in a healthy state certain amino acids, which are building blocks of proteins and a number of minerals and vitamins have to be provided in appropriate amounts from an external source. They cannot be generated within our bodies. Our diets, therefore, have to include carbohydrates, proteins, fats, vitamins and minerals in right proportions or in other words, it should be a balanced diet. Generally, an average per capita intake of about 60 gm/day of proteins is considered sufficient for an individual. With unbalanced diets, the working efficiency is reduced and individuals fall prey to deficiency diseases or infections, which are normally avoided.

(I) GLOBAL FOOD SITUATION AT THE BEGINNING OF THIRD MILLENIUM

Much of global demand for eatables is met with plant products of which cereals, legumes and pulses provide the bulk of energy and protein needed by the mankind. The world produces about two billion metric tons of cereals every year. The three major cereals, which constitute a large part of our food supplies, are wheat, rice and corn. Today more than 300 kg/year of grains are available to every person in the world if we equitably distribute the cereals produced annually – a quantity, which is considered more than sufficient (Fig 10.1). To this may be added course cereals, legumes and pulses as well as numerous types of fruits and vegetables, sugars and syrups, which are, produced all over the world to feed humans and his dependents. In addition to the supplies of eatables, which we obtain from plants, there are many animal products such as milk, meat, eggs, fishes and the produce from aquaculture that satisfy our hunger.

1. GLOBAL CEREAL PRODUCTION

Cereals include wheat, rise maize, barley, oats, millets and sorghum etc. With the exception few countries where the percentage foods derived from animals in daily diets is higher, cereals and pulses make up as much as 75-90% of the average daily per capita food intake of the majority of world population. Only three of these, namely wheat, rice and corn or maize constitute as much as 90% of world's cereal production. Majority of world's population relies heavily on cereals for the supply of energy and a little protein. In none of the earlier periods of human history global cereal production grew as fast as it did in the latter half of 20th century (1). It was a

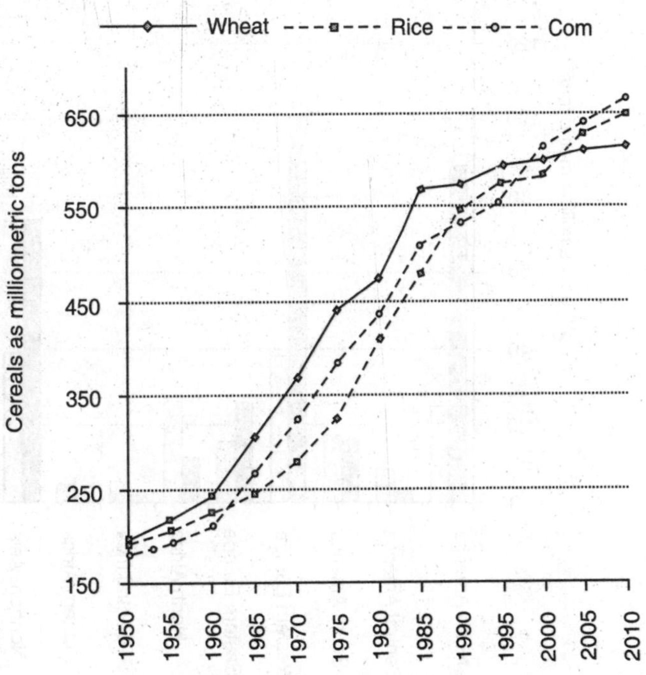

Fig. 10.2 *Growth of major cereal production*

meager 624 million metric in the year 1950 AD (Fig 10.1, 10.2). It now stands at about 2000 million metric tons – a growth of more than 3.5 times!

a. Wheat: Wheat is truly a world cereal. It has more than 650 varieties and is the most important and widely consumed cereal all over the world. It is mainly milled into flour. The whole grain contains plenty of carbohydrates, proteins, fats, minerals and vitamins. In regions of the world where wheat is the preferred cereal malnutrition and protein deficiency does not occur as the whole-wheat grains contain a relatively higher percentage of proteins with appropriate amino acids as compared other cereals. Wheat is unique as a world food grain because it contains a substance called gluten (a protein) with chemical properties most suited for the production of "risen" loaf of bread. Wheat and to a lesser extent rye are the only cereals that contain gluten. Today it is only in African countries, Middle East and Far East that wheat consumption exceeds its production (Fig 10.3).

It was the introduction of dwarf wheat from Japan into the United States which revolutionized wheat cultivation throughout the world. Dwarf wheat has stronger stems and can bear the weight of more tillers thereby eliminating the problem of lodging. An important derivative of dwarf wheats was Norin-10 which was crossed with American varieties to give high yielding dwarf American wheat. The spread of the *dwarfing genes* from Norin-10 has been meteoric by all standards. It constituted the very basis of Green Revolution. In Mexico, Norman Borlaug who was well aware of the necessity of short, thick and strong stemmed wheat plants in his breeding programme, made his first successful crosses in 1955 and developed two dwarf varieties Pitic-62 and Penjamo-62 which were released by the Mexican Government later (2;3). The possibility of raising wheat production with the help of dwarf varieties was realized in India as well. In 1965, 250 tons of seeds of both varieties were imported from Mexico and released to the farmers. Later another 18000 metric tons of seeds of these two species were obtained from Mexico and the Green Revolution commenced in India.

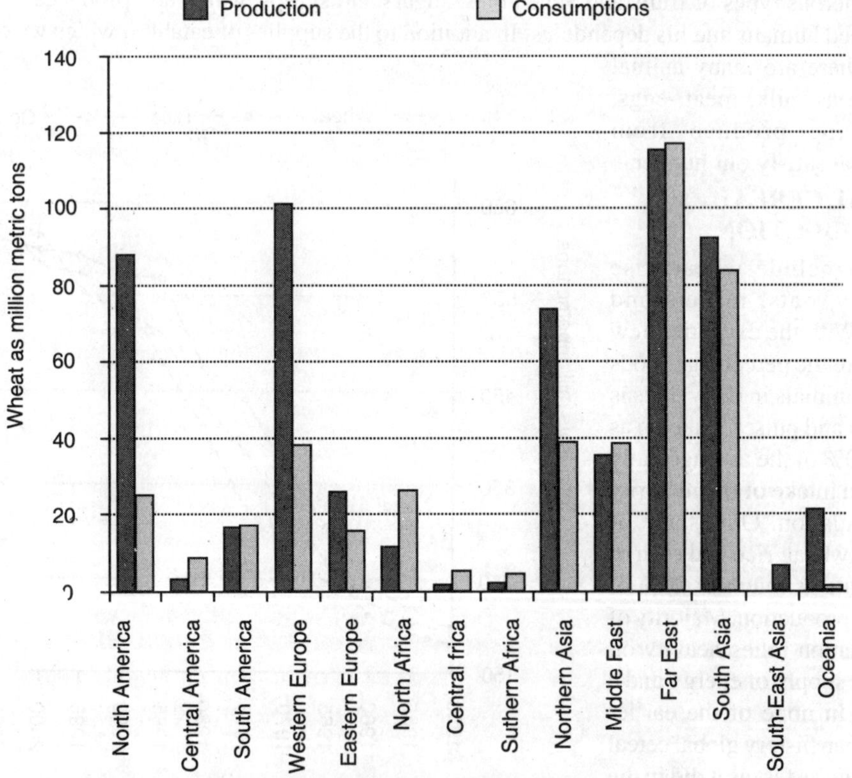

Fig. 10.3 *Productgion and Consumption of Wheat*

b. Rice: There are more than 2600 varieties of rice, which is preferred crop of the humid tropical regions of the world. Rice has a higher percentage of carbohydrates and as such a high-energy content. Rice crops generally yield more food per hectare of arable land than any other crop and is widely eaten in the Far East, South and South East Asian countries. However, it has rather a lower nutritive value as it lacks adequate proteins, minerals and vitamins. Removal of the outer husk of the rice grains deprives the rice of whatever nutritive value it has. Among the higher income groups, rice is invariably eaten along with pulses, vegetables or meat, which compensates for the protein deficiency. The daily diet of 2000 kcals or less consisting primarily of rice with nominal amounts of proteins, mineral and vitamins is the typical diet of millions of poors in East Asia, South Asia and South East Asia. As a result, chronic malnutrition and hunger is prevalent in these regions of the world.

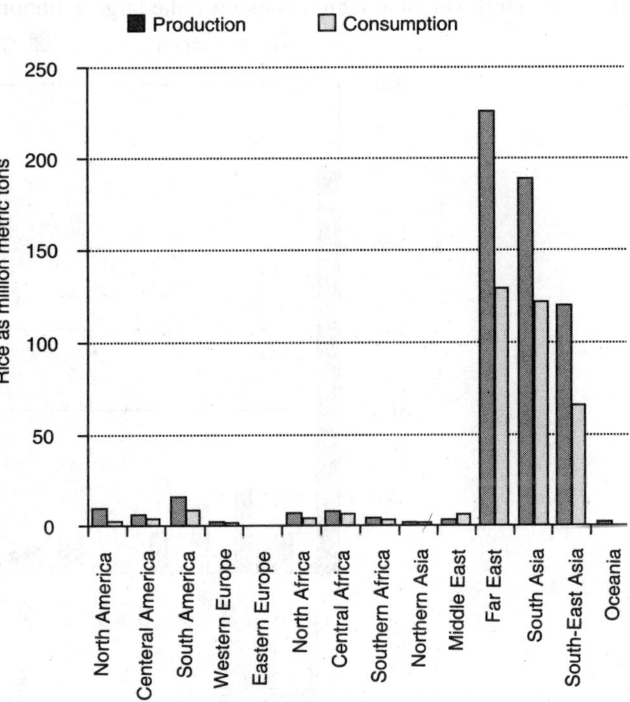

Fig. 10.4 *Global production and consumption of Rice*

A total of about 586.74 million metric ton of rice is produce globally while its worldwide consumption is only 353.86 million metric tons. A surplus of about 232.88 million metric ton of rise is therefore grown annually. A little more than 90% of total global rice production comes from just three regions of the world, which are Far East, South Asia and South East Asia (1). It is these regions of the world only which account for nearly 90% of global rice consumption. Outside these regions South America, North and Central America and North and Central Africa little rice is consumed much of which is produced locally (Fig 10.4).

It was introduction of semi-dwarf varieties of rice, such as Tai-chung Native-1, which was a significant event in the history of rice cultivation. It responded well to higher doses of fertilizers and was relatively insensitive to photoperiods. However, the semi-dwarf varieties were found to be susceptible to most of the diseases and pests and higher levels of fertilizer use further aggravated their intensity. Following the success in incorporation of dwarfing genes, attention was directed to introduce resistance to diseases, insects and pests. Genes for resistance to a number of diseases, insects and pests have been identified and resistant varieties have been developed. Attempts are being made to introduce traits like better taste and aroma in the high yielding, disease resistance varieties of rice being cultivated today (2;3).

c. Maize or Corn: Corn is also one of the most widely distributed food crop, which has a great nutritive value although it is not as nutritive as wheat. Corn is eaten as basic food in Mexico, some parts of South America and North Asia. Global production of corn exceeds those of wheat and rice. A corn crop matures somewhere in the world every month of the year. However, its worldwide consumption is much less than wheat or rice. About 602.5 million metric tons of corn are produced whereas only 106 million metric tons are consumed which results in a huge surplus of about 496 million metric tons (1). The largest producer of corn in the world is North America and the Far East.

It is in Central America, which consumes the largest amount of corn in the world (Fig 10.4).

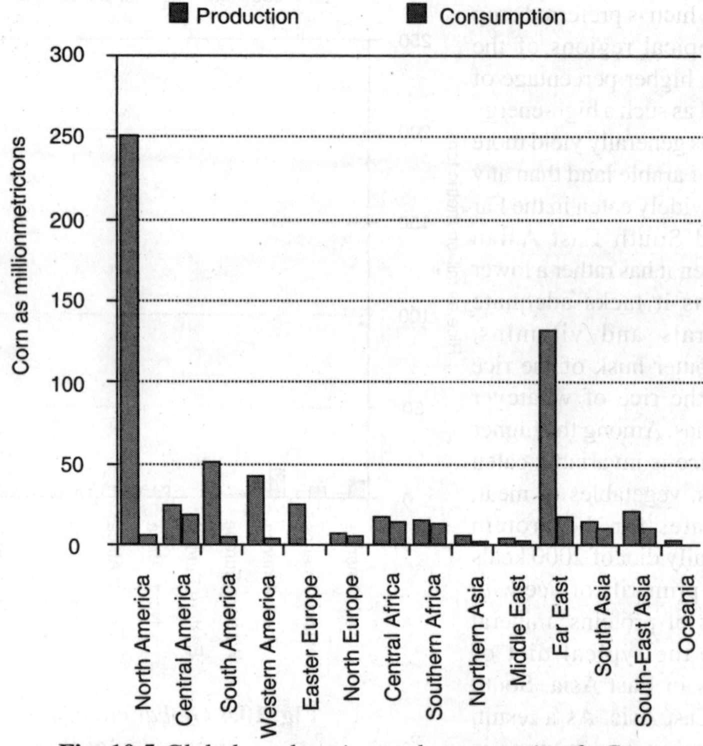

Fig. 10.5 *Global productgion and consumption fo Corn*

Corn is a cross pollinated crop. It was William James Beal, who probably made the first controlled crosses between varieties of corn for the sole purpose of increasing yields through hybrid vigour. Beal worked successfully without knowledge of the genetic principle involved. It was Donald F. Jones, of Connecticut Agricultural Experimental Station, who in 1917 developed the technique which later became famous as "Double Cross". The first hybrid corn using this technique, to be produced commercially, was marketed by the Connecticut Agricultural Experiment Station in 1921. Later it was Henry A. Wallace, who organized the first seed company devoted to the commercial production of hybrid corn. By 1960 nearly 96% of the total corn acreage was being planted with hybrid seeds. Now corn cultivation with hybrids seeds has been adapted all over the world. The technique of breeding hybrid corn has successfully been applied to other crops as well (2;3).

d. Course cereals: In addition to wheat, rice and corn there are a many cereals which are grown as human or animal feed. These cereals are referred to as course cereals because of their rough or course taste and a little lower nutritive value. Most of these cereals are grown in rough and rugged terrains, in soils which are poor in nutrient content and can withstand an appreciable degree of water stress. Important among course cereals are: oats (*Avena sativa*), barley (*Hordeum vulgare*), sorghum (*Sorghum bicolor*), pearl-millets (*Pennisetum typhoideum, P. americanum, P.glaucum* and *P.spicatum*), rye (*Secale cereale*) and pigweed (*Amaranthus* sp.) etc. Although exact figures are lacking the global production of course cereals has been estimated to be about 340-350 million metric tons (Fig 10.6). With an ever growing human numbers to feed the demand on our food grain production is going to multiply in future. With gradually rising rate of degeneration of our cropland and shortage of water for irrigating our fields it could be difficult to grow enough to feed the human numbers in times to come. Higher costs of fertilizers shall add to our miseries. As course cereals can be grown on rough, nutrient poor soils in dry, water stressed regions of the world these cereals could provide an attractive alternative supply of grains to the mankind (2:3).

2. GLOBAL PRODUCTION OF PULSES:

Legumes and pulses have high protein content and provide proteins, which are a very important constituent of human diet. In places where foods derived from animal sources forms only a small fraction of human diet, legumes and pulses can make up much of the protein deficiency. This is particularly important for people of the rice-eating regions of the world where the diets of much of the poors consist mostly of carbohydrates.

On world wide basis about 57.7 million metric tons of pulses are produce and the global consumption is about 36.3 million metric tons which leaves a surplus of about 11.6 million metric tons (Fig 10.7). The largest consumption of pulses is in South Asia where nearly 14.5 million metric tons of pluses are consumed. South America followed by Far East countries and Central America are also the regions where pulses

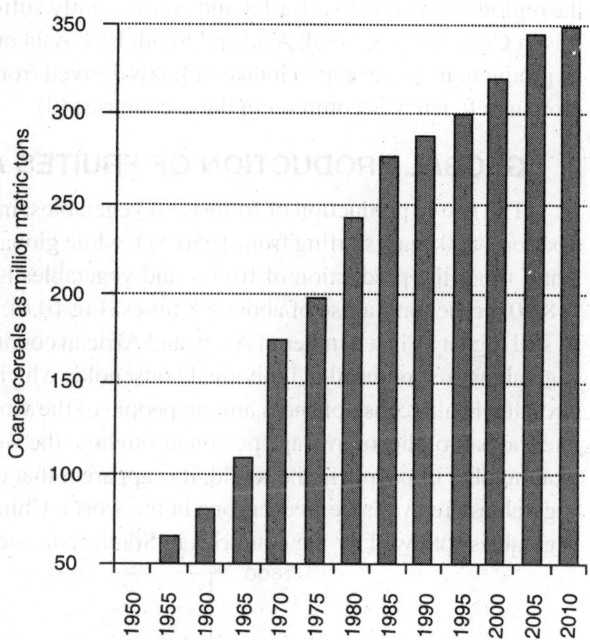

Fig. 10.6 *Global production of course cereals*

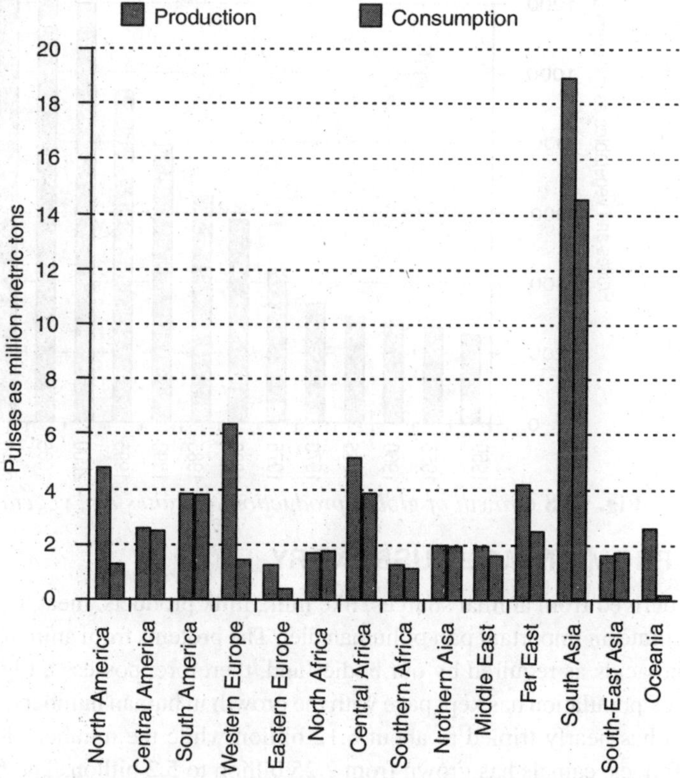

Fig. 10.7 *Global production and consumption of pulses*

form an important part of daily diets. Since pulses are more costly than other cereals many poors of the region are unable to afford it and consequently suffer from malanutrition and protein deficiency. It is in CebtralAfrca, South Asia and South East Asia only where the consumption of pulses exceeds its production. As per capita intake of foods derived from animal sources is very low malnutrition and protein deficiency is common in these countries (3).

3. GLOBAL PRODUCTION OF FRUITES AND VEGETABLES:

The rise in production of fruites and vegetables, including potato, has been phenomenal. Within a period of 50 years starting from 1950 AD, while global cereal production multiplied about three and a half times the production of fruites and vegetables has shot up from 254.8 million metric tons to 1486.0 metric tons, a rise of about 5.8 times (Fig 10.8). As a matter of fact the actual quantities could be still higher as in a number of Asian and African countries a surprisingly large amount of fruites and vegetables are produced in individual households which are usually not included in the reported data. Growing health consciousness among people of the world has made fruites, vegetables and salads an integral part of diet of average person throughout the world. If we divide the total production with the total number of people in the world, it is apparent that today more than 230 kg per year of fruites and vegetables are available to everyone in the world. China leads the world in production of fruites and vegetables followed by the countries in South Asia and Western Europe.

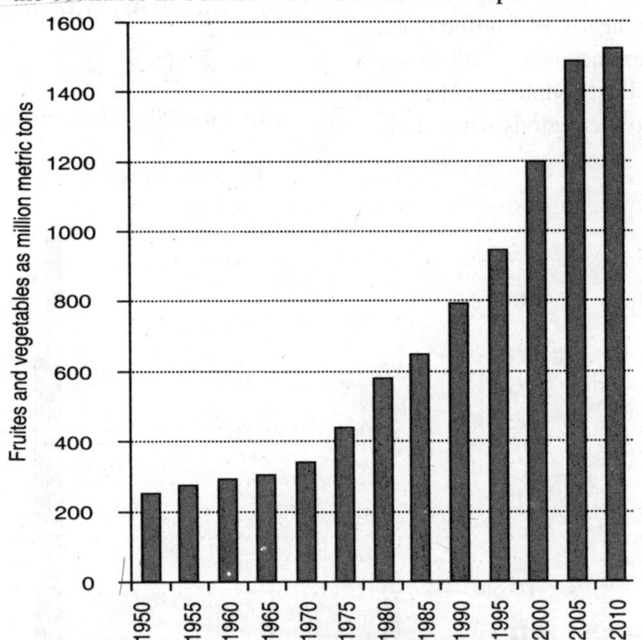

Fig. 10.8 *Growth of global production of fruites and vegetables*

4. PRODUCE FROM ANIMAL HUSBANDRY

Food stuffs derived from animal sources, like milk, milk products, meat, fish and other edibles from waters, constitute an important part of human diet. The proteins from animal sources contain the right mix of amino acids as required by our bodies and, therefore, possess a higher nutritive value. The rise in livestock population has kept pace with the growth in human numbers. Since 1950 AD, the human population has nearly tripled to about 6.12 billion while the number of cattle, pigs, sheep, goats, horses, buffaloes, camels has grown from 2.25 billion to 5.2 billion. The fowl population has multiplied even faster – from 3 billion in 1950 to about 12 billion in the year 2000 A.D.

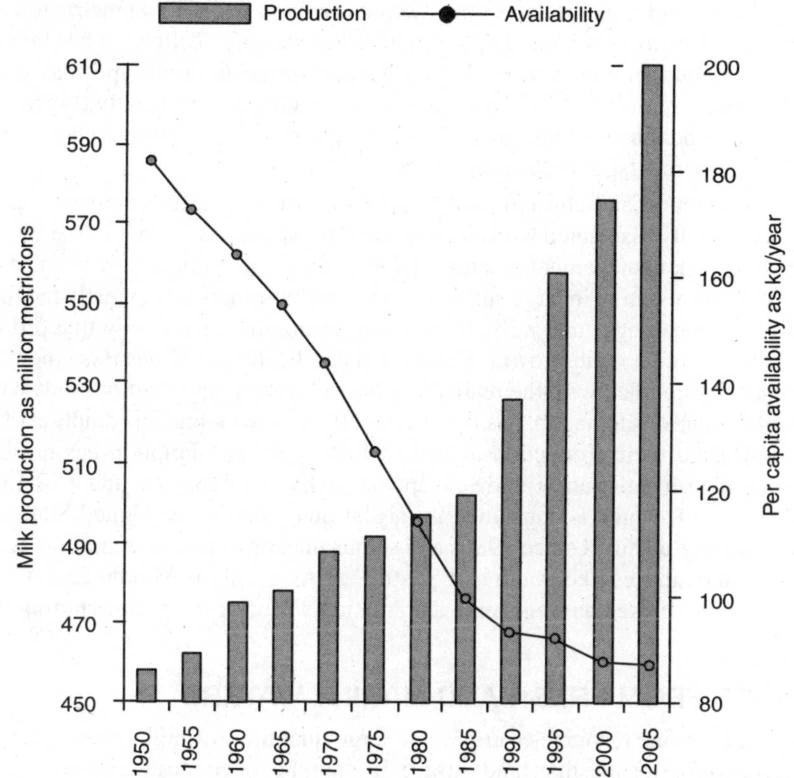

Fig. 10.9 *Global production and per capita availability of milk*

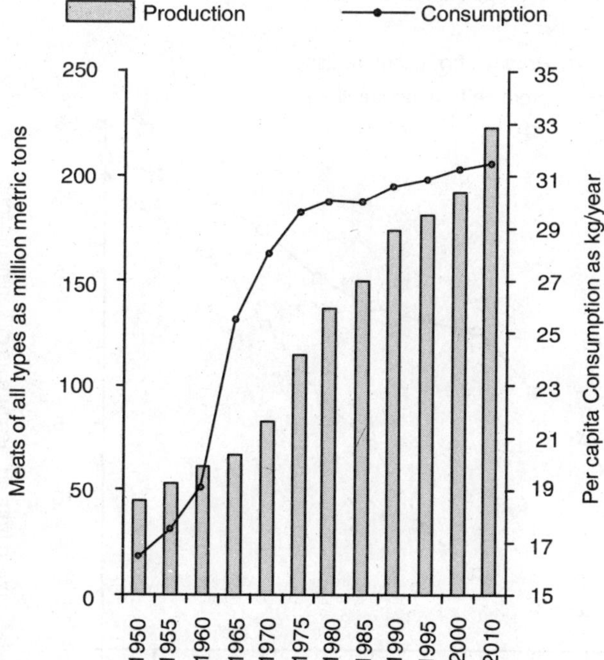

Fig. 10.10 *Global Production and Consumtion of measts of all types*

Global production of milk and milk products now it stands at about 610 million metric tons per year. However, per capita availability of milk and milk products has steadily declined. It has failed to keep pace with the growth in human numbers. Global production of meats of all types has grown steadily since 1950 AD. However, after 1980 its consumption has declined significantly (Fig 10.10). This is probably due to rising health consciousness among people of world. People today want to avoid diets containing higher percentages of saturated fats.

The rapid stride in production and consumption of meats of all type during the latter half of twentieth century, which is usually associated with affluence and prosperity, is a result of rising grain supplies and changes in livestock management practice. Global cereal production grew at a fast rate between the years 1965 to 1986 resulting in large surpluses. The surplus grains were mostly fed to the live stock population to raise meat, eggs and milk. It has been during this period only that produce from animal sources has registered a rapid growth. However rising health consciousness among the affluent sections of our society coupled with the realization that too much high quality foods with a high saturated fat content is injurious to health, has the overall effect of reducing consumption of red meats. This in turn has affected their production as well. Global pork consumption has not been growing as fast as it did during seventies and eighties. It appears to have stabilized at about 12-13 kg/head/year. Consumption of beef, which is consumed in largest quantities in the United States and European countries, has steadily declined since 1980s and so has the consumption of mutton, which is the most common meat in countries like South and South East Asia and the Middle East. Global production and consumption of chicken has registered a dramatic rise during the period starting from 1965.

5. GLOBAL PRODUCE FROM FRESH AND MARINE WATERS

Living fresh water and marine resources can provide large quantities of high quality proteins. Aquatic habitats are much more extensive than land surface. In a number of countries protein obtained from aquatic sources constitute a major portion of total per capita protein intake. For example, nearly 25% of dietary protein requirement of average Japanese is met with by fishes and other products from aquatic systems.

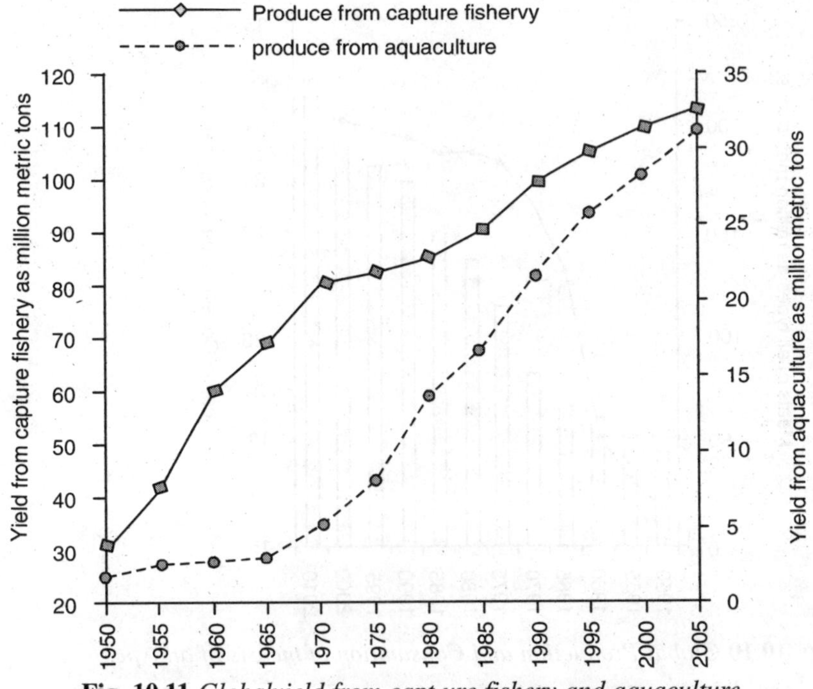

Fig. 10.11 *Globalyield from capt ure fishery and aquaculture*

The bulk of fish and other products from aquatic systems come from marine habitats (nearly 88-90%). Our oceans contain more than 97% of the total water free for circulation on this planet, covering an area of 361 million sq kms. Total annual productivity of our oceans has been estimated to be about 55 billion metric tons in terms of dry biomass. The world catch of fishes and other products from marine and fresh water grew at the rate of about 6% per year between the years 1950 to 1970. This growth nearly tripled the world production within a span of two decades (Fig 10.11). Following 1970 AD, there has been a steady rise at a somewhat slower pace. Today total produce from fresh and marine waters amounts to about 115 million metric tons. Global production of fish and other edible material from aquatic systems has suffered serious setbacks following 1970 AD. The oil crisis of 1973 and then 1978 caused many large vessels to stop fishing activity. Rising pollution of freshwater, bays, estuaries and coasts coupled with over fishing has drastically reduced our fish-stock. According to one estimate, we were already harvesting nearly 70% of the total available stock of fishes and other animals by 1972 A.D. And today this figure has reached nearly 94%. Over-harvesting or putting too much strain on the system could damage it for a considerable period of time (4).

Aquaculture is a labour intensive occupation. It is most suitable for densely populated poor countries of the world where labour is cheap and per capita income is low. Unlike capture fishing, where the size of the catch is determined by the magnitude and availability of populations, aquaculture is limited only by the size of the suitable area. The natural productivity of the water concerned is of little importance in aqua-culture as food for the developing population can be provided from outside sources (5). A major portion of the global produce from aquaculture comes from under developed countries where the labour is cheap. From a meager production of about 2-3 million metric tons per year in early fifties the total global production from aquaculture had jumped to about 32 million metric tons million tons by 2005 AD. In a number of developing countries, it provides the much needed high quality proteins, foreign exchange and employment to millions of people. Together with agriculture, aquaculture may also provide a desirable way to use of animal and human wastes.

(II) GLOBAL DIETARY ENERGY AND PROTEIN SUPPLY

The figures of global food production mask great anomalies. No doubt, today we are producing more eatables than is actually required on per capita basis. Yet, hunger persists. Malnutrition bogs down working efficiency, reduces life expectancy and millions fall prey to diseases, which they could avoid on an adequate diet. No doubt on per capita basis average global supply of dietary energy is about 2770 kcals/day. However, enormous differences occur from country to country and region to region (Fig 10.12). Only about 20% of world's population lives on daily-diets, which provide more than 3000 kcals of dietary energy on per capita basis. Majority of the world, nearly 42.2% consume diets which provide 2500-3000 kcals of dietary energy per capita per day, whereas about 32.8% of world's population lives on diet which give per capita 2000-2500 kcal/day. About 5% of world's population has to contend with diets, which provide less than 2000 kcals per capita per day. The capacity of an individual to do physical work is reduced with lower dietary energy provision. The total dietary energy supply should not be so low as to make him unfit to lead even the most sedentary life-style possible. Per capita per day dietary intake of about 2000 kcals may be taken as this limit beyond which hunger and starvation becomes evident. Signs of malnutrition and hunger also become apparent when people are put to hard physical labour but are allowed diets lower in energy content than that required for the work they have to do (6).

Adequate per capita protein supply, minerals and vitamins are important for the maintenance of good health. About 60 gm/day of proteins are necessary to built up and maintain one's body in healthy state. Malnutrition and health problems associated with unbalanced diets appear, as our bodies have to carry on its metabolic activity in absence of few or many essential amino acids, minerals and vitamins. In Central and South Africa, South Asia and South East Asia per capita per day protein consumption is less than the amount usually considered adequate. In Central Africa, per capita protein intake is poorest in the world, being only 47 gm/day. In Asian countries, it is about 56 gm/day. The diets of millions of poors in these countries consist largely of carbohydrate with very little proteins, vitamins and minerals. Malnutrition, infant mortality and various types of health problems associated with unbalanced diets are common in these regions of the world.

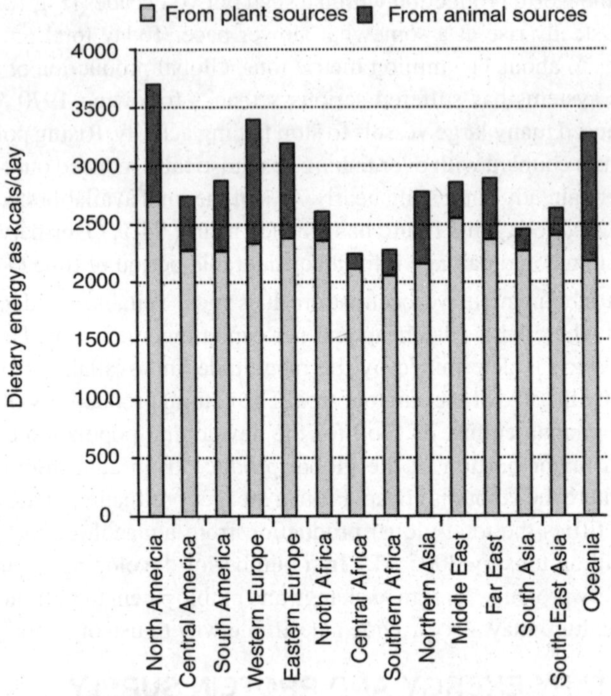

Fig. 10.12 *Global supply of dietary energy.*

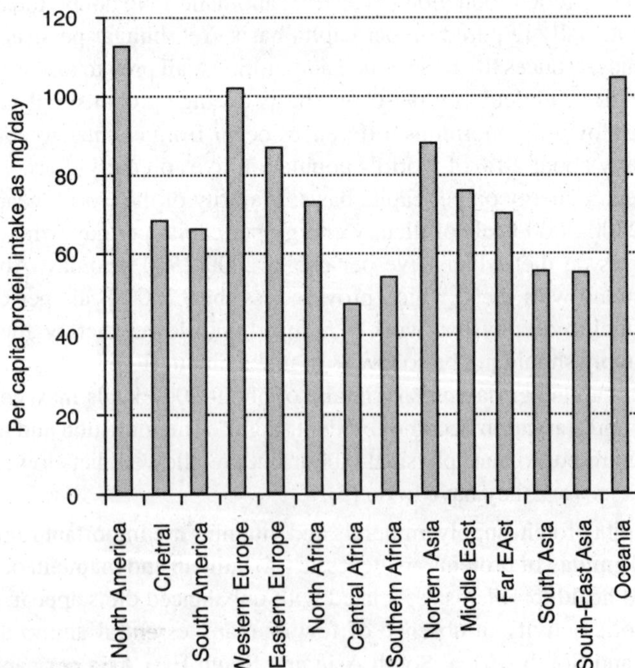

Fig. 10.13 *Per capita protein intake in different regions of the world*

(III) MALNUTRITION, SCARCITY AND HUNGER AMIDST PLENTY

The resources of arable land and freshwater are not equally distributed among countries of the world. Fertile land is a precious commodity. The area of arable land or the land, which can be used for growing food, varies from region to region (Fig 10.14). In China and Japan a single sq km of land supports more than 1000 people. In South Asia, which includes India, a single sq km of arable land has to produce eatables for more than 600 people (Fig 10.15). If we take into consideration the per capita availability of only major cereals consume by man (wheat, rice and corn) it is painful to realize that even today Central America, entire Africa and Middle east fail to produce enough (200 kg/head/year) cereals for their people (Fig 10.16).

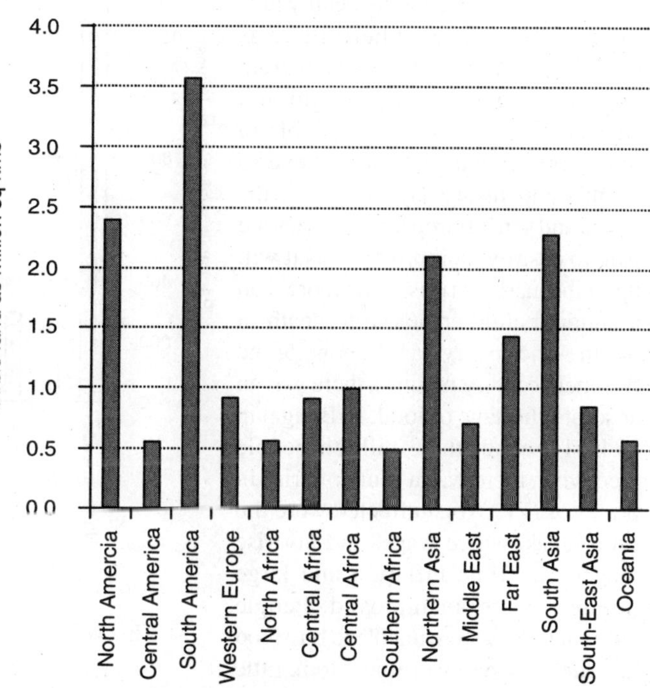

Fgi. 10.14 *Global availability of arable land*

We live in a divided world. If we examine region wise food production and its per capita availability it will be seen that there are some parts of the world in which food production is much higher than the demand whereas in a few countries the demand is much higher than the production. At many places, the surplus is simply wasted or destroyed or else it is fed to animals to raise beef, pork or chicken. Countries, which are unable to produce enough foodstuffs, have to depend on world market to feed their people. However, in a market one should have money to buy things or some commodity to trade in exchange. Many of such countries are, however, burdened with huge populations to feed – while resources are inadequate to meet the demands.

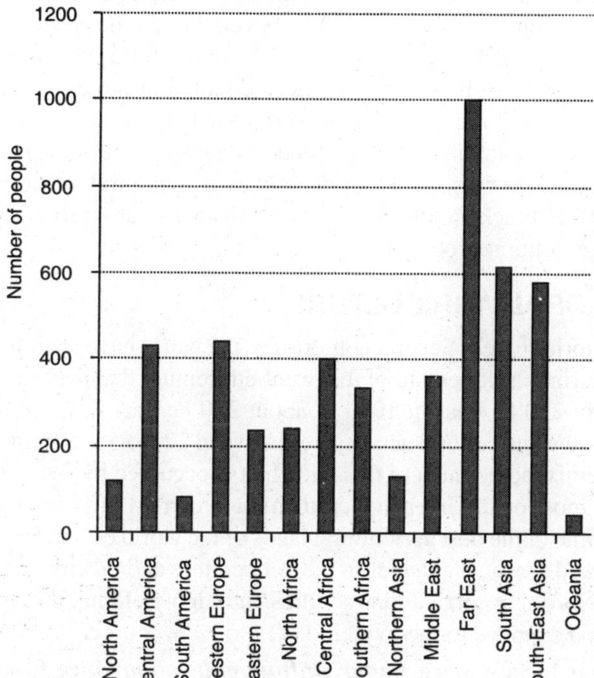

Fig. 10.15 *People supported by each sq km of arable land*

It is rather surprising to note here that hunger and malnutrition are often not due to absolute shortage of food. It has been rather an outcome of bad management. Today food is produced for the markets, not for needs of the people. Due to socio-economic limitations, political barriers and at times lack of communication and transportation facilities the surplus of one locality fails to reach the places where there is deficit. Nearly 40% of the global grain production goes to feed cattle and chickens while nearly 12% people in third world countries do not get even a single squire meal a day. Farmers in the States and some European countries are paid to destroy their produce, as it will upset the market prices while more than a hundred children starve to death in South Africa every day. In none of the twentieth century famines, there was an absolute shortage of food. In Bengal in 1943-44, about three million people died when there was a four-fold rise in prices of rice. Worst affected were the rural areas, where wages did not rise due to war time inflation and large numbers were unemployed. People without money were unable to buy food and the British Government took little action. One of the worst famines of modern times, therefore, took place when the amount of food per head in Bengal was actually 7% higher than that of the preceding year and food stocks were at record levels. In Ethiopia, in 1972-74, about 200,000 people died of hunger in the provinces of Wollo and Tigre even though the country's food production fell by just over 5% only. During the period, in spite of famine food was being exported from the affected provinces. In Bangladesh in 1974, rice prices doubled in three months after severe floods. Due to dislocation caused by floods, supplies could not reach at places where required, causing a drastic price rise. People could not afford to buy food. As a result, one and a half million people died of starvation. In fact, there was no shortage of rice either as total produce or on per capita basis. In South Africa around 50,000 black children starve to death each year - 136 every day. Yet South Africa is a net exporter of agricultural products.

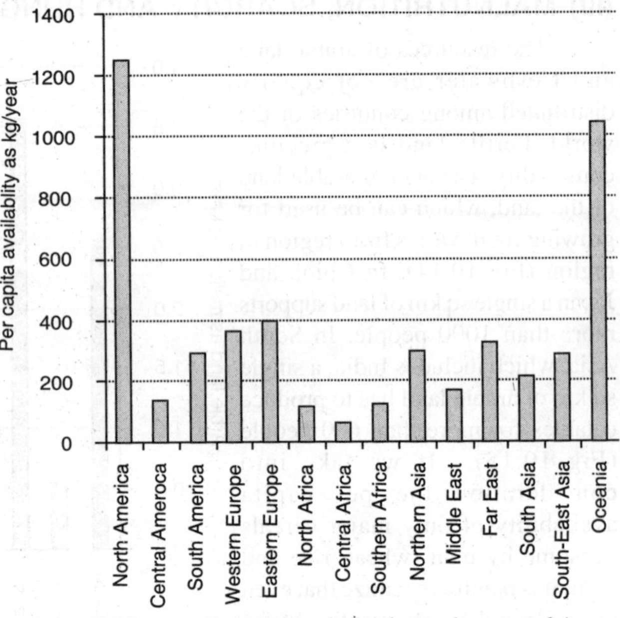

(Major cereals include wheat, rice and come only)

Fig. 10.16 *Global availability of major cereals in differentregions of the world*

(IV) THE CURRENT STATE OF GLOBAL AGRICULTURE

Global agriculture has many success stories to tell. Several countries of the world have acquired self-sufficiency in food production. It was during the latter half of the twentieth century that per capita global grain production shot up from nearly 250 kg/year in 1950 to about 370 kg/year in the year 1985 (Fig 10.1). The area subjected to intensive agricultural activity lies between 22° to 55° latitudes on either side of the equator. In Southern hemisphere, much of this wide belt is occupied by sea. It is primarily in the Northern hemisphere that most of the intensive cultivation is carried out in areas, which are well supplied with fresh water. Some of the best agricultural lands of the world occur in this region. The climate is characterized by marked seasonal changes in the temperature and precipitation patterns. Optimum use is made of the soil, fertilizers, irrigation facilities and high yielding, disease resistant varieties to provide the bulk of food supplies to the mankind (7).

The tropical region of the world has largely been under *shifting cultivation* since times immemorial as soils are poorer. It involves clearing of a patch of land for the needs of a few families. The vegetation is slashed and burned and the land is cultivated for a few cycle of crops. As the productivity declines and crops fail, the plot is abandoned. Farmers move on to a fresh patch. After few years, the abandoned plot recovers and patch is covered with forests again. However, today the pressure of demand forces people to continue cultivation even when the productivity has declined appreciably. Synthetic fertilizers and modern agricultural techniques are used which place additional strain on the system. The cleared patches are not allowed sufficient time to recover before fresh cultivation is taken up and an increasingly area of dense forests is lost in this way every year (8;9).

Modern technology has enabled man to solve most of the problems which reduce the agricultural output and raise food production. In the process, he uses means and methods, which are detrimental to maintenance of ecological balance while energy and material inputs in agriculture are steadily rising. Agriculture today is characterized by:

1. Over-expansion of our crop fields, deforestation and degeneration of our soils:

Modern agriculture has probably expanded beyond the sustainable limits of world's natural systems. Out of 18.25 million sq km of global arable land available to mankind, the area devoted to cereal cultivation was about 5.90 million sq km in the year 1950. By the close of the second millennium, it grew to about 7.12 million sq km. This nearly 20% rise in the harvested area has altogether no parallel in man's recorded history. Total global area devoted to cereal cultivation attained its peak value in the years 1975-76 when it was about 7.20 million sq km (Fig 10.17). Even after thirty years of sincere efforts it has not risen at all, rather it has dropped by about 0.08 million sq km. We have probably reached the utmost limits. In future, it will be very difficult though not impossible to expand our crop fields any more. Much of the expansion of our crop fields beyond the level of 1950 has been on marginal lands, land that is too fragile to support intensive agriculture. Most of these lands were earlier under agriculture, which depended on rains for irrigation.

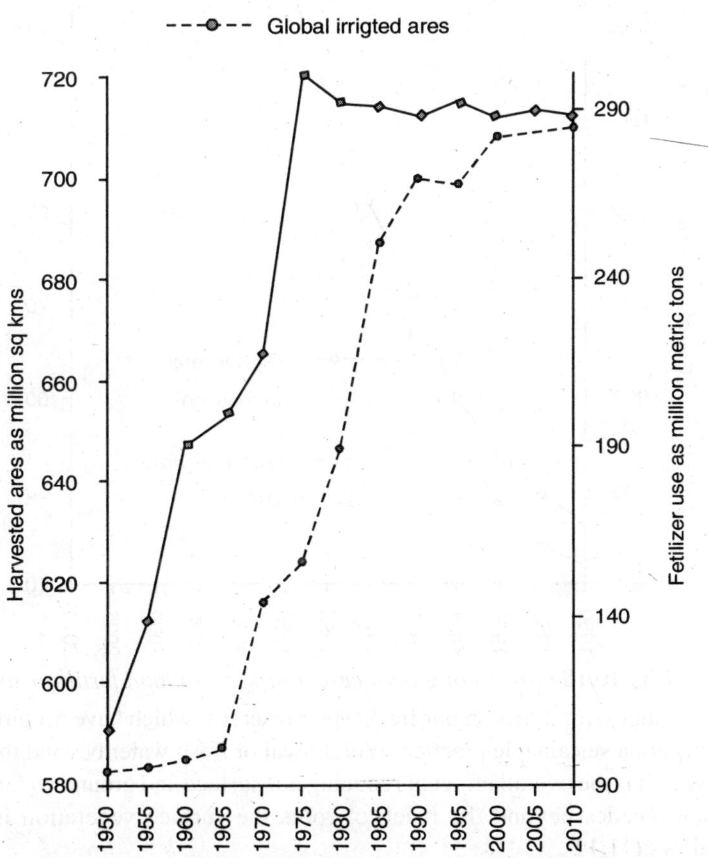

Fig. 10.17 *Growth of Gloabl harvested and irrigated area*

2. Set-backs caused by today's irrigation practices:

Giant irrigation projects which brought waters of streams and rivers to irrigate these fields were launched the world over. Subsidies were generously distributed to dig tube-wells to draw more and more water from underground deposits to irrigate the crop fields. No doubt global area under irrigation expanded by about 3.5 times since 1950 (Fig 10.17). But, these waters brought salts also and following one or two decades of boom in productivity, problems of salinity and alkalinity sprang up almost everywhere. Plants take up only those salts, which they require. The rest are left, which under conditions of poor drainage, accumulate in the soil. These salts are washed away or leached to deeper strata of the soil when drainage is proper and are remove via sub-surface drainage. Large-scale removal of tree cover helps the agencies of desiccation and erosion, which have a free play over the soils. An impervious hard pan develops underneath, which impairs subsurface drainage. Problems of salinity and alkalinity develop. Today, we are losing an increasingly large area of our land to salinization, desertification and degradation while encroaching upon the forests land for more crop fields – reducing our forest wealth (8;10).

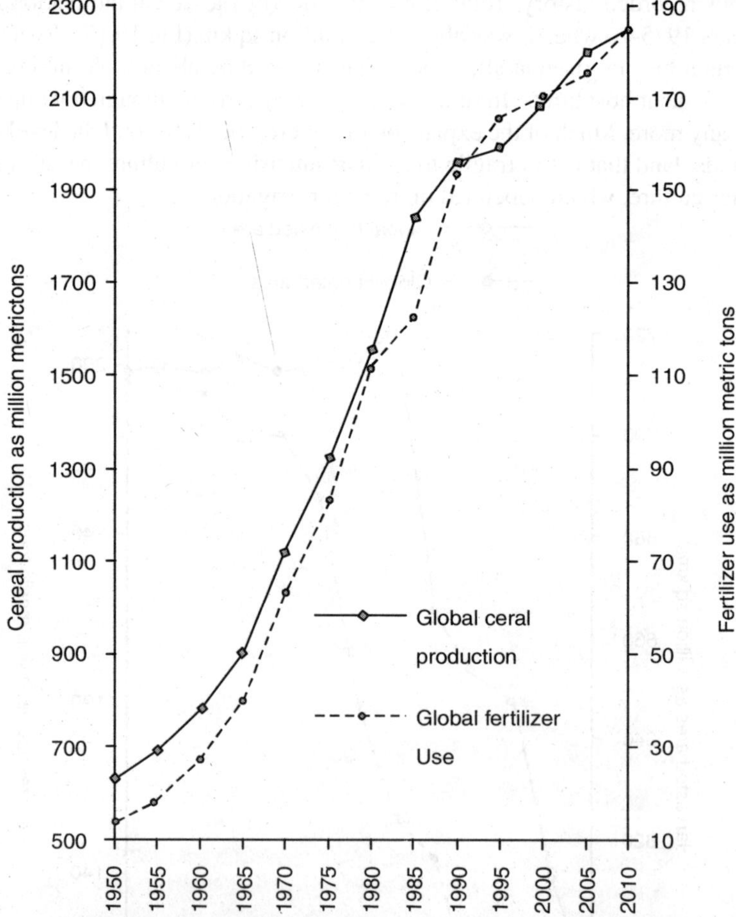

Fig. 10.18 *Growth of global cereal production and fertilizer use*

We are making huge over drafts on our fresh water reserves, which have accumulated over long periods of time. It is not a sustainable practice. Withdrawal of fresh water beyond the level, which is replenished each year, has the overall effect of reducing our surface and ground water deposits. As the ground water table recedes beyond the reach of roots the surface vegetation is destroyed and desertification follows (11;12).

3. Chronic dependence on chemical fertilizers and energy inputs in agriculture and loss of natural fertility of our soils:

Today's agriculture has as its basis a handful of high yielding synthetic varieties, liberal use of chemical fertilizers and pesticides and better irrigation facilities which we provide by huge overdrafts on our freshwater resources. Since 1975-76 AD the area devoted to cereal cultivation has not increased at all, it has declined a little. But global cereal production has been rising steadily. This is due to liberal use fertilizers, pesticides, intensive irrigation and energy inputs. If we divide the global cereal production with the global use of total fertilizers we shall find that the world used about 14 million metric tons of fertilizers to produce about 624.0 million metric tons of grains in 1950. At the beginning of the third millennium we used 170.0 - 175.0 million metric tons of fertilizers to produce a little over 2000 million metric tons of grains. Meaning thereby, that application of one metric ton of fertilizer yielded 44.6 metric ton of grain just fifty years ago. Today application of the same quantity of chemical fertilizer (1 metric ton) gives only 12.2 metric ton of grains. Each year we have to add a little more chemical fertilizer to obtain the same yield. Why? The answer is simple. We are ignoring, rather suppressing gradually, the natural nutrient generation capacity of our soils. The soil is a living entity. It thrives with life. The microbial life which occurs in soil regenerates nutrients, mineralizes organic matter, enhances porosity, and raises water retention capacity of the soil. If we allow the soil to lie fallow for sometime it is ready to yield almost the same quantity of grains the following season. We do not allow this and thus overstrain the microbial community. Repeated application of chemical fertilizers without any input of organic matter starves the microbial community. After all it is the dead and decaying organic matter which nourishes these microbes. As organic matter diminishes, the microbial communities disappear and soil is turned in to a dead heap silt, sand and clay (13).

4. Adverse consequences of liberal pesticide use in agriculture:

The use of chemical poisons, fungicides, herbicides and pesticides has become increasingly popular among farmers of today. It offers a very simple, easy and cheap means to reduce losses caused by diseases, insects and pests. However, extensive use of pesticides presents a lot of problems (14). These are summarized below:

1. The pesticides which we use in agriculture kill both target and non-target organisms. Many of the insects and pest so killed are useful to us. What is more damaging is that these chemical poisons are washed down to the soil where they damage the microbial community which affects nutrient regeneration capacity of our soils.
2. The pathogen, insects and pests develop resistance and cross resistance. We have to use increasingly larger doses of these chemical to obtain the same effect. The *Systemic fungicides* which we use enter the plants and make them resistant to pathogens. However, the practice is like poisoning the grains so that pathogens may not damage it. The same grain has to be eaten by us!
3. Apart from damaging the non-target organism many of the pesticides or their residues persist in toxic state in the environment for long durations of time. They get lodged within the bodies of the organisms exposed to it. During lean periods when the fats dissolve they are released into circulation to cause a number of health problems.
4. They are accumulated in ever growing concentrations by many organisms and passed on to higher trophic levels in the food chain. At each trophic level they are concentrated many hundred times which poisons the entire food chain. Thus the process of bio-accumulation and bio-magnification makes exceedingly small quantities of pesticides or their toxic residues in the environment available to the living organisms in a concentrated state.
5. Another serious consequence of entry, accumulation and bio-magnification of the pesticides within biological system is their widespread dispersal. Living organisms carry these poisons

wherever they go. Thus the pesticide applied in Florida may appear in Polar Regions and cause toxicity there (15).

5. **Dependence over a handful of synthetic varieties for food production and degeneration of biodiversity:**

The bulk of our food supplies, today, come from a dozen of high yielding, disease resistant, synthetic varieties. Man has slowly and consciously been reducing the number of plants, which he cultivates. For better taste and flavours, more yield, disease resistance, suitability to the conditions of environment etc. we have been ignoring our age-old, traditional varieties. Hundred of the indigenous strains, which have withstood the trials and tribulations of thousands of years of cultivation are disappearing one after the other. The loss of these traits or genes is an irreparable damage to our capacity to develop new cultivars in future. We do not know under what conditions we shall have to cultivate our food crops in future as major changes in the global environment are expected to unfold within a century or two. A pool of genetic resources as large as possible would be a very valuable asset to us in times to come.

An immediate consequence of declining biodiversity in our crop fields is the increased vulnerability of our crops to unforeseen changes in the environment. Our crop fields have now turned into mono-cultures which have made them increasingly susceptible to unkind environment or unforeseen pathogens. Being genetically uniform the entire population tends to behave in the same way. Inclement weather or a single virulent strain of a pathogen could wipe out the entire population. In 1970, American farmers lost nearly 80% of their maize crops when a virulent fungus attacked their genetically uniform crops. In the former Soviet Union the Bezoskaja winter variety of wheat which was grown over an area of about 15 million hectares during successively mild winters was destroyed by a single spell severe cold wave. This brought almost famine like conditions at both places. In crops, which have plants of several strains growing mixed together at least, some survive. The fields provide at least something to live on (16).

(V) THE CURRENT STATE OF ANIMAL HUSBANDRY

Traditional livestock management utilized agricultural wastes, grasses, other wild herbs and useless organic matter to feed the animal population. It was essentially supplementary to agricultural activity. Livestock served man in a number of ways. In an agrarian society or agriculture based society, domestic animals are still an essential component of domestic establishment in most of the developing countries of the world. To millions of poors livestock provides a bare means of subsistence. However, in the latter half of twentieth century animal husbandry changed considerably. Today it is no longer a supplement to agriculture. It has been transformed into a **full-fledged industry** where animals spend their time eating a carefully measured diet under carefully controlled conditions. About 40% of global grain production – includes corn, barely, sorghum, oats etc. – is now fed to the livestock while 12 out of every 100 people in the world go hungry. In USA livestock feed accounts for about 70% of the domestic grain consumption. The area planted to grow livestock feed has risen from 15 million hectares in 1950 A.D. to about 59.6 million hectares by the close of the second millennium (17).

The transformation of our traditional animal husbandry from small units attached to agricultural households into a full-fledged industry has created a variety of problems for mankind. These may be summarized as follows:

1. ***Heavy demands for energy and material resources:*** Livestock management on modern lines requires large amounts of energy and material resources. High quality livestock feed, water, energy and suitable space etc. has to be arranged to raise huge populations of animals in carefully controlled conditions.
2. ***The problem of waste disposal:*** The traditional practice of livestock management involved raising the animal population mostly on organic wastes from agriculture whereas the faecal

matter from live stock population provided excellent source of energy as well as organic manure for application to the agricultural land. This nutrient loop has been disrupted by modern animal husbandry which tends to isolate livestock from agricultural set ups. Large amounts of organic wastes produced by animal farms have to be disposed of on one hand whereas large quantities or organic debris produced from crop fields piles up and has to be disposed of on the other hand. The two fail to reach other. This multiplies the problems of organic waste disposal (18).

3. ***The problem of methane emission:*** It has been estimated that ruminating animals discharge about 80 million tons of methane gas in belching and flatulence every year. Microbial activity in organic wastes and animal feeds also releases about 35 million tons of methane per year on worldwide basis. This adds up to 115 million tons. A total of 15-20% of global methane emission is contributed by livestock population we breed and about 3% of global warming from all green-house gases (19;20).

4. ***Over Grazing:*** We have developed huge livestock population, which need something to eat, something to live on. This has resulted in enormous destruction of natural vegetation. Cattle and other ruminating livestock graze almost half of the planet's total land area. With pigs and poultry they also consume fodder and feed produced on about one fourth of our cropland. Though familiar and ubiquitous, the tremendous impact of our livestock population on the global environment has not been fully recognized. Over grazing, is a major threat to forests and wild land in a number of developing countries of the world. Every year vast areas of forested land are cleared to be used as pastures. In Central and South America, nearly 75% of the deforested land is put to use for growing grasses and fodder to be fed to cattle, which are slaughtered to feed the lucrative beef markets of the United States (21).

Therefore, instead of promoting large scale, industry like management, it should be better if we encourage the development of small animal husbandry units and maintain a closer co-ordination between animal husbandry and our agricultural set-ups. A management on co-operative style, which involves small farmers, the rural poors, should be promoted so that the benefits go largely to the masses instead of industrialist or capitalists. Prosperity at grass-root level shall automatically moderate many of the ill effects originating from the huge livestock population, which we have to breed. This should also keep the problems created by industry like animal husbandry units within manageable limits (22).

(VI) THE CURRENT STATE OF WORLD FISHERY

All the world over, fishery industry has been suffering from the adverse consequences of water pollution. Our freshwater systems, streams, rivers, ponds and lakes, are grossly polluted. Ever expanding deforestation and poor land use practices have resulted in our streams and river to be increasingly muddy. This has resulted in substantial decline in the harvest of freshwater fishes. In marine environments also our coastal water mangroves, continental selves, bays and sanctuaries are grossly polluted. This has degraded the habitats of most of the economically useful varieties of fishes which are rapidly declining in number. There has also been a decline in harvest from deeper water although they are relatively free from the effect of pollution. In fact, it is in the shallower parts of the sea, coastal waters, bays and sanctuaries that most of the fishes return for breeding where they suffer from adverse effects of pollution which largely results in decline in their numbers.

To this is added damages caused to the population by over-fishing. Generally smaller fishes have to left as such so that there is enough stock to harvest during the next season. However, in practice non does so and fishes of all sizes are captured which significantly damages the harvest during the following year. Many countries of the developing world simply do not possess enough resource to utilize the full potential available in their Exclusive economic zone. Adequate number of large costly fishing boats and trawlers are not available. The fish stock is left unharnessed and is finally lost as there is a limit to the crop biomass which the system can sustain. So while global fishery is suffering from damages caused by over fishing, at many places of the world full potential of the available resources is not utilized because of the lack of means to harvest the fish stock.

(VII) AGRICULTURE, ANIMAL HUSBANDRY AND FISHERY IN INDIA

A. Indian Agriculture:

Before 1950 food shortages were frequent in India. The poor state of Indian agriculture was highlighted by the famous Bengal famine of 1942, which took a heavy toll of human and cattle life. However, after independence top priority was accorded to the agriculture sector, which caused miraculous developments in Indian agriculture and made us self sufficient in food grain production by the year 1970 Fig.10.19. Now we are not only self sufficient but have enough surplus food grains to export to other countries as well. The record annual cereal production above 200 million tons in the beginning of the third millennium is a testimony to the resilience of Indian agriculture.

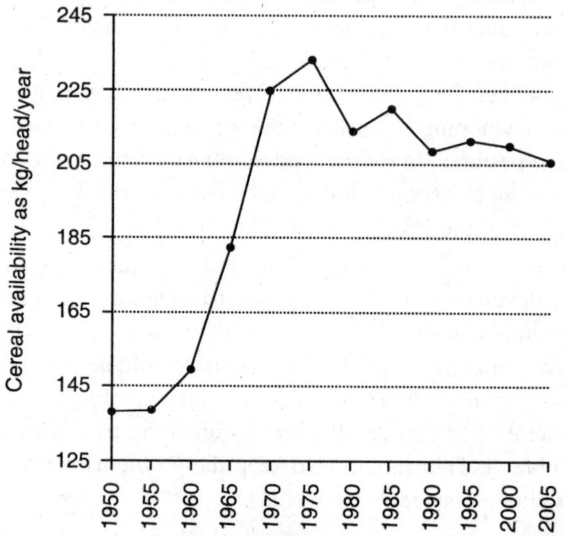

Fig. 10.19 *Per capita availability of cereals in India*

The rapid rise in total food grain production in India was made possible by the use of modern technology of intensive agriculture. For a period of about two decades, beginning in 1950 AD the rising productivity was largely due to rapid rise in overall area under cultivation of cereals. The total cropped area grew from 97.3 million hectares in 1950 to about 124.3 million hectares in 1970 AD, while the food grain production was doubled. From a meager 50.8 million tons in 1950 AD it grew to 108.4 million tons in 1970. After 1970, there has been insignificant rise in total area used for grain cultivation (Fig 10.20). However, the rising trend in production has been maintained at the same pace due to the Green Revolution, which revolutionized Indian agriculture during the period. It enabled us to produce more and more grains from the same fields. Total cropped area fluctuated around 125 million hectares but the food production shot up to 200 million metric tons 2000 A.D. In order to maintain the crops we had to expand irrigation facilities and use large amounts of fertilizers and pesticides (Fig 10.21). This was naturally an expensive proposition but we had to feed a rapidly growing population and acquire self-sufficiency in food grain production (23). These practices have slowly been degrading our resources. Today a big task for Indian agriculture is to conserve our resource bases while maintaining productivity at levels sufficient to feed the growing population.

B. Animal husbandry in India

Since times immemorial animal husbandry has been a vocation complimentary to agriculture in India. India is the land of **sacred cow**. There has always been more emphasis on maintenance of livestock for the recurring benefits which they provide instead of killing and consuming them *in-toto.* This is highlighted by the fact that in spite of having a huge livestock population, India has only about 3000 registered slaughter houses and an annual production of about 4.8 million tons of meat only.

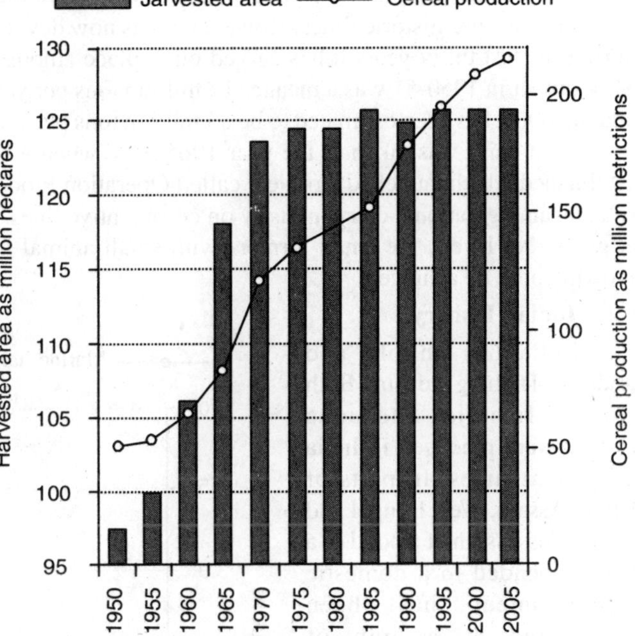

Fig. 10.20 *Growth of harvested area and cereal production in India*

In India goat's meat is the preferred meat with a contribution of about 60% or more to the total meat output of the country. Cattles are not reared for meat production. Most of the meat from large animals comes from their slaughter at the end of their productive life-span as milch or draft animals. Unlike Western Countries, in India, there is a little demand for processed meat which involves treatment of fresh meat in such a way as to make it more palatable to the consumers and require little kitchen work. There are few units in our country manufacturing processed meat products, most of which are confined to the larger cities. Export of processed meat from India started early in seventies and in 1994 A.D. It fetched an amount of Rs.30 billion only.

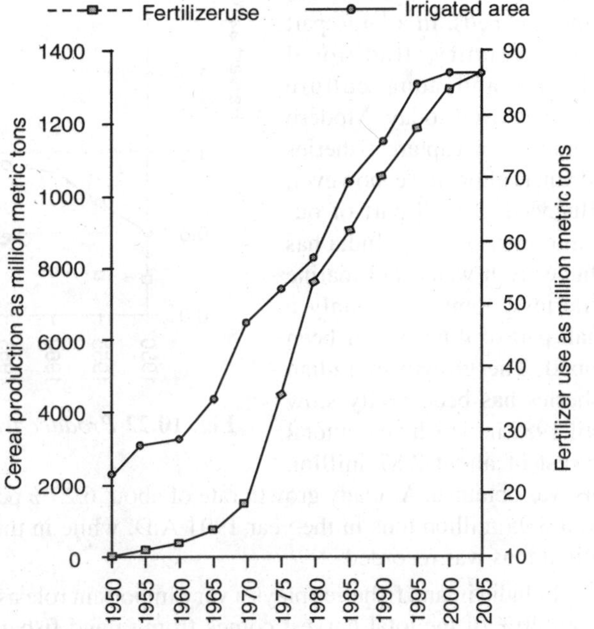

Fig. 10.21 *Fertilizer use and area under irrigation in India*

Poultry farming as commercial enterprise is only a recent development in India. Organized poultry farming was introduced only in the sixties. However, the growth of Indian poultry from a back-yard activity to an organized industry has been phenomenal. Today, in spite of many setbacks, it has achieved a production of about 3200 million eggs and 325 million broilers. By the year 2000 A.D., there has been more than fivefold increase in egg production and about seventy fold rise in broiler production over the figures of 1970 A.D.

In India dairy has always been a subsidiary to agriculture and a means of bare subsistence for the poor since pre-historic times. However, it has now developed into a well-recognized industry and during the last thirty years it has carved out a place among the major milk producers of the world. Milk output in 1950-51 was a meager 17 million tons per year which rose to 55 million tons by 1992 and in 2000 it has been estimated to be 85 million tons per year. An important step towards improving the Indian dairy was taken in the year 1965 A.D. when a National Dairy Development Board was established which launched a project called Operation Flood. The project aimed at building a viable self sustaining national dairy industry on co-operative lines. It was due to the co-operation of people at grass root levels, the small farmers with small animal husbandry units that such a high rate of growth could be achieved.

C. Indian Fishery

India has an old and traditional fishing industry. Both capture fisheries and aqua culture were practiced in India in ancient times. In parts of Bihar, Assam, West Bengal and Orissa possession of a pond or a tank appended to a domestic establishment has been considered as a sign of prosperity. Pond attached with individual households provided water as well as fishes for the domestic needs. In a large part of our country, traditional fisheries and aqua culture continues even today. Modern techniques of capture fisheries and aqua culture have, however, influenced a small part of our fishing industry only. India has a huge fresh-water and marine fisheries potential but only a small part of it has so far been tapped. The growth of Indian fisheries has been pretty slow until 1985 in which year a total harvest of about 2.87 million tons was obtained. A steady growth rate of about 6.25% per year, however, resulted into a harvest of about 3.95 million tons in the year 1991 A.D. while in the year 1995 a total production above five million tons was recorded.

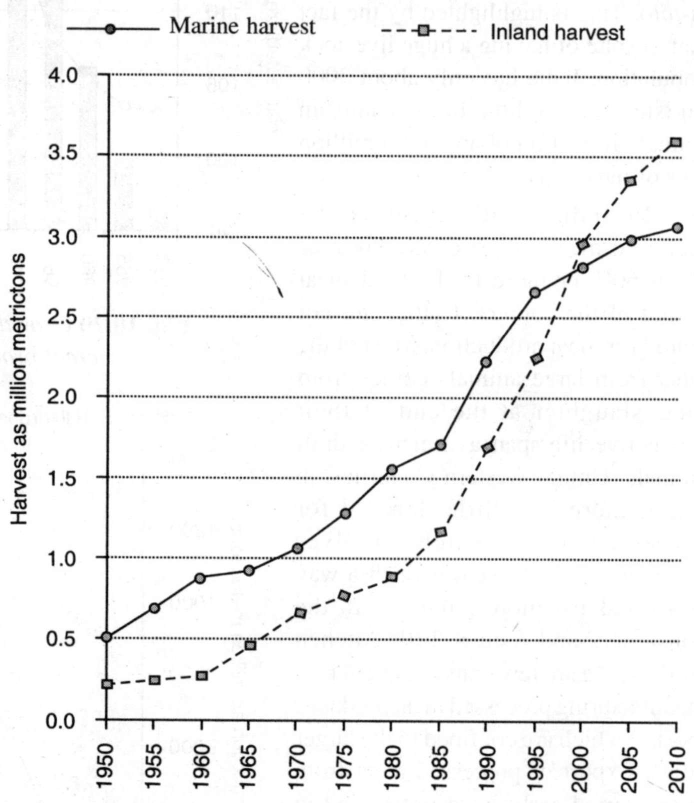

Fig. 10.22 *Produce from marine and inland fishery in India*

In India inland fisheries plays a very important role as compared to other countries of the world. About 40% of the total harvest comes from inland fisheries the bulk of which comprises of fresh water fishes, which inhabit ponds, tanks, rivers and water reservoirs spread over about 5.5 million hectares in India. The total production from fresh waters in 1991 A.D. was about 1.69 million tons which rose to 2.95 million tons by the year 2000 AD. In our country produce from inland fishery now exceeds the total marine catch. It has been estimated that the Indian waters can produce about 5 million tons of fresh water fishes. We are utilizing only a small part, about 35-40% of this huge potential.

The potential of capture fisheries which depends on the natural productivity of aquatic systems being limited the future shall probably witness a rapid rise in aquaculture. The farming of fishes and other economically useful animals utilizes natural productivity as well energy source provided from outside. Presence of suitable waters, space or location is the only requirement of aquaculture, which is a labour intensive industry. All we need is the technical expertise, the technology for growing different kind of sea foods in India. We have been successful in shrimp farming from which we earn and exchange worth Rs.3.27 million annually. We should expand our aquaculture to cover other economically useful organisms and cash in the lucrative international market of seafood, which occurs around us (24).

Questions

1. Discuss the resource bases of global agriculture.
2. Discuss the state of global agriculture. Global food production is not rising as fast as it did nearly fifteen years ago. Why?
3. What are the adverse consequences of modern agriculture?
4. How can we turn out agriculture into a sustainable system?

PART – III

Biodiversity

11

Chapter

Bio–Diversity and its Degeneration

The biosphere constitutes a vital life support system for man. Its existence in a healthy and functional state is essential for the existence of human race. It is the complex collection of innumerable organisms – the bio-diversity – which makes our lives both pleasent and possible. No one knows exactly how many species occur on our planet. Scientists believe that the total number of species on earth is in-between 10 million to 80 million (1;2). We have been able to enlist only 1.4 million species so far. Nature has taken more than 600 million years to develop this exceedingly complex spectrum of life on this planet. The existence of human race depends on health and well being of other life forms in the biosphere. However, we are losing this accumulated heritage of millions of years at a very fast rate. The very basis of our existence is being undermined. The onset of biological poverty or reduction in diversity of life forms is bound to have grave consequences for the entire living world.

(I) WHAT IS BIOLOGICAL DIVERSITY?

Complex beyond understanding and valuable beyond measure biological diversity is the total variety of life on our planet. The total number of races, varieties or species, i.e., the sum total of various types of microbes, plants and animals present in a system is referred to as biological diversity or simply as bio-diversity. Biological diversity is usually analyzed at three different levels, each of which has its own significance. These are:

1. Diversity of biotic communities and ecosystems: Depending largely upon the availability of abiotic resources and conditions of the environment an ecosystem develops its own characteristic community of living organisms. A small pond for example, constitutes an ecosystem and possesses a set of flora and fauna different from a river which is another type of ecosystem. Different types of forests, grass-lands, lakes, ponds, rivers, wet-lands etc. represent diverse ecosystems each with a characteristic biotic community.

2. Diversity of species composition within a community: The biotic component in an ecosystem may be composed of a few species only or a large number of species of plants, animals and microbes, which react and inter-act with each other and with the abiotic factors of the environment. The richness of species in an ecosystem is usually referred to as **Species diversity**.

3. Diversity of genetic organization within a species: Within a species there are often found a number of varieties or races or strains which slightly differ from each other in one, two or a number of characters such as shape, size, quality of their product, resistance to insects, pests and diseases, ability to withstand adverse conditions of environment etc. These differences are due to slight variations in their genetic set up. This diversity in the genetic makeup of a species is referred to as **Genetic diversity.** A species with a large number of races, strains or varieties is considered to be rich and diverse in its genetic organization.

A brief discussion is needed here to elucidate the meaning of the terms, species, variety or races. A species is usually the unit of classification in most of the taxonomic works and can be defined as a group of organisms genetically so similar to each other that they can interbreed and produce fertile off-springs. For example, Horses and Zebra are different species. Although they are genetically similar and can interbreed but the off-springs produced are infertile. A species is usually recognizably different from each other in appearance but sometimes the differences are so minute that it is difficult to distinguish a species from another. Genes, which are biochemical packages passed on by parents to their offsprings, are similar in organisms belonging to a given species. However, subtle differences occur which are expressed as differences in size, colour, or sometimes invisible characters such as susceptibility to some disease, resistance to drought or low temperatures *etc.* These differences are used to classify individuals of a species into different varieties or races. They do not affect the organism's ability to interbreed. In fact, greater diversity in the genetic constitutions within a species enables it to adopt and survive the adversities of a perpetually changing environment in a more effective way.

(II) SPECIATION, NATURAL LONGEVITY OF SPECIES AND OPTIMUM BIODIVERSITY

There are usually two ways in which a species is created. A large number of species have evolved simply by a single stroke in the evolutionary process – through polyploidy. Polyploidy involves multiplication in the number of gene bearing units – the chromosomes. This happens either within a pre-existing species or sometimes in a hybrid between two species. The polyploids are unable to form fertile hybrids with the parent species because polyploidy introduces enough differences in their genetic makeup. They have to interbreed among themselves and establish as a new species. The second important process is geographical speciation which takes a much longer time. When populations or group of organisms are isolated by some barrier such as a mountain range or extension of an arm of sea the differences in the environment which are inevitably there, cause the isolated populations to change. In due course of time, enough differences accumulate which make interbreeding between the two populations impossible. Thus, even when the barrier between the two populations is removed, they co-exist as separate species.

Nature has maintained species diversity at an approximately even level throughout the course of organic evolution, although this consistent level has been punctuated by accelerated extinction rates after periods 10-50 million years apart. What this approximately even level should actually be? How many species can occur in a definite area under ideal conditions? These questions cannot be answered with precision but an approximation or a guess founded on sound logic is possible. Within a cluster of islands the number of species of birds (or reptiles or ants or any other equivalent group) found on an island approximately rises by one fourth root of the area of the island in kms. Thus, the number of species present in an isolated habitat tends to rise by x $(\text{area})^{0.25}$ as its area increases. Here exponent 0.25 may vary depending upon local conditions between 0.15 to 0.35. According to this theory of island biogeography in a typical case (where the exponent is 0.25) there is a doubling in the number of species for every ten fold rise in its area (3). In a study of ant population of Hispaniola, 37 genera were discovered in amber from Miocene period – about 20 million years old. Exactly 37 genera occur on the island today also. However, 15 of the original genera have become extinct and are replaced by 15 others, which have invaded the island. Thus, an even level of diversity of ant populations has been maintained since last 20 million years on the island (4).

As fossil records reveal, in a particular taxonomic category, species have a remarkably constant life span. The probability that a particular species will become extinct within a certain span of time after it split off as a new species is roughly constant. Thus the percentage of species being lost in each period of time is approximately similar. This regular pattern has, however, been disturbed during the past 260 million years by major episodes of extinction, which have been estimated to occur rather regularly at an interval of 26 million years (5). The estimate of extinction rates in the Paleozoic and Mesozoic marine fauna ranges between one to ten extinctions for every million years, depending upon major taxonomic categories (5;6). As compared to these figures, the current rate of extinction of species would be about 1,000 to 10,000 time higher, which has no parallel in about a half billion years of the history of the biosphere (7).

(III) UNEVEN DISTRIBUTION OF BIOLOGICAL WEALTH ON OUR PLANET

All life has a common biochemical basis. All life activities such as growth, development, reproduction etc. depend on enzyme mediated biochemical reactions. There is a certain range of environmental conditions in which most of these complex reactions occur at their maximum ease and efficiency. This does not mean that all organisms require similar set of conditions for their optimum growth. However, it does suggest that within this optimal range a large number and type of organisms are likely to occur, grow and multiply. In habitats, which do not provide these optimum conditions, organisms shall have to adopt themselves to the prevailing adverse conditions. The process of natural selection shall automatically exclude delicate forms restricting the number of species to only those, which are able to exist under the adverse set of conditions.

The enormous diversity of life forms in the biosphere has evolved essentially through a process of trial and error during the course of organic evolution. Those changes in characters of a living organism, which confer some advantage to the species, are retained. Disadvantageous changes cause elimination of the organism or its entire population. Harsh conditions of environment quickly force the enfeebled and weaker sections of natural population to extinction, without giving them much chance of struggle, adaptation and evolution to a life form better adjusted to its surroundings. Regions of world with mild, warm and humid climates have developed a bewildering diversity of living organisms as most of the changes in the living stock are perpetuated and allowed to evolve into a new life form.

Warm tropical regions between the tropic of Cancer and Capricorn on either side of equator have since long provided the most suitable habitat for living organisms. The temperatures vary between 25° to 35°C a range in which most of metabolic activities within a living organism occur with maximum ease and efficiency. The rainfall is ample often more than 200 mm per year and the most remarkable feature of the tropical climate is its stability. Seasonal changes in temperature are usually less than the total variations between the day and night temperatures – a change which is hardly noticeable. There are only slight changes in humidity and photoperiods over the different seasons. In Singapore, a tropical city, the mean January temperature (30.5°C) differs from the July temperature (31.9°C) by 1.4°C only whereas the daily fluctuation in the day and night temperatures is about 10°C (from 25°C in the night to 35°C in the day). As we move towards the poles – the Sub-tropical, Temperate and Polar Regions on both sides of equator, climatic conditions gradually become harsh and severe. Temperatures drop and there is a gradual reduction in the mean annual precipitation. The climate is marked with strong seasonal fluctuations. There are drastic variations in temperatures, humidity, rainfall and photoperiods between different seasons. In Dawson city, Alaska, the mean temperature in January (–26°C) differs by about 47°C from the mean July temperature (21°C).

The changes in climatic conditions are reflected in the distribution of living organisms and the pattern of biodiversity on our planet. The number of species present per unit area declines as we move from mild tropics to the severe tundras and taigas. Polar Regions have no trees. Coniferous forests of Northern Canada, Alaska, Northern parts of Europe and Russia possess less than 12 species of trees. The temperate forests of the United States have 20-35 species of trees whereas the tropical forests of Panama have over 110 species of trees in a relatively small area. In almost every plant and animal groups a similar trend prevails. Canada has 22 species of snakes whereas 126 species of snakes are found in USA. From Mexico about 295 species of snakes are recorded. Peter S. Aston of Harvard

University has reported around 700 tree species from 10-selected one hectare of plots in Kalimanthan, Indonesia. This almost equals the total number of tree species of the entire North America.

Table 11.1 *Land areas, approximate number of flowering plants, and annual deforestation rate in mega diversity countries of the world.*

	Countries	Land area*	Number of species	Rate of deforestation**
1	Brazil	6.3	55,000	13,820
2	Columbia	0.8	45,000	6,000
3	Mexico	1.4	25,000	7,000
4	Australia	5.7	22,500	N.A.
5	Indonesia	1.4	20,000	10,000
6	Peru	1.0	20,000	2,700
7	Malaysia	0.2	15,000	3,100
8	Ecuador	0.2	15,000	3,400
9	India	2.2	15,000	10,000
10	Zaire	1.7	10,000	4,000
11	Madagasker	0.4	10,000	15,00

* Land area as percentage of total global area.

**Deforestation rate as hectares/year (closed forests only).

(*Source:* Ryan John., 1992)

(IV) THE MEGA DIVERSITY COUNTRIES OF THE WORLD

Warm and humid regions in between the tropic of Cancer and Capricorn are provided with a rich and diverse plant, animal and microbial life. In this wide belt around the globe occurs more than half of the total number of species present on our planet. Countries, which happen to lie in this zone, are referred to as **Megadiversity countries** as they possess a wide variety of plant and animal species. These include Brazil, Columbia, Mexico, Indonesia, Peru, Malaysia, Ecuador, India, Zaire, Madagascar, Australia etc. Table 11.1 provides the approximate number of flowering plant species recorded in these countries as well as the share of land surface as percentage of the total land surface available on this planet.

(V) ENDEMISM AND 'HOT-SPOTS' OF BIOLOGICAL DIVERSITY

Endemic species can be defined as those species, which are confined only to a particular locality. Such organisms are very important from the point of view of conservation as their disappearance means total extinction of the species as they are not found anywhere else. No doubt, endemism represents a unique step in the process of evolution, which could be perpetuated and sustained only in the locality concerned by imperfectly understood attributes of environmental quality. The diversity of life forms, which we see around us, has essentially evolved by a method of trial and error. In the process of natural selection, changes which confer some disadvantage to the organism are eliminated while those changes which are advantageous are retained. It is the environment only, which is instrumental in the operation of the process of natural selection. This makes the habitats in which endemic species thrive very important. The importance of the habitat or locality is further highlighted by the fact that in most of the cases such localities possess a number of endemic species distributed in several taxonomic categories or groups. The endemism in a particular taxonomic category is usually matched by more or less similar degree of endemism in other taxonomic groups.

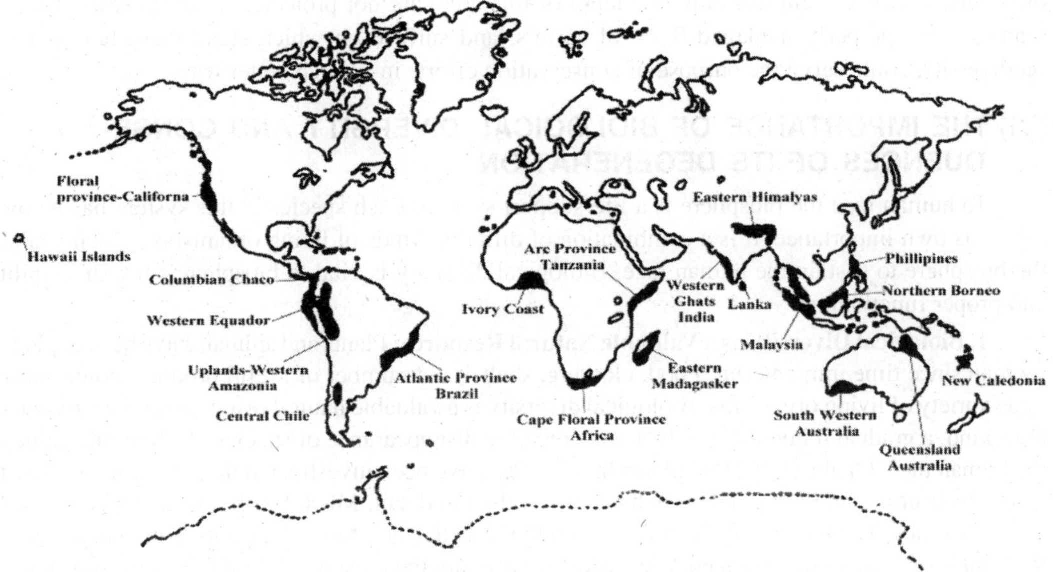

Fig. 11.1 *Hot-spots of biological diversity*

Naturally the endemic species and the habitats which are likely to be lost forever should receive urgent conservation attention. Once lost, there is no way to recover them. Based on the degree of endemism in species composition, Myers (1988) has identified 12 such localities in tropical regions of world which require urgent conservation attention (8). These localities are:

1. Hawai Islands
2. Columbian Chako
3. Western Ecuador
4. Uplands of Western Amazonia
5. Atlantic forest area Brazil
6. Eastern Medagasker
7. Eastern Himalyas
8. Peninsular Malayasia
9. Northern Borneo
10. Phillipines
11. Queensland Australia
12. New Caledonia

These areas are spread over 292000 sq kms only and represent barely 0.2% of the earth's total land surface. Of the world's 8.34 million sq kms of primary forests, these hotspots of biological diversity represent only 3.50%. However, they possess 34400 endemic plant species which is about 27% of all tropical forests or 13% of the total plants found on earth. In 1990, Myers identified another eight hotspots of endemic species diversity in other climatic regions of the world (9). They are:

1. California, Western Floral Province
2. Central Chile
3. Ivory Coast
4. Cape Floral Province, Africa
5. Western Ghats, India
6. Sri Lanka
7. South West Aistralia
8. Eastern Arc Province, Tanzania

These eight hot spots of biodiversity cover 454,400 sq. kms or only 0.3% of world's total land surface but possess 15,555 endemic species of plants representing 6% of the total number of species of plants world-wide. Thus about 49,955 endemic species of plants occur in an area barely 786,400 sq kms or only 0.5% of the world's total land area. On worldwide basis about 4.9% of earth's land surface has been set aside as nationally protected areas. However, much of earth's hot spots of biodiversity, which comprise only one tenth of this area, are not protected at all. Instead of badly managed 5% properly managed 0.5% of earth's land surface on which stand these hot spots of biodiversity, could serve the purpose of conservation efforts in a much better way.

(VI) THE IMPORTANCE OF BIOLOGICAL DIVERSITY AND CONSEQUENCES OF ITS DEGENERATION

To human race the biosphere is a life-support system. Each species in this system has its own role – its own importance. It is a combination of different kinds of living organisms, which enables the biosphere to sustain the human race. Biological diversity is vital to biosphere's health, stability and proper functioning.

1. Biological Diversity as a Valuable Natural Resource: Plants and animals have been exploited by man since time immemorial. Food, clothing, shelter and number of useful products come from a wide variety of living organisms. Biological diversity is a valuable natural resource for the survival of man-kind, a gradual reduction of which may result in disappearance of species of economic value to the human race. Of about 250,000 plants hardly 10% have been investigated in a very cursory way to assess their utility and about 1% only has been studied in detail. Much less has been done in case of animal species. The biological wealth so far on the records, therefore, represents a huge resource for the man-kind waiting to be tapped. If we take into account the huge number of life forms which have not yet been discovered and documented *this pool of biological resources becomes much larger.* Imagine the plight of man-kind if *Penicillium* or *Cinchona* would have disappeared before their curative properties were discovered. The very basis of life support system on this planet shall be gradually lost if this huge largely unknown and untapped biological wealth is allowed to degenerate (10).

Repeated cultivation of a variety over long periods of time under human care gradually reduces its vigour and productivity. Breeders have been using breeding techniques, essentially to pool together useful genes into the varieties cultivated so as to match them with quality requirements and the prevailing conditions of cultivation. However, to do so, we should possess an assortment of traits or gene pool as large as possible for breeding and improvement in domesticated strains of plants and animals. The genetic organization of wild plants and animals has been evolving for thousands of years. It has withstood the tests and trials of a perpetually changing environment since time immemorial. Old traditional varieties and the wild relatives of domesticated plants and animals constitute a vital **genetic resource** for us. No one can anticipate the traits or the genes which shall be needed in future to improve the cultivars of the time as our environment has been undergoing rapid changes during the past two or three decades. Under such conditions we should have a collection of traits or a gene-pool as large as possible. It is only from this pool that we can synthesize the cultivars of future. Reduction in biological diversity shall inflict irreparable damage to future breeding and improvement activities, which are necessary to support the human society (10).

2. Biological Diversity as Instrument of Maintaining a Stable and Healthy Ecosystem: In an ecosystem, everything is related to everything else. A system of complex interactions exists between various components of a healthy ecosystem, which occur in a state of dynamic equilibrium. To every action there is an equal and opposite reaction so that the outcome is nil. This system of checks and balances is of a fundamental importance in an ecosystem, which is maintained in a functional state by the activity of a large number and types of organisms. So in a complicated ecosystem with several trophic levels each of which is composed of several species, elimination of a single species or few species does not create any problems. There are several species or alternatives, which can take over and keep the system in a functional state. But in a simple system loss of a single or few species could be catastrophic because of the lack of alternatives. Thus diversity imparts stability to an ecosystem.

3. Biodiversity as an Instrument of Regulation of Ecosystem services: Ecosystem Services which are broadly defined as the benefits provided by ecosystems to humanity. Nature works incessantly to provide these free benefits to mankind through the enormous collection of living organisms which it maintains – the biodiversity. We are often not even aware of the benefits we derive from a healthy ecosystem (Please see Chapter 2 also for Ecosystem Services). Functions like soil formation and its maintenance in state fit for plant growth, decomposition of organic wastes and recycling of nutrients, detoxification of harmful chemicals, resistance to invading exotic species, biological nitrogen fixation, regulation of local and global climate, pollination and seed dispersal etc are essential for mankind which are regulated by biodiversity, the complex collection of plants, animals and microbes.

1. A large quantity of the top soil is lost every year due to erosion and denudation much of which is compensated by new soil formation. Our food production system shall collapse if the process of soil formation and its maintenance in healthy state is disrupted, both of which are carried out by innumerable microbes in the soil – the biodiversity.
2. Each year about 88.06 billion metric tons of organic wastes are produced worldwide. Worldwide about 100,000 chemicals are used which when discarded create hazardous waste dumps. Maintaining biodiversity in soil and water is essential for improved effectiveness of the process of biodegradation and detoxification.

Table 11.2. Number of species in major taxonomic groups known to us

	Taxonomic groups	Number of species
Invertebrates	Protozoa	50,200
	Porifera	10,000
	Coelenterata	9,000
	Platyhelmenthes	12,000
	Nematoda	10,000
	Mollusca	100,000
	Annelida	8,200
	Arthropoda	920,000
	Echinodermata	6,000
Vertebrates	Pisces	23,000
	Amphibia	2,400
	Reptilia	5,200
	Aves	8,300
	Mammalia	4,200
Total number of animal species		**1,168,500**
Lower green plants	Algae	20,175
	Fungi	80,000
	Bryophytes	14,900
	Pteridophytes	11,000
	Conifers	650
Flowering plants		300,000
Total number of plant species		426,725
Total number of plants & animals		1,595,225
(*Source:* Various taxonomic works)		

3. Seventy seven million metric tons of nitrogen is applied to our fields as commercial fertilizers which costs about 38.5 billion US dollar. Worldwide biological nitrogen fixation carried out by diverse micro-organisms provides about 150 -180 metric tons of nitrogen. Replacing it with commercial fertilizers could cost 85 billion US dollars to mankind.
4. Higher species richness or the biodiversity in a healthy ecosystem increases the biotic resistance against invasion by exotic species. Invasive species threaten biodiversity, change ecosystem functioning and cause economic losses. The threat of invasion by exotic species can be eliminating if a rich and diverse biota is maintained in the system.
5. It is the precise composition of green house gases in the atmosphere, regulated by cumulative activity of plants, animals and microbes, which maintains the mean global temperatures at about 14°C enabling life to flourish on our planet. Albedo which is the proportion of solar radiation reflected back to the atmosphere has been shown to have a critical role in warming of earth's surface. Surfaces with low albedo reflect a small amount of sunlight, those with high albedo reflect a large amount. This causes earth's surfaces with low albedo to get heated more than places with higher albedo. The differential heating of earth's surface determines the pattern of precipitation and local climate.
6. In natural ecosystems a process of checks and balances (prey-predator interactions) is operative which automatically controls the populations of diverse insects, pests and pathogens. In a natural ecosystem, rich biodiversity controls nearly 99% of insects, pests and diseases.
7. A number of studies have demonstrated that failure of fertilization due to absence of pollinators is the cause of fruiting failure and thereby low fruit yield in a number of fruit crops. Pollination limitation increases inbreeding, reduced genetic fitness, and increased susceptibility to environmental stresses. Nearly one third of world's food production could be lost if this natural pollination service is disrupted. In order to maintain adequate population of pollinators, therefore, a rich and diverse natural system is needed.
8. Seed dispersal is an important process in natural ecosystems and population dynamics. Absence of an effective population of animal agents of seed dispersal causes the plant population to be confined to localized areas which results in crowding, wastage of seeds, inbreeding depression and loss of genetic diversity in the species concerned.

(VII) REDUCTION IN BIOLOGICAL DIVERSITY – THE SITUATION TODAY

Millions of years of organic evolution has handed over to us a vast variety of plants, animals and microbial species. We have been able to describe only about 2.5 to 12.0% of the total number of species present on our planet which Scientists believe is in between 10-80 millions (11). The approximate number of species in different taxa of animals and plants as compiled from a number of taxonomic studies adds up to about 1,595,225 only (cf. Table 11.2)

The glaring inadequacy of our knowledge is apparent from the fact that still about 200-250 species new to science are described from Tropical Africa alone. Plant collecting expeditions in Tropical South America reveal one new species of plants among every five-collected (12). We have spent billions of dollars on moon's exploration but have done little to understand our own planet. Our knowledge about the forests of tropical and sub-tropical regions is alarmingly poor. After-all the moon is going to stay in its place much longer than the tropical forests of our planet.

Ordinarily if a particular life form is not located for a considerable period of time (say two or three decades) it is considered extinct. But an element of uncertainty always persists. It is not possible to rule out the possibility of occurrence of the organism in some unexplored corner of earth's surface. There are several instances of organisms believed to be extinct, being recorded after a long period of time.

The general pattern of extinctions, which emerges from a study of about 500 cases since 1600 A.D., indicates that nearly 75% of extinctions have occurred on islands instead of main continents. These were mostly associated with expansion of European influence. As European colonizers with their exotic animals and plants settled in colonies and agricultural activities expanded a systematic destruction of natural habitats started. On islands surrounded by water there was no alternative for wild life to go to and they perished one after another. A large proportion of these extinctions involved birds and amphibian and the least affected were mammals. Of those extinctions, which occurred on the continents, more than 65% involved aquatic organisms. There is little information about invertebrates and lower green plants. However, their number should also have been considerable as extinction of one species usually brings about the disappearance of a chain of species dependent upon the species lost (13).

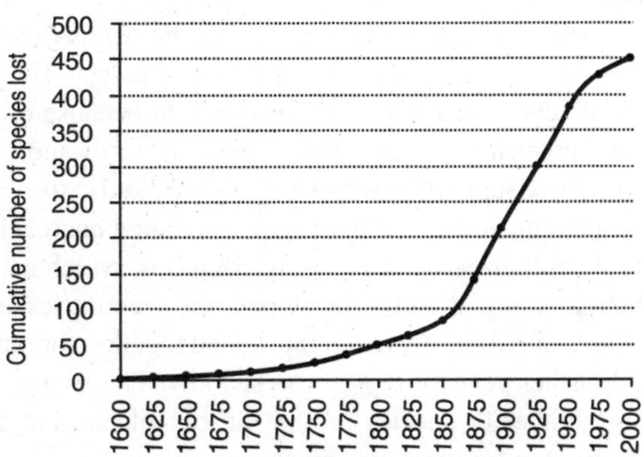

Fig. 11.2 *Exlictions since 1600 AD.*

During the course of organic evolution, a number of species have appeared simply to disappear. Biological extinction is a natural phenomenon, which has been taking place rather in a balanced way ever since life came into existence. It is actually the fast rate of extinction of live forms, which is causing serious concern to us.

Before man's appearance on this planet the rate of extinction was one species per thousand years. However, the pressure of human activity has drastically changed the picture. In a span of three hundred years between 1650 A.D. to 1950 AD about 30 species of higher animals were lost and now we are probably losing one species every year (Fig. 11.2). The estimate of endangered wildlife usually understates the actual magnitude of the problem because it deals with higher life forms known to us in general. All fishes, amphibians, reptiles, birds and mammals put together constitute about 2-3% of the known species of living organisms. Nearly one third of a million species are contributed by green plants alone while a little over a million species of invertebrates have been recorded to occur on our planet – let alone the enormous number of species which we have not yet been able to enlist or describe. An assessment of the gravity of situation should take into consideration the entire spectrum of life on our planet.

(VIII) MAJOR CAUSES OF REDUCTION IN BIOLOGICAL DIVERSITY

Usually, disturbances of any type in an ecosystem tend to reduce its biological diversity. As human population rises, an ever-increasing demand for raw material, food and space is placed on natural ecosystem while enormous quantities of wastes and spoils are introduced into the environment. The pressure of human demands and pollution of environment collectively damage the biotic component of natural systems either partially or completely. Major causes of reduction in biological diversity can be summarized as follows:

(1) DESTRUCTION OF NATURAL ECOSYSTEMS:

The requirement of space, food and raw material for expanding human establishments is the most important singular cause of such a rapid decline in biological diversity. Much of the surface area of our globe where agriculture or cattle ranching are possible has been brought under human use. All over the world the process of biological impoverishment is taking its toll of species after species.

The northern most belts skirting the Arctic Sea is probably the only habitat in the world where losses of biological diversity are modest as yet. This region is very thinly populated and possesses little diversity in its species composition because of the harsh and severe climatic conditions. The zone between the coniferous forests of north and the tropical zone in the south in the Northern hemisphere supports extensive agriculture and cattle ranching. Isolated patches of forest and woodland left standing amidst the plains are also being encroached upon. The tropical regions of our globe are largely colonized by developing countries, which are often designated as mega-diversity countries. Though poor and under-developed, these countries are endowed with richest flora and fauna of the world. Annual deforestation rate added together for mega-diversity countries amounts to about 61,520 sq. kms per year for closed forests only. This is alarming. Tropical forests house nearly half of the total number of species present on our planet and destruction of their natural homeland shall be catastrophic for the entire biosphere.

Wetland ecosystems are important store houses of biological diversity. They are often world's most productive systems and provide suitable habitats for a large number of species. They are of much help in regulating water flow and eliminating silt, sediments and pollutants from moving waters. In many parts of the world the wetland ecosystem are being drained and dried for agriculture use or for human settlements or are converted to aquaculture ponds. Mangroves and other coastal wet-lands form an important buffering zone between the land and sea. Mangroves possess a rich flora and fauna unique in their own way and like terrestrial wetlands are highly productive systems. It is this zone which protects the coastline from erosion by ocean waves while clears the silt, sediments and pollutants from streams before the water flows into the sea. Mangroves have suffered extensive losses in Asia, Latin America and Western Africa. An important consequence of elimination of mangroves is the destruction of coral-reef ecosystems. Coral-reefs are marine equivalents of a tropical forest. They are highly diverse and productive ecosystems. An increasing quantity of silt, sediments and pollutants are washed into the sea as there is none or very little mangrove vegetation to check, and clear out the inflowing waters. Although it is difficult to study under water communities directly, it is suspected that these rich repositories of bio-diversity are in no better condition that than wetlands and mangroves.

(2) ADVERSE CHANGES IN BIOTIC OR ABIOTIC ENVIRONMENT

Unfavourable changes in the biotic or abiotic factors of the environment of an ecosystem may be brought about by:

1. Environmental pollution.
2. Over-exploitation of selected species by humans
3. Habitat fragmentation.
4. Introduction of exotic species.
5. Natural calamities.

1. Environmental Pollution: Pollution involves introduction of undesirable and harmful material or energy in the form of a gas, liquid, solid heated effluents or radiations in an ecosystem. Most of these pollutants, even seemingly harmless materials also, adversely affect the biotic community. The hardy, tenacious and resistant forms survive. Weak and susceptible species are eliminated. With each eliminated species, follows a chain of reactions, which eventually disturbs the structure and function of the entire ecosystem and a number of species suffer.

There is little poisonous or toxic material in ordinary biodegradable pollutants (the sewage). However, many species cannot tolerate organic enrichment and disappear. The microbial demand on oxygen creates oxygen deficit, which suffocates a number of species. Most of the biodegradable materials, however, are broken down into simpler constituents which over-fertilize the aquatic systems which promote some species while suppressing others. The inevitable outcome is reduction in the biological diversity of the system. More damage is done by substances which persist in toxic state for long durations. Enormous quantities of gaseous material and particulates, which are discharged into the air, undergo reactions in the atmosphere and are disseminated over long distances. They are finally brought down to earth's surface as dry precipitation or with rains to cause slow but extensive damage to plant and animal life. Only a few plants survive around nickel smelters in Sudbury, Ontario, USA while acid fallouts have destroyed plants and fish populations in lakes almost 65–70 kms away. Within a period of about 40 years smelters at Trail, British Columbia, killed nearly all conifers within an area of 20 kms – stunting plant growths at places almost 65 kms away. Such dead zones are still being created all around the world, taking a heavy toll of both plant and animal life.

2. Over-Exploitation of Selected Species: Merciless hunting or collection of a select group of living organisms for food, profits or recreation is an age-old cause of extermination. There is considerable evidence to suggest that the disappearance of large Pleistocene animals of North America -- the woolly mammoth, horses, camels and mastodons – was caused by the Stone Age hunters who came to the continent through Bering Strait, which was once a land bridge between Asia and America. Even today, hunters and collectors are significant threat to a number of species. Whale, elephants, most of the wild cats, rhinoceros, many species of snakes, crocodiles and a number of other animals are mercilessly hunted for their hide, tusk or horns.

Over exploitation is one of the main causes of disappearance of plants of scientific and medicinal value. The pitcher plant, *Nepenthes khasiana*, *Drosera sp.*, *Gnetum sp.*, *Psilotum sp.*, *Isoetes sp.* are ruthlessly sought and collected for teaching and laboratory work. They have already become rare. Medicinal plants like *Podophyllum* sp., *Coptis* sp., *Aconitum* sp., *Rauwlfia* sp., *Saussura lappa*, *Atropa acuminata* etc. are also disappearing rapidly as a consequence of merciless over-collection. Similarly, the natural populations of a number of economically important trees like *Pterocarpus santalum*, *Dysoxylon malabaricum*, *Santalum album* that yield valuable timber are fast dwindling. In the category of over-exploited plants may also be placed a number of orchids, which produce world's most showy flowers. Plants like *Paphiopedilum fairieyanum*, *Cymbidium aloiflium*, *Aerides crispum* etc. are in great demand but their natural populations have almost disappeared.

Table 11.3 Exotic weeds common in India

	Weed	Family	Probable origin
1	*Agrostemma githago*	Caryophyllaceae	Eruope
2	*Argemone mexicana*	Papavaraceae	Mexico
3	*Lantana camara*	Verbenaceae	Tropical America
4	*Parthenium hysterophorus*	Asteraceae	North America
5	*Tribulus terrestris*	Zygophyllaceae	Africa
6	*Verbena lonariencis*	Verbenaceae	South Africa
7	*Xanthium strumarium*	Asteraceae	America
8	*Alternanthera philexeroides*	Amaranthaceae	South America
9	*Eichorinia crassipes*	Pontederiaceae	Brazil
10	*Salvinia molesta*	Salviniaceae	Tropical regions

Among the domesticated organisms, a general neglect of traditional varieties and their substitution by a few selected strains for immediate advantages has caused a significant damage to biological diversity. Man has slowly and consciously been reducing the number of species and varieties which he cultivates. Those which give better yield, desirable tastes and flavours are preferred. All over the world traditional varieties, which together constituted a diverse mosaic, are being dropped one by one being replaced by a few high yielding strains. A victim of cultivators neglect and apathy old traditional strains are disappearing at an alarmingly fast rate while many of the wild relatives of the cultivated species are facing extinction due to pressure of human activity on natural ecosystems. The reduction of genetic diversity among the cultivated species and the disappearance of their wild relatives, drastically limit possibilities of creating new cultivar in the future, which could be disastrous for human race.

3. Habitat Fragmentation: Today, the pressure of human activity, population and development are fragmenting wild biological communities into small patches surrounded by urban or agricultural land. Railway tracks and highways across forests have the over-all effect of cutting it into small segments. These patches virtually become small islands in the sea of man-dominated landscape. When a habitat is fragmented into small patches its capacity to support or sustain viable populations of different species declines drastically and consequently the excess numbers of species are removed. Small forest patches suffer from rapid changes in microclimate, which normally affects only the margins of a large forest. Winds and encroachment by various agencies, damage trees and progressively reduce the size of forest patch or fragment. The immediate consequence of habitat fragmentation is movement of most of the mobile forms from margins of the patch to deep interior. The margins, which are frequently subjected to encroachments, lose much of their species. Deeper zones of the patch are subjected to crowding where competition for space, food and water, causes the weaker organisms to perish. Crowding often results in stepped up predation or overgrazing which destroys vegetation and the organisms are forced to live in a stressful environment. The rate of reproduction slows down or at times organisms fail to reproduce altogether while life under stressful conditions shortens their life span.

4. Introduction of Exotic Species: Organisms adapted to life in a stressful environment become tough, tenacious and resistant. When introduced into locality with gentle, mild and hospitable living conditions such organisms thrive well. The original inhabitants being unable to compete are gradually suppressed and forced out of existence. Invasive species threaten biodiversity, change ecosystem functioning and cause economic losses.

While economically useful plants are deliberately introduced a large number of exotic weeds are transferred from one locality to another accidentally. The wheat imported to India from USA under PL-480 scheme was contaminated with seeds of *Parthenium hysterophorus*, the congress-grass and *Agrostemma githago*, the corn cockle. Both of these plants have spread throughout India as a pernicious weed in wheat fields. *Lantana camara*, probably one of the ten most common weeds of the world was introduced in India as an ornamental hedge plant from Sri Lanka in nineteenth century. After cutting trampling or burning, this plant regenerates itself quickly to form dense thickets. Prolific growth and adaptability has enabled this plant to overrun large tracts of land including pastures, cultivated fields, wasteland and forests. Table 11.3 provides a list of exotic weeds which are common in India.

Water hyacinth, *Eichornia crassipes*, was introduced for the first time in 1914 at Narayanganj in West Bengal. The first appearance of Alligator weed, *Alternanthera philexeroides*, was reported near Calcutta airport in 1965, while *Salvinia molesta* was brought to India by an aquarist. These plants vigorously produce stolons and regeneration of detached parts is quick which result in development of thick mats floating on the water surface. The damages caused by these plants are extensive. They impede run off in streams and promote waterlogged conditions. The decay and decomposition of dead plant part places a heavy demand on dissolved oxygen in the aquatic system. Mat-like spread cuts of air and sunlight and sub-surface biota is almost completely eliminated. Due to active transpiration, both alligator weed and water hyacinth accelerate water loss by 130–250% as compared to the evaporation from clear water surface. In West Bengal alone the annual loss of fisheries due to

Some workers notably Paine (1966), Terborgh (1986), have coined the term **'Key-stone species'** to distinguish those species which influence the ability of a number of other species to survive in a community. Some of these Keystone species, like top predators, which often control herbivore populations, may be quite conspicuous. Others may not be so. A large number of inconspicuous plant species possess specialized set of insect fauna. Their disappearance leads to disappearance of those insect species as well. Naturally, Keystone species in a community should be accorded top priority in conservation efforts because their loss entails a loss of other species as well (14;15).

(IX) BIOLOGICAL DIVERSITY AND FUTURE CHANGES IN CLIMATE

Most of the scientists of the world now agree that our planet will warm up significantly during the next hundred years as a result of rising concentrations of green house gases (CO_2, CH_4, NO_2, CFCs etc.) in the atmosphere. Molecules of green house gases absorb infrared radiations preventing them from radiating back into the space. This causes increase in the average global temperatures. Exactly how much and how soon these changes shall occur is, however, still not clear. One estimate predicts a warming of 3° ±1.5°C by the end of the next century. Another estimate predicts a similar magnitude of change in average global temperatures within next 50 years. A rise of 3° ±1.5°C may seem small. Actually it is not so. A change of 3°C in average global temperatures shall leave our planet warmer than at any period in the preceding 100,000 years (16). The degree of warming which different regions of the world will experience shall vary according to their altitude and latitude. Polar and sub-polar regions shall experience more warming than tropical and subtropical zones of the world (Fig. 11.3)

Associated with warming will be changes in precipitation pattern causing some places to receive more and others less rainfall than they do now (Fig. 11.4). A rise in mean global temperatures shall melt more ice deposits, which along with thermal expansion of seawater shall cause a rise in mean sea level, inundating many low-lying areas. The physiological effects of rising CO_2 concentrations on plants could affect interspecific relationships as responses of different species to higher levels of CO_2 and their water requirements vary. The soil chemistry could be significantly altered by changes in both precipitation and elevated CO_2 concentration in the atmosphere.

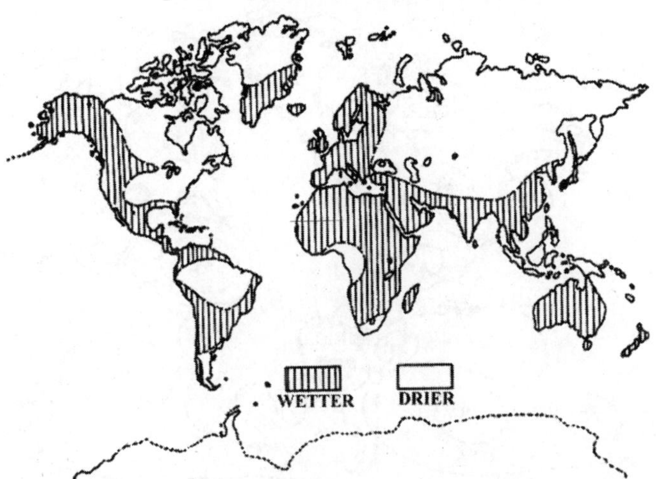

Fig. 11.4 *Changes in pattern of precipitation*

An inevitable consequence of global warming shall be expansion of tropical belt, displacement of sub-tropical and temperate zones northward and southward away from the equator while Polar Regions shall diminish in size. As inferred by distribution of fossils during past ages, the most important consequence of climatic change shall be the expansion of the range of distribution of different species towards regions, which are more suitable, or the **Climatic optima** and withdrawal exotic weeds has been estimated to be about 8.5 million tons. Mats of water hyacinth break loo are pushed by winds to adjacent paddy fields, which are rendered unproductive. A number of ι water plants are displaced by these vigourous but useless weeds. There is an overall reductic biological diversity wherever these exotic weeds migrate.

5. Natural Calamities: Natural calamities, such as floods, drought, forest fires, earthquak volcanic eruptions, epidemics etc., sometimes take a heavy toll of plant and animal life. Floods a frequent in moist tropical regions of the world, which inundate much of the ground vegetation, trap large number of animals while leaching away soil nutrients to make the already poor tropical soi poorer. Failure of monsoon, for example, in succession for two or three years dries up ground vegetation and as the subsurface water table recedes, trees are affected. With plant life animals also suffer. In such localities, heavy seasonal downpour also causes extensive damages. Epidemics sometimes destroy large portions of a natural population. Sporadic diseases are quite frequent in a natural community. However, at times virulent strains appear, cause diseases, which put on the dimensions of an epidemic. A large number of organisms are killed. In nature such episodes are usually confined to specific plant or animal populations as the pathogen is often specific to particular species or group of species.(3).

(3) Chain Extinctions

In an ecosystem, everything is related to everything else. The component species are maintained by a complex set of interactions among the enumerable life forms as well as between the life forms and the abiotic factors of the environment. Extinction of a component species, therefore, results in a chain of events affecting a number of other species adversely. Those forms, which are, obligately dependent on the extinct species have to die out. There are several instances of animals starving to death following the disappearance of species, which constituted their main source of food. The chain of events following near extinction of genus *Hibiscadelphus* which resulted in disappearance of nany Hawaiian honeycreepers and its pollinators has been well illustrated (13). Extinction of each tropical plant species leads to a loss of about 10–30 species of insects because many tropical insect species are highly specialized in their feeding habits and behaviour.

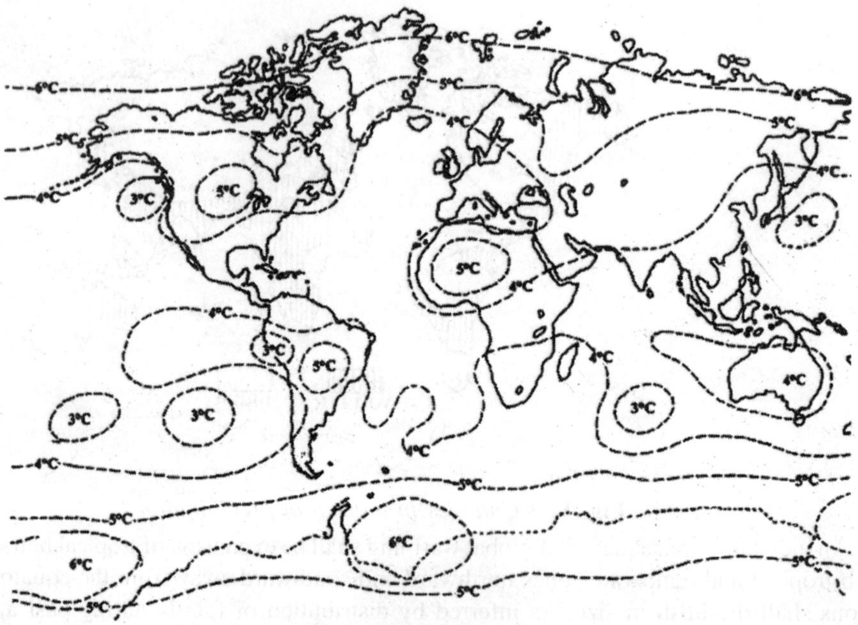

Fig. 11.3 *Future trends in global warming*

from localities, which have become unsuitable for their existence. In the past episodes of warming, fossil records reveal that many species shifted towards both higher latitudes and higher elevations (17;18). Thus large-scale migrations shall occur. Species incapable of migrating according to dictates of the climatic change shall be caught in unfavourable conditions and shall disappear.

During the last spell of warming when the average temperatures of North America were only 2°–3°C higher than what they are now, *Maclura* sp. (Osage oranges) and *Asimina* sp. (pawpaw) occurred many hundred kilometers north of their present day distribution. *Trichechus* sp. (manatees) could be found of the shores of New Jersey. *Tapirus* sp. (*Tapirs*) and *Tayssus* sp. (peccaries) foraged in Pennsylvania. A forest similar to the present day forest of North Carolina occurred at Cape Cod. As far as altitudinal shifting is concerned, many species of plants, like hemlock (*Psuda canadensis*) and white pine (*Pinus strobus*) were found about 350 meters higher up the mountains than they occur now. In general, a cooling, which is observed by a short climb of about 200 meters in altitude up a mountain, is obtained by a shift of about 100 kms in latitude towards the poles. As each species disperses itself at different rates, the climatic change, which forces various species to shift their range of distribution, also results in development of new plant and animal associations. Those species which are localized in distribution – such as endemic species – may not be able to reach and colonize new habitats. Even some animals, which are capable of swift movement, do not travel far because of behavioral reasons. The rate of movement of several species of dears for example, has been observed to be less than two kilometers per year. Such species could be adversely affected if the climatic changes are faster.

Based on the evidences of shift in range of distribution of species due to warming in the past and the physiological tolerances of the species it has been estimated that many suitable ranges for occurrence of a number of species shall move 500–1000 kms northward while Boreal Coniferous forests shall be reduced by nearly 40% of their present extent. This shall happen within a period of hundred years or so. Meaning thereby that the species shall have to shift northward at a rate of at least 500–1000 kms per century. This is too fast a rate for most of the species to move if we compare it with the migrations that have occurred following the last ice-age about 10,000–12,000 years ago. When the last glaciers retreated and the climate warmed up by 3°–5°C, a number of species had to move 25–30 kms northward per hundred years. Analyses of fossil pollen grains indicate that the fastest migration was by spruce forests (*Picea* sp. mainly) which migrated at a rate of about 180–200 kms per hundred years while Beach forests (*Fagus* sp. mainly) could migrate at the rate of 20 kms only.

Today, natural habitats are rapidly shrinking being replaced by human settlements, agriculture fields or much of land is rendered barren and degraded, of no use to many of the living organisms. Wilderness has become more or less like islands in the sea of man-dominated landscape, which many species cannot cross to reach the locality with suitable conditions or their climatic optima. Natural barriers such as a mountain range or an arm of the sea may also check the movement of many species. No one exactly knows how the biosphere shall react to the climatic changes brought about by global warming. Many species could become extinct, many others could survive in reduced numbers while many may flourish with a changed geographical distribution. As different species shall respond to changes in climatic patterns in different ways, the entire biotic spectrum over the globe is expected to change and an inevitable outcome of this upheaval could be drastic reduction in the biological diversity.

(X) REPURCUSSIONS OF THE CURRENT DEGENERATION OF BIOLOGICAL DIVERSITY ON FUTURE OF EVOLUTION

What will be the effect of such a rapid degeneration of biological diversity and destruction of natural ecosystems, on the future course of evolution? The average extinction rate in Paleozoic and Mesozoic extinction episodes involving the marine fauna varied from one taxonomic group to other and ranged between one to ten extinctions (of species) per million species per year. The current extinction rates have been estimated to be 1,000 to 10,000 times faster than these rates of extinctions. Scientists believe that such a fast rate of extinction, if allowed to continue for a hundred years or so, shall leave the biosphere on our planet crippled for millions of years to come.

If we compare the current scenario with that of the episodes of extinctions, which have, occurred in past geological era, we find that in these episodes for at least about 5–10 million years the biosphere limped along in crippled state. The taxonomic groups, which were dominant earlier, were gone or were represented by a few individuals only. The few species, which survived the onslaught of extinction, flourished without competition and became surprisingly abundant. The process of evolution slowed down or was suspended altogether as the biotic environment, which makes a species adopt and adjust and thus shapes a species, became ineffective. It was only after several million years that the process of evolution could gather momentum and this was followed by development of new forms – the evolutionary radiance of innovations. But the earlier forms which were abundant at one time were gone and new taxonomic groups appeared as dominant organisms. For example, it was only after the disappearance of the giant reptiles – Dinosaurs etc – that mammals, which were already there in Jurassic period, could gain enough strength to emerge as dominant forms.

The current spell of extinctions is, however, different from the extinctions which occurred in the geologic past. In the past there was no one to destroy, slash and burn habitats and clear vast stretches of land of natural vegetation. There was no man made pollution, no global warming or ozone hole. In the past episodes much of plant species and their habitats survived and provided the vital resource and environment for the process of evolution to start its course. The evolutionary radiance of new life forms which we find in our fossil records within a few million years bears ample testimony to the fact that in spite of the spell of extinctions, enough, particularly the habitats, was left. The current episode of extinction in the evolutionary history is characterized by massive destruction of plant species as well as their habitats. More species of plants are expected to vanish as natural habitats are cleared, tropical forests are destroyed, wetlands are drained and ocean beds are covered with toxic debris and sediments. With these habitats gone, the storehouse of resources from which evolutionary process may commence shall disappear. It will take much time and effort for the evolutionary mechanism to begin. It will be temporarily (if not permanently) suspended.

QUESTIONS

1. What are Hot–Spots of Biological diversity? Why should we accord top conservation priority to these sports? Discuss.
2. What do you understand by Biodiversity? Discuss its importance and uneven distribution over earth's surface.
3. Give brief account of the current state of degeneration of Biodiversity.
4. Enumerate major causes and consequences of degeneration of biodiversity.
5. Discuss in brief the likely impact of future changes in climatic conditions on Biodiversity, How the current degeneration of Biodiversity shall affect the future course of evolution of life on this planet.

12

Chapter

Ex-Situ Conservation

The existing system of protected areas, national parks, sanctuaries and biosphere reserves is not adequate for conservation of biodiversity. Many protected areas and habitats are too small and subject to frequent changes to sustain the viable populations of species they seek to preserve. As human enterprise expands and wild life habitats shrink numerous life forms will be lost. Large land vertebrates, which come in conflict with man, shall be ultimately confined to wild life preserves. Whatever we do species shall continue to disappear and biosphere shall continue to lose its constituents one after the other in near future. With wild life failing to survive in protected areas and natural habitats, conservation under human care appears to be the only solution. No doubt, it is impossible to preserve all of the species thus affected but prudence demands that we should intensify our efforts for ex-situ conservation as well since it shall save at least some of the fragments of once extensive wild life for the future.

(I) WHAT IS EX-SITU CONSERVATION

Ex-situ conservation involves maintenance and breeding of endangered plant and animal species under partially or wholly controlled conditions in zoos, gardens, nurseries and laboratories. There is nothing like 'golden freedom' in wilderness. Wild animals have always to be alert, compete for food, water and space and have to die a lingering death due to diseases, injury, starvation or thirst. They are often unable to breed due to absence of a mate. Plants growing in wilderness experience equally difficult living conditions. Grazing animals, diseases, forest fires and other calamities often destroy entire plant populations while there is a tough competition for nutrients, space, light etc. Human care eliminates the stresses which living organisms experience in natural habitats and provide conditions necessary for a secure life and breeding (1;2).

(II) THE STRATEGY OF EX-SITU CONSERVATION

Like *in-situ* conservation, *ex-situ* conservation is an age-old practice. Man has been breeding plants and animals under his care from times immemorial for various purposes, a practice which is basically the strategy of *ex-situ* conservation. However, these captive breedings were rarely done for the purpose of wild life conservation. Today captive breeding and maintenance of wild animals and plants have become a very important method of conserving some of the characteristic and important life forms. The application of advance scientific methods and technology has transformed the process into a well-organized discipline of science (1;2). The strategy of *ex-situ* conservation involves:

A. Identification of species to be conserved.

B. The methods of *ex-situ* conservation.

(A) Identification of species for ex-situ conservation

With a multitude of species likely to be lost in near future, we shall have to pick and chose the important ones which we can preserve as we cannot conserve all of them. Naturally, we shall have to devise some criteria for the selection of species to be placed under *ex-situ* conservation efforts. These criteria may include:

(1) Vulnerability of the species to extinction.

(2) Economic, ecological or aesthetic importance of the species.

(1) Vulnerability of Species to Extinction

The expansion of human enterprise in near future will inevitably result into fragmentation of natural habitats, which shall in turn cause disappearance of one species after the other. There is considerable evidence to suggest that the number of species in an isolated patch of habitat decrease over time. Therefore, with so many species likely to be extinct in near future, a crucial issue for conservationists is to identify those species, which are most at risk of extinction after habitat fragmentation from the knowledge of their biology and ecology. A number of attributes or traits of life cycle have been proposed as factors determining the species vulnerability to extinction (3;4). Important among these include:

1. **Rarity:** Abundance of species before fragmentation is an important criteria for determining the vulnerability of the species to extinction. It has been found that after fragmentation, rare understory birds occupy fewer forest fragments per species than common ones. Thus, fewer individuals of rare species as compared to the common species are likely to occur in habitat fragments. Such species are prone to extinction sooner than other species.
2. **Dispersal ability:** Species capable of migrating between fragments of habitats or between mainland areas and the fragments are able to nullify the effect of small population size. Thus mobile species which are capable of migration with ease and swiftness are not at much risk of being lost.
3. **Degree of specialization:** Specialized species usually use and subsist on resources which are distributed in small patches in time and space. These resources are likely to be effected more frequently than others. The reason for rarity of specialized species is the rarity of these resources. Such species are, therefore, more prone to extinction. Specialized species may also be vulnerable to successional changes in the habitat fragments and to the break-down of co-evolved mutualism and food web.
4. **Niche location:** Species adopted to or able to tolerate conditions at the interface between different types of habitats are less affected by habitat fragmentation than others. These species are actually benefitted by habitat fragmentation. Species intolerant to conditions of life at the interface between two types of habitats are more prone to extinction in a relatively shorter time.
5. **Population variability:** Species with relatively stable populations are much less vulnerable to extinction than those, which undergo large fluctuations in population densities. The population of those species with stable population is unlikely to fall below the critical threshold from which it is difficult to recover.
6. **Trophic status of the species:** Animals at higher trophic levels are usually less abundant and thus more vulnerable to extinction because of rarity.
7. **Adult survival rate:** Species with naturally low adult survival rates are usually more vulnerable to extinction than those species, which have higher adult survival rates.
8. **Longivity:** Usually long-lived animals are less vulnerable to extinction than those species which have shorter life span.

(2) Economic, ecological and aesthetic importance of the species

Once the likelihood of extinction of species has been assessed, the choice of species for *ex-situ* conservation rests on its economic, ecological and aesthetic importance. As the resources available to us are limited and cannot be applied to all the endangered species, the species which are or are likely to be more important to man in near future among the endangered species shall be preferred for *ex-situ* conservation. There is little logic in conserving and wasting our limited resources on species, which has no use for man in future.

(B) The methods of ex-situ conservation

The practice of *ex-situ* conservation involves techniques, which are essentially meant to maintain, multiply or help the species to survive under natural conditions. These include:

(1) Long Term Captive Breeding

The method involves capture, maintenance and breeding in captivity on long-term basis of individuals of the endangered species. Captive breeding and propagation on long-term basis is usually under taken for species, which have lost their habitats permanently, or there are present certain such factors in the habitat, which shall force it to extinction again. In majority of cases where human interests – such as poaching, excessive hunting etc. – come in direct conflict with existence of the species long-term maintenance and breading in captivity is resorted to.

There are a number of species which are being helped by *ex-situ* conservation technology, *i.e.*, are being maintained and bred in captivity under human care while their original homes have been so changed that they cannot survive therein. These species include: Siberian tiger, Pere Davids deer, Lion tailed macaque, European bison, Przewalski's horse, Brown eared pheasant, Edwards' pheasant, Bali Myna, White naped crane, Addax, Slender horned gazelle, Scimitar horned oryx, Gaur, Grevy's Zebra, Puerto Rican horned toad, Chinese aligator, Mauritius pink pigeon, Madagascar radiated tortoise, Aruba island rattle snake etc (4).

(2) Short Term Propagation and Release

Short-term maintenance, captive breeding followed by release of the animal in their natural habitat is a method, which is usually resorted to when the population of a species declines due to some temporary set-back in their living conditions. There is no irreversible change in the habitat and the animal concerned can survive in its natural home after the factors causing the set back are eliminated. For example, if the decline in the population of the species is due to over hunting, it may be banned by law. The endangered animals maintained and bred in captivity under human care are subsequently released in the wild habitat. Thus, a little help enables the species to re-establish itself in its natural home (4).

There are a number of species which man has helped to survive by using *ex-situ* conservation technology. Important among these are: Cheetah, wolf, Red-wolf, American bison, Arabian oxyx, onager, Andean condor, Bald eagle, Peregrine, Hawaiian goose, Lord Howe Island wood rail, Guam rail, European eagle, Owl, Guam kingfisher, Galapagos giant turtle, Galapagos land iguana, etc.

(3) Animal Translocations

Translocation involves release in a new locality of animals, which come from anywhere else other than the place in which they are being released. In some cases translocation may also involve removal of the animals from a natural community. The capture, transfer and release of animals from one locality to another usually involve maintenance of the animal in captivity for some time. However, care should be taken that this period of captivity should be as short as possible (5).

Manatee (*Trichechus manatus*), a huge blimp-like gentle creature is ruthlessly hunted for its fine quality meat, hide and oil and is the most endangered of all aquatic mammals. Among other methods used to conserve the species translocation to new habitat has also been successfully tried. The animals have established itself in its new home. For Golden lion Tamarins (*Leontopithecus rosalia rosalia*) a strikingly beautiful, highly endangered squirrel sized monkey, endemic to a small part of forests of Rio de Janeiro state, all conservation methods are being tried. In 1983, a small population of these animals was transported to a safer habitat where they appear to have established themselves (5). Though there may be several reasons for carrying out the translocation of animals from one place to another – like compassion educational and commercial reasons – the main objective behind undertaking transfer of animals from one habitat to another is their conservation. Translocation is usually resorted to under following circumstances:

1. The wild population may be threatened with prospects of extinction due to habitat destruction, uncontrollable hunting or some other reason.
2. The population within a locality may have become surplus, i.e., quick multiplication may have caused the number of individuals to rise beyond the capacity of the habitat to sustain them.
3. There may be present apparently suitable empty habitats which can be used to accommodate the additional species which is in need of conservation. Thus threatened species may be conserved by providing it with a new home by translocation.
4. Small isolated populations decline due to continued inbreeding as the individuals have no other choice but to breed with the available mates. In such cases introduction of co-specific types through animal translocation could ensure survival of the population.
5. In cases where a particular species occurs in a restricted locality, or is endemic to a particular place, its disappearance due to some local cause may result in its extinction. Successful translocation to other places could extend the range of distribution of the species and save it from prospects of extinction.
6. Translocation may be resorted to as an alternative to culling. Individuals of the species may be removed from the community to release the pressure on another species which is rarer and thus more important, or to improve the population by correcting structural imbalances, or to save healthy individuals of the population by removing the unhealthy ones.
7. Translocation may also be used if the animal populations become a nuisance to humans, threatening life, health, property or livelihood. The animal population has to suffer if man retaliates. Under such circumstances it is often in the interest of conservation to transfer the excess population to some other place where it does not come in conflict with man.

(III) ANIMAL REINTRODUCTIONS

Animal reintroduction involves release of animals either born in captivity or caught in infancy from the wild and grown in captivity, into an area from which they have either declined or disappeared as a result of human pressures (such as hunting) or due to some natural causes (like an epidemic). These reintroductions may also involve rehabilitation of the species as well. When an animal is born in captivity or is captured as an infant and raised in captivity, it is ignorant of the *modus vivendi* in the wild habitats. Such animals are naturally deprived of the process of learning which enables the individual to survive in the wild and have to be trained before being introduced into their natural habitats. The process of training naive animals to live in their natural habitats is referred to as *rehabilitation*. Rehabilitation is a very important step in the reintroduction procedure involving intelligent animal such as Primates in which learning plays an important role in the development from infancy to adulthood. It is less significant or even unnecessary in case of animals in which instinctive behaviour pattern predominates–like crocodiles (6).

It is for purpose of reintroduction of the animal that enormous efforts and expenditure involved in captive breeding are accepted by the Society. Indeed many of reintroduction efforts have been fruitful. American Bisdon (*Bison bison*), White tailed sea eagle (*Haliacetus albicilla*), Goshawk

(*Accipiter gentilis*), Bald Eagle (*Haliacetus leucocephalus*) *and Osprey* (*Pandion haliacetus*), the Harris' Howk (*Parabuteo unicinctus*), Griffon vulture (*Gyps fulvus*), Peregrin falcon (*Falco peregrinus*) have been successfully bred in captivity and reintroduced in their natural habitats. However, there have been many costly failures. The Action Plan of Species Survival Commission (IUCN) recommended only 68 species for captive breeding and reintroduction, majority of which are being maintained and bred in captivity, out of about 2409 species listed in the Red List of IUCN, 1990 (6). Only 145 known cases of reintroductions of animals borne in captivity have been documented. These include 13,275,295 individuals of about 126 species of which 32% are mammals, 45% birds, 15% reptiles, 6% fishes and only 1% invertebrates (7). One of the main reasons for recommendation of such a small number for reintroduction is that most of the species have declined because of habitat destruction or alteration. The species existing in their natural habitats now, are occupying fully what little habitats remain, i.e., the habitats are already saturated. So reintroduction is not likely to be a successful tool for the conservation of species threatened with prospects of extinction unless enormous efforts are undertaken for the ecological restoration of degraded parts of the ecosystems.

a. Circumstances Under which Captive Breeding and Reintroduction are Undertaken: It is important to realize that the species recommended for reintroduction are those species which have suffered from human pressures at the species level only and not at the habitat level and large tracts of empty habitats suitable for the existence of the threatened species still exist. If the original causes of their decline are identified and eliminated, reintroduction could be an effective solution for their survival (7). The circumstances under which captive breeding and reintroduction may be undertaken are:

1. When species come in direct conflict with man, such as grey wolf, mained wolf, big cats, large and potentially dangerous animals such as Rhinoceros etc.
2. When species is ruthlessly hunted as a food source, or other products like hide, fur, horns medicines etc., such as Addax, Gazelles, Oryx, Tortoises, Rhinoceros, etc.
3. When the species is vulnerable to some disease such as black footed ferrets.
4. When the species is a victim of pet trade such as parrots.
5. When the species exists in the form of small fragmented population which are vulnerable to natural disasters.

b. Problems Associated with Successful Reintroductions: The basic aim of reintroduction attempts is to re-establish self-sustaining, genetically viable populations of the introduced animal in natural habitats, which have lost them. Repeated failures and few successes have demonstrated that translocation attempt shall have greater chances of success under following conditions:

1. When the number of founder organisms released is large.
2. When habitats are suitable.
3. When the environmental conditions within the habitat are stable and there are little variations in the conditions.
4. Herbivores carry greater chances of success than carnivores.
5. There is a high degree of genetic diversity among the founder members.
6. When we are well aware of the patterns of behaviour of the organisms under different set of conditions and a proper care is taken to train and acclimatize the individuals of the population to be released.

It is often difficult to comply with many of these criteria. The release of a large number of individuals at a time shall naturally carry greater chances of success of the reintroduction attempt. However, it is often difficult to breed a large number of individuals of a species as it may involve more manpower money and resources which are difficult to obtain.

The carrying capacity and suitability of the habitat in which the animals are being released are very important for the reintroduction attempt to succeed. Careful investigation should be conducted to identify the factors behind the decline or disappearance of the original population, which should be moderated or eliminated altogether. This is often difficult. Reintroductions involving populations consisting of herbivores have been found to be more successful in establishing themselves than carnivores. This is because the organisms at higher trophic levels are usually more specialized in their food and habitat requirements. But as both herbivores and carnivores alike are threatened there may be times when conservationists have to deal with carnivores also (8).

There should be as much genetic diversity as possible among the individuals of the introduced populations for the success of reintroduction attempts. At the same time genetic makeup of the introduced species should also be suitable to the local conditions. Usually full spectrum of variations present in the original population of the species in the habitat is rarely available. To maximize the genetic diversity in the animals to be released the population is interbred with other sub-species or con-specific types in hope that natural selection after the reintroduction may produce correct locally adopted genotypes. This is like taking a chance, which in itself has more probability of failures than success. Unusual selection pressures under captivity even though we start with a large number of founders may cause loss of wild genetic traits. Captive breeding tends almost invariably to result into domestication. Therefore, it is desirable to keep the number of generations between the capture from wild and the reintroduction as low as possible – as the higher the number of generations in captivity, the greater would be decline in the wild traits within the species. The interbreeding done in an effort to maximize the genetic diversity in the organisms to be introduced may also result in appearance of such traits, which are undesirable for the locality. This may result in an ecological disruption rather than restitution of the species in its original habitat (8).

We have little understanding of behavioral traits of many of the threatened species. The knowledge of habitat requirement, food preferences, the attributes of the suitable environment, which the animals require for mating, breeding etc. have to be known for a successful reintroduction attempt. How can we help a species if we do not exactly know what it actually needs? A thorough study of these attributes of behaviour will require time and in fact it is gigantic task to carry out these studies in the case of so many species which are at present declining or are likely to be lost in near future. This lack of understanding about wild life is a major bottleneck in the process of successful reintroduction.

(V) ZOOS AND BOTANICAL GARDENS – INSTITUTIONS IN TRANSITION

While settling on earth's most productive habitats and changing them to agricultural and other human uses, mankind has tremendously reduced wild organisms, both plants and animals and this process of wild life degeneration is now taking place at an accelerated rate. Most of the people are becoming urbanized with the result that a vast majority of them will never see the significant diversity of wild life which occurs on our planet except in zoos and botanical gardens. Thus, zoos and botanical gardens are living museums of natural history. Today they are undergoing a transition from institutions, which simply exhibit wild life to organizations, which maintain breed and conserve wild life as well. The role of conservation of biodiversity has come to zoos and botanical gardens automatically as it is these organization alone which have been dealing with wild life and have some means to keep, multiply and thus conserve the species disappearing in nature. Many species maintained in zoos and botanical gardens have already become rare and some even extinct in the wild.

(1) Zoological gardens or zoos

Maintenance of wild animals in captivity is a very ancient practice – almost as old as the civilization itself. It was by 16th century AD, that zoos were established at Cairo, Karlsberg, Dresden, Prague and Versailles. The oldest zoo, the Schonbrunn zoo, which exists today also, was established in Vienna in the year 1759. In 1794 AD, the first public zoo was founded in Paris. London zoo came into existence in 1828. The zoo at Bristol was established in 1836.

In India the first menagerie was setup by Lord Wellesly at Barrackpore in 1800 AD. The first Indian zoo was established by Raja Rajendra Mullick Bahadur at Marble palace in the heart of Calcutta in 1854. This zoo happens to be the oldest existing zoo in India. By 1860 AD, the zoo-movement took roots and their management became more scientific. In 1907, Carl Hagenbeck of Hamburg developed the concept of barless moated enclosures and with it attempts were made to maintain animals in more natural settings. The primary objectives behind maintenance of wild animals in zoos and menageries were exhibition and entertainment. The realization of research and educational values of zoos provided most of the zoos of the world their modern shape and the scientific basis for their management. By 1960 AD, scientists and naturalists the world over were faced with the reality of an alarming rate of extinction of wild animal species and zoos of the world were looked upon to help in the conservation of endangered species. Today the zoos maintain well over 500,000 individuals of terrestrial vertebrates, representing about 3000 species of mammals, birds, reptiles and amphibians apart from a large number of fishes and other less conspicuous forms in aquaria and other collections. There are about 1140 zoos all over the world with an average area of 55 hectares which are visited by about 2.88 million people each year (9).

The earlier objectives with which zoos were initially founded, that are exhibition and entertainment, have paled out of significance and more important objectives have replaced them. The primary goals behind maintenance of zoos today are:

(I) Participation in wild life conservation efforts

1. Maintenance and breeding of endangered species.
2. To provide populations of endangered species to be reintroduced in natural habitats.
3. Maintenance of *ex-situ* population for exhibition or research purposes from where these animals can easily be obtained.

(II) Study and research.

1. For acquisition of scientific knowledge which will ultimately benefit conservation efforts.
2. To obtain scientific knowledge on captive populations which could provide an insight into the basic biology of the species?

(III) Educational purposes: Zoos and animal exhibitions can creat public and political awareness about the conservation of wild life and natural resources.

Zoos of the world have contributed significantly to wild life conservation efforts. Pere David deer (*Elaphurus davidianus*), Przwalskii's horse (*Equus przwalskii*), Oryx (*Oryx dammah*), Adax (*Addax nosomaculatus*), Alpine Ibex (*Capra ibex-ibex*), Ospreys (*Pandian haliacetus*), Harris' Hawk (*Parabuteo uncinctus*), the douc langur (*Pigathrix nemaeus*) survive only because of the enormous efforts put in by world's zoos. At one time these species had completely disappeared from their natural habitat. The few specimens which happened to be in captivity in zoos of the world were carefully maintained bred and most of them, with few exceptions like *Pigathrix nemaeus*, have been introduced back into their natural homes. Attempts are being made to reintroduce the remaining species as well (10).

Similarly zoos have done excellent work in restocking, replenishing and re-strengthening the natural populations of a large number of species which occur as small fragmented populations. But for these restrengthening attempts many of these species would have disappeared. The Siberian tiger (*Panthera tigris altaica*), Lion tailed Macaque (*Macaca silenus*), Golden lion tamarin (*Leontopithecus rosalia rosalia*), American bison (*Bison bison*), Black footed ferret (*Mustela nigripes*), Eagle owl (*Bubo bubo*), White tailed sea eagle (*Haliacetus albicilla*), Goshawk (*Accipiter gentilis*), Bearded vulture (*Gypaetus barbatus*) Philippines eagle (*Pithacophaga jefferyi*), etc. are some of those species which were survived by dangerously low populations. Individuals of these species were painstakingly captured maintained and bred in captivity. When sufficient populations of these species were built up they were released in wild habitat to strengthen the natural populations which were on verge of extinction. They are flourishing in most of the cases now (10).

(2) Botanical Gardens

A botanical garden can be described as a place where flowers, fruits and vegetables are grown. Since times immemorial, man has been cultivating plants for various purposes. It is a general practice to maintain a small garden in every well to do household and kings and courtesans maintained large gardens where a variety of plants were grown. The basic purpose for the maintenance of these gardens was beauty, aesthetic sense and the calm and quite environment, which a secluded place full of greenery and colourful flowers provide.

With the advent of the era of scientific development, most of botanical garden started keeping interesting exotic plants for the purpose of exhibition and education. A number of botanical gardens became institutions of scientific research and undertook the job of cataloguing and classification of plants. Many of them have contributed significantly to our knowledge of taxonomy and systematics of plant life. As early as 1545 AD, the botanical garden of Padua, Italy was established. The natural history museum of Paris, France, came into existence in 1635 AD. In 1646 the botanical garden and museum of Berlin, Germany was established. The Royal Botanical Garden of Edinberg, Scotland, U.K., was founded in 1670 AD. The Botanical garden and institute of Vienna, Austria was established in 1754. The Indian botanical garden, Sibpur, Howrah, India was founded in 1787 AD. The Royal Botanical garden of Peradiniya in Sri Lanka was founded in 1810 and that of New South Wales, Sydney, Australia was developed in 1816 AD. The Royal Botanical garden, Kew, England was established by merging the private gardens of Sir Henry Capel and Dowager Princes of Wales, Princes Augusta in 1841. The New York Botanical garden, New York was established in 1895.

Apart from maintaining live specimens of plants from all over the world, many of these gardens also maintain preserved samples and herbaria along with record of their geographical distribution ecology and life cycles. The task of cataloguing and classification of plants which botanical gardens undertook in addition to upkeep and maintenance of plant life have resulted into substantial addition to our knowledge of systematic and taxonomy of green plants. Hooker J.D. in association with George Bentham published their Genera Planaterium during the years 1862 to 1883. These authors have contributed a phylogenetic system of classification of green plants, which represents a landmark in the field of systematics and taxonomy. Both of these authors were associated with Kew Botanical Gardens England. A number of research journals and magazines are published by Botanical gardens of world. Notable among these are the Kew Bulletin and Index Kewensis started in year 1887, Annals of Missouri Botanical gardens, Annals of Royal (Indian) Botanical Gardens, and Annual reports etc.

Though something was done to preserve the germ-plasm of crop plants and their wild relatives earlier, it was only in the eighties that plants growing in the wild received some attention. Now a wealth of rare and endangered plants thrives in about 1848 botanical gardens and arborea in 148 countries of the world, which have taken up the job of conservation of plants seriously. They maintain more than 4 million living plant accessions among which there are representatives of more than 80000 species, Many of these botanical gardens have been equipped with latest technology and are doing excellent work to preserve, propagate and reintroduce plants in their natural habitats (11;12).

Plants like *Diospyros hemiteles* known from a single female tree and *Olax psittacorum* known only from old collections which is lost in the wild but is survived by a single tree, are protected and work is going on to breed them through tissue culture techniques. *Tambourissa tetragonia* is known to us only by two specimens and tissue culture techniques are being tried to propagate it at Kew

Botanical gardens. *Limonium tuberculatum* has disappeared from the wild. The botanical garden at Canary Island, Spain, had a few specimens of this plant from which a population of about a dozen plants was established. It is now being further multiplied to be introduced in its natural home. *Ramosmania heterophylla* a plant whose parts are in great demand because of its medicinal properties has disappeared from the wild. A few specimen of this plant are maintained in a botanical garden in Mauritius. Attempts are being made to cultivate this plant in tissue cultures at Kew Botanical gardens. Similarly, *Lotus kunkellii*, *Onopordon nogalesii*, *Euphorbia handiensis*, *Senecio hadrosomus*, *Clianthus punieus*, *Hibiscus columnaris*, *Badula crassa* etc. have been brought under human care in various botanical gardens and are being multiplied for reintroduction into their natural homes. A sanctuary has been established for *Nepenthes khasiyana* the famous insectivorous plant badly sought after for scientific study in educational institutions. Natural populations of many orchids, which produce world's most showy flowers, like *Paphiopedium fairieyanum, Cymbedium aloiflium, Aerides crispum,* have already declined to few individuals in isolated patches. The Indian Botanical garden, Sibpur, Howrah is doing a remarkable work by preserving these plants and multiplying them in its Orchid house (11;12;13).

(VI) CONSERVATION OF BIO-DIVERSITY IN SEED-BANKS, GENE-BANKS OR GERM PLASM RESERVES

A large number of plant species form seed with variable periods of dormancy following which they can be germinated to yield daughter plants. Most of such plants, therefore, can be preserved in the form of their seeds in small packets for long durations. Places where the seeds are stored are known as **seed-banks** or **gene-banks** or sometimes **germ-plasm banks**. The germ plasm of a plant is any of its living organ or a part of it from which new plants can be generated.

(1) Early Conservation Efforts:

In fact, it was in response to the rapid erosion of genetic diversity among, pulses, cereals, fruits and vegetables that serious efforts were started to conserve the genetic resources of crop plants early during the present century. As early as in 1899 A.D., the Government of United States of America setup a separate section for collection, preservation and introduction of plants under its Department of Agriculture. In 1899 A.D., American Embassies all over the world were asked to collect and send samples of plants likely to be useful to the country. It was in 1947, that the Regional Plant Introduction station was setup at Ames, Iowa, USA. In 1959, the first Cold-storage centre for grains was established at Fort Collins, Colorado, USA. This centre is now one of the most important gene banks in the world for wheat, sorghum and soya (14).

Since 1920 A.D., Russian workers have also done pioneering work in the field of conservation of genetic resources of crop plants. It was V.I. Vavilov, a geneticist and agronomist, who observed a number of highly resistant wheat varieties when he was a student. Later as the director of the Leningrad Institute of Plant Industry, he sent out collection expeditions all over the world. It was through his efforts that the Russian germ-plasm bank has now become one of the biggest collections in the world. Following the Second World War, a number of seed banks, gene banks and germ-plasm reserves have appeared all over the world. The task of conservation of genetic resources of fruit trees goes on in France under the auspices of the Public Services Department and the National Research Institute. Conservation of plants like banana, coffee-shrub and cocoa-tree has been undertaken at Napo Research Station, Ecuador, under the auspices of the National Institute of Agronomic Research (14).

(2) The Global Strategy:

For several decades, the efforts for conserving plant resources were taken up independently by developed countries alone. Developing countries were not even consulted. However, it was around 1960, that the rapid diminution of genetic diversity among crops of agronomic importance forced the advanced countries of the world to prepare plans for conservation of plant resources at global level.

The Food and Agriculture Organization (FAO) of the United Nations convened the first technical meeting for exploration, collection and conservation of plant resources of the world in 1961. The United Nations Conference on Human Environment convened in 1972 adopted a resolution concerning the International Programme for the preservation of genetic resources of plants cultivated in the tropical regions of the world. Almost simultaneously, a global strategy for preservation of seeds was developed at Beltsville, USA. The International Board for Plant Genetic Resources was framed in 1974 with its head quarters in Rome. By 1985, working on rather a modest budget it had setup a chain of 43 gene banks all over the world of which 21 are situated in the developing countries. An International Gene-bank controlled by the Food and Agriculture Organization was also proposed in another conference of Food and Agriculture Department in November 1983 (15). Some of the important centres for the preservation of plant resources around the world are:

1. Institute of Plant Industry, Leningrad, Russia for wheat, sorghum and barley.
2. National Seed Storage Centre, Fort Collins, Colorado, USA, for Sorghum, wheat and soya.
3. Plant Resources Conservation Centre, Beltsville, USA, for barley and wheat.
4. International Rice Research Institute, Philippines, for rice.
5. International Centre for Agricultural Research in Dry Areas, Aleppo, Syria, for wheat, peas, chick pea, beans and lentils.
6. International Centre for Improving maize, wheat and barley, Mexico.
7. International Crop Research Institute for Semi-arid Tropics, Hyderabad, India, for sorghum, ground nut, millets and chick pea.
8. Napo Research Centre, Ecuador, for cocoa, coffee and banana.

(3) Maintenance of Plant Germ-plasm for Long Durations:

Seed banks or gene banks earlier used simple techniques for preservation, which involved storage of dried seeds under ambient or lower temperatures. General types of seeds such as those of many pulses, cereals, some vegetables and fruits possess natural dormancy and can be easily preserved for long durations by any of the following methods:

1. **Short-Term Storage:** Seeds which are to be stored for a few years are sun dried and kept in sealed containers at a temperature of 5°C or even at room temperatures.
2. **Mid-Term Storage:** Seeds which are to be stored for 15–20 years may be sun dried and maintained in sealed containers at temperatures between 0°–5°C. If these seeds are dried to about 4.5% of their moisture content, they can be maintained in viable state for longer durations.
3. **Long-Term Storage:** If seeds are to be maintained for several decades or more they are previously dried to about 4.5% of moisture content and kept in sealed containers at temperatures between –10° to –20°C.

Some of these seeds, of tropical plants, bushes or fruit crops do not possess a natural dormancy. They die quickly if not allowed to germinate immediately. Little could be done to preserve these plants as simple techniques available at the time failed to preserve them. They had to be constantly cultivated. Similarly, a number of plants, which reproduce by means of their vegetative parts such as tubers, corns, rhizomes etc, had to be regularly cultivated and multiplied. Recent techniques of preservation have greatly extended the scope of germ-plasm banks. Under liquid nitrogen, which lowers the temperature to about –196°C, seeds dried to a low moisture content can be preserved for almost indefinite periods – several hundred years or so. Through the application of tissue culture and cryo-preservation techniques almost any type of plant can be maintained in viable state for long durations. Plants, which cannot be preserved by traditional methods, are preserved in the form of tissues or meristem cultures under liquid nitrogen. Small pieces of simple vegetative tissues or better still few meristematic cells of plants to be preserved are introduced in synthetic medium under aseptic conditions and are allowed to grow into small clumps – called the **callus.** These cultures or calluses are maintained under liquid nitrogen for any length of time. They can be multiplied and young plants differentiated from the tissues whenever required (16).

(4) Short Comings and Controversies:

Conservations efforts aimed at the preservation of germ plasm of plants of agronomic importance suffer from the following basic shortcomings and controversies:

1. Plant conservation activities involving seed collections and preservation usually revolve round few plant species of agronomic importance – plants that provide food to humans and his dependents. A large number of plants, which are otherwise of known economic importance, are ignored. It is obvious that we cannot provide refrigerated rooms for each and every plant. The procedure is costly and requires a lot of expert attention. Howsoever, efficient and versatile the approach may be, a number of agronomically important plants are expected to be left out.
2. At seed or germplasm storage centres, a large number of plants are preserved in small packets or petridishes. Due to high costs, it is not possible to maintain several of these centres. In the event of some unforeseen calamity or a War, the entire collection of rare plants may be destroyed. We shall be left with nothing if the plant-variety is not available anywhere else.
3. Collection and preservation of germ plasm in seed or gene-banks tend to interrupt the vital process of evolution. It is not only the preservation, which is necessary, but also the adaptation and evolution of the plants to an ever-changing environment is important. Plants developed from seeds stored for long durations may entirely be misfit weaklings because by the time they are germinated the conditions of environment could have changed drastically. Thus a variety valued as resistant to a particular fungus may not be of any use as the fungus might have learned to over-come the resistance while the host plant has been lying dormant.
4. It is possible to improve the cultivated plants by using the desired genes from other varieties. Through genetic techniques, these genes can be pooled together in a single cultivar, which improves its commercial value enormously. Seeds or germs containing useful genes, therefore, acquire immense importance so also are the seed banks and germplasm collections, which possess a large assortment of varieties. Huge profits can be made by cornering the desired germplasm and selling it at a high price to farmers and plant breeders.

The lure of huge profits has caused a number of big Multinational companies like Sandoz (Swiss), Imperial Chemical Industries (U.K.), Pfizer (USA), Ciba-Geiegy (Swiss), Up John (USA), to enter into the venture. Many of these multinationals have their own seed-banks with worldwide collections and elaborate facilities for conducting research in the field. These companies not only monopolize plant resources but also the techniques, expertise and innovations necessary to develop improved plant varieties. Huge sums are charged for providing the germplasm and the necessary knowledge by these companies (14).

The developing countries of the world, from where the desired germplasm might have originally come, are unable to obtain it or are asked to pay enormous sums of money for the resources of their own land. Many developing countries of the world have resented the establishment of these germplasm banks and do not co-operate fully in the venture. As almost two third of plant resources of the world happen to lie in these countries the situation has resulted in a serious set-back to the conservation efforts for plants of agronomic importance. The matter has been taken up at the 22nd and 23rd sessions of FAO conferences but no consensus has so far been achieved, as the advanced countries of the world are intent on protecting the interests of the multinational firms (17).

(VII)ADVERSE CONSEQUENCES OF EX-SITU CONSERVATION

A number of disadvantages are inherent in the practice of ex-situ conservation. The wild life, which we painstakingly breed, may not be exactly the same which it would have been if allowed to continue in nature. Some of these disadvantages are:

1. In order to prevent genetic drift the populations in captivity have to be larger than we have been maintaining in most of the cases. We do not have provisions, space, funds and manpower to keep the required number of individuals of each species. Out of 629 species considered as threatened only 20,628 specimens belonging to 140 species are kept in zoos of the world (9). This is too small a number to avoid genetic drift and the eventual loss of a number of useful genes.
2. Animals in captivity may undergo genetic adaptation to their artificial conditions and thus become useless for the natural habitats. With time homogenization of their genetic set up occurs and captive parents are able to pass only a limited assortment of traits to their offsprings which tend to become more or less carbon copies of each other. The lack of genetic diversity makes them unfit to adjust and adapt to the persistently changing conditions of the environment in nature.
3. Animals in captive population may lose their knowledge of the natural environment. This makes them incapable to live in the wild to the extent that no amount of acclimatization or training can rehabilitate them in the wild.
4. The *ex-situ* populations may represent only a limited portion of the gene pool of the species. This is particularly true of wide ranging species with many scattered populations each of which has its own genetic makeup. To sample and breed the complete genetic spectrum, representative specimens from many populations are required which is a difficult task.
5. *Ex-situ* conservation efforts require a regular supply of resources, funds and institutional support etc. any interruption in the course of which may result into considerable losses.

(VIII) JUSTIFICATION FOR CONTINUING EX-SITU CONSERVATION

Ex-situ conservation efforts are usually under-taken in cases of those endangered organisms, which do not have any chances of survival in the wild. In-situ conservation efforts even if carried out with all earnestness shall not be able to preserve them. If we have to preserve such species, there is no other option but to bring them under human care. Once they have been multiplied and infused with vigour by interbreeding with con-specific types they can be reintroduced into their natural homes. This will enable us to save the species. Thus, ex-situ conservation is the resort of last choice. Today a number of species are surviving because of our ex-situ conservation efforts. For example, Pere David's deer, Addax, Oryx, the douc langur, przwalski's horse etc. live today because ex-situ conservation methods were applied for their preservation. The main objections to the practices of ex-situ conservation are:

1. Wild life is a vast and diverse assemblage of plants, animals and microbes. We can at best bring under human care only a very small fraction of the vast and diverse assemblage now facing the prospects of extinction.
2. *Ex-situ* conservation strategy has so far acquired rather a limited success. Animals and plants can be maintained in captivity. Breeding and multiplying them present a little difficulty but has been done in a large number of cases. However, problems appear when reintroduction of the species into their natural habitats is attempted. There have been frequent and costly failures, though in a few cases reintroduction attempts have been remarkably successful.
3. In *Ex-situ* conservation efforts, all the money and resources are spent on the same species, which is to be conserved. In *in-situ* conservation a number of species, known or unknown to us are benefited when we look after the entire habitat for the species to be conserved.

4. A major objection to *ex-situ* conservation efforts is the heavy expenditure incurred in the process. There is, no doubt, a great disparity in the money spent on captive maintenance, breeding and reintroduction as compared to that spent on similar *in-situ* programmes (4). It has been calculated that keeping African elephants (*Loxodonta africans*) and black rhinos (*Diceros bicornis*) in captivity shall cost 50 times as much as protecting and managing the same number of individuals in their wild habitats.

We agree that natural selection cannot be simulated in ex-situ populations and for all practical purposes ex-situ conservation leads to domestication and disappearance of wild genes. However, ex-*situ* care of breeding populations can effectively serve as means to conserve unique or useful traits rather than the entire species as such (18). Individuals with a set of useful genes can be maintained in captivity for those genes only. We may let the captive population suffer all the adverse consequences which follow when the species is kept imprisoned for long periods of time. As long as the useful genes persist the individuals of the population are useful to us. These genes may be used to improve domesticated organisms and even the wild ones, which are threatened because of the loss of wild genes, may be compensated by these preserved genes. Thus ex-situ conservation offers an additional means to conserve the threatened in-situ populations as well.

QUESTIONS

1. What is ex–situ conservation? How shall we identify species vulnerable to extinction?
2. Discuss in brief various methods commonly employed for ex–situ conservation of species.
3. What do you understand by Captive–breeding? Discuss its utility for the conservation of threatened species.
4. Discuss the merits and demerits of Captive–breeding and re–introduction as a commonly adopted method for ex–situ conservation.
5. Discuss the problems associated with captive breeding and successful reintroduction of a threatened species.
6. Under what circumstances animal translocation and re-introduction is usually undertaken. Discuss.
7. Discuss how advanced science and technology has been helping ex-situ conservation efforts.
8. Write a brief essay on seed banks or gene banks. What are controversies associated with it?
9. Discuss the role of zoological gardens play in conservation of Biodiversity.
10. Discuss the role Botanical gardens play in conservation of Biodiversity.
11. Discuss adverse consequences associated with the ex-situ conservation. Why should we continue ex-situ conservation efforts.

13

Chapter

In-Situ Conservation

In situ conservation is an age-old practice. Establishment of protected areas, parks, sanctuaries or reserve forests, where wild life could grow and multiply, has been the common response of societies and governments to the threat of diminishing wild life since times immemorial. Kings and courtesan often carved out large areas of wilderness for preserving game animals. Traditions and taboos of indigenous tribes have preserved many patches of forests as refuge of spirits and deities, which have remained undisturbed since prehistoric times. Though these preserves were intended for an entirely different purpose, they have been instrumental in preserving much of wild life in the vicinity of human establishments.

(I) WHAT IS IN-SITU CONSERVATION

In-situ conservation involves conservation of species in its natural habitat – in places where the species normally occurs. The natural surroundings or the entire ecosystem is protected and maintained so that all the constituent species, known or unknown to us are conserved and benefited. We do not have to isolate a few individuals of the species, create artificial habitat for their maintenance, feed them and provide a mate for their reproduction. We simply have to eliminate factors, which are detrimental to the existence of species concerned, and the rest is left to the nature, which takes care of its own self.

(II) ADVANTAGES AND DISADVANTAGES OF *IN-SITU* CONSERVATION

In-situ conservation of wild life offers many advantages. In fact, the method involves promotion of natural system to take care of its own self – we simply provide conditions to let life- forms do it themselves. The advantages of *in-situ* conservation can be summed up as follows:

1. *In-situ* conservation is a cheap and convenient way of conserving biological diversity as we play a supportive role only. Factors detrimental to the existence of the species concerned are eliminated and the species is allowed to grow in its natural environment in which it has been growing since a long time. This reduces the cost of conservation efforts enormously.
2. In order to ensure the survival of the species we protect the entire natural habitat or the ecosystem. Thus, a large number of organisms are protected and maintained in the process. In-situ conservation offers a way to protect to a large number of organisms simultaneously known or unknown to science.
3. In a natural system organisms not only live and multiply but evolve as well. A natural ecosystem allows free play of natural agencies – like drought, storms, snow, fluctuation in temperatures, excessive rains, fires, pathogens etc – which provide an opportunity to the organisms to adjust to the diverse conditions of the environment and evolve into a better adopted life form. By

isolating a species in artificial environment under human care we exclude these agencies. This freezes the vital process of evolution. The germplasm tends to stagnate.

Probably the only important disadvantage of *in-situ* conservation is that it requires large areas of earth's surface if we have to preserve the full complement of biotic diversity of a region. This involves minimizing or excluding human activity and interference from that locality which is often difficult in the face of growing demand for space. After all humans, also need space to live.

(III) THE STRATEGY OF IN-SITU CONSERVATION

The strategy of *in-situ* conservation revolves around establishment of small or large protected areas which are set aside exclusively for wild life. Human activities like, hunting, firewood collection, timber harvesting etc. are restricted in these areas so that wild plants and animals could grow and multiply in a natural but protected environment. The strategy though not adequate or fully successful has been of much help in preserving a considerable variety of life forms. Today about 7,000 protected areas, parks, and sanctuaries, nature reserves all round the world cover more than 650 million hectares of earth's surface representing about 5% of the total land area of our planet. These include a variety of national parks, sanctuaries, biosphere reserves etc. that differ from one another in their size, purpose and the degree of human interference or management efforts accorded (1).

National parks and **Sanctuaries** are usually meant for the protection of one two or more species and their habitats. They are usually small reserves, which on an average range between 100 sq kms to 500 sq kms. The boundaries of National parks are usually well marked and circumscribed where as those of a sanctuary are often not well defined. Controlled biotic interference is permitted in sanctuaries, which allow tourist activities as well. In National parks, tourist activity as well as a limited biotic interference on the outskirts or the buffer zone may be permitted but no interference is allowed in the deeper interior or the core zone of the park. The **Nature reserve** or **Biosphere reserve** on the other hand are usually large protected areas with boundaries circumscribed by legislation and are usually more than 5,000 sq kms in area. They are not meant for any species in particular but are intended to preserve a representative sample of entire biotic spectrum of the locality or the climatic zone. Except for a limited degree of biotic interference in the buffer zone or the out skirts, no exploitative human activity or tourism is permitted in a Nature reserve or Biosphere reserve. They are usually scientifically managed and due attention is accorded to Research and Conservation of gene pool within its confines (1;2).

(1) Requirements for an ideal protected area

The basic aim of all conservation efforts is to protect and conserve the biotic spectrum present on our planet for the benefit of the present and future generations of mankind. Today an overpopulated world, with its population still growing, is faced with the dilemma that expanding human enterprise could swallow the entire wilderness at some point of time in near future. This could result in the destruction of the vital life support system on our planet. However, it is difficult to stop the expansion of human enterprise. What we can do at best is to mark out biologically significant localities on this planet and pay sincere attention for their maintenance and conservation (3). These biologically significant areas should:

1. Should include representative sample of the biotic spectrum of the locality.
2. Should be capable of sustaining viable populations of all the constituent species within the system.
3. Should be able to allow free play of natural agencies, such as fire, drought, floods, pathogens, insects and pests etc. which make an organism grow, adapt to the prevailing conditions of environment and evolve into a better adopted life form, so that the process of natural selection and evolution may continue uninterruptedly.

(2) Selection of site for protected areas

A certain degree of biodiversity occurs within all habitats because genetic diversity has allowed life to adapt to varying conditions of environment. However, species are not evenly distributed over the earth's surface. Biodiversity is greater in some areas than others. Therefore, it becomes important to identify important centres of biological diversity before undertaking any conservation project. Which areas should be given priority? Which are the areas, which require immediate conservation efforts? An obvious answer to these questions is that we must choose those areas for immediate conservation efforts which possess the richest and the most diverse plant and animal communities under some kind of threat from internal or external sources (3). Identification of important areas to direct the conservation efforts can be achieved at national or global levels by following methods:

1. **Observations on Over-all Species Diversity:** The simplest method of targeting areas for conservation is to select countries, which possess the largest number of species – countries with greatest species richness. A very small number of countries occurring in tropical climate possess the largest proportion of world's species diversity. These countries are referred to as 'Mega-diversity Countries (4;5). It has been suggested that these countries should receive special international biodiversity conservation attention. Mc Neely *et al.* (1) have used the lists of vertebrates, swallow tail butterflies and higher plants of different countries to identify twelve such megadiversity countries which together possess the largest number of species (nearly 70% of all) in a relatively small area (See Table 11.1, in Chapter 11).

   This approach is relatively simple as it needs only lists of species, which occur within a geo-political zone, and emphasizes that the conservation effort should be managed at country level. However, the approach suffers from the disadvantage that it fails to take into account the uniqueness of the flora and fauna of the region. By focusing our attention on habitats, which possess the greatest number of plant, and animal species we may leave out unique and important organisms, which happen to occur in regions with a very low diversity. Moreover, there may be considerable degree of overlap – the same species may occur in a number of countries. Thus, relatively unimportant, widely distributed species, which may not be threatened at all, may receive more conservation attention than the species, which actually need it.

2. **Observations on Endemic Species Diversity:** An endemic species is that species which is restricted to a given habitat or locality. The fact that an endemic species is confined to a particular location and is not found anywhere else makes it important. Areas with largest number of such species should receive more attention as endemic species once lost from its original abode cannot be obtained from anywhere else. Based on the degree of endemism in species composition Myers (6) has identified 12 such localities in tropical regions of the world, which require urgent conservation attention. In 1990, Myers (7) identified another eight hot spots of endemic species diversity in other climatic regions of the world (See Fig. 11.1. in Chapter 11). Thus, nearly 49,955 endemic species of plants occur in an area barely 786,400 sq. kms or only 0.5% of the world's total land surface. Although there may be a few exceptions, in majority of cases a greater degree of endemism is met with in those regions of world, which have a richer and more diverse flora and fauna. Moreover, it is usually observed that a greater degree of endemism in one taxonomic group corresponds with a similar degree of endemism in other taxonomic groups as well. Endemism is, therefore, a valuable parameter for the identification of areas of greater biological significance.

3. **Critical Faunal Analysis:** The concept of critical faunal analysis was first introduced by Ackery and Vanc Wright in 1984 (8). It is now being increasingly used to determine conservation priorities. The entire spectrum of life forms within a group under consideration in a locality is called a 'compliment'. The portion of forms not included in this compliment is

designated as 'residual compliment'. After the selection of a compliment which represents the maximum number of endemics the priority for the second site is determined by the residual compliment which adds the largest proportion of life forms to the initial choice. The third site is the site, which adds the maximum residual compliment to the second site. In this way the process may be continued in the selection of fourth and fifth site in step wise fashion till the entire biotic spectrum is brought under conservation efforts. The process offers the advantage of providing an objective and optimized sequence of priority for the conservation efforts to be carried out and the performance of any other sequence can be judged for its relative efficiency in representing total biodiversity of the locality.

For example, Serengiti National Park (Tanzania), is one of the earliest site, protecting a population of 24% of the antelope species. Kufu National Park (Zambia), if including, shall bring under conservation efforts the highest number of antelope species. Together these two parks contain 38% of all antelope species. If we add two more reserves, *i.e.*, Haut Dod Faunal Reserve (Cote D'Ivoire) and Quadi Reme-Quadi Achim Faunal Reserve the four protected reserves shall represent about 56% of the antelope species diversity. Thus just these four protected habitats can bring under conservation efforts about 56% of all antelope species of Africa.

(3) Critical size of protected areas and populations

In a world with rapidly rising population, space is usually a sparse commodity. Man needs space for agriculture, industries, housing, military establishments, railways and roads etc. and at times it becomes difficult to set aside large areas of biological significance exclusively for wild life conservation. The question asked in this connection is what should be the minimum critical area which is just sufficient for the conservation of a given set of flora or fauna?

When a large forest is fragmented into small patches its capacity to sustain viable populations of different species declines drastically. A number of species either are forced to flee away or are pushed to extinction. The number of species, which a small isolated patch of forest can support, is a function of its productivity, habitat requirement of the species concerned and the conditions of environment – both abiotic and biotic. According to the bio-geographic principles, a tenfold decline in the area of the patch could reduce the species to about half of their original number (9;10). It has been observed that size of the habitat has much influence on the number of viable population of many species inhabiting the system. In Illinois, Graber and Graber found that the number of bird species in small patches of grassland decline at a much faster rate than the simple reduction in the total area of the patch would suggest (11).

Thus, a protected area should be large enough to support the population of the species, which we intend to conserve. This population should be a viable population – populations that are capable of growing and multiplying actively. If the population in the system is unable to multiply or multiplication occurs at a rate slower than the mortality, the entire conservation effort will be wasted, as at some point of time in future the species shall be eliminated. Moreover, in order to strengthen chances of survival, the populations should be large enough to encompass as much genetic diversity as possible. A population consisting of individuals with a homogenous genetic makeup is usually more susceptible to chance variations in environmental conditions and attacks by pathogens as compared with populations having heterogeneous genetic makeup. Thus, in order to contain, a greater number of species with populations, large enough to have a secure future, the habitat should be large enough to feed all individuals of all viable-populations of all species and to meet the habitat size requirements of the species to be preserved.

(4) Management of protected areas

The basic objectives of management of protected areas, be it a national park or a sanctuary or a nature reserve or a biosphere reserve, is to ensure that:

1. The viable populations of species present in the system co-exist, grow and multiply without any interference from outside factors.
2. Free play of chance variations in the environment, like floods, drought, hailstorms, various pathogens, fires etc. are permitted to enable the organisms within the system to adjust and adapt to the chance variations in the conditions of the environment. These changes should not be so extensive as to wipe out the entire population. In brief the process of natural selection and evolution should be allowed to continue uninterruptedly.

The very objectives of the management of protected areas suggest that the managerial interference in the affairs of wildlife should be as little as possible. However, excluding detrimental human influence is a big task in itself and so is the elimination of detrimental biotic influence both of which call for an intensive monitoring of protected areas and an expert human attention (12;13). The management practices generally adopted for a protected habitat may, therefore, be grouped into following:

1. Excluding detrimental human interference.
2. Excluding the exotic species or the detrimental biotic interference.
3. Using limited corrective measures.
4. Intensive monitoring of protected areas.

1. Excluding detrimental human interference

Probably there is no place left on this planet uninfluenced by human activity. Protected areas are often surrounded by human establishments and many of them have these establishments within their boundaries. Man uses the forest around for collection of fire wood, timber, fruits and nuts, hunting game animals etc. These regular encroachments by man are injurious to the wild life, which has to move to other parts of the protected areas where crowding and other undesirable changes may occur. A gradual thinning of the margins of protected areas takes place and in cases where human establishments are situated within the confines of the protected area this thinning occurs in gradually enlarging circles around the human establishments. Indigenous tribes living in the locality have been using the resources of forests since times immemorial. Establishment of protected area snatches away their traditional resource base, which is strongly resented. It becomes very difficult to exclude the detrimental human encroachments in the face of hostile people living around and within the confines of the protected area. Many inhabitants of these establishments trade in wild life products which is an important subsidiary source of their livelihood. It is often very difficult to deal with such people and at times war like encounters are required to check the merciless killing of the prized animals.

Complete elimination of human influence from natural ecosystems is often questioned. Man is just another species in the biosphere and organisms have been evolving in his presence ever since he appeared on earth's surface. Wild life has to co-exist with man in future as well. Though man's influence has been more on the detrimental side, a number of organisms have adopted themselves to his presence and have thus been benefited. Some scientists believe that complete elimination of man's influences may not be as wise as we assume. The problem of checking pollution of air and water is more severe today. Human establishments, industries, waste disposal sites situated near the protected area pollute the air with harmful gaseous effluents and streams flowing through the area may carry polluted waters. Sulphur dioxide emitted by smelters may cause acid rains over large areas. Too much of shoot, ashes and smoke emitted by power plants may cause problems of toxicity when brought down to the area by precipitation. All this is injurious to wild life in general (12;13).

2. Excluding detrimental biotic interference or Controlling exotic species

With rapid strides in communication and transportation technology, extensive migrations have occurred during the past few centuries. Wherever man migrated, he carried with him, his pets and plants and introduced them to new localities. Original flora and fauna of the locality had to bear the onslaught of invading species and compete with them. A number of forms, which failed to do so, could not survive. Those, which survived, had to adapt to life in presence of alien organisms. This brought about changes in the entire biotic spectrum and could also affect the course of future evolution of the life forms of the locality.

The case of Hawaiian Parks, Hawaii islands, in the middle of Pacific Ocean, illustrates the true dimension of the problem of exotic species. Due to its position in the middle of Pacific Ocean and distance from any of the mainland around, Hawaiian Islands had a unique flora and fauna, which had evolved in virtual isolation from the biotic communities at other places. When Polynesians arrived on these islands about 1,760 species of plants existed out of which 95–99% were endemic and unique (14). Polynesian colonists introduced about 30 species (15). The advent of distant people from diverse cultural backgrounds, which followed from 18th century onwards, introduced their own distinctive food, medicinal and ornamental plants etc. This has caused the addition of about 4,600 species of vascular plants alone. Of these, nearly 700 are reproducing successfully and maintaining themselves as conspicuous populations. This has caused the extinction of nearly 200 endemic species (recorded cases) and another 1,000 or so are endangered and completely changed the indigenous flora and fauna of Hawaii group of islands(16;17;18).

Therefore, if we have to protect original habitats along with their indigenous biotic spectrum in a protected area, we shall have to identify the problem species or the exotic species and concentrate our efforts to exclude them from the protected habitats.

3. Using limited corrective measures

Chance variations in the conditions of environment, such as drought, floods, fires, pathogens, epidemics etc. often affect the wild life in a protected area adversely. These trying situations or rough times are useful for the development of populations, which become better adapted to their environments and thus have greater chances of survival in future. However, the adverse changes may at times be so severe as to decimate the entire population – leaving us nothing to conserve. Under such circumstances, it becomes necessary to apply limited corrective steps to ensure that viable populations of as many species as possible survive through these rough and trying situations. These steps vary from area to area and depend largely on the nature of adversities faced by the biotic community. They may include provision of space, food and water, treatment of diseased animals, vaccination, elimination or removal of exotic species, elimination of diseased individuals if they are likely to infect others in the population, displacement of threatened populations to safer places, etc.

No doubt, the corrective steps stated above cannot be resorted to in the case of every individual or every population within the biotic community because of inadequacy of our means. Only a few selected species or group of species can be helped in this way. These methods tend to interfere in the vital process of natural selection, which is undesirable. Therefore, such methods should be applied only under conditions of dire necessity – when the entire population is likely to vanish, it is better to save at least some of it from which new populations may be generated when the conditions normalize.

4. Intensive monitoring of the protected areas

Intensive monitoring of the protected area is essential in order to keep a close watch over the biotic spectrum, which we want to protect. This monitoring usually involves:

1. **Observations on air, water and soil quality:** Food water and air are the basic requirements for all life activity. With human establishments all around, it becomes necessary to keep a close watch on air, water and soil quality of the protected habitat. Adverse changes in these components of the environment may induce changes in the biotic spectrum of the protected area. Susceptible species shall be suppressed, hardy and resistant forms shall come up and multiply. This may cause both biotic and abiotic environment within the system to change making life difficult for the organisms we wish to protect. Various physical and chemical

parameters of air, water and soil quality should be regularly monitored (19).

2. **Observations on changes in population densities, frequencies, migrations and health of the biotic community within the system:** The physico-chemical parameters of environmental quality provide information about changes in the abiotic component, while observations on changes in the biotic community reveal the effect of these changes. Physicochemical methods are of little use when the system is subjected to some type of biotic stress, such as rise in poaching activity or entry of some exotic organism. Many changes in environmental quality are so subtle that physical and chemical methods available today are unable to detect them. However, these changes are reflected in the biotic components of the system. Thus, a close observation on the biotic component (Biological monitoring) of the system may reveal changes, which are not detected by various physical and chemical methods. Biological monitoring also provides an early warning system so that preventive or precautionary measures may be taken before the changes appear in its full intensity (19).
3. **Observation on behaviour, growth pattern, growth requirements and life cycles of organisms occurring in a protected area:** We know very little about wild life, which occurs on this planet. Detailed studies on growth pattern, growth requirements and life cycles have been carried out only in the case of a few selected forms. The vast and heterogeneous collection of millions of species, which are important to us in one-way or other, has never been subjected to intensive scientific study. Collecting naming and listing the species in the flora or fauna of a locality is one thing but to conserve them and help them to survive under the stresses induced by habitat destruction or pollution is another. How can we ensure the species' survival if we do not know what it requires and under what conditions it grows and multiplies? The detailed observations, which we have on the bamboo eating Pandas or different species of whales, the lions, tigers and different cats, crocodiles, vultures, the hippos and rhinos etc., are relatively recent ones. It is these studies, which have helped us to conserve them successfully. Thus, research and scientific study of various component species of an ecosystem be it an insect or an algae or a higher plant is a must. This knowledge shall be extremely helpful in our future conservation efforts (20).

(IV) SHORTCOMINGS IN THE EXISTING SYSTEM OF PROTECTED AREAS

Though an impressive proportion of surface of our globe has been earmarked as protected areas, there exist great anomalies in the system. Many of these parks and sanctuaries simply exist on paper and in majority of cases where some efforts are undertaken to manage and protect the wild life they are hardly adequate. Major drawbacks in the existing system can be summarized as follows:

(1) In a number of countries though a large proportion of land has been set aside as protected area for wild life, little wild life actually occurs in these places. For example, about one fifth of the total land area of Bhutan has been classified as protected area. Most of Bhutan's protected areas are high up in the Himalayas where little wild life occurs. The protected habitats of Chile are largely concentrated high up in Andes – a mountain range – and much of the country's unique vegetation types are not protected at all. Only less than 2 kms of Belize's 220 kms long barrier reef – the longest in the western hemisphere and known to possess exceedingly rich variety of marine forms is covered as a protected habitat. Though there are about 7,000 protected areas spread over about 5% of earth's surface much of the biologically significant localities, which are rich in rare and endemic species and high biodiversity, are not covered in the existing system. All over the world high altitude areas have received a greater share of protection efforts whereas areas of greater biotic significance, low land forests, wetlands and aquatic system with enormous diversity of forms remain unprotected (21).

(2) Moreover, the size of the protected area set aside for many habitat types is awfully insufficient to conserve the entire biodiversity of the locality in most of the cases. Of about 178 biogeographic regions represented in the current system of conservation areas only thirty or so have area nearing 1,000 sq. kms. Thus even if we have marked nearly 5% of worlds total land surface for wild life, the current system of conservation in protected areas may not forestall the extinction of many species which need large habitats and are susceptible to chance variations in climate and environmental factors (21).

(3) Many of the existing protected areas are simply on paper as they lack adequate staff or budget for their maintenance. They are simply carved out as protected areas on maps and many even lack a proper boundary so that it is difficult to demarcate them in field. A survey in 1988 found, that only sixteen out of a hundred Caribbean marine parks outside USA has some management plans, staff and funds to carry out the conservation work. Even in wealthy nations, like Japan, similar problems exist – Shiratoka National Park, Japan, a biologically rich protected area has never had more than a single wildlife ranger for more than 37,000 hectares (22).

(4) A number of parks and sanctuaries encourage profitable activities, which are injurious to the wild life. Logging operations are carried out with Government permission in many parks of Canada, Czechoslovakia, and Indonesia etc. In many developing countries, the inefficient and corrupt administration allows logging and fuel wood collection surreptitiously, causing extensive damage to the wild life. Many of the wild life reserves in Europe and Canada are oriented primarily for tourism and recreation. Conservation of biodiversity has assumed secondary position or is often absent altogether in these areas (21).

(5) Human settlements present around or within the protected areas are often detrimental to the cause of wild life conservation. It is very difficult to find such large areas of wilderness devoid of human settlements or interference as required for proper and adequate wild life protection efforts. Natives and indigenous people who largely depend on wild plants and animals for their subsistence usually frequent even those parts of wilderness, where modern societies, agriculture, rails and roads have not reached. Wild life conservation efforts tend to rob these people of the vital resources on which they have been living since times immemorial. This is resented. The local people do not co-operate and at times develop hostile attitude, which defeats the very purpose of the conservation efforts.

(6) Although local people and natives are fully capable of abusing land and hunting wild life to extermination, it is interesting to note that many of the world's healthiest and richest ecosystems of today occur in regions under their control. Farmers of South East Asia traditionally honour sacred groves – patches of wilderness amidst agricultural fields and rural landscape – as abodes of powerful deities. Indigenous Indians of Panama leave patches of forests as super natural parks for the refuse of the wild life and spirits. Tukano Indians of Brazil guard forests and waterways and their traditions and taboos have protected as much as 60% of the streams of the locality as sanctuaries for fishes and other aquatic life. Taboos and traditions have preserved the only orange-tang population now left in upper reaches of Batrang-Ai River in Southern Sarawak (23). Indigenous people, their costumes and traditions, are rarely involved in wild life conservation efforts. These customs and traditions are disappearing at a fast rate. There is no harm in adapting modern scientific ways and methods but in the process healthy customs and traditions which represent the accumulated wisdom of past generations are also being forgotten.

(V) CONSERVATION BEYOND PARKS, SANCTUARIES AND RESERVES

Whatever we do, it is very difficult or rather impossible to stop expansion of human enterprise, reckless over exploitation of natural resources and pollution of the environment within the time frame available to us. This becomes clear when we scrutinized the progress made in this direction during the past fifty years. This will have serious negative impact on our society. The number of species currently available for our use shall be reduced. Essential ecological services, such as regulation of water quality and quantity, recycling of nutrients, biological control of insect, pests, pathogens and the buffering action of forests on climatic extremes could be impaired or terminated. Mankind should face this issue and make concerted efforts to minimize the projected loss of biodiversity. No doubt we shall lose a significant part of our biological wealth but what is important today is the part which we are able to save. The future of mankind shall depend on what is left after the current episode of extermination passes away (24;25).

Saving a few species in our zoos and botanical gardens, freezing a fragment of the germplam in liquid nitrogen or maintaining a part of once extensive natural habitats in protected areas or refuge for wild life are no solutions to the problem. Even if we do it with sincerity putting in all the resources which we can spare, species shall continue to disappear. The principal cause of the loss of biodiversity lies in the rapid destruction of natural habitats of species, which is in turn due to:

1. Rising population pressure and requirement of land for cultivation.
2. Unjust and inequitable distribution of wealth which forces people to over exploit the natural resources.
3. Industries, logging and human settlements etc which either encroach upon or introduce harmful wastes to the natural systems damaging an ever increasing part of wild life habitats.
4. Commercial logging operations, extraction of fuel wood and other products for which an increasingly larger area of natural ecosystems is searched rummaged and damaged.

The main burden of facing this challenge happens to fall on the poor and developing countries of the world. It is they, who possess the largest share of biological wealth of our planet. Possibly 50-70% of all species have their homes in nearly 6-7% earth's surface possessed by the developing countries of the world. These countries are, however, burdened with large populations to feed, want poverty, malnutrition and diseases. Natural forests, timber, fuel wood extraction and various wild life products, shifting cultivation etc. constitute an important part of their livelihood. To a person who labours all the day simply to earn enough for a squire meal for himself and his family, wild life conservation makes little sense. It is abstract poverty which forces people in the developing countries to over exploit the wild life and natural habitats. If we have to conserve natural habitats we shall have to:

a. Intensify agricultural productivity and provide the people of the area with alternatives so that further encroachments on the natural habitats are checked.
b. Intensify forest management and develop compensatory plantations which can provide an alternative to continued exploitation of natural forests.
c. Peruse a vigorous forest conservation policy so that substantial areas of remaining natural habitats are protected from all types of exploitation.
d. Developmental projects affecting wild land or natural habitats of wild life adversely should be discouraged. In cases where they are essential, they should be accompanied by compensatory management practices.

(1) Intensifying agricultural productivity

Introduction of land reforms, just and equitable distribution, and scientifically managed high productivity sustainable agriculture could go a long way to remove the pressure of over-exploitation from forested wild land. If the people living in the locality are provided with viable alternatives for their needs, natural systems shall no longer be subjected to frequent encroachments. The Malaysian experience of Jengka Triangle has demonstrated that well managed tree crops could provide an attractive alternative for settled farmers and help them to reduce dependence on shifting agriculture. Such tree plantations also protect soil and water resources. A well chalked out plan for the region was started in

1960 which demarcated available land for these plantations and set aside nearly 60% of the area as forest reserve. After 20 years, the village area and the forest land has remained relatively stable and the wild habitat enjoys freedom from persistent pilferage and encroachments (26).

A high priority area for future research is the trial of new technology for sustainable agriculture on soils which are poor in nutrients and on which much of our biologically significant habitats occur. The system of growing crops interspersed with leguminous plants which has promise of reducing dependence on fertilizers or the zero tillage technology which is helpful in retention of organic matter in the soil are some of the approaches which may be tried in future. It is only a very small number of plant species which man has been using as food crops. There are a number of plants which can also serve as source of food for mankind. However, their potential as food plants has not yet been properly explored. A little diversification of food habits so that these plants may be used as a food source shall be helpful to widen the range of food crops available for indigenous consumption. The winged bean *Psophocarpus tetragonolobus*, for example, has been known for centuries to the indigenous tribes of New Guinea, which use it as food. It has now been shown that the plant is nutritionally as valuable as Soybean with 40% proteins and 17% edible oils.

(2) Intensifying forest productivity and reforestation

Timber and fuel-wood are the two common commodities for which forests are usually ravaged and depleted. A better alternative for meeting the timber and fuel wood demand could be fast growing species of woody plants which can be grown on degraded land and wood used to produce ply-wood and straw boards which can be substituted for commercial timber. Plants like *Gmelina arborea, Albizzia falcataria, Leucaenia species, Eucalyptus species, Pinus carribes* etc. grow at a rate 4–5 times faster than typical timber species most of which take 60–85 years to mature. This will reduce the demand on slow growing timber species which occur in natural habitats. It has been estimated that in the Asian region all the needs for industrial wood by 2000 A.D. can be met by growing fast growing trees over 25 million hectares which is less that 10% of the remaining forest area (27). Thus, support for agroforestry programme outside forests could provide the people of village communities with fire wood for structural works and other forest products – thereby reducing the pressure of demand on natural forests.

(3) Persuing a vigorous forest policy

After the World War II, the concern about environment, natural resources and biodiversity took almost two or three decades to emerge, from the minds and writings of scientists, as a general public concern. In a democratic society once the public is motivated it does not take much time to translate the general concern into effective action. Many governments all over the world have now enacted laws to stop reckless over exploitation of natural resources and degeneration of biodiversity. These measures are not yet adequate, as the situation appears to demand. Moreover, in many localities though the conservation laws have been brought into force they are enforced only half-heartedly. About 1.8 million hectares of the forests of all tropical Asia have already been set aside as nationally protected parks and nature reserves. However, this area, which amounts to about 6% of the Tropical Asia's forests, is not sufficient to ensure the preservation of the unique germplasm and wild life resources, which occur in the region. At the same time, another 50 million hectares of deforested land lies useless in tropical Asia in urgent need of rehabilitation. This land if intensively utilized can provide plenty of resources to the people of the area. This will not only relieve the pressure from natural forests but the area allocated to the wild life can also be expanded (27).

Raising political awareness and mobilizing support for conservation of natural habitats (forests) and wild life shall achieve little unless and until the fundamental policy needed to ensure the conservation is well defined. People may be prepared to act but they should know what to do. The more important needs for immediate future are:

1. A strong political commitment to land reforms, and redistribution policy, particularly in areas adjacent to the threatened habitats. This will encourage shift to more sustainable agriculture and relieve pressure of demand from natural ecosystems.
2. A strong commitment by agricultural sector to the diversion of resources to the development of intensive agriculture in buffer zones around threatened forest land and rehabilitation of degraded water shed.
3. A strong commitment for curbing logging operations, fuel wood collection and providing alternatives by planting fast growing species on degraded landscape. This can be done by imposing higher taxes on timber export which will in turn encourage the use of alternative materials.

(4) Scrutinizing developmental projects effecting natural habitats and wild life

Developmental projects, which encroach upon natural habitats and wild life, should be carefully screened to assess the degree of adverse affects they cause. The following types of economic development projects may be included in the category of projects, which require a close scrutiny and environmental impact assessment:

1. Agricultural and animal husbandry projects.
2. Projects for transportation, such as building of roads railway tracks etc.
3. Hydroelectric projects.
4. Industrial and mining projects.

These projects affect natural habitats and wild life directly or indirectly. Human needs are going to get priority over the concern for habitats and wild life. Expansion of agriculture and animal husbandry shall occur in future only if the food supplies from existing sources fall short of the human requirements. If we intensify our agriculture and animal husbandry, modify our food habits, distribute and share the surplus judiciously in such a way that the produce from the existing sources is enough for our needs there will be little necessity of agricultural expansion. Just and equitable distribution of resources is equally important in this connection. To-day the world produces enough to meet the requirement of all people living on earth's surface. However, about 39% of the global grain production is fed to live-stock while 12 out of every 100 people in the world go hungry.

Transportation projects, which involve the construction of highways, roads, railways or canals, pierce through natural habitats slicing them into pieces. A busy highway or railway track has somewhat the same effect on a natural system as caused by habitat fragmentation. This results in reduction in the capacity of the natural habitat to sustain the viable populations of some species which are either driven away or eliminated altogether. There is an over-all reduction of biodiversity. The roads, railways or canals also make the area accessible and there is an influx of human establishments, which further damages the natural habitat. Thus as far as possible transportation projects should by-pass the biologically significant habitats. Where they are absolutely necessary care should be taken to provide compensatory management components.

Hydroelectric projects involve development of large reservoirs and water diversion schemes, which submerge large area of land or drastically transform aquatic as well as terrestrial habitats. The programme of water shed protection and construction of power lines etc. may disturb natural localities. These projects are usually developed in rock-filled areas where mountainous terrain provides suitable damming site for storage of water. The rock strata beneath the reservoirs have to bear the pressure exerted by the standing mass of water table. The stored water percolates down raising the water table of the area, which also adds significantly to the transformation of the locality. Silting and sedimentation shortens the life span of the hydroelectric project, which has to be protected by a proper watershed management programme. The large hydroelectric projects, which are usually multipurpose projects, bring about tremendous changes in a natural habitat. These require a very careful risk-benefit or cost-benefit analysis before they are implemented.

Industries and mining projects involve chemical and thermal pollution of surrounding localities and are never friendly to the environment. They should be located as far as possible away from biologically significant areas and proper measures should be taken to treat and minimize the harms caused by the effluents discharged from the industry or mining project.

(VI) IN-SITU CONSERVATION: RESTORATION OF DEGRADED HABITATS

Today the foundation and health of agriculture depends largely on the rich genetic diversity. The centers of genetic diversity, ironically, have shifted from natural systems and traditional agriculture to germplasm collections or the gene banks – with all the liability they carry. The existing preserves of wilderness are mostly inadequate in size and most of them are occupied fully, i.e., saturated with whatever wildlife that is there. Numerous species are at the end of their tether, likely to be lost in near future because their habitats are drastically reduced or destroyed. For such species, the only hope of survival lies in recreation of new homes by human beings. Moreover, the ex-situ conservation efforts shall be meaningless in the long run if we fail to provide suitable habitat for the species thus saved.

There are two basic approaches, which are currently being tried for the restoration of degraded habitats (28). These are:

1. Recreation of authentic replicas of lost ecosystems.
2. Recreation of useful production oriented ecosystems.

1. Recreation of authentic replicas of lost ecosystems

This type of restoration activity is more of an academic venture to recreate true copies of ecosystems, which once existed at the place. There is emphasis on selection of the correct mix of species, exotics are shunted out and environmental conditions are so recreated that the diversity, which occurred in the system earlier, may be reinstated. The enormous labour and skill was involved to bring back organisms, which had disappeared from the locality, to drive away the exotics, which had come to stay in the area, to recreate conditions of soil, water and other environmental factors as they existed earlier. However, is it really worth the effort and money put in these projects? It is like returning to the good old times. However, we live in a rapidly changing world. By forcing the system to continue in a state in which it was many years ago, we could be denying the wild life chances of responding to fluctuations in the conditions of environment. Such systems, therefore, tend to become extensively managed systems, with all attendant shortcomings. They do not cater to the immediate needs of humanity, which it probably did in good old times.

2. Recreation of useful production oriented ecosystems

In this category of restoration, the objective behind conservation efforts is to enable the ecosystem to do things that cannot be done by degraded systems. Forests can built up soils, control the capture and release of moisture, regulate nutrient cycles, withstand climatic changes and disturbances and above all can support a large spectrum of life forms. The organisms constituting the restored systems may not be the same, which occurred there earlier – before the degradation of the system. Often such species are introduced in the system, which have never occurred in locality but are economically important. The environmental crisis on our planet is to deepen. The task of restoration of degraded habitats, it is likely, may have to be undertaken on a much larger scale. This means that compassion or charity for the disappearing wild life cannot be the only reason for restoration, as we simply cannot afford it. For a meaningful change, the restoration of degraded habitats shall have to be linked with economic activity, which provides something to the mankind.

Creation of production-oriented ecosystems is a development-linked conservation. It is a type of agriculture blended with ecology. It does not have the destructiveness of monoculture which modern agriculture uses to enhance the productivity. Elements of diversity are maintained with specific economic objectives in the background and both have to go together. The wild life has to be content with whatever environment these restored ecosystems provide. Integrating nature's restoration requirement with economic needs of people may be the only way on which the terrestrial fabric of our planet can be rebuilt.

QUESTIONS

1. What is in-situ conservation? Discuss its advantages and disadvantages.
2. The strategy of in-situ conservation revolves round protected areas. Discuss. Also enumerate the short-coming of the existing systems of Protected Areas.
3. What are the requirements for an ideal protected area? How shall we select the site for establishing a protected area?
4. An ideal protected area does not need much managerial attention the nature takes care of its own self. Discuss.
5. Discuss the common methods adopted to manage a protected area.
6. Why conservation in parks, sanctuaries and nature reserves is considered inadequate and the only sensible way is to conserve the entire ecosystem or habitats. Discuss.
7. Discuss the role of developing countries in conservation of biodiversity.
8. What major strategies should be adopted for habitat conservation in developing countries?
9. Should we recreate authentic replica of degraded ecosystems or re-create useful production oriented ecosystems on degraded land?. Discuss.

PART – IV

Environmental Pollution

14

Chapter

The Pollution of Environment

Human activity generates a tremendous amount of waste materials. These are discharged in various components of the environment in which they bring about undesirable changes. The phenomenon is termed as Environmental pollution which has been defined as an undesirable change in physical, chemical or biological characteristics of air, water or land that will be or may be harmful to human and other life, industrial process, living conditions and cultural assets or cause wastage of our raw material resources.

(I) MAJOR CATEGORIES OF ENVIRONMENTAL POLLUTION

Entry of pollutants causes disturbances in an ecosystem which manifest themselves into a chain of adverse reactions often very complex in nature. The degradation of the environment thus brought about can be grouped into various categories based either on the nature and type of environment being affected or the nature and the type of pollutant causing the problem. Based on the nature and type of environment affected the pollution may be of the following three types:

1. Air pollution
2. Water pollution
3. Soil pollution.

Both air and water are components of the environment, which are dynamic entities. They are an effective means of transfer and transport of the waste gases and other materials. Soils on the other hand are stationary entities, which are indirectly affected by the pollution of water and air. Toxic wastes dumped on a soil may not cause any harm to the plants growing nearby if there is no means (such as water or air) of transfer of the toxicant to the plants body or the roots. Soils are affected only when water leaching down the waste dump enters the soil or gases and particles afloat in air are adsorbed on surface of soil particles. The ecological effects of pollution on soils or soil pollution unfold only when the water or the air within the soil environment becomes polluted. The problems caused by pollution of water and soil are generally similar in nature whereas the air pollution presents a radically different set of problems. Therefore, two broad categories of pollution have been distinguished here which are:

(a) The pollution of earth's surface: Water and Soil

This category includes pollution caused by pollutants, which are discharged on earth's surface – both on land and in water. As the biosphere forms a thin crust over land surface as well as in water, these pollutants interact with the living constituents of the environment. While the constituents of the biosphere are adversely affected, the pollutants are also biologically degraded into simple, harmless compounds and are finally eliminated. The following four types of pollutants may be distinguished:

1. **Simple Biodegradable Pollutants:** These include simple wastes, which are easily and quickly degraded into harmless constituents. Domestic wastes, organic matter of plant and animal origin, faecal matter, blood, urine etc. are placed in this group.
2. **Complex Biodegradable Pollutants**: These include wastes, which are resistant to degradation by biological agencies. They are degraded very slowly due to which they persist in the environment for long durations. It is due to their persistent nature that they cause problems different from those caused by simple biodegradable wastes. Synthetic chemicals, persistent pesticides, plastics, various polymers, polymeric resins, plasticizers, petroleum crude etc are placed in this group.
3. **Non-Degradable Pollutants:** These include wastes, which are not degraded by biological agencies. They may undergo a change of form or chemical combination but are indestructible in nature. Heavy metals and toxic trace elements are placed in this group.
4. **Pollution Caused by Physical Agents:** These include the pollution caused by harmful and irritating physical agents such as heat, noise etc. Thermal, Noise and Radioactive pollutants are placed in this group.

(B) The pollution of earth's atmosphere

This category includes gases, vapours and fine particulate material, which escape into the atmosphere. They pollute the air around us and come in contact with biosphere only when they are unable to go high up in the atmosphere. At higher altitudes, where the biosphere is absent they cause problems of a radically different nature from those caused by pollutants of earth's surface. Most of these pollutants are finally brought down to earth's surface along with precipitation to be converted to harmless form on land surface or in aquatic systems.

(II) ENTRY OF POLLUTANTS IN THE ENVIRONMENT

Pollutants when released in the environment gain entry into the soil by adsorption, into the water by dissolution and into the air by evaporation. To cause environmental degradation and evince its undesirable effects a pollutant has to enter the system. If they are unable to do so they stay as a separate entity in the medium concerned, such as a heap of garbage lying in waste dumps or the particulate matter, which floats up, stays suspended or settles down in an aquatic medium. As such, they form inert packages of pollutants in the environment.

(III) TRANSFER, TRANSPORT AND DILUTION OF POLLUTANTS

Pollutants released into the environment may remain at the point of their discharge or be transported to other places. The distance up to which they are transported depends mainly on the mobility of the agent carrying them. Water and air currents are effective media for transfer and transport of waste materials. Soil depends largely on moving water and winds for dispersal and dissemination of pollutants.

The mobility of a pollutant, however, is not confined to their dispersion within a single medium. A pollutant released into water may get distributed in soil and air as well. Pollutants released into the atmosphere may contaminate water as well as the soil. Waste discarded on soil may contaminate water, which seeps through it or the air through vaporization. The transfer of pollutants from one medium to another may be of two main types:

1. Bulk transfer: Bulk transfer involves massive movement of pollutant from one compartment to another irrespective of concentration differences. For example, the entire quantity of Sulphur dioxide in the atmosphere dissolves in rain and is transported in bulk to the soil or water underneath. In cases where the pollutant is so much that the carrying capacity of the medium is saturated, only the quantity which dissolves in the medium is transported.

2. Interfacial transfer: Interfacial transfer involves transfer of pollutants from one phase to another caused by concentration differences on either side of the interface. Such transfers are greatly influenced by environmental conditions such as pressure temperature etc. Three such transfers occur in nature. A substance may be transferred from air to soil and water, from water to air and soil and from soil to air and water. These transfers involve the process of adsorption, dissolution and vaporization. The nature and properties of the pollutants, their concentration, solubility, vapour pressure etc. play an important part in interfacial transfers.

(IV) ENTRY OF POLLUTANTS INTO BIOSPHERE

Pollutants can enter a biological system from any of the three media, such as water, soil or air, which carry them. In lower plants and animals, plasma membrane is the main barrier between the outside environment and the biological system inside the cell. In higher plants and animals, which possess much better protective devices such as coating wax, cuticles, multilayered epidermis etc. pollutants have to take other routes. These routes may be through the roots, root hairs, rhizoids, stomata etc. in plants and through nasal mucosa and mucosal lining of mouth, respiratory track and the gastro-intestinal track in animals. In many cases, even the epidermis is pervious to certain types of pollutants.

Nature has provided higher animals including man with an efficient device, which helps to guard themselves against pollutants, which regularly enter through intestinal track. Materials absorbed from intestines are routed through *hepatic portal system* to the liver wherein harmful constituents are reacted upon by liver's enzymatic machinery. A number of harmful material absorbed from the intestinal track are degraded or converted to harmless state before reaching the blood circulation. However, this is not so in the case of pollutants entering the system through the respiratory track. The air in alveoli of lungs is in contact with capillaries across a thin membrane. The material absorbed from gases inside the lung enters straight into the blood circulation. This becomes more significant in the light of the fact that we breathe in by weight about 16 times more air than the total foodstuff and water consumed orally.

Plasma membrane, through which all substances entering a cell must pass, is a lipoprotein membrane being made up of a thin layer of lipid or lipid-like substance covered on both sides by a layer of proteins. The basic structure of the plasma membrane is strikingly uniform throughout the living world. The rate of entry of pollutants into a biological system is influenced by the size of molecules, its concentration in the medium, the nature and solubility of the substance in water as well as in lipids and fats. Substances soluble in lipoid material (lipophilic ones) are carried across the plasma membrane simply by dissolution and diffusion through the lipid layer. Lipoid insoluble but water-soluble pollutants (hydrophilic ones) may require an active transport mechanism. Smaller molecules such as those of water simply pass through the pores in the plasma membrane.

(V) BIO-ACCUMULATION AND BIO-MAGNIFICATION

Inside a biological system, non-essential and harmful substances may either be stored or are reacted upon by the enzyme system and excreted. The phenomenon of accumulation of pollutants inside cells and tissues of living organisms at concentrations above those of immediate environment is known as Bio-accumulation. The concentration of a foreign substance within a biological system is dependent upon the rate of uptake, the duration of exposure and the rate at which it is being eliminated or reacted upon by the system. The extent upto, which a chemical shall be absorbed and bio-accumulated by an organism, depends on its solubility in fats or the lipophilicity. A higher degree of correlation has been found between octanol water partition co-efficient (Log P^{ow}) and the extent of bio-accumulation (1;2). Pollutants soluble in lipoid material or capable of forming complexes with macromolecules within the cell may be stored for long duration of time. By virtue of their complex formation and dissolution in lipids and fats which are storage products of cellular metabolism, these substances stay away from enzymatic reactions which act on them (3). Residue of fat-soluble insecticides may be stored in fatty tissues of living beings for years together (4;5).

The term **Bio-magnification** denotes enhancement in the concentration of various substances along the food chain. Harmful and toxic substances, which enter the trophic structure at the level of primary producers, get concentrated many times at each trophic level as they move up the food chain. It is only a small portion of organic material received as food from the lower trophic level, which is retained and added to the body mass of an individual. A large fraction is respired and consumed. If the organic material (food) contains a small amount of toxic constituent, which is neither excreted nor metabolized, the whole amount present in the entire mass of the organic material taken up from the lower trophic level shall persist in the organism consuming it as it cannot get out. As much of the material taken in is oxidized and eliminated the toxic constituent shall occur in the fraction of material added to the body weight of the organism and hence its concentration.

An indication of the occurrence of the phenomenon first came from Illinois (USA) where elm trees were sprayed with DDT. A large number of American robins (*Turdus migratorius*) were found killed near these trees for no apparent reason. It was later discovered that these birds were killed due to DDT poisoning. The lethal dose came from earthworms, which they consumed. Earth-worms had concentrated DDT residue by feeding on fallen elm leaves. In a study, Barker (1958) found that soil under the elm trees contained 10 ppm while earthworms contained 86 ppm of DDT and 33 ppm of DDE a derivative of DDT. Robins died after consuming only 11 or 12 such earthworms. An analysis of dead robin's body revealed 250 ppm DDT in brain and 744 ppm DDT in the liver (6).

In Lake Clear, California, DDT was sprayed to kill gnats (*Cheoborus astictopus*) on three occasions in 1949, 1954 and 1957. Following the last application, grebes a much desired water bird (*Aechmophorus occidentalis*) started dying. It was found that the concentration of pesticides in the planktonic algae in the lake water was 0.02 ppm, in crustacea 5 ppm, in fishes 100 ppm and in the gonads of dead grebes 1600 ppm. The pesticide was bio-magnified 250 times in the primary consumer (Crustacea), 5000 times in the secondary consumer (Fishes) and about 80,000 times in the tertiary consumers, *e.g.*, grebes (7;8). In waters of East Coast estuary, Woodwell et al (1967) have reported 0.0005 ppm of DDT, while zooplanktons in the same waters contained 0.04 ppm, needle fish 2.07 ppm and gulls 75.5 ppm. DDT is thus biomagnified several thousand times in fish eating birds where its toxicity manifests itself (9). It is mainly because of the adverse effects of DDT on bird life and its global distribution that DDT pollution has received worldwide attention (10;11;12).

(VI) BIO-DEGRADATION OF POLLUTANTS

The process of bio-degradation can be defined as decomposition or destruction of pollutant molecules by the action of enzymatic machinery of a biological system. There is hardly any complex organic molecule present in the environment or any chemical which man's ingenuity and synthetic skill may synthesize which cannot be decomposed by this excellent and versatile machinery of enzymatic degradation.

(1) Agents of Bio-Degradation: Though bio-degradation can be brought about by most of the living organisms, tiny microbes such as bacteria, algae, fungi, small animals etc. play a very important role in decomposition of various pollutants in an ecosystem. This is due to their catabolic versatility, the diversity in species composition and their ability to perform reactions at a much faster rate per unit weight as compared to other organisms. These microbes perform specific biochemical reactions which degrade macro-molecules into small units in step-wise fashion. Usually a series of organisms are required to bring about the complete degradation of a large molecule or a group of molecules.

(2) Conditions Necessary for Bio-degradation: Though there may be some microbes which perform a number of metabolic reactions, in most of the cases only specific type of microbes are required to perform specific reactions. In many cases a number of species have to act one after the other – each performing its own specific function – to bring about the complete degradation. The presence of microbes which specifically perform the task of decomposing specific structures in the pollutant molecule is, therefore, one of the basic necessities of bio-degradation. Another necessity of the biotic decomposition reaction is that contact has to be established between the microbes and the pollutant molecules which should be in the form suitable for the reaction. It is only as a sequel to the establishment of this contact that the enzymatic machinery required for the purpose is usually brought into existence or in other words induced into being. Pollutants present in such a state which is unable to induce this machinery escape the process of bio-degradation. Availability of favourable conditions of environment such as water, suitable pH, temperature etc in which the microbes may proliferate and the enzyme systems may operate, further speeds up the process of bio-degradation.

(3) The Nature of Bio-degradation Reaction: Bio-degradation reactions break large molecules into smaller units so that they are rendered harmless and are eliminated from the system. These reactions include oxidation, reduction, hydrolysis and different types of structural re-arrangements. Simple carbohydrates, proteins, fats, nucleic acids etc. are quickly degraded. Molecules having highly branched chains, an increased degree of substitution, simple or fused aromatic rings, cyclo-paraffins etc. are decomposed with considerable difficulty. Degradation of these molecules involves step-wise removal of substituted groups or atoms, destruction of ring structures and progressive alteration of constituent units etc. This explains the persistence of poisonous pesticides, petroleum crude and other products, plastics and plasticizers etc.

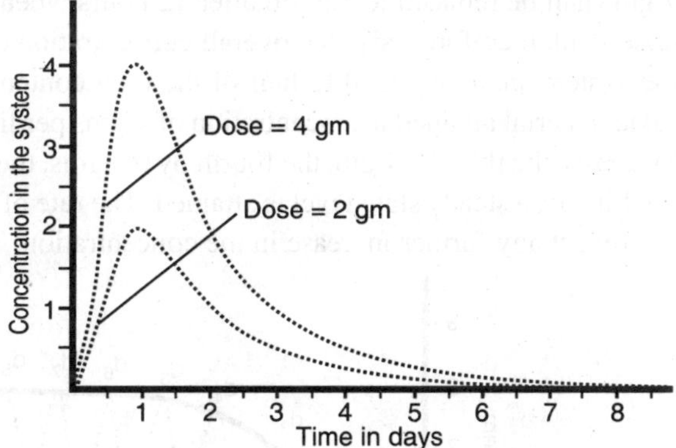

Fig. 14.1 *Degradation of pollutants in a biological system*

In general most of the lipophilic compounds which pose a greater hazard to a biological system as they are capable of diffusing through biological membranes, are converted to water soluble or hydrophilic ones. This results in restriction of their movement within a biological system and in due course of time they are excreted. To bring about this conversion, synthetic reactions are also employed by the biological system. Conjugation with glucuronic acid, sulphates, amino-acids, acetylation and methylation are some of the common reactions which transform a pollutant molecule into water soluble compound. Once transformed in this way the pollutant is excluded from the biological system.

(4) Unique Dynamism of Bio-degradation Reaction: The process of bio-degradation has a remarkable capability to adjust its activity according to the amount of pollutant present in the system. The rate of bio-degradation depends upon the ratio between the concentration required to saturate the system and the concentration of pollutant within a biological system. Below the saturation level, degradation reactions decompose a constant fraction of the total content of pollutant per unit time. Therefore, if a concentration of 200 mg of a pollutant is reduced to half of its original value in an hour, implying a rate of 100 mg per hour, a concentration of 20 mg shall take the same time to get reduced to 10 mg which however shall occur at a rate of 10 mg per hour. Larger loads of pollutants, which pose a greater toxicological risk, are decomposed at a faster rate, which slows down as the concentration of the pollutants within the system diminishes (Fig. 14.1). It is for this reason that the term half-life is used to measure the period of retention of pollutants or any foreign substance within a biological system. Half-life of a foreign substance in a biological system denotes the duration of time within which its concentration is reduced to half of its original value.

Small quantities of pollutants regularly entering a biological system are efficiently decomposed and eliminated. Larger amounts which the system is unable to decompose completely before the entry of fresh dose tend to accumulate within the system. As the concentration of foreign substance rises, the rate of degradation or elimination is also speeded up. This results in a state of dynamic equilibrium between the intake and rate of decomposition, which adjusts itself in such a way as to keep the concentration of the pollutant at a reasonably constant level.

For example, a pollutant which has a half-life of 12 hours is introduced in a biological system at a rate of 4 gms after every twelve hours (Fig. 14.2. d1, d2, d3...). A concentration of 4 gms per litre shall be attained in the system after 12 hours when the second dose of 4 gm is introduced. Samples collected after 24 hours show a concentration of about 6 gms only. This is because in the preceding 12 hours (the half-life) the initial concentration of the pollutant was reduced to 2 gms to which the next dose of 4 gms was added. The third dose shall raise the concentration to about 7 gms only as by the time it reaches the system, earlier concentration of 6 gms would be reduced to 3 gm per litre. Similarly 7 gm shall be reduced to 3.5 gm after 12 hours when the next dose of 4 gms enters the system. This dose shall, therefore, raise the overall concentration of pollutant to 7.5 gms only. Each dose entering the system shall be added to half of the total content of the pollutant present 12 hours earlier. The toxic material attained a concentration of 4 gms per litre after the first dose. The second dose raises it by 2 gms, the third by 1 gm, the fourth by 0.5 gms, the fifth by 0.25 gm, the sixth by 0.125 gms and so on. Finally, a steady state level is attained. The rate of bio-degradation reaction rises after a few doses to prevent any further increase in the concentration. The input equals the output.

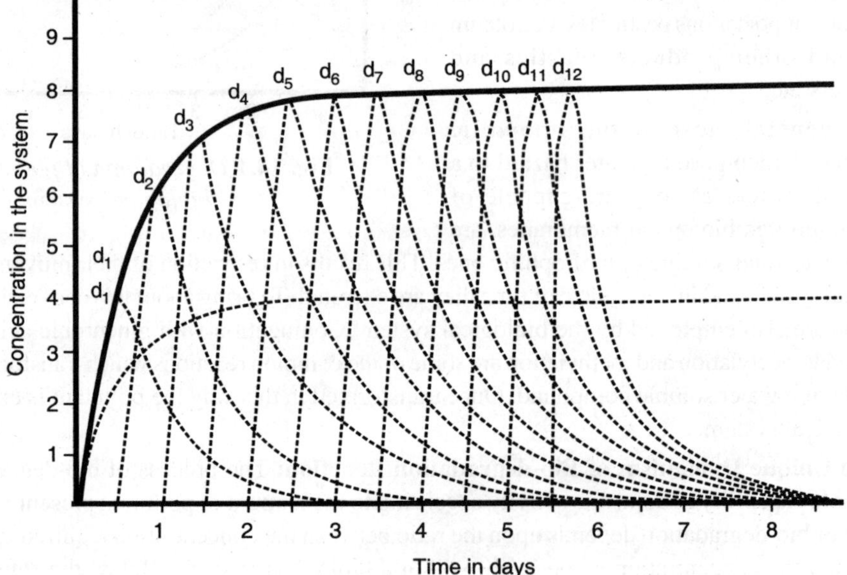

Fig. 14.2 *Acquisition of steady state concentration of pollutants regularly entering a biological systgem is sub-acute doses*

The steady state concentration of pollutants regularly entering a biological system depends largely on the rate of uptake and the half-life of the pollutant concerned in the system. If the rate of input is the same, substances, which have a longer half-life, tend to attain a higher concentration. In natural environment, a biological system may receive small exposures, which are spread over long durations. A constant rate of bio-degradation could have caused simple arithmetic addition of pollutant, which at some point of time could reach lethal toxic level. However, it is the remarkably versatile machinery, which works insidiously to keep the continuous inflow of pollutants at a uniform concentration by making adjustments in its rate of bio-degradation, no matter, howsoever, long may be the duration of exposures. If this machinery is disrupted even small amounts which pass for unnoticed in a healthy biological system add up to cause acute pollution.

However, with a continuous rise in the rate of input, a condition is reached when the biodegradation reactions are forced to occur at their maximum pace to prevent further increase in the concentration of pollutants in the system. The system is strained to work at its fastest possible rate. The corresponding concentration is the critical concentration level. It is only below this level that recovery or self-purification is possible. Any further rise in the concentration beyond this level affects the bio-degradation machinery adversely and the system loses its self-purifying capacity. With enfeebled pace of decomposition or elimination reaction, a phase of biologically unlimited pollution sets in and recovery is not possible.

(VII)THE FATE OF POLLUTANTS IN THE ENVIRONMENT

There occurs a bewildering variety in the physical nature and chemical composition of waste materials. Almost anything, which is useful at one time, may become useless at another. However, almost all types of pollutants or waste materials have a similar general fate. They are degraded or changed into their simple, harmless constituents and recycled back to the channels of their respective bio-geo-chemical cycles.

(1) The Fate of Pollutants of Earth's Surface: Land and Water

The pollution of earth's surface is mainly caused by biodegradable wastes, wastes resistant to degradation, non-degradable wastes and surplus or waste energy. Biodegradable wastes can be decomposed rather easily by abiotic and biotic agencies. The final products of decomposition of these wastes are : carbon dioxide, water, nitrogen, mineral matter etc. which are recycled back into their respective bio-geo-chemical cycles. Complex biodegradable wastes, which are resistant to degradation, may cause some problems. However, eventually these are also degraded into their simpler constituents and disposed of in the same manner as decomposition products of other biodegradable wastes.

Waste material, which cannot be degraded by abiotic or biotic agencies but can be subjected to a variety of physical or chemical transformations, consists of the inorganic components, the constituent elements of our planet. They cannot be destroyed. In nature they are transformed from one form to another, converted to their most stable state and are finally carried to oceans through streams and rivers. It is in the sea where they lie forming a part of silt and sediments, which come up only when the ocean floor rises to expose the sedimentary rocks thus formed.

Loud noise deliberately produced or caused by moving objects, vibrating machinery etc. is often a source of great nuisance. So is heat energy, a large amount of which is wasted in the process of its utilization. Whatever may be the amount or the nature of surplus energy, it is ultimately dissipated in the environment in a highly diluted state. Nuclear energy may be considered as a special type of energy, emission of which is a consequence of fusion, fission or disintegration reaction in unstable nuclei. Many of these unstable isotopes have a very long life. Nuclear energy is also dissipated like any other form of energy while the nuclei emitting the energy return to their stable state. However, some radioactive materials have so long half-lives that they continue to emit radiations for millions of years. These materials also behave and are handled by nature in the same way as non-degradable pollutants *i.e.* they are recycled back to their respective bio-geo-chemical cycles.

(2) The pollutants of the atmosphere

All gaseous effluents, vapours and fine particulate material, which escape into the atmosphere, have to come down to earth's surface at some point of time. They may react, re-react and form harmful and irritating products. Some of these may enter the stratosphere and cause ozone depletion. However, the entire burden is finally brought down to earth's surface in the form of dry deposition or precipitation to be disposed of and recycled back to their respective bio-geo-chemical cycles.

QUESTIONS

1. Enumerate major categories of pollution of environment. What is their general fate ?
2. Discuss the phenomenon of bio-magnification? How does it differ from bioaccumulation ?
3. What do you understand by biodegradation? How does it influence the concentration of pollutants entering a biological system in sub-acute doses? What will happen if the mechanism fails?
4. Why does a pollutant regularly entering a biological system in small sub-toxic doses attain a steady state concentration? What are the factors, which influence this steady state concentration?

15 Chapter

Pollution of Earth's Surface: Water and Soil 1. Simple Organic Wastes

Pollution which can be minimized or eliminated and its components degraded into simpler, harm-less constituents by biological activity is termed as bio-degradable pollution. Dead and decaying remains of plants and animals, products of their metabolic activity, faecal remains, excreta, waste waters and materials which originate from day to day activity in a domestic establishment, agricultural wastes, wastes from agro-based industries, tannery effluents etc, come under this category. Various lipids, fats, oils, proteins, carbohydrates, nucleic acids, urea, cellulose, hemi-celluloses, lignin, pectin etc. along with ash, dust and detergent bearing waters containing plenty of phosphates form the major part of simple bio-degradable pollution.

(I) PROBLEMS ASSOCIATED WITH SIMPLE BIO-DEGRADABLE WASTES

Biodegradable pollution, which mainly involves water and soil, causes little long range problems. Actually, it is the enormous quantity of sewage and organic wastes which causes difficulties. Large towns and cities discharge a tremendous volume of waste material for handling and treatment of which adequate facilities are often not available. Smaller amounts of such pollutants pose no problems as the biological agencies of degradation quickly decompose most of the organic matter forming carbon dioxide, water and mineral matter. Problems associated with biodegradable pollution can be summed up as follows:

1. The problem of odorous gases and volatile substances derived from organic wastes.
2. The problem of abundance of infectious microbes.
3. The problem of oxygen deficit caused by decomposition of organic matter.
4. The problem of generation of plenty of plant nutrients in water bodies.

(1) Odorous gases and volatile substances

Presence of such compounds which produce smell or odour in the environment are often undesirable. There are a number of biodegradable substances, which continuously lose molecules to the atmosphere, *i.e.*, are odorous or volatile, Such substances are known as 'osmogen'. and the study of odours or smell is termed as 'Osmics'. These compounds, osmogens, in gaseous state are taken in the lungs through respiratory track and may cause irritation, loss of appetite, giddiness, nausea, anorexia etc. Through lungs, these chemicals enter straight into blood stream and the general circulation and could affect the living system adversely if inhaled in higher quantities (1). The presence and concentration of odour in the atmosphere depends upon a number of factors such as the rate of emission, the distance from the source, the wind velocity and its direction etc. There are three main sources of odour in the atmosphere:

1. Odorous Exudates, Excreta and Secretions of Living Plants, Animals and Microorganisms: A number of bacteria, fungi, algae, plants and animals produce odours of various types. Many gymnosperms and higher plants release hydrocarbons (mainly terpenes) which have a strong aromatic smell. Heteroptrohic bacteria *Micromonospora, Nocardia, Streptomycetes* etc. are known to produce a variety of odours. *Anabaena circinalis,* a blue-green alga, produces earthy smell. *Synura petersonii* emits fishy or cucumber like odour while *Trichoderma viridae* a fungus is known to produce coconut like odour. Free living protozoans, belonging to genus *Uroglenopsis,* pruduce an aromatic oily secretion which has an objectionable odour. A considerable amount of various volatile organic compounds, each with its characteristic odour, is discharged in the atmosphere by plants. Total natural hydrocarbon emission from the state of Pennsylvania, USA, alone was estimated to be about 34,000 tons per day in the month of August (2).

2. Domestic and Industrial Discharge of Odorous Material: A large number of odorous substances are used in modern domestic and industrial establishments. Wastes from these units contain hydrogen sulphide, ammonia, amines, alcohols, aldehydes, phenols, marcaptans, esters, chlorine, and chlorinated organic compounds etc., which are released into the atmosphere to create problems of odour pollution (1).

3. Microbial Decay and Decomposition of Organic Matter: Decay and decomposition of organic matter is probably the most important source of odorous material in the atmosphere. The process of bio-degradation involves decomposition of large macromolecules into simpler components in gradual steps. It requires a series of microbes and favourable conditions for the activity of these organisms. Some of the microbes or the conditions under which they act may not be available. This disturbs the entire chain of bio-degradation reactions. Undesirable intermediate products collect and cause problems.

Majority of heterotrophic microbes require oxygen to respire. In its absence or short supply, their growth is suppressed. Those microbes, which are capable of growing under oxygen deficient conditions, thrive. Aerobic process is replaced by anaerobic decomposition. Instead of carbon dioxide and water, which are the end products of aerobic degradation, various acids, alcohols, methane, hydrogen sulphide, ammonia and a number of organic compounds are produced due to anaerobic activity. These escape into the atmosphere and cause odour problems or odour-pollution. This often happens when solids or semi-solids decompose. Oxygen is usually in short supply or absent altogether under the surface layers or the crust. In still waters containing plenty of organic matter a similar layer of froth and particulate forms on the surface, preventing gaseous exchange from air above. Inside this layer oxygen deficient conditions develop and anaerobic decomposition quickly replaces aerobic activity. Foul smelling gases and vapours are emitted causing odour pollution when outermost crust is disturbed (1;2).

(2) Abundance Of Infectious Microbes

Organic wastes, sewage effluents, excreta, exudates and faecal matter etc. support a rich population of microbes. Numerous viruses, bacteria, algae, fungi, protozoans, helminths, annelids, larval stages of various insects, pests etc. thrive on the organic matter. Some of these are responsible for causing dangerous diseases of man, animals and plants. The survival time of various pathogenic microbes in sewage effluents and organic wastes depends upon a number of factors. These include the nature of microbial species, conditions of temperature, pH, etc., the chemical nature of the environment and the composition of microbial species in the medium. The capacity of some of pathogenic organisms to form highly resistant, encapsulated stages, spores and cysts plays an important part in their prolonged survival. Some pathogens may survive only for a few hours while others may persist for several months or years together. The problem of infectious microbes, which arises because of accumulation of large amounts of bio-degradahle wastes, has two important aspects (3). These are:

1. Contamination of Water Supplies, Food and Edible Material Caused by Direct Contact: Chance contamination of water supplies food and other edible material with sewage effluents and organic wastes is often responsible for the outbreak of a number of diseases. Typhoid caused by *Salmonella typhii* and cholera, caused by *Vibrio cholerae,* are mainly spread through contaminated waters. Violent air currents may form virus of bacteria bearing aerosols. Resistant spores and cysts of many pathogens persist even when the sewage effluents dry out completely. Winds blow them around as dust. With air currents, disease-causing organisms are carried to distant places.

2. Development of a Population of Vectors or Carriers of Serious Diseases: Dead and decaying organic matter is the starting point of the detritus food chain. A series of microbes, insects, pests and rodents feed on organic wastes or organisms, which live on these wastes. Abundance of biodegradable wastes, therefore, causes a substantial rise in the population of aphids, mosquitoes, houseflies, rodents and such other animals. These become a nuisance in day-to-day human activity. Many of these organisms are carriers or vectors of serious diseases of man, animals and plants. These insects and pests physically carry a number of pathogenic organisms from one place to another. Some viruses, bacteria and protozoans may enter the body of the insect or pest, feed, breed, multiply and persist therein for long periods. The pathogen is transferred to other animals or man with the saliva when the insect bites or feeds upon them. Or else such carriers simply continue to excrete the pathogen for long duration of time.

Malarial parasite is tranferred by the bite of the female mosquito of genus *Anopheles.* African sleeping sickness, caused by *Trypanosoma gambiens and T. rhodesiense* is spread through the bite of Tse-tse flies (*Glossina palpalis* and *G.morsitans*). Likewise, sandfly (*Phlebotomonas sp.*) disseminates Cutaneous Leishaminasis and Kala-azar both caused by *Leishmania* spp. The mosquitoes of genus *Aedes* (*A. aegippti, A.allopicus and A. acutellaris etc.*) transmit viruses causing Dangue fever, Yellow fever and the Japanese Encephalitis (3).

(3) Creation of Oxygen Deficit in Aquatic Systems

The problem of oxygen deficit caused by organic wastes concerns aquatic systems. In a water body organic matter is decomposed to harmless constituents by the activity of series of heterotrophic organisms. These include bacteria, fungi, protozoans and other small animals which live on dissolved organic material and organic debris. Most of these organisms need oxygen to respire. There are two main sources of dissolved oxygen in an aquatic system:

1. **Oxygen absorbed from atmosphere above:** Atmospheric oxygen continuously dissolves into water at air-water interface.
2. **Oxygen produced by green plants within the system:** Green plants as well as phytoplanktons carry on photosynthesis which results in oxygen production.

Organic matter serves as a source of energy to microbial life and other animals, which feed on these microbes. Over-abundance of organic matter causes rapid rise in microbial activity and quick decomposition occurs. This in turn places a heavy demand on oxygen in the aquatic system. However, amount of oxygen present in the system is limited which is rapidly consumed and oxygen deficient conditions develop. Very low levels of oxygen are injurious to a number of organisms which as a consequence perish or if capable of movement leave the place. Massive fish-kills reported in 1972, 1975 and 1977 in Gomti River near Lucknow, U.P., India, were caused mainly due to the development of oxygen deficiency in the river water. There was no toxic or poisonous material present in the distillery effluents which were discharged in the river. Due to oxygen deficiency aerobic activity is replaced by anaerobic activity. Various organic acids, alcohols, phenols, gases like ammonia, methane, hydrogen sulphide etc. accumulate in the system and harm the aquatic life. The environment becomes congenial to many pathogenic organisms. The water body starts emitting foul smell. Its waters acquire an undesirable taste.

(4) Generation Of Plant Nutrients and Accelerated Eutrophication

The decomposition of biodegradable pollutants generates rich plant nutrients, which often threaten the very existence of the water body. Mineral elements, which go into the structure and composition of a living being, are released when their dead bodies, exudates and excreta disintegrate. The products of decomposition of organic matter precisely match the mineral requirement of aquatic plants. The growth of green plants is, therefore, stimulated which in turn supports a rich crop of consumers. The plant and animal biomass, which develops in the aquatic body often, becomes a nuisance (4;5).

1. Eutrophication: The term eutrophication means nutrient enrichment. A freshly formed body of water has a very low concentration of plant nutrients. Little plant life develops in such waters. Low primary production limits animal communities as well. Surface runoffs, wind borne dust and organic debris, excreta and exudates of animals, which use the water, slowly raise the nutrient content. Bacteria and blue green algae fix atmospheric nitrogen. Phosphates present in rocks and detritus at the bottom are solubilized by the microbial activity. In gradual stages, the nutrient status of the water improves. A moderate population of plants, animals and microbes now develops in the system. With passage of time further nutrient enrichment occurs. Dense population of plants, phytoplanktons and animals now appears. At this stage, the aquatic system becomes highly productive in terms of fish yield and other produce.

Based on nutrient status and productivity an aquatic system can be classified into the following three types:

a. **Oligotrophic:** water with poor nutrient status and productivity.

b. **Mesotrophic:** water with intermediate or moderate nutrient status and productivity.

c. **Eutrophic:** water with rich nutrient status and high productivity.

Oligotrophic waters slowly turn into mesotrophic waters and then to eutrophic waters. Further aging causes over-abundance of nutrients. This in turn gives rise to such profuse growth of rooted and floating green plants that the water body loses its aesthetic and economic value. Its waters become useless. Organic debris and silt settles at the bottom. The margins of the aquatic body turn into a marsh with a small shallow pond in the centre. Organic debris and silt finally fill the depression and what was once a lake becomes a dry land.

2. Accelerated eutrophication: It is the human activity, which has speeded up the process of eutrophication of our water bodies. Natural eutrophication is a very slow process. Huge amounts of mineral nutrients and organic matter are added in water reservoirs in the form of sewage effluents, organic wastes, agricultural run-offs, excreta and exudates of living beings etc. These contain plenty of phosphates, which are used in large quantities in soaps, detergents and as water softeners. Water flowing through agricultural land picks up a lot of nitrates and phosphates applied as fertilizers. The aquatic body acquires more nutrients in a day than it would do in a year under undisturbed natural conditions. All this promotes luxuriant growth of algae and other water plants. A rich microbial community and animal population also develop. Expensive cleaning operations have to be undertaken if the water has to be used for any domestic or industrial purpose. The process of natural eutrophication is accelerated. Silt and organic debris collects in the bottom and the system turns into a shallow and muddy pond, then to a marsh and finally to a dry land. The water body, which could have served us as a reservoir of fresh water, fishes and other products for hundreds of years, becomes useless within a span of a few years only (4;5).

(3) Algal blooms or water blooms:

In some eutrophic water dense population of planktonic algae develops. The water turns green within a span of a few hours. The phenomenon is called water or algal blooms because of the fast growth rate of algal population. Such waters are useless for human use as it is very difficult and costly to remove the microscopic green plants. The entire mass of planktonic algae may die, often abruptly. This gives rise to large quantities of dead and decaying organic matter which causes taste and odour problems (7).

Presence of plant nutrients in concentrations sufficient to support the huge population of planktonic algae is the primary requirement for the formation of algal blooms. In their absence blooms cannot develop. High levels of dissolved organic material in water in which bloom formation occurs, which yield plant nutrient following decomposition, probably give rise to a highly specialized environment. Only one or two species are able to live and thrive under such conditions. This explains the usual abundance of one or two species in a water bloom.

Association of water blooms with hot, humid and conditions of high light intensities appear to be an important aspect in water bloom formation. Hot humid conditions promote intense microbial activity. The abundant quantity of dissolved organic material present in the system would be quickly decomposed. This should produce plenty of nutrients for algal growth. In tropical and subtropical regions, hot and humid conditions also cause thermal stratification. With sunrise and the heating effect, upper layers of water, get heated. Being lighter they stay above the cooler layers which are below. Oxygen dissolves in the upper layers from the atmosphere while intense sunlight promotes active photosynthesis. Oxygen saturation results. The presence of large population of planktons in upper layers prevents light to go deeper down. Lower layers thus suffer from oxygen deficit and anaerobic activity is stepped up. This suffocates other aquatic life. The system turns foul and loses its utility (8).

(4) Role of nitrogen and phosphorus in eutrophication: Life as it exists on this planet is composed of about 40 elements. However, the bulk of this organic world is made up of about nine essential elements. These are: carbon, hydrogen, oxygen, nitrogen, sulphur, phosphorus, calcium, magnesium and potassium (cf. Table 15.1). In addition to these nine elements a number of other elements are required in very small quantities which are called micro-nutrients. These elements go into various enzymes, co-enzymes and co-factors without which metabolic activities of a living cell are not possible.

Most of the natural waters contain enough calcium, magnesium, potassium and sulphur in amounts sufficient for growth of luxuriant plant life. Both oxygen and hydrogen come from water. Carbon is also present in plenty as dissolved carbon dioxide, bicarbonates and carbonates. Micronutrients are required in very small quantities. Most of them are present in quantities enough for plant growth. The deficiency of these elements may occur but is rare. This leaves nitrogen and phosphorus as the most important elements, which are required by plants in amounts large enough, which may not be present in natural waters always (8).

Table 15.1 *Elemental composition of green plants.*

	Elements	Concentration		Element	Concentration
1	Carbon	45.00%	9	Sulphur	0.10%
2	Oxygen	45.00%	10	Molybdenum	0.1 ppm
3	Hydrogen	6.00%	11	Copper	6.0 ppm
4	Nitrogen	1.50%	12	Zinc	20.0 ppm
5	Potassium	1.00%	13	Manganese	50.0 ppm
6	Calcium	0.50%	14	OIron	100.0 ppm
7	Magnesium	0.20%	15	Boron	20.0 ppm
8	Phosphorus	0.20%	16	Chlorine	100.0 ppm

After Epstein 1965.

As early as 1840 A.D. Leibig formulated the idea that growth of plants is dependent on the amount of the nutrient, which is available in minimum quantity. In an aquatic system, the development of plant life is often restricted either by the lack of phosphorus or of nitrogen. Only that much plant growth occurs in the system, which is permitted by the availability of phosphorus or nitrogen as the case may be. If there is no import of phosphorus in the form of dissolved salts or organic material, the plant life in aquatic body has to depend on the phosphorus solubilized from the bottom strata and phosphorus derived from the mineralization of organic matter present within the system. There is no other source of this element in such waters. Both of these processes are very slow and are influenced by a number of factors such as pH, temperature, microbial activity, the nature of the bottom strata etc. In absence of input from outside sources phosphorus enrichment and thereby growth of aquatic life is a very slow process (7;8).

The other element, nitrogen is not so critical. It is available in mud, detritus and rock strata at the bottom. It is released when organic matter decomposes. A number of bacteria and blue green algae can fix atmospheric nitrogen whereas some of it may come from the process of abiological nitrogen fixation. Therefore, unlike phosphorus natural water bodies have their own devices to obtain nitrogen from external sources. Internal sources are not the only source. Given enough phosphorus, the system acquires nitrogen required for plant growth and in time rich plant life develops in the system (9).

(II) TREATMENT AND DISPOSAL OF BIODEGRADABLE WASTES

Biodegradable pollutants can be effectively treated with microorganisms and decomposed to simple harmless constituents. The biological treatment is rather an inexpensive treatment. It not only eliminates the pollution but may also provide economically useful products. Little human efforts are required in process of disposal of biodegradable wastes. Given enough time, the excellent microbial agencies of nature can efficiently degrade these wastes. However, problems arise when large amounts of organic wastes are produced. The microbial machinery is unable to handle such huge quantities in the given space, time and conditions. We often have to expedite the decomposition process and confine it to a limited space so that its intermediates or end products or the microbes themselves may not contaminate human establishments (10).

Human skill has tamed nature's microbial machinery to work at its maximum efficiency within a confined space. This has enabled man to dispose of large quantities of biodegradable wastes quickly, efficiently and with very little expenditure. Biodegradable wastes produced in a domestic or industrial establishment are usually of the following two types:

1. **Liquid Wastes:** Such as sewage, waste waters, industrial discharges etc. These may also contain plenty of semi-solid and solids in suspended state or in colloidal solution.
2. **Solid Wastes:** Such as trash, garbage, pulp, peelings, packing materials, rags etc. Industries involved in canning, processing and marketing meat, poultry, dairy and agricultural products discard plenty of solid and semi-solid organic wastes.

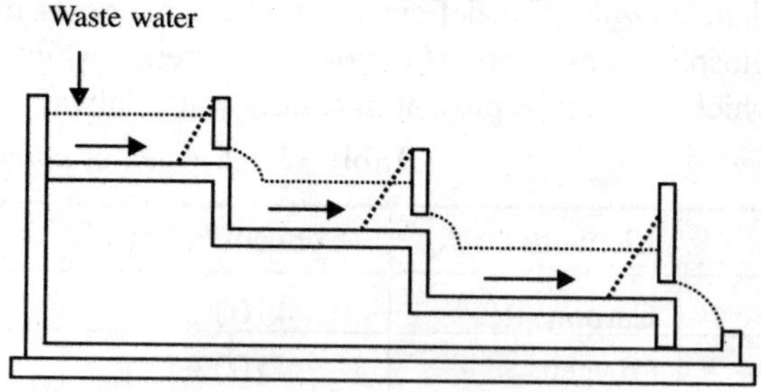

Fig 15.1 *Screening Chambers*

The bio-degradation of organic wastes involves two processes: anaerobic and aerobic. Liquids and semi-solids are usually subjected to both aerobic and anaerobic degradation. Solid wastes are on the other hand more conveniently digested and mineralized by anaerobic processes.

(1) Treatment And Disposal of Liquid Wastes

The biological treatment of sewage and wastewaters is done in three stages. These stages are:

1. Primary Treatment

Sewage and wastewaters contain plenty of solids in suspension or in colloidal solution or simply as floating objects. These have to be removed before the biological treatment. To separate large floating objects like pieces of vegetables, rags, dead animals and other junk the sewage is passed through screens or nets in chambers which are called screening chambers (fig 15.1). Larger suspended or floating objects are held back in the screening chambers. After screening wastewaters are run through a series of grit chambers (Fig 15.2). These chambers contain large stones or brick ballast. Passage through these chambers removes much of coarse particulate material. Special mechanical devices are also available to remove coarse particles from wastewaters. These are known as hydrocyclones or centrifugal separators. The sewage is subjected to rapid circular motion in hydrocyclones, which generates strong centrifugal force. The coarse particles are pushed towards the periphery of the circular chambers. Clear water is drawn out from the centre of the tank.

Wastewaters still contain plenty of fine particulate material. To remove these they are stocked in large tank for a variable period of time, which depends upon the nature of suspended material. Lighter particles float over the surface. Heavier ones settle down. Later the liquid is drawn out from the middle layers, which leaves the upper and lower layers back in the sedimentation tank. Oils, fats and greases are also retained in the upper layers as they are lighter. These tanks, therefore combine, the process of flotation as well as sedimentation.

In order to make the process of flotation more effective some emulsifying or foaming agent may be added to the tank and the contents of the tank agitated with the help of compressed air or mechanically. Particles are adsorbed on to the surface of the bubbles and come up on the surface. This process is known as Froth floatation. Certain chemicals and coagulants may also be used to raise the efficiency of the process of sedimentation. A coagulant is a finely powdered material, which is mixed with wastewaters. Its particles add on to the lighter particles present in the sewage and make them heavier which facilitates their sedimentation. The process is commonly known as Coagulation. Salts of iron, aluminium, activated silica, polyacrylamides etc. are usually used as coagulants. The practice of adding coagulants or flocculants improves the efficiency of processes in primary treatment. However, it suffers from the basic disadvantage that such substances are added in wastewaters, which later interfere with the process of biodegradation in the secondary phase of the treatment. Various soaps, oils and other flocculants may clog the filter beds, which are extensively used in the secondary treatment (10;11).

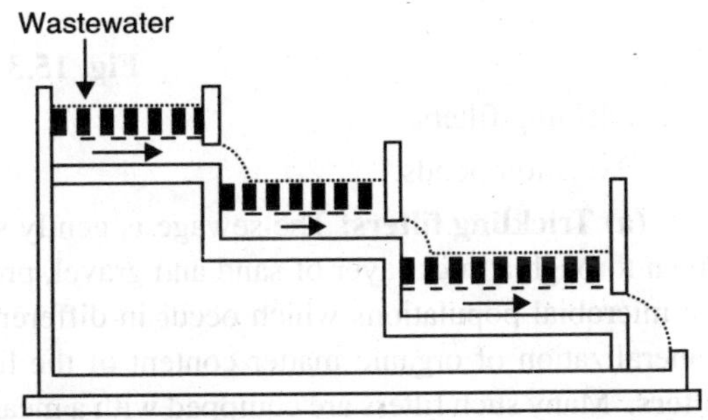

Fig 15.2 *Grit chanmbers*

2. Secondary Treatment

Primary treatment removes most of the solids and particulate materials. Wastewaters now have plenty of dissolved organic matter. It may also contain a little amount of very fine organic debris and particulate materials. It is now subjected to secondary treatment, which is in fact the biological treatment. There are two main kinds of secondary treatments:

A. Aerobic treatment

B. Anaerobic treatment.

A. Aerobic Treatment

Aerobic treatment involves bio-degradation in presence of oxygen. Conditions suitable for rapid growth and multiplication of micro-organisms are created in especially designed filters or tanks. Proper temperature and availability of oxygen are the two important requirement of the treatment. Low temperatures slow down growth and activity of the microbes. Lack of oxygen inhibits aerobic activity. Aerobic treatment is usually carried out in:

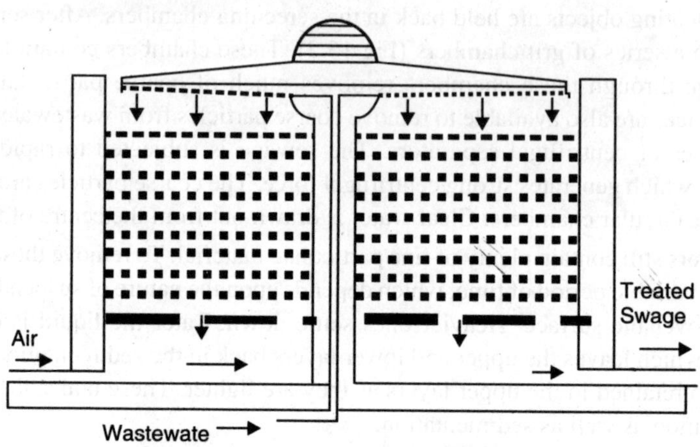

Fig. 15.3 *Tricking filters*

(a) Trickling filters.

(b) Oxidation ponds.

(a) Trickling filters: The sewage is gently sprayed over the filter beds and allowed to trickle down through a thick layer of sand and gravel, prepared over a perforated plate. As it moves down, the microbial populations which occur in different zones of the filter bed cause rapid oxidation or mineralization of organic matter content of the liquid. These giant filters are called the **Trickling filters.** Many such filters are equipped with a means to force compressed air through pipes under the filter bed. The air provides oxygen to the microbial population.

The microbial populations grow and multiply in between the sand and gravel particles. These include various bacteria, green and blue green algae and diatoms in the upper zones where plenty of light is available. In the lower zone various fungi bacteria, protozoans, and other small animals, which live and thrive on the dead and decaying organic matter are found. Passage through biologically active layers oxidizes much of the organic matter present in the wastewaters. After treatment in trickling filters, wastewaters are usually subjected to further oxidation in a series of tanks, which are referred to as oxidation ponds (11).

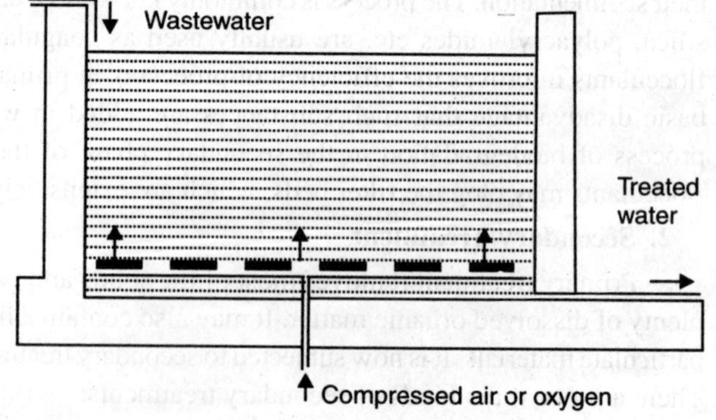

Fig. 15.4 *Compressed air or oxygen*

(b) Oxidation ponds: Treatment in an oxidation, also known as aeration tanks, involves slow passage of wastewaters, through a series of large tank in which a large number and variety of microorganism are present. Wastewaters are mixed with activated sludge in these tank, Activated sludge is a solid or semi-solid sediment derived from the settlement of particulate matter in waste waters. It comprises of a complex mixture of microbes present in the sewage. It also includes pathogenic organism. When dry it forms a dark coloured powder, one gram of which contain as many as 10^{14} bacteria. The specific composition of microbes in sludge depends chiefly on the composition of wastewaters. These microbes rapidly decompose most of the organic matter present in water. Compressed air is blown through the bottom of the tank, which provides oxygen necessary for the oxidative decomposition.

The liquid from the oxidation tank is then taken out to the settling tanks where activated sludge and other particulate material is allowed to settle down and the clear supernatant is taken out for disinfection and final disposal or subjected to the tertiary treatment (11).

B. ANAEROBIC TREATMENT

Anaerobic treatment involves degradation of organic wastes under conditions of oxygen deficit or its total absence. Many of the common heterotrophs of the soil, water and intestinal track are anaerobes or facultative anaerobes. These microbes acting together possess a remarkable capacity to decompose proteins, fats, carbohydrates, cellulose, woody materials, phenols and many other complex substances. The products of anaerobic degradation are reduced substances such as hydrogen sulphide, ammonia, methane and other simpler compounds (10).

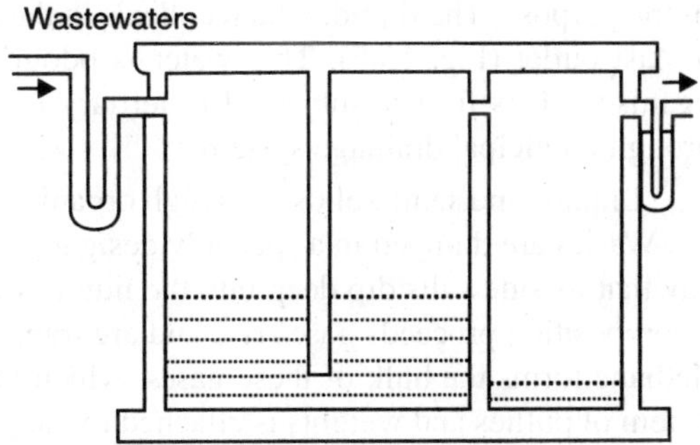

Fig. 15.5 *A closed septic*

A number of products of anaerobic degradation are useful. Many organic compounds like organic acids, alcohols etc. can be obtained by the activity of microbes under anaerobic conditions on suitable substrates. However, sewage effluents are a complex mixture of organic substances, dissolved salts, particulate material, pathogenic organisms etc. It is not possible to obtain useful products with the present state of our knowledge and technology. Methane is a notable exception. It is a common product of anaerobic degradation of organic matter. It can be used as a cheap fuel gas as is done at present in many domestic establishments. The sludge or the solid material left after anaerobic decomposition is exceedingly rich in plant nutrients. It can be used as manure to fertilize our fields.

Septic tanks are employed for anaerobic treatment of wastewaters. A septic tank is usually a large elongate reservoir with a curved or funnel-like bottom. Sewage is allowed to flow slowly through the tank. Sludge precipitates and settles at the bottom. Lighter part of it rises to the surface to form a continuous floating layer. Rising gases trap sedimented particles and bring it to the surface where a thick crust develops. The oxygen supply from the atmosphere is cut off due to this crust of solid and semi-solid material. The mass of wastewater moving in between the upper crust and bottom sludge undergoes rapid decomposition under anaerobic conditions. The products of microbial activity accumulate in the middle layers. Much of the gaseous material produced is methane, ammonia, hydrogen sulphide etc which slowly escape to the atmosphere through the upper crust and the smell turns foul. After passage through the septic tank, which may take few days to a week, the liquid is drawn out from the middle layers of the tank. This leaves back the upper crust and the bottom sediment in the tank. Wastewaters from a septic tank are subjected to further treatment in the tertiary phase (10;11).

The semi-solid and solid sludge is left in the septic tank. It is retained in the tank for long durations. During this period, much of it is decomposed and its volume is greatly reduced. In order to ensure satisfactory mineralization of organic matter the entire decomposing mass should be periodically stirred and diluted with fresh sewage. Quite a large number of pathogenic organisms, spores, cysts, eggs of helminths etc. are present in the bottom sludge. Though quite rich in plant nutrients, it cannot be used as fertilizer. It is usually spread out on wasteland. Microbial activity continues while drying occurs. As there is little organic matter left now pathogenic organisms are slowly reduced and the dried scum becomes fit for application in agricultural fields as manure.

A slight variation of the simple septic tank discussed above is the conventional covered septic tank used in many households, which lack sewer lines. Faecal material and wastewaters are dumped into a covered chamber. Microbes decompose it under anaerobic conditions. Semi-solid faecal matter being light collects in the chamber while liquids pass through an aperture near the bottom to equalize the level of the tank. It takes a long time for the matter in the first two chambers to attain the height of aperture to reach the third tank. By the time it reaches this chamber most of the organic matter has been decomposed. Most of the gases collect in the first chamber, as it is in this chamber that much of the faecal matter decomposes. These foul smelling gases escape through a water seal and pipes provided for the purpose. The liquid, which collects in the third chamber, flows out when it attains the level of the last outlet (Fig. 15.5). This water is odourless, rich in mineral nutrients and disease causing organisms. It is further subjected to tertiary treatment. In practice, however, it is allowed to flow through municipal drainage system.

Liquids and semi-solids with high organic matter content may be used to obtain methane, a fuel gas. Wastes are dumped in a specially designed compartment. A large jar is inverted over it in such a way that its sidewalls dip deep into the liquid wastes along the walls of the chamber (Fig. 15.6). As decomposition proceeds gases rise and are trapped over the liquids under the top of the inverted jar. Methane forms the bulk of these gases, which is taken out through a tube to be used as fuel gas. A system of pullies and weights is attached to the jar to adjust the pressure under the jar. The scum left in the chamber after decomposition of organic matter is rich in plant nutrients. It is dried and finally applied to the fields as fertilizer.

3. Tertiary Treatment

Most of the organic matter present in wastewaters is decomposed during secondary treatment. Wastewaters now contain plenty of nitrates, phosphates and ammonium salts, which are the products of bio-degradation. In addition, these waters contain large population of microbes – bacteria, algae, fungi, protozoans etc. Phosphates, nitrates and ammonium salts are essential plant nutrients. If released into natural waters, such waters cause the problem of eutrophication or excessive plant growth. The microbial population present in these waters has to be eliminated before these waters are discharged in a river or a lake as many of the microbes are dangerous disease-causing organisms.

After the secondary treatment, wastewaters are subjected to tertiary treatment. It is during this phase of the treatment that much of the plant nutrients and microbes are

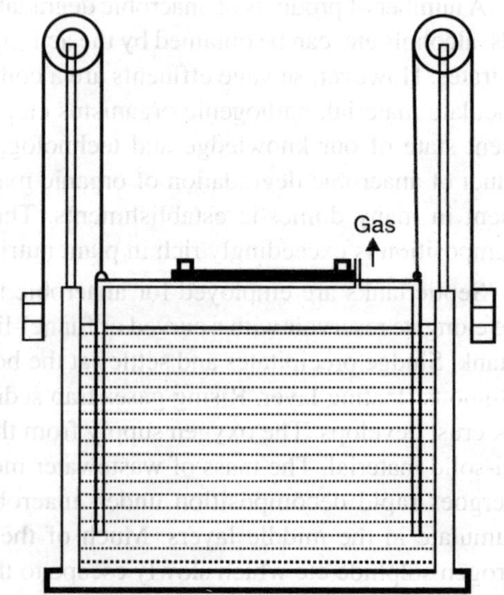

Fig. 15.6 *A digester equipped with device to recover combusitible gases*

eliminated. This is done in biological ponds in which wastewaters are allowed to stay for some time. Biological ponds are a series of simple shallow tanks about 1 meter in depth, which are interconnected with each other. Rich population of algae develops in these ponds as plant nutrients are absorbed. Primary producers (algae) support a series of consumers. The mineral content of the water passes on to higher trophic levels and is eliminated. Profuse algal growth generates enough oxygen in the system to support a diverse community of aerobes which consume much of the harmful pathogenic forms.

After lagooning in biological ponds the waste waters are subjected to the process of disinfection which further eliminates the pathogenic organisms still present. This may be achieved by chlorination, treatment with potassium permanganate, ultra-violet irradiation or any such physical or chemical means. After the tertiary treatment the waste waters are finally fit for being discharged in a natural water body such as a lake or a river (11).

(2) Treatment and Disposal of Solid Bio-Degradable Wastes

Agriculture, industrial or domestic activities of man produce huge quantities of solid wastes. Production of about 50 million tons of rice creates about 17 million tons of paddy husk and 2.8 million tons of rice bran. A cigarette making industry using about 7.8 million kg of tobacco leaves produces about 0.39 million kg of solid wastes. About 10,000 kg of cotton turned into cloth creates about 800 kg, while tanning and dyeing 1000 hides generate about 1200 kg of solid refuse. The amount of waste solids produced per person per day in Indian cities varies from about 300 to 600 gm. On global scale all this adds up to an enormous quantity, disposal of which is a tough task.

Bio-degradable wastes pose little problems. Much of agricultural wastes is used to feed the live stock population. In India about 410 million tons of manure are produced every year from organic waste produced by bovine population. Large compost pits are dug and organic wastes are dumped in it to stay there for a long time. Pits are covered with a layer of soil. Organic wastes, if dry may be soaked with water which helps in their rapid decomposition. The process of degradation under anaerobic conditions changes the physical structure of the waste material. Its mass is drastically reduced. The time required for adequate decomposition depends upon the nature of organic wastes, moisture content, temperature etc. The residual matter is exceedingly rich in mineral nutrients. It is used as manure in agricultural fields (10).

Problems arise when the mass of organic wastes is enormous. As biodegradation is a slow process, land once used as dumping ground or compost pits is engaged for a long time. Lower temperatures and scanty moisture content delay the process considerably and in very cold climates the decomposition of organic wastes is negligible. In temperate and tropical regions as well, the growing amount of organic wastes is causing concern. More and more land has to be used to accommodate the dumping grounds for solid wastes as human establishments expand. Useful land has to be converted to ugly looking waste-land.

A rather quick method of getting rid of the enormous bulk of solid wastes is to burn it under controlled conditions. The heat produced may be utilized for some useful purpose. Ashes left after burning are dumped in water or disposed of in fields. They represent only about 8-10% of the total mass of waste material thus burnt. When organic wastes are subjected to high temperatures in absence of oxygen, large molecules are broken down into smaller ones. The process is known as pyrolysis. Gases like CO_2, CO, CH_4, C_2H_2, C_2H_4, black tar like liquid and charcoal is produced. These can be used as fuel. However, heat treatment yields plenty of gaseous pollutants, which adds to the atmospheric burden of gaseous wastes. Rising concentration of these gases in the atmosphere shall have far reaching consequences such as global warming, stratospheric ozone depletion etc. Hence, the practice of burning organic wastes is not a very sound practice.

The combustion of organic matter releases, quickly, mineral elements which appear as ashes. These are either dumped on open land or disposed of in streams and rivers. Flowing waters carry the precious mineral nutrients down to the ocean from where little returns to the land. Natural biogeochemical cycles are cut short. If the period of residence of mineral elements in the biosphere is shortened, the soil becomes poorer as it receives little recycled mineral nutrients and has to obtain it from virgin sources. Quick depletion of nutrients occurs. The practice of burning cow-dung cakes has resulted in wide spread depletion of nitrogen and phosphorus in many Indian fields. The mineral content in the faecal matter produced by cattle could have helped to maintain proper soil structure and nutrient status. Instead it is burned to produce heat and ashes which are quickly leached away during rainy seasons. The soil becomes poorer after every rain (10).

QUESTIONS

1. Discuss problems associated with simple biodegradable wastes. How can you expedite the decomposition and elimination of these wastes?
2. What is accelerated eutrophication? Discuss the role of nitrates and phosphates in eutrophication.
3. What are algal blooms? What problems do they cause? Discuss the role of nitrogen, phosphorous and carbon in their formation.
4. What do you know about the primary, secondary and tertiary treatment of biodegradable wastes? How can you use these wastes as a source of energy?
5. What happens when large quantities of non-toxic, biodegradable wastes are discharged into a lotic or a lentic system?

16
Chapter

Pollution of Earth's Surface: Water and Soil II. Complex Biodgradable Wastes

In contrast to simple biodegradable substances, which decompose quickly, there are some compounds, which have a remarkable degree of resistance to the natural agencies of decay and decomposition. These are often harmful substances, which persist in the environment for long duration of time during which they are taken up in the biosphere, accumulated and bio-magnified to concentration potentially toxic to organisms at higher trophic levels in the food chain. Many of these chemicals are carcinogenic, teratogenic and mutagenic in nature. Pollutants grouped together under the category of wastes resistant to degradation, are organic substances produced naturally or are synthesized by man. Many of these are such chemicals, which have caused much concern due to their widespread use and dissemination. Broadly speaking these chemicals can be grouped into the following three categories:

(A) Pesticides and allied chemicals
(B) Crude petroleum and its derivatives.
(C) Polymers, plastics, plasticizers and other wastes.

Though most of these chemicals are referred to as being persistent, the versatile and efficient machinery which Nature employs for decay and decomposition of pollutants is capable of decomposing even these recalcitrant and refractory chemicals. But to do so, suitable conditions and abiotic as well as biotic processes have to be put together for which Nature requires time on its own scales and not on scales dictated by human needs. These chemicals, if left to Nature's devices alone are degraded rather slowly, slower than the rate of their entry into the environment. They tend to accumulate within the system causing toxicity to those living organisms, which happen to get exposures. Grave ecological problems appear where the magnitude of pollution is considerably large.

A. Pesticides and allied chemical

Under the name pesticides are grouped a large number of chemicals which are used to suppress or eliminate undesirable organisms. Most of these chemicals are poisonous substances capable of damaging one type of organism drastically while causing none or only nominal damage to the desired one even if the two are in a close association. It is due to this property of exerting a selective toxic action on different species or group of species that these chemicals have acquired a very important role in a modern society.

It was after the discovery of the remarkable insecticidal properties of DDT by Mueller in 1933 that a chemical war has been waged against various insects, pests and other undesirable organisms. In his endeavour to feed a rapidly growing population man had to develop sophisticated technology for raising vegetables, fruits, cereals and live-stock of which the control of various pathogens, insects, pests, nematodes, rodents etc. forms an integral part. In good old days the problem was not so severe. Sprawling human establishments and practices of intensive agriculture have created additional resources for subsistence of various pathogens, insects, pests etc while the disturbed prey predator relationship in a drastically altered ecosystem has weakened or eliminated the natural means of population regulation. The number of disease causing organisms and pestilent insects has been estimated to be well over 86000 apart from a large number of species of fungi, ticks, mites and nematodes. On global scale, these could cause a loss of over 50% of total world foodstuff production. (1).

(I) PROBLEMS ASSOCIATED WITH THE USE OF PESTICIDES

Pesticides are unique in position among toxic substances as they are deliberately added to suppress or eliminate some form of life. Under ideal conditions, the injuring action should be specific and affect only the target organisms. The toxicity should disappear after the purpose for which it was applied has been achieved. However, none of these features are met with in most of the pesticides which are in common use these days. This has made the use of these synthetic organics extremely hazardous pollutants of the environment and the biosphere. Some of the characteristic features of pesticide application may be summarized as follows:

(1) The use of synthetic pesticides has become a necessity in various branches of our economy.

(2) Most of the pesticides are poisons and their handling is hazardous. The selective action of pesticides is never perfect and many non-target organisms are affected by their toxicity - some of which may be useful organisms.

(3) It is difficult to prevent the circulation of these chemicals in the environment. They are usually applied with the help of some aerial or surface spraying device or simply dusted manually. Air, water and living organisms carry them to far off places.

(4) Most of these synthetic organics or their decomposition products persist in toxic state in the environment for long durations. Thus, once applied they continue to harm the non-target organisms for long periods. They are bio-accumulated and bio-magnified, features, which cause problems at higher trophic levels in an ecosystem. Persistence of DDT has been recorded for periods as long as twenty-five years (2).

(5) It is impossible to reduce the rate of application of pesticides. Their application should be in adequate doses to eliminate the entire population of unwanted organisms without providing them any chance to develop resistance to the chemical employed. However, in actual practice, uneven distribution of insecticides over the area under operation usually exposes many of the undesired organisms also to lower concentrations (sub-toxic or sub-lethal concentrations). Thus, resistant populations are developed and subsequent applications have to be in higher and higher doses. Decreasing the rate of their application could defeat the very purpose for which they are applied.

(6) Immunotoxicity of Pesticide and Allied Chemicals: The effect of various pesticides on immune system has become an area of great concern. A number of pesticides which include atrazine, captan, dinitro-ortho-cresol, hexachlorobenzene, 2-4-5-T have been shown to have marked effect on immune system of laboratory rats (3). About forty different pesticides have been cited to influence the host defense mechanism by Street (4). With an impaired immune system, the organism becomes susceptible to any type of infection howsoever mild.

(7) Mutagenic, Carcinogenic and Teratogenic Action: Mutagenic, carcinogenic and embryo-toxic action of foreign chemicals in biological systems is due to the interference which they cause in function, synthesis and structure of DNA and RNA molecules. A number of pesticides like DDT, Aldrin, heptachlor, Hexachlorobenzene, Toxaphene, Captan, Mirex etc. all of which are in common use these days have been shown to induce cancerous growth and tumours in mammalian systems (5). Likewise Carbaryl, Captan, Folpet, Difolatan, Organo-mercury compounds, 2-4-5-T, Pentachloronitrobenzene and Paraquat has been demonstrated to cause embryo-toxic effects in experimental animals (6). Similarly many pesticides have been shown to cause mutations in a series of tests by Epstein and coworkers (7).

(II) ECOLOGICAL EFFECTS OF PESTICIDE POLLUTION

Differential toxic action of pesticides on different species results in an unbalanced biotic community. Species susceptible to the toxic action of the pesticide are eliminated while the resistant ones multiply without competition. Organisms on which the eliminated species depended for its food requirement are also affected. Absence of predation causes these forms to multiply in numbers. Thus a chain of events is initiated which alters the species composition of the ecosystem beyond recognition. Insignificant pest may become so numerous as to cause another pestilence.

Pesticides often kill a series of organisms owing to their non-specific action. System diversity and complexity is, therefore, drastically reduced which make it highly vulnerable. The population, which emerges to replace the older one, is often characterized by a faster rate of growth at the level of primary consumers while the predator population takes longer time to build up enough strength to keep the primary consumers in check. This may result in development of population of other organisms as serious threat to productivity (2;8).

(III) PERSISTENCE OF PESTICIDES IN THE ENVIRONMENT

The persistence of pesticides in the environment depends upon their chemical and physical properties, dose and formulations (e.g., liquid, powder or granules etc.), type of the soil, its moisture content, temperature, physical properties of the soil, composition of the soil microflora and the plant species present. As far as the duration of existence in the environment is concerned various pesticides may be divided into the following categories :

1. Organochlorine insecticides with persistence in the environment of more than 18 months. Some of these compounds may persist for periods over 10 years.
2. Derivatives of triazine, urea, picloram etc. with persistence in the environment of about 18 months or so.
3. Derivatives of benzoic acid, amides and various other acids with persistence in the environment of about 12 months.
4. Phenoxyalkylcarboxylic acids, nitriles and derivatives of toluidines with persistence in the environment of about six months.
5. Organophosphate pesticides and derivatives of carbamic acid with persistence in the environment of about three months or below three months.

It should be noted here that the persistence of these chemicals in the environment depends on a number of factors. Under favourable conditions of decomposition, the degradation may be brought about rather quickly while adverse conditions may delay it for considerable period. For example, DDT may be eliminated within two years only, under favourable conditions while its persistence for periods ranging between 20-25 years has also been recorded (12;19).

(IV) BIO-ACCUMULATION AND BIO-MAGNIFICATION OF PESTICIDES

Pesticides, even though present in exceedingly low concentration in the soil or the surrounding water are taken up by various microbes, plants and animals which may accumulate and concentrate them several thousand times. The concentration of 0.00001 ppm of DDT, for example, may be magnified to almost 70,000 times in oysters within a period of 40 days. One kg of soil may contain only 0.0001 mg of an organochlorine pesticide whereas carrots grown on this soil may contain as much 2-6 mg per kg and the rabbits feeding on these carrots may contain as much as 22-35 mg per kg of the toxicant. Toxaphene may occur in lake waters only in the concentration range of 0.0002-0.0006 mg per litre but the water plants growing in the lake may contain as much as 0.2-0.4 mg per kg, invertebrates 0.5 - 1.5 mg per kg and trout and salmon may contain as much as 3.0 - 6.0 mg per kg. Toxaphene is decomposed very slowly and even after a period of six or seven years there is little significant change in its concentration. Similar bioaccumulation and biomagnification have been recorded for a number of other pesticides as well (2).

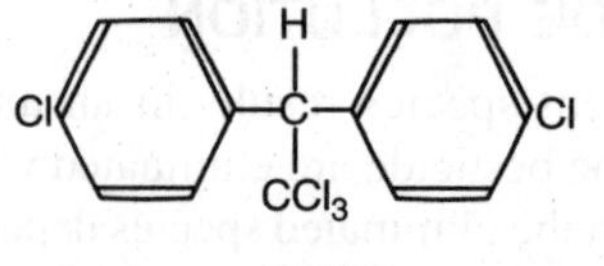

DDT
LD_{50} in male rates: 113 mg/kg

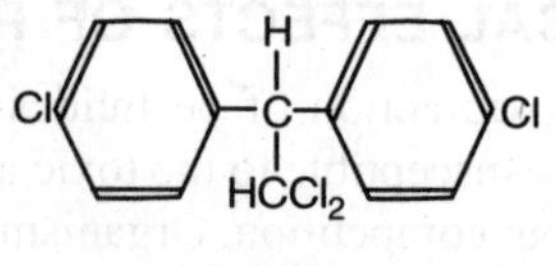

DDD
LD_{50} in male rates: 3400 mg/kg

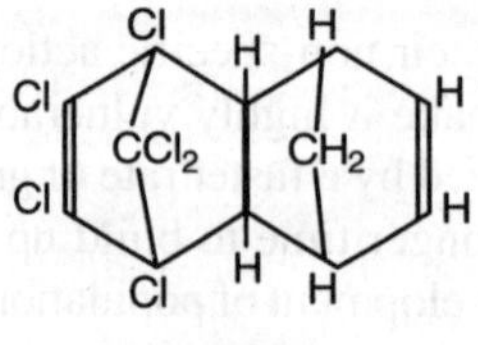

ELDRIN
LD_{50} in male rates: 39 mg/kg

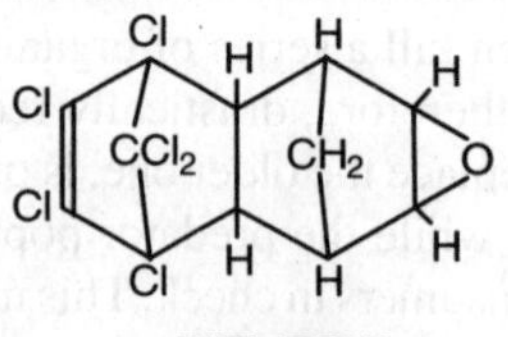

DIEL DRIN
LD_{50} in male rates: 46 mg/kg

$C_{10}H_{10}Cl_{10}$
TOXAPHENE
LD_{50} in male rates: 70 mg/kg

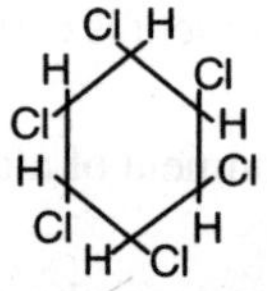

LINDANE OR GAMMAXANE
LD_{50} in male rats: 88 mg/kg

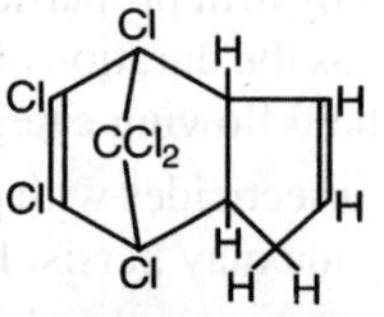

HEPTACHLORE
LD_{50} in male rats: 100 mg/kg

Fig 16.1 *Common organochlorine insecticides.*

The process of bioaccumulation and biomagnification makes exceedingly low quantities of pesticides or their toxic residues available to the living organism in a highly concentrated state. Either the symptoms of toxicity appear in the animal accumulating the poisonous material or it makes the animal poisonous to other organisms, which feed on it. The worst sufferers are animals at the top of the trophic structure. For example, a very low concentration of DDD, an organochlorine insecticide was bio-magnified 5000 times in fishes in Lake Clear of California soon after it was sprayed in 1957. Each fish possessed about 100 ppm of the pesticide and grebes, which represented the top level of the trophic structure, succumbed to its toxicity after eating 20 such fishes (9;10).

Another undesirable consequence of the phenomenon of entry, accumulation and biomagnification of highly persistent pesticides is the transport and dissemination of these poisons to far off places in a highly concentrated state, within the bodies of living organisms. If a bird carries 100 mg per kg of DDT within its body, with thousands of birds migrating from one place to another a large quantity of the pesticide, about 1 kg per 10,000 kg of bird-biomass, shall get transported to new regions where it is released after the birds die and their bodies decay. Thus, the pesticide applied in Florida may appear in Polar Regions and cause toxicity there (11).

(V) CLASSIFICATION OF PESTICIDES

In general, pesticides are classed into several groups, which are named after the type of organisms against which they are used. As far as the chemical nature and structure of these substances are concerned, there occurs a bewildering variety. Various organo-chlorine, organo-phosphate derivatives, carbamates, thio-carbamtes, triazines, bipyridyls, urea, nitriles, nitroanilines, amides, phenols, derivatives of aresenic, copper and mercury, even diesel and kerosene may be employed as pesticides

(2,8). Major groups of these chemicals which are employed on such large scale as to cause problems of poisoning, carcinogenesis, mutagenesis and environmental degradation are:

A. Insecticides.

B. Herbicides.

C. Fungicides.

A. Insecticides

Insecticides are chemicals which are used to suppress unwanted insects. The selective action of most of the insecticides in common use these days, stems from the fact that it implicates such vital systems in the body of the organism to be eliminated which are altogether absent in organisms to be protected. In most cases the target is the nervous system. It is the damaged nervous system which is responsible for the death of the insect (13). As this system does not occur in plants no harm is done to its productivity. Insecticides in common use these days can be grouped into three main categories. These are:

1. Organochlorine insecticides.
2. Organophosphate insecticide.
3. Carbamates and others.

(1) Organochlorine Insecticide :

These insecticides include chlorinated derivatives of ethane such as DDT, DDD, Methoxychlor etc., cyclodiene which include Aldrin, Dieldrin, Heptachlor, Toxaphene etc, and hexa-chloro-cyclohexanes such as Lindane. In general these are very stable chemicals which can withstand the action of various environmental factors, like temperature, solar radiations and moisture for long duration of time. In a biological system also they are degraded with considerable difficulty. It is stability and persistence of these synthetic organics, which is responsible for their prolonged toxicity to insects and pests. Most of these chemicals dissolve well in organic solvents and fats while their solubility in water is very poor. They have a broad spectrum of action, which implicates the nervous system of the insect causing disruption of transmission of nerve impulses. The animal dies of malfunctioning of the nervous system. Fig 16.1 lists some of the common organochlorine insecticides.

(2) Organophosphate insecticides

Organophosphate insecticides are one of the most important groups of pesticides in common use these days which are preferred over organo-chlorine derivatives. They are effective over a wide range of pestilent insects and are quickly degraded into harmless metabolites within a living system as well as in the environment. Within a period of three months after the application they are almost completely eliminated from the system. Though fat soluble and capable of rapid penetration into living organisms, they are never deposited to an appreciable extent within a living system. They are required in much lower amounts than most of the organochlorine derivatives.

$(C_2H_5O)_2P(=O)-O-P(=O)(OC_2H_5)_2$

TEPP

$(C_2H_5O)_2P(=O)-O-C_6H_4-NO_2$

PARATHION

LD50 in male rats: 13 mg/kg

$(C_2H_5O)_2P(=O)-S-CH(COOC_2H_5)(COOC_2H_5)$

MALATHION

LD_{50} in male rats: 1375 mg/kg

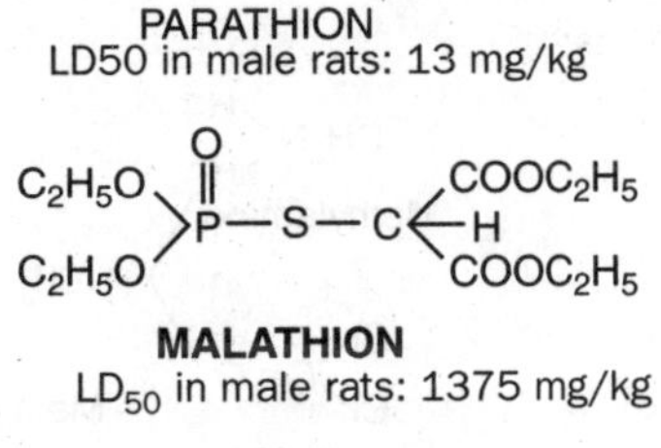

Fig 16.2 *Common Oranophosphate insecticides*

However, an extremely toxic nature and relatively rapid development of resistant populations following repeated use are the main negative features of these insectides. Most of the cases of accidental or intentional insecticide poisoning involve organophosphate insecticides such as parathion or malathion. Moreover, due to the rapid decomposition, the insecticidal action of these substances is not as long lasting as those of organochlorine insecticides. This necessitates their repeated use and makes them costlier.

Like organochlorine derivatives, most of the organophosphate insecticides implicate the nervous system of the target organisms. They are capable of inhibiting the activity of the enzyme acetylcholine esterase which catalyses the removal of acetylcholine from the synaptic cleft after an impulse has passed through the junction. The inactivation of this enzyme causes acetylcholine to accumulate in the synaptic cleft, which as a consequence remains in a charged state blocking further transmission of nerve impulses through the cleft. Some of the important organophosphate insecticides are given in Fig 16.2.

(3) CARBAMATE INSECTICIDES

These insecticides are analogues of carbamic acid, which are very effective against many insects and pests. As compared to organophosphate derivatives, most of these compounds have a lower toxicity to mammalian systems.

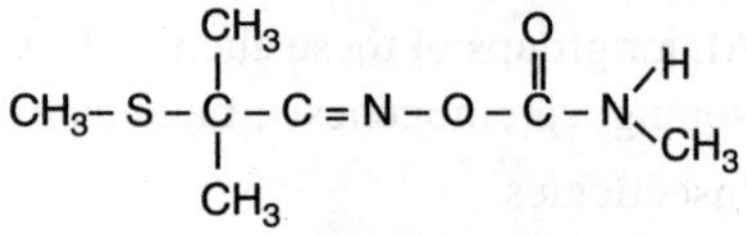

$CH_3-S-C(CH_3)_2-C=N-O-C(=O)-N(H)CH_3$

ALDICARB

LD50 in male rats: 0.8 mg/kg

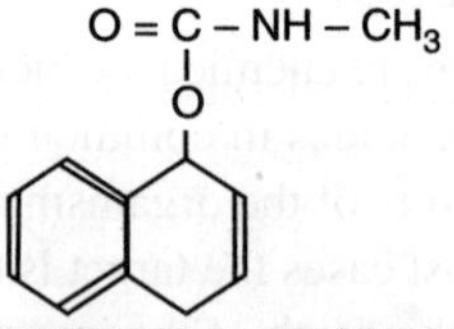

$O=C-NH-CH_3$ | O (naphthyl)

CARBARYL OR SEVIN

LD50 in male rats: 850 mg/kg

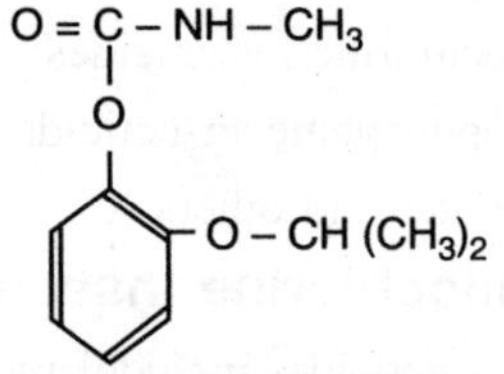

$O=C-NH-CH_3$ | O (phenyl) $-O-CH(CH_3)_2$

BAYGON OR PROPOXUR

LD_{50} in malke rats: 83mg/kg

Fig 16.3 *Common carbamate insecticides*

Carbamate insecticides are readily degraded in the environment and in a living system, also they are easily detoxified and excreted. Apparently, the insecticidal activity of most of these compounds is due to competitive inhibition of enzyme acetylcholine esterase, which catalyses breakdown of acetylcholine in the synaptic cleft between two nerve endings and at the neuromuscular junctions. Like organophosphate insecticides, these also block the conduction of nerve impulses through the synaptic cleft. Important insecticide of this group is given in Fig. 16.3.

The episode of toxic gas leakage in Bhopal, M.P., on December 3, 1984 involved methyl isocyanate, which is used in the manufacture of Carbaryl or Sevin. Though the cause of the tragedy is still shrouded in mystery, scientific explanation suggests that the accidental addition of water in MIC storage tanks could be the cause of the tragedy. The manufacture of Carbaryl involves the following reactions:

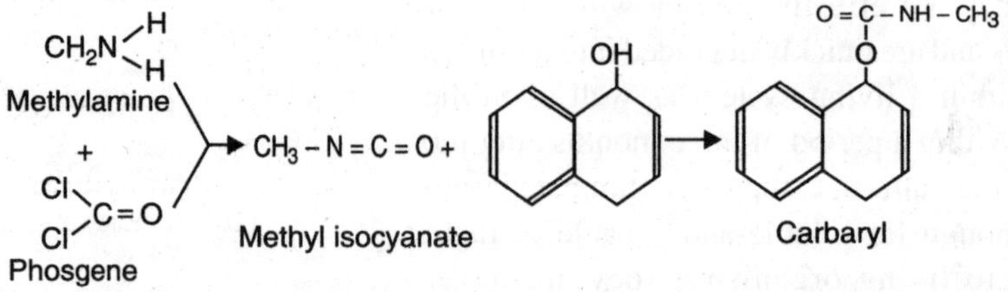

CH_2NH_2 (Methylamine) + $Cl_2C=O$ (Phosgene) → $CH_3-N=C=O$ (Methyl isocyanate) + naphthol (OH) → Carbaryl ($O=C-NH-CH_3$ on O)

MIC is stored with phosgene, which prevents its polymerization. Addition of water causes phosgene to react with water molecules and produce HCl. HCl catalyses polymerization of MIC, a reaction which is strongly exothermic. Accidental addition of water triggered the reaction. An enormous amount of heat was produced due to which things went out of control. About 40 tons of MIC was consequently released. MIC is an extremely reactive, poisonous and volatile chemical with a capacity to penetrate living systems (13). The damages caused in the living organisms, particularly, in biological membrane-systems are irreversible. The tragic incidence left about 2890 people and 3000 cattle, goates and other animals dead while 2,00,000 were affected with its chronic poisoning.

B. HERBICIDES

Herbicides are chemicals, which are used to suppress unwanted plants. The use of these chemicals has increased markedly during the last twenty years. The production of herbicides now rivals or even exceeds those of insecticides in quantity, the volume of application and the total area under treatment. Therefore, these poisonous substances possess almost similar potential to cause environmental problems, contamination of our food and water supplies, as do various insecticides. Fortunately, in India, herbicides are not used on such a large scale as they are in developed countries.

There are only forty or fifty species of plants, which provide 90% of the world's food supply, whereas about 7000 species are considered undesirable since most of them compete with plants cultivated on large scale for nutrition, water and space. The unwanted plants or weeds can cause considerable damage if allowed to grow uninhibited under the highly favourable conditions, which we provide to plants we grow. In smaller establishments such as are frequent in India, it is possible to weed out mechanically most of the useless plants but in larger ones it is not possible to do so. Herbicides have, therefore, become a necessity.

Herbicides prepared with intentions to eliminate unwanted herbs should present little problems of chemical toxicity to vertebrates because of the obvious morphological and physiological dissimilarities between the two. Indeed many herbicides have very low toxicity to mammals. However, there are some herbicides, which are extremely toxic to higher animals including man. Some common herbicides are listed in Fig 16.4.

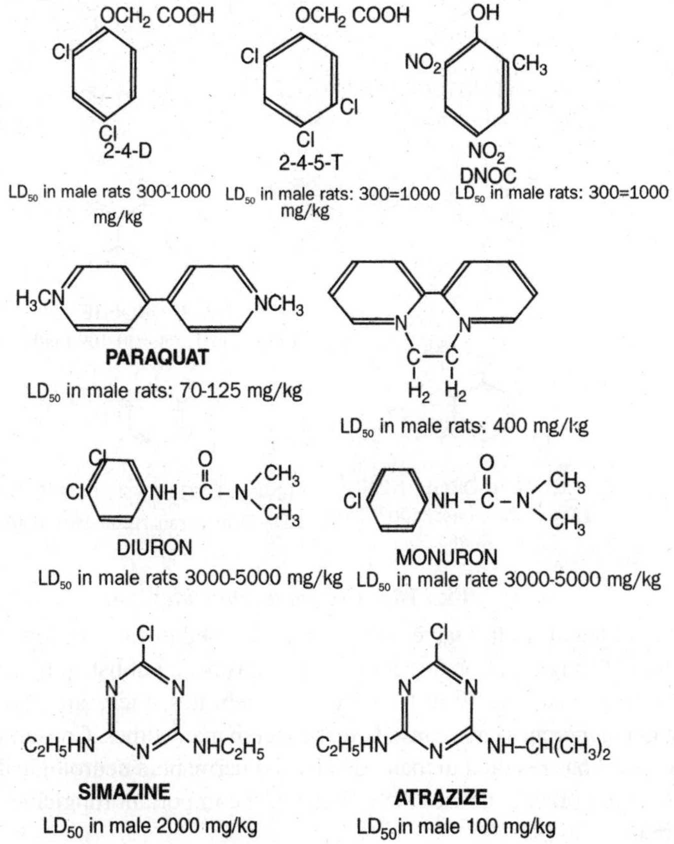

OCH2 COOH
Cl
Cl
2-4-D
LD50 in male rats 300-1000 mg/kg
OCH2 COOH
Cl
Cl
Cl
2-4-5-T
LD50 in male rats: 300=1000 mg/kg
OH
NO2
CH3
NO2
DNOC
LD50 in male rats: 300=1000
H3CN
NCH3
PARAQUAT
LD50 in male rats: 70-125 mg/kg
N
N
C—C
H2 H2
LD50 in male rats: 400 mg/kg
Cl
Cl
NH – C – N
O
CH3
CH3
DIURON
LD50 in male rats 3000-5000 mg/kg
Cl
NH – C – N
O
CH3
CH3
MONURON
LD50 in male rate 3000-5000 mg/kg
Cl
N
N
C2H5HN
N
NHC2H5
SIMAZINE
LD50 in male 2000 mg/kg
Cl
N
N
C2H5HN
N
NH–CH(CH3)2
ATRAZIZE
LD50in male 100 mg/kg

Fig 16.4 *common herbicides*

A great deal of public attention has been drawn to the application of about 44 million lbs of **Agent Orange** by the Americans on about 1.4 million hectares of rich fertile land and forests in Vietnam War during the years 1961-71. It consisted of a mixture of 2-4-D and 2-4-5-T with varying degrees of contamination by tetrachlorodibenzo-p-dioxan (TCDD) which develops during the manufacturing of the herbicide concerned. TCDD is an extremely toxic compound (oral LD_{50} in guinea pigs being 0.0006 mg per kg only) and has been shown to be carcinogenic as well as teratogenic (14;15). The entire stretch of land subjected to the application of Agent Orange was rendered barren. Famine like conditions ensued with hundreds of claims of carcinogenesis and teratogenesis among people exposed to the herbicide mixture (16). There is no conclusive evidence about 2-4-D or 2-4-5-T but the contaminant, TCDD, has been shown to be definitely carcinogenic and teratogenic. This was an incidence of havoc caused by biological warfare, which was more disastrous for the people than a conventional war.

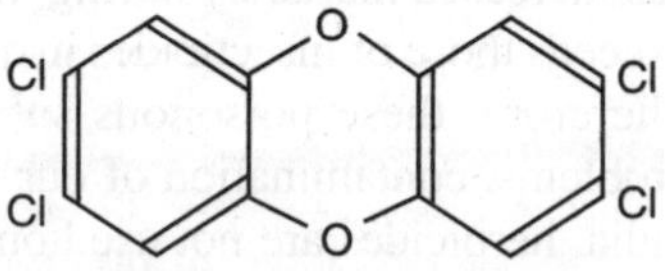

2-3-7-8-tetrachlorodibenzo-p-dioxin

C. FUNGICIDES

Fungicides are chemicals used to kill fungi. These chemicals may be used before the onset of fungal infection in which case they are termed as protective fungicides. The fungal infection is unable to enter and develop in the host plant in presence of these fungicides. Eradication or curative fungicides are applied to cure the fungal infection which has already set in.

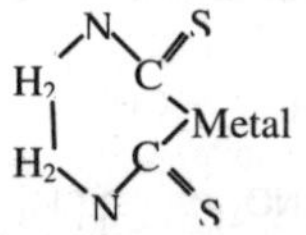

DIETHYLDITHIOCARBAMATE
Mildly toxic to mammalian system

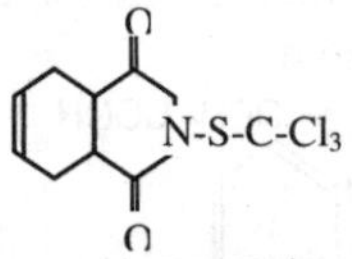

CAPTAN
LD_{50} in male rats: 9000-12500 mg/kg

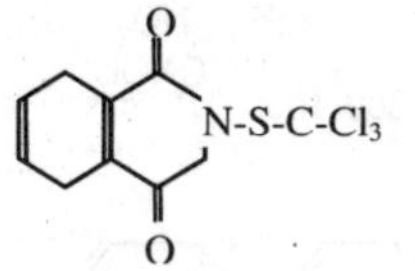

FOLPET
LD_{50} in male rats: 10000 mg/kg

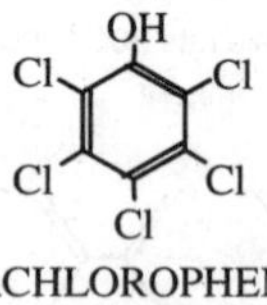

PENTACHLOROPHENOL
LD_{50} in male rats: 30-100 mg/kg

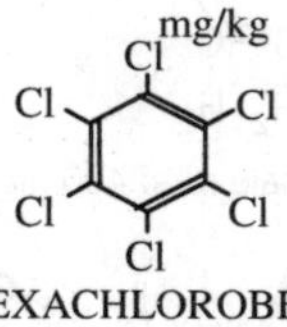

HEXACHLOROBENZNE
LD_{50} in male rats: 3500-4000 mg/kg

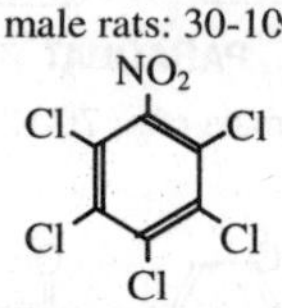

PENTACHLORONITROBENZENE
LD_{50} in male rats: 1200-1650 mg/kg

Fig. 16.5 *Common fungicides*

Detailed toxicological studies have only been performed in cases of few fungicides. A detailed review of the action of fungicides on target organisms has been published by Lukens (17). Although many of these compounds have little toxicity to vertebrates, there are few notable exceptions. Compounds containing mercury have caused great concern many times. Consumption of grains treated with organo-mercurials has resulted in many deaths and permanent neurological disability in humans in Pakistan and Iraq and other countries (18). Some of the important fungicides which are in general use these days are listed in Fig 16.5.

(VIII) DETOXIFICATION OF PESTICIDES AND ALLIED CHEMICALS

The process of detoxification involves abiotic or biotic transformation of pesticides into relatively harmless substances. it has often been observed that toxicity of substances depends on its chemical structure and a slight change in which may result in radical changes in its properties. Loss of toxicity naturally follows if active groups are detached and in many cases a little change in structure of the compound detoxifies it completely. Substances possessing highly branched structure, increased substitution, aromatic compounds, cycloparaffins etc. are degraded with difficulty and hence their persistence (12;13). Degradation of pesticides in environment involves:

1. **Abiotic transformation:** Abiotic transformation includes mainly photolysis, hydrolysis, cleavage of various types of bonds, adsorption by humic material and other colloids and formation of stable complexes within the medium. Solar radiations, presence of moisture, temperature, the chemical nature of the humic material and other substances present in the medium etc. play an important role in abiotic transformation of poisonous substances in the environment.
2. **Biotic transformation:** Biotic transformation involves biochemical reactions brought about by enzymatic machinery of living organisms. Most of the pesticides are lipophilic compounds which are excreted with difficulty. They have to be acted upon by the enzyme systems in such a way as to convert them into hydrophilic ones so that they could be excreted out. The process of biotransformation, however, does not always result in detoxification. In certain cases products much more toxic than the parent compounds are also formed. Import biochemical reactions, which occur in a living system and bring about a variety of transformation, may be:
   a. **Degradation reactions:** These reactions involve oxidation, reduction, hydrolysis, cleavage of important bonds etc, which result in displacement of important functional groups the pesticide molecules by groups like –OH, COOH, NH_2, etc. Many pesticides lose their toxicity in the process.
   b. **Conjugation reactions:** Conjugation reactions, which usually follow degradation reactions involve synthesis reactions in which the pesticide molecule now carrying groups acquired during the earlier reaction, are conjugated to other molecules within the living system to form highly ionized water-soluble substances. These are quickly excreted from the living system.

Although all living organisms are capable of biodegradation, tiny microbes such as bacteria, fungi etc play a very important role in decomposition of pesticides in the environment. They are able to perform diverse biochemical reactions at a much faster rate per body weight as compared to higher plants and animals while the diversity in their species composition ensures that a wide variety of pollutant shall be subjected to the degradation activity. Usually a number of species have to act, each performing its own specific task to bring about the complete decomposition of pesticide molecules. In most of the living organisms, a battery of enzymes is located on endoplasmic reticulum, which is a network of lipoprotein membranes within a cell. Microbial decomposition is brought about mainly by these enzymes. In higher animals including man, the cells of the liver carry most of the functions of degradation though other cells are also capable of performing a certain degree of biodegradation activity (12).

B. Crude petroleum and its distillates

Crude petroleum or mineral oil is the term usually applied to the complex mixture of gases, liquids and associated solids, which occur deep inside earth's crust. The composition of crude petroleum is highly variable. However, all of them contain hydrocarbons both aliphatic as well as aromatic, cycloparaffins, naphthenes etc. In addition to these, crude petroleum also contains a number of compounds of sulphur, nitrogen and a number of metallic constituents like nickel, chromium, cadmium, vanadium iron etc. Crude petroleum is distilled to separate various useful components, which include light oils such as petrol and benzene, medium oils such as kerosene and diesel, heavy oils such as lubricating oils, greases, hard wax and coal tar etc.

(I) Oil spills

Petroleum crude is a precious substance and its distillates drive millions of vehicles around the world, light countless number of homes and provide energy for domestic, industrial and agricultural use. As its distribution is limited to certain places and rock formations, huge amounts of crude petroleum has to be transported across the sea and land from one place to another and so are its finished products. In the process, some of this precious material spills out. In marine environment these oil spills cause enormous damage. Each litre of spilled oil may spread to cover an area of about 4000 sq meter over the water surface while its oxidation requires about 3.3 kg of oxygen, which usually occurs in four hundred thousand litres of seawater.

Oil spills have been occurring ever since large scales transportation of crude petroleum and its distillates started. However, probably the first oil spill which drew enormous public attention, occurred on March 18, 1967 when a Liberian tanker, Terry Canyon, ran aground near the entrance of English Channel spilling about 60,000 tons of crude petroleum into the sea. In January 1969, an offshore oil well blew off near the coast of Santa Barbara in the United States and discharged about 4500 litres of petroleum crude per hour causing extensive damage to littoral and sub-littoral life. In 1978, the much-publicized Amoco Cadiz disaster dumped about 68 million gallons of crude oil along the French coast. The Alaskan oil spill from the 30,000-ton-super-tanker, Exxon Valdez occurred on March 24, 1989 when the tanker plied up on a reef, causing the release of about 11 million gallons of crude petroleum into the clean waters of Prince Williams Sound. As the oil hit 2000 kms of the shoreline about a hundred thousand birds, which included some rare species as well, were killed. A countless number of sea animals died and a couple of bears and dears, which lived on sea animals and plants, were found dead. The largest oil slick, which made history, occurred in 1990 during the Gulf war in which Kuwait's oil installations dotted along the shore were blown up. About 330 million gallons of petroleum crude was discharged into the waters of the gulf which spread over an area of about 700-800 sq kms. A countless number of sea animals were affected and an ecological catastrophe followed. Apart from these major oil spills, minor leakages of crude oil keep occurring all along important trade routes in sea through which oil is transported. While other type of pollution is being minimized, the frequent presence of oily waters and floating tar balls along important trade routes and ports testify to the fact that oil pollution is slowly rising all around the world (20).

(II) Natural oil seeps:

Man made oil spills are not the only source of crude petroleum and associated substances in the sea. Natural seeps developed in submarine rock strata have been discharging crude petroleum into the environment since millions of years. More than 200 submarine oil seeps have been identified around the world and the amount of crude discharged annually by them is far greater than the total contamination from off-shore oil production and transportation activities. Most of these oil seeps occur deep inside the ocean where the slow discharge of petroleum crude is taken care of by natural agencies of degradation. Man-made oil spills, however, tend to be disastrous because they introduce suddenly enormously quantities of crude petroleum on ocean surface which causes problems of toxicity to the organisms exposed to it and an ecological catastrophe.

(III) PROBLEMS ASSOCIATED WITH CRUDE PETROLEUM POLLUTION

The problems associated with pollution caused by petroleum crude and their distillates can be grouped into the following categories:

(1) Light and medium fraction of crude oil:

Light and medium fraction of crude petroleum are mixtures of various hydrocarbons. Lighter of these are gases and volatile substances vapours of which are poisonous.The toxicity of hydrocarbons is usually indirectly proportional to their boiling point and viscosity. As the viscosity and boiling point rise the toxicity decreases. Saturated hydrocarbons or paraffins like methane and ethane are asphyxants, gases that cause suffocation. Some of these hydrocarbons are depressant of central nervous system. Liquid paraffins can remove oil and coating of wax from exposed skin and caused dermatitis and pneumonitis. Inhalation of vapours can cause acute intoxication and anesthesia. Benzene is particularly toxic and long-term exposures even at low concentrations can cause anaemia and leukopenia. Oral ingestion of these chemicals causes symptoms of mucous membrane irritation, vomiting and central nervous system depression. In acute cases albuminurea, haematuria and derangement of hepatic enzyme system may occur. Due to low boiling point and highly volatile nature, light and medium fractions of petroleum crude evaporate quickly and as such pose no long term problems.

(2) Heavier fraction, greases, waxes, tar etc:

Heavy oils, greases, waxes and solid fraction consists of 16-40 carbon joined together in various ways forming complicated chains, rings, polycyclic and heterocyclic structures and fused ring systems. It is this structural complexity, which makes these compounds remarkably resistant to natural decay and degradation. These compounds have a considerably low toxicity but their recalcitrant and refractory nature makes them important environmental pollutants. Heavy oils, lubricants take months to oxidize and be decomposed.

Aromatic, heterocyclic and polycyclic hydrocarbons, which form an important constituent of heavy oils, greases, wax asphaltic bitumen etc. and the smoke derived from their combustion have been implicated in many instances of carcinogenesis. Most of these chemicals are strongly lipophilic in nature and this result in their unrestricted entry into a biological system. Teratogenic (Embryo-toxic) and mutagenic properties have also been attributed to these complex organic molecules. They have been suspected to be involved in affecting adversely the reproductive processes and immune system of living beings. On contact, skin irritation, eczema, cancer of skin etc. are caused by many of these chemicals. Nitro-, chloro-, and sulphur derivatives of these compounds are even more dangerous.

(3) Ecological problems caused by crude petroleum:

The problems caused by crude petroleum and its associated chemicals in marine ecosystems such as land locked bays, main oil tanker routes, ports and harbour areas etc. have been causing us much concern. Important aspects of crude petroleum pollution can be summarized as follows:

1. Crude oil floats over the water surface and forms an insulating layer between the water and the atmospheric air above. This stops exchange of gases, which regularly occurs at the air-water interface. This layer also restricts light penetration, which retards photosynthesis. Bio-degradation of crude oil also consumes plenty of oxygen and adds to the oxygen deficient conditions. Thus oxygen deficient conditions rapidly develop. Oxygen deficit causes large-scale death of aquatic organisms.
2. Oils form a layer around eggs, larvae, smaller plants and animals thus encapsulating them from the surrounding environment, which prevents exchange of gases and other materials. This kills most of the smaller organisms. Even fishes are affected as oil sticking on the gills interferes with their normal respiratory process.
3. Sea birds exposed to petroleum crude are unable to fly when their feathers get soaked in oil. Oil dissolves away the protective oil or wax-like coating as consequence of which feathery coat loses its insulating capacity and thereby ability to withstand low temperatures.
4. A large proportion of compounds present in petroleum crude are lipophilic. They can cross the lipo-protein membranes by the process of simple diffusion and accumulate in lipid and fats present in the body of the organisms exposed. Thus, fishes and other sea-produce may be tainted with oil, which imparts an unpleasant flavour and render them unpalatable to consumers.

5. Water soluble fraction of petroleum crude contains a number of toxic substances, which exert their toxic effects. Sulphur compounds, such as carbonyl sulphide present in the crude can be highly poisonous and so are a number of other compounds containing nitrogen, phosphorus and heavy metal ions almost invariably present in small amounts in the crude oil.
6. Crude oil almost invariably contains a small amount of a number of carcinogenic, teratogenic and mutagenic substances, which pose a grave threat to those exposed to it. However, definite evidence of these dreaded properties is available in case of only few of these chemicals but the evidence so far available should be enough to caution us to avoid such exposures.

(iv) Oil spill cleaning operations

Natural degradation of huge amounts of crude oil takes its own time. However, as long as the oil continues to float over the water surface, it harms the aquatic life and as such cleaning up of oil spills becomes very important. This is done as follows:

(1) Confining the Oil Spill to a Limited Area: It is easier to deal with an oil spill if it is confined to a smaller area. Attempts are also made to prevent the oil slick from reaching the shoreline or shallow areas as it is these regions, which are much affected, and it is in these regions that maximum diversity of life forms occurs. For this purpose, barriers or booms are placed around the oil slick. Once confined to a limited area away from the shoreline, spilled oil may even be left as such for the nature to take care of. In turbulent sea, however, crude petroleum may splash out or pass through from under the floating booms to contaminate fresh areas. When the volume of the spill is large, it is often a stupendous task to confine it to a limited area.

(2) Mechanical Removal of Oil from Water Surface: Once the spill has been contained by the barriers or mechanical booms, it is usually sucked up with the help of skimmers. Skimmers are mechanical devices, which remove the top layers of the oil-contaminated water.

(3) Burning the Crude Petroleum on Sea Surface: Burning of oil on the surface of water has generally been found to be less successful because more volatile fraction evaporates quickly while water below removes heat faster and the fire is extinguished. Burning of oil slick also leads to extensive air pollution.

(4) Use of Absorbents: Absorbents also facilitate an oil slick cleanup operation. They absorb the oil and prevent it from spreading further. Cheap natural materials like saw dust, pine bark, peat moss, straw or synthetic absorbent like polyethylene, polystyrene, polyurethanes etc. are spread over the oil layer. A large quantity of oil can be removed from the sea surface when these materials are removed from water.

(5) Treatment of the Crude Oil with Dispersants: Dispersants are chemical mixtures, which cause oil to spread and disperse in the same manner as soap removes oil and grease from one's hands. It contains a surfactant, a solvent and a stabilizer. The solvent enables surfactant to mix with, penetrate the oil and turbulence causes the formation of emulsion while the stabilizer prevents the emulsion from coalescing once it is formed. Dispersion, therefore, increases the slick surface and is effective in diffusing the crude oil. The process of microbial degradation is also speeded up. However, dispersants contain chemicals, which may be harmful to marine life while emulsified oil droplets may sink down carrying toxic substances with them, which may cause further damage.

(6) Microbial Treatment of the Crude Oil: There are a number of bacteria almost found universally which live on hydrocarbons and are responsible for bio-degradation of petroleum crudes. They occur in large numbers at places, which are rich in oil. A major drawback in these naturally occurring microbes is that their genetic make-up is capable of decomposing only a limited number of hydrocarbons as raw material and that too at a very slow rate. *Pseudomonas* represents such a group of bacteria, which decompose a number of esoteric compounds which most of other microbes are incapable of degrading. As the genes coding for enzymes, which attack hydrocarbons are located not on the main bacterial chromosome but on plasmids, it has been easier for the biotechnologists to pool together genes from different strains of bacteria and produce a super-bug. This super-bug is capable of decomposing a wide range of hydrocarbons at a faster rate than most of the natural strains. Though the super-bug has not been used commercially the day is not far away when such microbes, cultivated in laboratory mixed with straw and dried would be ready to be sprinkled over the oil spill. The microbe would quickly decompose the petroleum crude.

Microbial products can serve as an efficient replacement for chemical dispersants and surfactants. A strain of *Pseudomonas aeruginosa* has been developed which produces large quantities of surface-active agent capable of reducing viscosity and surface tension at the oil water interface. Being a product of biological activity, these chemicals are highly biodegradable. Chemical surfactants on the other hand are highly toxic substances with a very poor bio-degradability and hence could do more damage than good.

c. Polymers, plastics, plasticizers and other wastes resistant to degradation.

Under this category are grouped a variety of substances which form paper, synthetic yarn, the versatile plastic and associated material and a number of other synthetic compounds which are not easily destroyed by natural agencies. Much of these materials appear as solid wastes, garbage, discarded daily in a city's garbage dumps and their quantity has been growing day by day. As the rate of disposal of these solid wastes is slower than the rate of generation, huge quantities of these materials tend to accumulate and create problems. In U.S.A. alone about 410,000 tons to solid wastes accumulate per day while U.K. discards about 25,000 tons of solid stink daily. In Delhi alone about 3000 tons of solid garbage accumulates in a day, to pile up in about a dozen of landfills around the city.

(1) Problems associated with solid wastes resistantto degradation:

Major problems associated with solid wastes consisting of polymers, plastics, plasticizers and other such wastes stem from their highly resistant nature. They are usually compounds made of repeating units of one, two, or few small molecules (called monomers) linked together in straight chains or cross-linked to form complicated giant molecules. The problems associated with solid wastes resistant to degradation can be summarized as follows:

1. Most of these solid wastes persist in the environment for long periods. They do not decompose even when buried deep inside a landfill for periods as long as 20 to 30 years. With rapid strides in the synthetic resin and polymer technology, the volume of such persistent chemical in the solid wastes is increasing at a very fast rate. With much of the space available around rapidly expanding cities already occupied by solid waste dumps for long duration, we have to look for newer sites every day.
2. Many of the synthetic polymers are composed of highly reactive toxic monomers such as nitrogen mustard, epoxy, ethylene imine groups, which are known for their biological alkylating action even at very low concentrations. Though polymers as such are inert and harmless compounds, there is always a distinct possibility that traces of their highly injurious monomers and catalysts used in the manufacturing process may persist in the final products. Moreover, the degradation of these recalcitrant molecules could also yield highly toxic residues. The smoke and fumes, which are produced when the synthetic polymers are burned, are also very poisonous in nature.
3. Many synthetic polymers and plastics contain plasticizers. These compounds impart plasticity and mouldability to the material. Phthalate esters like di-2-ethylhexyl phthalate and di-n-butyl- phthalate are important plasticizers, which are produced in large amounts for the purpose. Some quantity of these plasticizers migrates into the product from the bags and

containers used as packing material from where they are released into the environment. Phthalate ester plasticizers are relatively stable, recalcitrant compounds with a low acute toxicity. Teratogenic and carcinogenic properties have also been shown for these chemicals. Their persistence and bio- accumulation at all levels in the food chain is also a distinct possibility. A concentration ranging from 5.0-130.0 mg per litre has been reported from marine environment in U.S.A. and their occurrence in a number of plants and animals has been recorded. They occur complexed with fulvic acid component of humic substances in the soil as well as aquatic bodies.

4. One of the most important aspect of solid waste pollution is the nuisance caused by leachates. Waters flowing through or percolating down the pile of partially decomposed or undecomposed solid wastes dissolve away plenty of harmful substances, organic remains and carry down associated microbes. These waters may contaminate surface water bodies. Waters carrying these leachates may percolate down the soil strata and spoil sub-surface waters as well. This could render wells, tube-wells and hand pumps etc. useless.

(II) Management of solid wastes resistant

Management of the problem of pollution caused by solid wastes resistant to degradation involves a two-pronged strategy. Firstly, the nuisance caused by leachates has to be dealt with effectively so that the dissemination of harmful ingredients present or generated from the wastes is restricted. Secondly, means have to be devised for decomposition of the garbage safely, quickly and economically. So the management of solid wastes involves:

(1) Handling of the problem of leachates: If there are any serious ill effects associated with ever-expanding solid waste dumps, it is the odour and contamination of our surface or underground reservoirs of water by the leachates, which are responsible for many public health problems. If the leachates flow over land surface, the soil may be affected.

The nuisance caused by leachates can be eliminated largely by a slight modification of waste-disposal practices. If the walls and the floor of the dumping pit are lined with some material like plastic or cement concrete leachates shall remain confined to the pit. These may be drained out later and suitably treated. The practice of controlled tripping as used in Delhi dumps is also helpful in reducing the volume of leachates from escaping the dumping site. The waste are spread in thin layers and compacted with bulldozers after which a layer of soil is laid. In this way, alternate soil and solid waste layers pile up to fill the shallow depression which converts the dried out lake into a re-usable ground. Much of the leachates generated are retained within the confined area due to compact stratification and with passage of time, the trash decomposes on its own.

(2) Disposal of solid wastes resistant to degradation: Quick safe and economical disposal of resistant wastes has become a necessity today. The enormous volume of stinking, ugly looking, perpetually expanding garbage dumps occupy large area of otherwise useful land for a considerable period. Even compost pits dug with costly labour fail to accommodate them. The only solution to the problem is to reduce the volume of the garbage to manageable amount. This can be done by:

1. Incineration of Wastes: Incineration of solid wastes involves the combustion of garbage at a temperature usually above 200°C. It reduces the bulk of solids to ashes while the heat produced may also be utilized in some useful way. Burning of solid wastes, however, produces smoke and harmful vapours like hydrogen peroxide, carbon dioxide and monoxide, sulphur dioxide, various hydrocarbons, and a number of other toxic fumes which further pollute the atmosphere. Moreover, wastes to be incinerated should contain enough material which burns. The solid wastes discarded in most of Indian cities have a very low calorific value as most of the combustible trash like paper, plastics, rubber, polythene etc. is picked by rag-pickers which throng the streets. The solid wastes which accumulate in Delhi have a calorific value as low as 600-800 per kg while many of the solid waste incineration plants in India are designed for burning trash with a calorific value higher than 1500 per kg.

2. **Pyrolysis and Vitrification of solid wastes:** Pyrolysis involves heating of solid wastes at high temperature (600°-800°C) in oxygen deficient conditions so that the trash does not burn but only smolders. There is little change in the bulk of solid wastes and the product of pyrolysis is almost like charcoal, which can be further used as fuel, if burning it does not produce any poisonous waste gases. As we have to provide heat for pyrolysis, the process is expensive. Vitrification often done in situ is a process which may be used for a mixture of recalcitrant and refractory organic wastes which are unaffected by any other type of treatment. Two electrodes are inserted in the heap of solid wastes and a very powerful electric current is passed through them. The heat generated melts mud, glass, plastics and other materials to produce a glass like fusion product, which can be, dumped anywhere since it produces little leachates and creates no problems. All these processes require large investments and high grade of technical expertise so that the entire trash is dealt with effectively without creation of any further problems.

3. **Microbial degradation:** Probably the best and the most economical approach to the problem of solid waste disposal involve microbial degradation of solid wastes. Biodegradable wastes can be conveniently handled in this way. Recalcitrant and refractory solids like plastics, polymers and many plasticizers create problems. These materials are synthetic substances and as such are often too new to the biological system. The microbial system simply lacks the enzymatic machinery needed for their decomposition.

   Recent strides taken in the field of biotechnology has come to be of great help in this direction. Genetically engineered microbes have been produced which can decompose a number of organic compounds which were considered to be non-degradable earlier. It was only an Indian born American Scientist Chakrabarty A.N., who has patented for the first time a genetically engineered strain of bacteria, *Pseudomonas*, which decomposes a number of complex and toxic hydrocarbons. Efforts are also being made to produce biodegradable plastic and polymers through genetically engineered microbes. Biologically produced plastic and polymers shall create much less problem as microbial system shall decompose them quickly and effectively into simpler constituents.

4. **Sorting and recycling of solid wastes resistant to degradation:** As much of the solid wastes generated these days consists of materials resistant to degradation, recycling and re-use appears to be the best solution to the problems. Recycling of paper, plastics and materials resistant to degradation and non-degradable wastes has been in practice since long. Sorting out of the re-usable stuff from heaps of waste material constitutes the first step in the process, which may often involve manual labour. This often renders the cost of the operation exorbitant and makes the final product if not costlier at least as much expensive as the fresh one. In a country like India, poor raddi-wallahs or the garbage collectors make a living from the discarded solids and in the process do a commendable job by removing much of recalcitrant solid trash from the garbage dumps. Pieces of metal, glass, rubber, plastics of various types, etc. are thus removed to be recycled again as finished product.

However, the products derived from recycling operations are often not of the same quality as original ones. Paper, for example, derived from recycled stuff is of a rough quality and has to be used as packing material, in cartons, in corrugated boards etc. With little improvement in the processing and manufacturing technology, a better quality paper can be made from the recycled stuff as has been done in Japan, which recycles as much as 40% of its unwanted paper into new high quality material.

Similarly recycling of trash consisting of polyvinyl chloride (PVC) after retrieval from waste dumps does not yield good quality products. The thin polythene bags or the material of which polythene buckets and containers are made of can be re-used but the result is hard brittle material, which has to be discarded soon. The problem is, therefore, not solved but only delayed. A solution to this problem has been found in compatibilizer molecules which stick on these different types of plastic molecules making the recycled material sturdier, long lasting and having as much gloss and finish as the original one. The material of the transparent plastic bottles is, however, different from those of thin polythene bags. It consists of Polyethylene tetraphthalate (PET) which is capable of picking up poisonous substances present in waste dumps. Recycling wastes consisting of polyethylene tetraphthalate, though easier, is not without any risk. The poisonous material absorbed in waste dumps may persist in the final product. It may be released slowly in the material stored in the containers made of recycles material.

QUESTIONS

1. Give a brief account of the history of use of synthetic pesticides.
2. Discuss the problems associated with the use of synthetic pesticides. What are the ecological effects of their excessive use?
3. What are persistent pesticides? Discuss the problems associated with their persistent nature.
4. Briefly describe the major categories of pesticides in common use these days. How do these pesticides act?
5. What are the special features associated with exposure of mammals to synthetic pesticides. Give a brief account of the chemical basis of their toxicity.
6. How are the persistent pesticides decomposed and eliminated from the environment?
7. What are oil spills? Discuss problems associated with crude petroleum pollution.
8. What are the ecological problems caused by crude-petroleum pollution? Briefly describe the fate of crude petroleum released into the environment.
9. Describe briefly oil spill cleaning operations.
10. Discuss the problems associated with persistent solid wastes. How would you manage, dispose off and eliminate these wastes?

17

Chapter

Pollution of Earth's Surface: Land and Water III. Heavy Metal and Toxic Trace Elements

Pollution caused by substances, on which biotic and abiotic agencies of decomposition are ineffective, is a unique type of pollution. Chemicals causing it are a part and parcel of inorganic matter, which constitutes this planet. Toxic trace elements and heavy metals come under the category of non-degradable pollutants. The problem caused by these elements is in fact due to their concentration in the environment in bio-available state. They cannot be destroyed. Dispersal and dilution in such a manner that these toxic elements are no longer available to the biosphere in toxic state and quantities eliminate this type of pollution.

Hydrogen, carbon, nitrogen, oxygen, sodium, potassium, magnesium, phosphorus, sulphur and calcium collectively constitute about 99.5% of a living organism by weight. Fourteen other elements, which include boron, cobalt, copper, fluorine, silicon, vanadium, chromium, manganese, iron, selenium, molybdenum, tin, iodine and zinc make up the remaining 0.5% of the mass of a living organism. These elements are considered essential trace elements for the biosphere. The rest of the elements, which occur on this planet, are either non-essential for the growth and development of living beings or their function within a biological system has not been demonstrated so far. Some of these trace elements whether essential or non-essential when present in the environment in bio-available state above a certain concentration become harmful to the living organisms.

(I) NATURAL CYCLING OF TOXIC TRACE ELEMENTS AND HUMAN ACTIVITY

Trace elements, essential or non-essential are distributed and redistributed naturally in the environment by both geologic and biologic cycles. Weathering and disintegration of rocks bring them into the soil, streams and rivers. Ultimately they are carried to the oceans to be deposited as sediments. Geologic disturbances in earth's crust finally bring these deposits up as sedimentary rocks. From the soil or water these elements are taken up by plants in the living system and passed on to higher trophic levels. When living beings die, decay and decomposition release them in the environment. Some of these elements may become air-borne in the process from where they are returned to the soil or water underneath as dry fall-outs or along with rains.

The natural cycling of trace elements has, however, been disturbed by human activity in two ways. Firstly, man has caused a rapid increase in the concentration of trace elements in various components of the environment and the biosphere. Secondly, he has altered the speciation – the chemical or biochemical form of many of these elements so as to make them available to the biosphere. Man's thirst for minerals from which these elements come is enormous. Extensive use and combustion of fossil fuels bring out a number of toxic trace elements which are added to the environment. Naturally occurring rather inocuous minerals are mined, refined and processed to be used by technologically advanced societies. In the process a lot of the non-biodegradable stuff is converted to harmful state or is simply left over as wastes to be added, ultimately to the environmental load of pollutants.

(II) TRACE ELEMENTS IN A BIOLOGICAL SYSTEM

Trace elements including the heavy metals play a very important role in a living organism. These elements serve as bio-stimulants, catalysts, co-factors etc. and form indispensable structural component of a number of macro-molecules. This is demonstrated by the fact that a number of trace elements occur regularly in a living system in a definite amount and proportion. If they are deficient or absent altogether, the organism fails to grow normally. However, even those trace metals which are essential for growth of an organism may become harmful if their concentration is raised a little. There is rather a thin line demarcating the deficient, sufficient and the condition of excess supply.

The toxic action of most of trace metals stems from the fact that they are capable of inter-acting and forming strong bonds with metabolically active groups within a living system. Under normal conditions these bonds serve to provide important linkages which keep an organic molecule in proper configuration or shape to perform some specific function. But under conditions of an over-supply these bonds attach at random and cause symptoms of toxicity. The injurious action may be due to:

1. Distortion of configuration or shape of important structural or functional organic molecules.
2. Replacement of metabolically essential elements.
3. Occupation of or saturation of important reactive sites on organic molecules to render them metabolically defunct.
4. Formation of organometallic complexes or clumps involving a number of macromolecules at the same time rendering them useless for metabolic activity.

The toxicity of a trace element is a function of its concentration at the site of its action. To reach it trace elements have to cross a system of biological membranes. Uptake of these elements within a living being across the lipoprotein membrane system usually involves an active transport mechanism which is often impeded as toxic concentrations are reached within the cell. However, the capacity of many of these elements to form organometallic bonds imparts lipophilicity to these toxicants. The lipophilic nature enables them to diffuse through the lipoprotein membrane system and thus gain free access within a living being. Many of the problems of metal toxicity stem from the formation of organometallic compounds in the environment, which makes these elements more active and mobile within a living system.

(III) THE SOURCE OF TRACE ELEMENTS IN TOXIC CONCENTRATIONS

Being a part and parcel of the inorganic matter which constitutes this planet, toxic heavy metals and trace elements usually occur dispersed in very small amounts in soils, aquatic systems or in the biosphere. The concentrated deposits of these elements generally occur in insoluble state in rock deposits and sediments, which are usually not available to the living systems. The natural rate of weathering, decomposition and dissolution is normally pretty slow and as such there is hardly any threat of trace element pollution. Volcanic activity, may indeed contribute substantially to the load of trace element pollution. All over the world there is one volcano or the other which is in a state of active eruption all round the year. However, the problems created are usually of a local nature, which are confined to areas in the immediate vicinity of the emissions. It is mainly human activity which has added enormous quantities of these elements in various compartments of the environment. These may be summed up as follows:

1. Combustion of Fossil Fuels: An important source of heavy metals and trace elements in the environment is combustion of fossil fuels and organic matter. In coal, petroleum crude and dead organic matter most of these elements occur in small amounts. Ash or fly-ash or unburned scum left after the combustion is rich in a number of toxic trace elements which are usually added to the environmental burden of pollutants. Average trace element content of Indian coal (Gondwana coal) as analyzed from a number of samples is given in Table 1.1.

At the present rate of coal consumption in power generation units, which is more than half of the total coal production in India, we are adding almost 15 million tons of ashes in the environment annually. Almost one-third of this enormous mass goes to the atmosphere and the rest is dumped on land or in water. Ashes from combustion of Indian coal consist mainly of silica, alumina, iron oxide, carbon, potassium, calcium, magnesium, sodium, sulphur and traces of titanium, phosphorus and significant levels of trace metals and heavy metals. Cumulatively ashes represent about 15-40% of the total weight of Gondwana coal used in India, South Africa, Australia and Brazil. The term Gondwana refers to the rock formations which contain coal-deposits.

Table 17.1 *Trace element concentration of Indian coal and fly-ash.*

	Element	Content in coal*	Content in Fly-ash*
1	Lead	118.0	416.0
2	Zinc	38.0	112.0
3	Copper	145.0	90.0
4	Chromium	50.0	121.2
5	Nickel	38.0	112.5
6	Cobalt	22.5	3.5
7	Cadmium	8.5	10.5
8	Uranium	6.4	9.2
9	Thorium	16.0	32.5

*as ppm.

As a consequence of consumption of large quantities of fossil fuels urban localities around the world possess a higher concentration of heavy metals and trace elements in their soil, atmosphere as well as the plants, animals including man. For example, atmospheric concentration of 0.03-0.65 mg of lead is present per cubic metre of air sampled in busy streets of Kanpur while in samples taken from about 30 kms away it was barely detectable. In soil samples from the city 350-400 mg per gm of lead was detected while soils collected from fields 30 kms away from the city only 100-85 mg per gm of lead could be found. *Casia tora* plants growing in the city were found to contain about three times as much lead as was found in plants collected from rural areas about 30 kms away.

2. Contributions from Industrial Activity: A number of toxic trace elements are introduced into the environment as a consequence of industrial activity. Many industrial establishments burn huge amounts of fossil fuels for energy and extensively use a number of compounds of toxic trace elements. For example mercury finds wide spread application in the manufacture of plastics, chloralkali units, electrical and electronic industries etc. Effluents from these industries contribute substantially to the mercury content of the environment. In fact the tragedy of Minamata Bay, Japan, was due to the industrial discharge of mercury from plastic and paints manufacturing units. The episode of cadmium poisoning in Japan, Itai-itai, has been attributed to the industrial discharge of cadmium. Similarly much of the chromium present in the environment arises from leather tanning, explosive, photography, ceramics, pigments and paints producing units. Effluents from textile mills and electroplating establishments may contain as much as 25-40 ppm of chromium.

3. Mining and Processing Wastes: Mining of minerals and their processing to obtain the required metal have created enormous ecological disaster areas at many places around the globe. Most of the mineral deposits occur as complex mixtures of a number of elements. In the process of mining and extraction of a particular metal the entire mass is excavated, laid bare and exposed to environmental agencies of weathering, degradation and transportation. This results in extensive contamination of surrounding areas. During concentration and processing of the ore for extraction of the metal plenty of useless finely powdered stuff or tailings are left which are disposed of in the vicinity of the mining and processing establishments. Moreover, smelting and subsequent treatment of ores results in emission of arsenic, lead, cadmium, mercury, sulphur dioxide etc. in large quantities which is discharged into the atmosphere. Sulphur present in tailings or in exposed deposits cause enormous amounts of acid-mine waters which damage streams and rivers and even threaten under-ground aquifers.

Clark Fork Basin, Montana, USA, has been site of more than a century of mining and smelting activity. It includes what was earlier known as the largest open pit in the world – the Berkley pit copper mine. This pit and the network of underground mines contain more than 500 billion litres of acid-mine waters which contaminates Clark Fork river and has spoiled adjoining ground water acquifers. The water and sediments of the river are contaminated with arsenic, lead, zinc, cadmium and other metals while the soil throughout the valley is contaminated with smelter emissions (1). Another dramatic example of the effect of mining waste disposal is Pangua Copper mine at Bougainville, an island in Papua, New Guinea. Before its closer mining operations dumped about 130,000 tons of metal contaminated tailings per day, amounting to a total of about 600 million tons which cover 1800 hectares in Kawarang Jeba river system. No aquatic life survives in the river. It was the local anger at the destruction of the area which precipitated a civil war ! (2).

Worldwide smelting of copper and other non-ferrous ores only, release an estimated six million tons of sulphur dioxide in the atmosphere each year. This is almost 8% of the total emission which is the primary cause of acid rains. Uninhibited smeltings have caused the formation of dead zones where little or no vegetation survives at many places in the world. One such zone is around Sadburry, Ontario, USA, where acid fallouts have destroyed fish population in lakes about 65 kms away. In Japan about 6700 hectares of rich cropland has become too much contaminated with heavy metals and trace elements for rice production. In U.K., about 400,000 hectares of agricultural land have been lost to metal smelters since the Roman times (1). However, in spite of all this new dead zones are still being created all over the globe to meet the ever growing demands of humanity for metals and minerals.

(IV) PROBLEMS ASSOCIATED WITH TRACE ELEMENT POLLUTION

Trace elements in general cause only local pollution problems. Environmental significance of the enhanced levels of these elements is judged in terms of the degree of toxicity, the extent of exploitation of the element, their application and consequent mobilization into the air, water and soil. Problems caused by trace elements may be summed up in brief as follows :

1. Higher levels of trace elements are injurious to plants, animals and microbial component of the biosphere.
2. Chronic and sub-lethal effects of trace element pollution at low concentrations may at times evade detection and general non-specific type of ailments follow which in turn evade redressal.
3. Their persistence in the environment and subsequent transformation into more toxic state are distinct possibilities.
4. Many trace elements and heavy metals are rendered lipophilic in nature as a consequence of formation of conjugates with organic molecules. This provides them a free access into a biological system and makes them more dangerous as they can be readily taken up by living organisms.

5. Bio-concentration and magnification in the biosphere may confront us with highly toxic levels of trace elements. This may substantially damage our food supplies, water resources and agricultural land.
6. Children are particularly susceptible to higher levels of trace elements. With chronic or sub-lethal concentrations present in food, water or air, a higher burden of these elements may gradually be built up in their bodies. This in the long run may reach toxic levels or else it may interfere with therapeutically administered drugs given for the benefit of the individual concerned.
7. Synergistic effects, when two or more than two such elements are involved may greatly enhance the trace element toxicity.
8. Carcinogenic, teratogenic and mutagenic effects may occur even at low concentrations which often evade detection.

(V) SOME TOXICOLOGICALLY IMPORTANT TRACE ELEMENT AND HEAVY METALS

During the last five or six decades an increasing amount of trace elements including heavy metals has been discharged into the environment by industrial establishments, transport activities and power generation units etc. From time to time an increasing concern has been expressed about contamination caused by trace elements like lead, mercury, cadmium, chromium, arsenic, and nickel. Some of the toxicologically important elements are being discussed below:

1. Arsenic

Common sources of arsenic are its sulphide ores such as Arsenopyrite (FeAsS), Mispickel (FeSAs), Realgar (AsS), Cobaltite (CoAsS) etc. The element is produced as a by-product from metal mines in Australia, Canada and Brazil. Smelting of lead, copper, gold and iron ores yields volatile oxides of arsenic, which gets deposited in flues from where they are collected and refined. Leading producers of arsenic are Sweden, Mexico, France, South West Africa (Namibia) and some Russian states.

Arsenic is ubiquitous in distribution. In limestones and siliceous deposits, its concentration ranges between 0.5 to 2.0 ppm while in volcanic rocks as much as 20 ppm of arsenic could be found. The highest concentration of arsenic almost 10,000 ppm has been recorded from Waitapu Valley in New Zealand and from Buns in Switzerland. Natural processes such as weathering and degradation of rocks annually release about 80,000 tons of arsenic into the environment. Global emission of this metal due to human activity has been estimated to be about 240,000 tons per year.

The total daily intake of arsenic from all sources in India by humans ranges between 0.20-0.35 mg per day. Cases of arsenic induced dermatitis, jaundice and damaged liver have been reported from West Bengal where people were forced to drink underground water with high arsenic levels. Trivalent compounds of arsenic are known to be the main toxic forms of this metal. Absorption of trivalent arsenic through lungs depends upon the particle size, but its absorption from intestinal track is almost complete nearly 95%. It tends to accumulate in nails and hairs. Excretion of absorbed arsenic is predominantly through sweat and urine.

A number of proteins and enzymes containing sulphhydril groups have been found to be altered by exposure to arsenic. Mitochondrial enzyme system are particularly affected which disturbs tissue respiration. Ingestion of large doses of arsenic is followed, after a brief period, by a feeling of constriction at the throat, difficulty in swallowing and severe pain in stomach. This is followed by vomiting and diarrhea. Weak pulse and breathing is usually the symptom preceding respirocardiac arrest in acute cases.

Chronic arsenic poisoning is characterized by a general feeling of weakness, nausea, loss of appetite, often vomiting and diarrhea. Copious secretions of mucus in respiratory track, hoarseness, bronchitis and liver dysfunction or jaundice follow. Nerves of limbs are affected and loss of sensation in hands and feet occurs. Peripheral neuritis, which results in loss of muscle power, leads to difficulty in walking and carrying out co-ordinated movements. There may be local effects on skin causing inflammatory, eczema like changes characterized by blackening and mottling of skin. Though oral administration of arsenic in experimental animals has failed to provide any conclusive evidence, carcinogenic properties of arsenic as a sequel to prolonged exposures to low doses have been well documented (3). Small concentrations of arsenic are known to stimulate plant growth, larger concentrations present in irrigation water may cause reduction in overall yield of crop plants.

2. Cadmium

Cadmium belongs to the same family of elements as zinc and mercury. The metalloid mainly occurs with zinc sulphide ores and in lesser quantities with lead and copper ores. The metal is obtained as a by-product of mining and smelting of zinc and lead. Cadmium is a rare element, which was discovered only in the year 1817 and its industrial use was rather insignificant till the Second World War.

Cadmium gains entry into the environment from mining and metallurgical operations, electroplating industry, units manufacturing polyvinyl chloride, plastics, Ni-Cd batteries, several paints, pigments and dyes and the combustion of fossil fuels etc. Cadmium contamination of water may also come from use of metallic and plastic pipes while superphosphate fertilizers, sewage sludge and automobile tyres also contain some amount of this toxic metal. As cadmium is found associated with zinc, copper and lead in mineral deposists, it is released in the environment during mining and processing operations carried out for the extraction of these metals and this constitutes the most important source of local cadmium pollution. As estimated in 1980, about 2100 tons of cadmium are released into the environment annually.

Respiratory absorption of cadmium is about 15-20% whereas upto 5% of cadmium ingested orally is absorbed from the gastro-intestinal track. About 50-70% of cadmium content of the body is present in liver and kidneys. Since the excretion of this metal is appreciably slower than the rate of its entry into a biological system, there is a progressive accumulation of cadmium in soft tissues of the body, particularly in kidney and liver. Due to its potential for gradual accumulation, much concern has been expressed about the regularly growing levels of cadmium in cereals, vegetables and other materials (4). Autopsy studies done on dead bodies or carcasses belonging to earlier periods have shown that the body burden of cadmium is gradually rising with growing mobilization and industrial use of this metal from its deposits.

Cadmium is a potent enzyme inhibitor. It interacts with sulph-hydryl groups of several enzyme systems and forms metal protein complexes within a cell, a property that is responsible for its accumulation in kidneys and other tissues without any obvious symptoms of toxicity. A number of studies conducted on experimental animals as well as on tissue culture models confirm the protective role of cadmium-protein complexes or metallothineins, as they are commonly called. It has been suggested that toxic responses occur only when exposures exceed the capacity of the cellular system either to provide enough proteins to bind the excess cadmium or to synthesize the metal protein complexes. Much of the cadmium, which is excreted via urine in cases of chronic exposures, occurs bound to proteins. It is free cadmium or cadmium ions, which are responsible for causing whatever damage, which occurs due to this metal. Acute toxicities of cadmium are rare. It may result from accidental exposures or deliberate ingestion of relatively high concentration of cadmium in drinks or food. Nausea, vomitting, abdominal pain could occur following consumption of drinks containing only 15-20 mg of cadmium per litre (5). Inhalation of cadmium fumes may produce acute chemical pneumonites and pulmonary oedema.

Low-level exposures spread over long durations produce chronic pulmonary and renal tubular diseases. Skeletal and cardiovascular systems may also be affected in such cases. Obstructive lung diseases result from chronic bronchitis, progressive fibrosis in lower portions of lungs leading to emphysema. Cadmium damages kidneys as a consequence of which proteins, amino-acids, glucose, phosphates and other minerals are excreted out with urine. Morphological changes are rather non-specific consisting of tubular cell degeneration, progressive interstitial inflammatory reaction and fibrosis. Adverse effects of cadmium on calcium metabolism result in increased excretion of calcium and renal calculi. Associated skeletal changes include pain in bones, which become fragile and in severe cases, skeletal deformities may ensue. Hypertension and incidence of cardio-vascular diseases consequent upon low-level exposures of cadmium has also been verified. Though the potential for carcinogenic properties of cadmium on many mammalian systems is well documented, observations on human subjects are rather inadequate. Similarly, teratogenic effects of cadmium have been observed in a number of animal studies but little conclusive observations have been made on human subjects.

3. Chromium

Chromium is generally an abundant element in earth's crust, which occurs in oxidation states ranging between Cr^{2+} to Cr^{6+}. However, the trivalent and hexavalent compounds of chromium are of a greater industrial importance. An important source of chromium is the mineral chromite, which contains about 68% of this element.

Chromium metal and its salts are used in production of stainless steel, ferrochromium and other alloys, chrome pigments, in tanning of leather, mordant dyeing, wood preservation and as an anti-corrosive agent in cooling systems, boilers, oil drilling muds etc. Important sources, which add chromium to the environment, are ferrochrome production units, refining of ores, chemical industries and the combustion of fossil fuels. Combustion of fossil fuels alone releases annually about 1450 metric tons of chromium into the environment. In rural areas, chromium content in ambient air is usually less than 0.1 ng per cubic metres while in industrial areas it may range from 0.01 to 0.03 mg per cubic metres. Particulates ejected from coal fired power generation plants may contain 2.3 to 32 ppm of chromium. Waste waters from tannaries may contain 10-50 ppm, textile wastes up to 32 ppm while spent chrome liqours may contain up to 4500 mg per litre of chromium. (6;7;8).

Only trivalent and hexavalent forms of chromium are of biological significance. Trivalent chromium is the most common form of the metal in nature and it is in this state the metal almost always occurs in a living system. Hexavalent chromium is capable of crossing the cell membrane and is rapidly reduced to its trivalent state inside a living cell.

Small quantities of trivalent chromium are essential to carbohydrate metabolism in mammals while it is also a co-factor for action of insulin. A little amount of chromium is helpful in improving glucose tolerance in diabetic patients, weak and old individuals. No conclusive evidence of harmful effects of trivalent chromium is available. Most of the mammals can tolerate almost a hundred times more chromium in trivalent state than their usual body burden without apparent toxic effects.

Acute chromium toxicity causes serious renal tubular necrosis. Exposure to hexavalent chromium has been found to cause dermatities, allergic skin reactions, chronic ulceration and injury to nasal septum, gastrointestinal ulcers, etc. (9). Chronic chromium toxicity has been associated with incidences of cancers of respiratory track in occupationally involved workers. Teratogenic properties have also been attributed to this metal. It is not certain as to which form of chromium is a more potent carcinogenic agent. As, it is the trivalent form which is metabolically more active, it is probable that the trivalent chromium bound to nucleic acids should be responsible for the teratogenic and carcinogenic action. At present, however, both forms are considered equally potent carcinogenic agent (10).

4. Fluorine

Fluoride pollution may become a serious threat to mammals, as it is capable of distorting the skeleton, which may change an individual into a useless hump of flesh and blood. Minerals like Fluorspar (CaF_2), Fluor-apatite [CaF_2, $3Ca_3$ $(PO_4)_2$] and Cryolite (Na_3 AIF_6) are the main source of fluorine in nature. In India, fluor-apatite and cryolite are common. The disintegration and decay of these minerals bring fluorides into the soil. From the soil fluorides are absorbed by plants which may contain about 0.1% of fluorine in their ash content as fluorides. Fluoride particles may get air-borne as well.

Fluoride pollution in India is generally caused by waters which permeating through detritus and alluvia or fluorine bearing strata may acquire as much as 15-22 ppm of fluorine content. Such waters if supplied to the city may cause mass scale fluorine toxicity.

Fluorides are taken up by living organisms through lungs and intestines. Only a fraction of fluorides absorbed is excreted – most of it is retained inside. Fluorine is an important constituent of harder portions of bones, such as ball and socket joints, enamel of teeth etc. However, it is present in very small quantities about 0.1 to 0.3 ppm in human body. Its deficiency induces osteoporosis, dental caries and bones lose their rigidity. However, there is a very thin line demarcating deficient, sufficient and conditions of excess supply as the presence of about 5-12 ppm of this element becomes toxic. Plants take up fluorides more readily from water than from the soil and accumulate it forming fluoracetate and fluorcitrate. It enters livestock through the vegetation. In fact, fluoride toxicity has done more damage to livestock on worldwide basis than any other type of pollution.

Higher fluoride concentration has deleterious effect on teeth and bones. It interferes with calcium and phosphorus metabolism. Teeth acquire a dull white coating and brown patches. Enamel becomes brittle and may break away. Bones lose their elasticity and arthritic conditions follow. Bones become prone to fractures. Fluorine is an effective inhibitor of many enzymes, notable among them are enolases, which are important constituent of the glycolytic pathway. Its presence in toxic concentrations impairs glycolysis, which disturbs the overall metabolism of an individual.

5. Lead

Lead sulphide or galena is the main source of lead. About four million tons of this metal is produced in the world each year. It is the ubiquitous toxic metal and can be detected in practically all components of the environment as well as of the biosphere. Lead is toxic to most of the living beings while it has no known function within a biological system. Lead finds wide application in industries and effluents from these units contain a good quantity of this element. Lead may come from pipes carrying water, combustion of fossil fuels and from exhausts of automobiles run on petrol containing lead compounds, which are added as anti-knock agents.

Plants growing near busy highways are regularly exposed to fumes and smoke discharged from automobiles containing plenty of lead. It has been estimated that almost 450 tons of lead are emitted annually into the atmosphere in Bombay alone through automobile exhausts (11). There is little absorption of lead through root system of plants. It gains entry through stomata and is absorbed by the tissues within. Leafy vegetables, therefore, are likely to possess a higher lead concentration as compared to others. Lead is also accumulated by green plants and is passed on to higher trophic levels.

In animals including man, gastro-intestinal absorption of lead ranges between 5-15% of the total amount ingested. Only about 4-5% of the lead, thus absorbed is retained. The rest of it is excreted. However, children have been shown to absorb much more, about 40% of lead ingested with food and water and retain almost 30% of it (12). Absorption of lead through lungs is rather more efficient and almost complete. Vapours of tetra-ethyl-lead, used as an additive to petrol, may be absorbed through intact skin as it is strongly lipophilic in nature.

More than 90% of the lead absorbed goes to blood where it can be detected in red blood cell associated with its membrane and haemoglobin (13). Later it is distributed to liver, kidney and bones including teeth. It is finally deposited in bones. About 90% of the metal, which enters into a biological system, is locked up in bones, which causes no immediate harm to the organism. Its concentration in bones increases with age of the individual. Any subsequent disturbance resulting in osteolysis, tends to mobilize the deposit of lead in bones leading ultimately to its toxic action. Acute toxicities of lead are of rare occurrence. It may occur only in cases where the intake is deliberate. Ingestion of large quantities of lead salts such as lead acetate produces burning pain in mouth, throat and stomach. This is followed by abdominal pain accompanied by constipation or diarrhoea and often bleeding in severe cases. Finally, there may be failure of blood circulation and termination of liver and kidney functions. The patient passes into coma followed by respiro-cardiac failure.

Chronic lead toxicity is marked by a general feeling of unwellness, fatigue and pain in limbs. Anaemia is usually the first symptom in animals including man following low-level lead intake over long durations. It has been shown that lead inactivates aminolevulinic acid dehydratase and porphyrobilinogen decarboxylase, which cause suppression of haem-synthesis. However, the toxic effects of lead are probably more significant in terms of human agility and performance. Degenerative changes in motor nerves and ganglia occur which result in decreased speed of conductance of nerve impulses to and from the muscles. The individual becomes dull, slow and inactive. In growing child, degeneration of intellect and mental retardation may occur (14;15). Excretory mechanism of kidney is affected by this metal, which may cause discharge of proteins and even blood through urine. Irreversible lead induced chronic kidney pathology is characterized by degeneration of tubular cells, interstitial fibrosis, glomerular schlerosis etc. Lead reduces uric acid excretion and chronic lead toxicities have been associated with development of gouty conditions in the hapless patients. Hypertension has also been found to be associated with a high percentage of lead in blood stream.

Severe lead toxicities have long been associated with sterility and gametotoxic effects in both male and female animals. Lead is capable of passing through the placenta to foetus to cause developmental anomalies and still births. Some clinical studies have found increased chromosomal defects in workers with high levels of lead in blood. Suppression of immune system has been observed in laboratory animals even at such low doses, which do not cause any apparent symptoms of toxicity (16).

Lead poisoning usually causes the presence of characteristic nuclear inclusions, which appear as dense, homogenous eosinophilic bodies under light microscope while cells containing them are usually swollen. Under electron microscope, these bodies are seen to possess a dense core and an outer fibrillary region. They are composed of lead-protein complex and contain a large amount of aspartic and glutamic acids and a little cysteine (17). Experimental studies have pointed out that nuclear inclusion bodies are earliest evidence of lead exposure, which may be observed before any of the functional changes of intoxication in the system are detectable. It is thought that the sequestering of lead in these bodies is helpful in protecting other susceptible organelles.

6. Mercury

Main source of mercury is cinnabar. However, there are at least thirty different minerals in which mercury occurs in more than trace amounts. In chloralkali plants, plastics, electrical, electronic industries and in units manufacturing organomercurial compounds, mercury finds wide spread usage. Effluents from these industries and the subsequent use, abuse and refuse of their products constitute an important source of mercury in the environment. Another source of mercury pollution is the combustion of fossil fuels. Coal on an average contains upto 0.5 ppm of mercury while in fuel oils mercury levels range between 0.1 - 0.35 ppm. It has been estimated that on global scale about 5000 tons of mercury are released from the combustion of fossil fuels alone.

Elemental mercury and some of its organic derivatives are regularly released from earth's crust into the atmosphere – a process often referred to as 'natural de-gasing' of earth's crust. It has been estimated that about 25,000 to 150,000 tons of mercury are discharged by this process while man made sources contribute only 8,000 to 10,000 tons of mercury into the environment.

Based on toxicological characteristics there are three important forms of mercury: elemental mercury, inorganic and organic mercury. Certain anaerobic bacteria in bottom sediments of water bodies convert inorganic mercury to more toxic organo-mercury compounds. For example, *Clostridium cochlearum* under anaerobic conditions transform mercury to highly toxic compound, methyl mercury, while the same is done under aerobic conditions by some bacteria belonging to genus *Psuedomonas*, and *Neurospora crassa*, which is a fungus. Under alkaline conditions, monomethyl mercury could be changed to dimethyl mercury, a volatile compound which can contaminate the atmosphere. Under acidic conditions, dimethyl mercury is converted back to mono-methyl mercury. It is mostly methyl mercury, which is absorbed by aquatic biota. Organo-mercurials are converted to inorganic mercury under oxidizing conditions. Ultra-violet light hastens the process. Reaction with hydrogen sulphide under anaerobic conditions causes the formation of poorly soluble mercuric sulphide, which is mobilized to produce mercuric sulphate under aerobic conditions. Natural processes are, therefore, responsible for inter-conversion of one form of mercury into another (18;19).

Mercury vapours readily diffuse across alveolar membrane and are lipid soluble. Mercury has an affinity for red blood cells and central nervous system. Absorption of metallic mercury from the gastro-intestinal tract (about 0.01%) is not of any toxicological significance. The absorption of inorganic salts of mercury from intestines has been shown to be about 7% only. Organo-mercurials are, however, efficiently absorbed – almost 90-95% absorption has been recorded. Since the rate of excretion of methyl mercury is considerably slower than the rate of its uptake, it is bioaccumulated and biomagnified along the food chain. Kidneys possess largest mercury content following exposure to inorganic forms while organic mercury has a greater affinity for brain. Within a cell, mercury may bind to a number of enzyme systems including those of microsomes and mitochondria resulting in general non-specific cell injury or death. Mercury has particular affinity for ligand containing sulph-hydryl groups and in liver cells it forms soluble complexes with cystein and glutathione which are excreted through bile and are reabsorbed form the intestines.

All forms of mercury cross placenta and reach the foetus, which may possess a much higher concentration as compared to the maternal tissues. This is particularly true of alkyl-mercury compounds, which attain almost twice as much mercury content as found in maternal tissues. Thus, alkyl mercury exposures are very dangerous in case of pregnant females. Although mercury is excreted from the body predominantly through urine and faeces, almost 5% of mercury concentration in blood may appear in maternal milk. Thus, nursing may greatly enhance the neo-natal exposures.

In halation of mercury vapours produces an acute corrosive bronchitis, interstitial pneumonitis and in higher doses central nervous system may be affected which results in increased excitability, tremours, impaired vision, muscular convulsions, madness and paralysis. Ultimately kidney failure occurs. Deliberate or accidental ingestion of divalent mercury salts causes corrosive ulceration, bleeding and necrosis of gastro-intestinal tract. Severe abdominal cramps, bloody diarrhoea and suppression of urination follows. If the patient survives the gastro-intestinal damages, kidney failure occurs within twenty-four hours owing to necrosis of proximal tubular epithelium. Mercurous compounds are less damaging in their action than mercuric salts probably because of their low solubility. Possibly the most important form of mercury in terms of toxic effects is methyl mercury. The main site of injury, which is of a permanent nature, is the central nervous system. The neurons in focal areas of cerebral cortex undergo degenerative changes and necrosis. These results in tremours, inability to co-ordinate voluntary movements, paralysis, impairment of vision, loss of hearing, speech and coma in severe cases. There is no effective treatment known for alkyl-mercury poisoning. Mercury has teratogenic action and is capable of inducing abortions and embryotoxic effects (19).

Two major episodes of mercury poisoning have occurred in Japan, in **Minamata bay and Naigata** because of industrial discharge of mercury compounds. Mercury was absorbed, bioaccumulated and bio-magnified to high levels. Fishes collected from Minamata bay and Agano River contained as much as 10-12 mg of mercury per kg of their flesh and bones. The largest epidemic of methyl mercury poisoning, however, took place in 1971-72 in Iraq when scarecity forced people to eat wheat grains treated with methyl mercury based fungicide, which were intended to be used as seeds. The average methyl mercury content of these grains was about 4.8-14.6 mg per kg. About 6000 people were affected with the poisoning and some 500 deaths occurred (20). Mothers exposed to sub-lethal doses showed only slight symptoms of mercury poisoning but gave birth to retarded off springs and continued to secrete mercury with their milk for long periods. We do not have to stretch our imagination too far to visualize the fate of the unfortunate child who is fed mercury-contaminated milk straight from mother's breasts.

7. Nickel

Nickel is a known carcinogen of respiratory tract. Important sources of nickel are its sulphide and oxide ores. The silvery white metal is used in the preparation of many alloys, such as german silver, stainless steel etc, in electroplating industry as mordant in ceramic industry and as catalyst in many synthetic reactions. Apart from these sources, it is the combustion of fossil fuels, which contributes the largest amount of nickel to the environment. About 70,000 tons of nickel is discharged into the environment from combustion of fossil fuels alone on global scale. About 0.03 to 0.12 mgs of nickels is present per cubic metres of air of our crowded cities. Estimated daily intake of nickel in human diet averages around 165 mg per day (21), while in an earlier report the estimated daily intake in USA was found to range between 300-600 mg per day.

During the past few years, evidence has accumulated that nickel is nutritionally an essential element. It is associated with the synthesis of Vitamin B_{12}. Urease from Jackbeans has been shown to be a nickel-metallo-enzyme. Nickel requirement for urea metabolism has also been demonstrated for soyabean cell cultures. Nickel deficiency in rats is correlated with retarded body growth and anaemia, which, howevor, could be due to nickel's interference with iron absorption from intestinal tract.

People occupationally involved with refining and use of nickel are predisposed to nasal cancers. Nickel carbonyl [$Ni(CO)_4$] was found to be the principal carcinogen. However, it was later concluded from epidemiological as well as experimental studies that Nickel sulphide (Ni_3S_2) could also be responsible for the carcinogenic action as it is capable of producing local tumours at injection sites and by inhalation in rats. In vitro mammalian cell tests show that both nickel sulphide and nickel sulphate give rise to mammalian cell transformation (22).

Most common toxic effects produced as a sequel to nickel exposures in large amounts include dermatitis and respiratory disorders. Nickel inhibits the activity of a number of enzymes such as maleic dehydrogenase, cytochrome oxidase and isocitrate dehydrogenase while its powder or dust is carcinogenic. Nickel carbonyl, formed because of reaction of nickel with carbon monoxide is a volatile compound, which is the most toxic of all the forms of nickel. It is a carcinogen and half an hour of exposure to about 30 ppm could be lethal to humans. The illness begins with headache, nausea, vomiting and epigastric or chest pain, which are accompanied by fever and leukocytosis. More severe cases proceed to pneumonia, respiratory failure and eventually cerebral oedema and death. Post-mortem studies show the largest concentration of nickel in lungs with lesser amounts in kidneys, liver and brain (23).

QUESTIONS

1. What do you understand by natural cycling of toxic trace elements? In what ways man has disturbed these cycles ?
2. Enumerate briefly the source of trace elements in toxic concentrations in the environment. Discuss the role played by these elements in a biological system.
3. What are mining and processing wastes ? Discuss the problems associated with trace element pollution.
4. Write a brief account of toxicity of any two of the following elements: Lead, Mercury, Cadmium, Chromium, Copper, Arsenic.
5. Discuss the problems associated with leachates or tailing and other wastes from mining and processing industry.

18

Chapter

Problems Caused by Harmful Physical Agents: Noise, Thermal and Radioactive Pollution

Heat, sound and radioactive pollution are consequences of excess amount of energy used or generated for some purpose or produced as a by-product of some form of activity. While the generation of heat and sound energy involves conversion of one form of energy into another, nuclear energy is a consequence of disintegration of unstable nuclei, fission or fusion reaction. The pollution caused by various physical agents is not of the usual type involving addition of unwanted or harmful factors to the environment, instead it is the excess amount of energy released in the environment, which causes the troubles. Though apparently harmless in moderate amounts, heat, sound and radioactivity can cause severe damages to the biosphere if the system is exposed to them for long durations or their intensity is increased.

Many industries use heat for various purposes and need cooling systems, which employ water. This results in production of heated waters in large quantities, which may be discharged as such in the environment. Various activities in our cities and most of the industrial establishments generate high levels of noise, which not only interfere with hearing but also affect the peace of mind, irritability, health and behaviour of an individual. Radiations from man-made sources threaten to damage not only the existence but also the very future of mankind. Physical forces cause mainly the following three types of pollution:

(A) Noise pollution.

(B) Thermal pollution.

(C) Radioactive pollution.

(A) NOISE POLLUTION

Sound is a special kind of wave action, which is usually transmitted through air in the form of pressure waves. These waves are received by hearing apparatus of animals, including man, transformed into electrical impulses in the ear and carried to the brain, which enables us to hear. Sound waves are generated in a number of ways, such as explosive expansion of gases, turbulent movement of liquids, vibrations of solid objects etc, which start a series of pressure waves in all directions. The intensity of these waves diminishes as the distance from the object producing them increases. Sound waves are reflected and deflected by objects, which happen to come in their way. Porous materials, such as perforated board, absorb these waves. When two or more than two such waves superimpose they reinforce each other and in opposite phase they may cancel each other out.

(I) THE MECHANISM OF HEARING

The hearing machinery of our body consists of three parts: the outer, middle and the inner ear.

The outer ear comprises of two curled shells each on either side of the head and a tissue thin membrane, about the diameter of a pencil, tightly stretched like a drumhead, which lies in-between the outer and the middle ear.

The middle ear consists of a hollow cavity barely large enough to hold five or six drops of water. A set of three tiny bones, the hammer (malleus), anvil (incus) and stirrups (stapes) span across the empty space. At one end of the chain, the handle of hammer is connected to eardrums while towards the inner side stirrups are fastened to another membrane, an oval window about the size of a pinhead. This membrane separates the middle ear cavity from the fluids and the canals, which form the inner ear and control our sense of balance. Immersed in the fluid are numerous hair cells or ciliary cells lining the inner-ear chamber (1).

Table 18.1 *Loudness of Some Common Sounds on Decibel Scale.*

	Source of noise	Approximate loudness
1	Threshold of hearing	0 dB
2	Churches, hospitals etc.	10 – 10 dB
3	A room in a quiet house	20 – 30 dB
4	Public library	30 – 40 dB
5	Offices	40 – 50 dB
6	Normal conversation	50 – 60 dB
7	Normal city traffic	60 – 70 dB
8	Alarm clock	70 – 80 dB
9	Heavy city traffic	85 – 95 dB
10	Jetliners 150 metres overhead	100 – 115 dB
11	Running motor cycle	115 – 120 dB
12	Threshold of pain to human ears	120 – 140 dB
13	Jet planes – taking off	140 – 150 dB
14	Launching of space rocket	160 – 180 dB

In the process of hearing sound waves are funneled in ear canals, strike the ear drums and make it vibrate. These vibrations travel along the bony chain (hammer, anvil and stirrups) to the oval window which transmits them to the fluid in the inner ear. Therefrom, nerve endings whose magic is still little understood, these vibrations are translated into electrical impulses which are flashed along nearly 30,000 fibres of auditory nerve to the brain and we hear a sound. The hair cells or ciliary cells which line the inner ear canals and chambers play an important part in picking up these vibrations. Obviously with such a delicate mechanism many things can go wrong, if we subject it to too much strain (1).

(II) THE MEASUREMENT HEARING ABILITY AND THE LOUDNESS OF SOUND

Hearing ability of an individual is monitored by Audiometric tests, the most common technique of which is referred to as the Threshold technique. It is based on the determination of the minimum sound level, which an individual can hear. Starting from a zero value the sound level is gradually raised. The level of sound, which an individual can just perceive, is noted. This is followed by gradual lowering of volume of a louder sound and the level of sound, which is no longer audible to the subject, is recorded. An average of the two values is taken as the threshold of hearing of the individual concerned (2).

The intensity or loudness of sound is measured on a scale called decibel scale or dB-scale. It measures the loudness of sound in terms of relative units of energy or power on a logarithmic scale in accordance with the response of human ear. The scale starts from O dB, which is considered as the threshold of hearing – the faintest sound which human ear can hear. A sound of 10 dB is 10 times louder than 0 dB. A sound of 20 dB is 10 × 10 or 100 times louder, of 30 dB 10 × 10 × 10 or 1000 times, 40 dB is 10 × 10 × 10 × 10 or 10,000 times while a sound of 50 dB is 10 × 10 × 10 × 10 × 10 or 100,000 times louder than the threshold of human hearing. The loudness of some of the common types of sounds on dB scale is given in table 18.1.

(III) NOISE POLLUTION – THE SITUATION TODAY

Table 18.2. *Permissible Noise Levels in Different Types of Localities.*

	Locality	Noise level	
		Day	Night
1	Industrial area	75	65
2	Commercial area	65	55
3	Residential area	55	45
4	Silence zones for hospitals, educational institutions etc	45	35

In urban localities all over the world, noise pollution has been recognized as a major factor affecting public health and well-being. It is an ever-growing nuisance. Man-made sources are mainly responsible for increasing the ambient noise level in urban localities. Automobiles, industrial units, low flying aircraft and loud speakers have been recognized as a major source of noise affecting a large number of peoples. Thickly populated, poverty-stricken third world countries are however, the most affected ones, wherein the ill effects of noise pollution are rarely given serious attention. A survey in Delhi, Bombay and Calcutta conducted in 1988 revealed daytime noise levels ranging between 60 dB to 90 dB in residential localities. Many times during the day, this level exceeded 100 dB. Near aerodromes, railway tracks, busy highways and industrial establishments a noise ranging between 95 to 105 dB has been recorded during most of the daytime. On the basis of extensive studies, World Health Organization has recommending permissible noise levels for residential, industrial, commercial localities and silence zones. These are given in Table 18.2.

(IV) THE ILL-EFFECTS OF NOISE POLLUTION

Noise is unwanted sound which is often a source of great nuisance. It is the duration, pitch and loudness of sound, which determine the discomfort, and other ill effects, which noise causes. Human ear is sensitive to sound levels ranging from 0 dB to 150 dB, however, sound levels beyond 70-80 dB cause plenty of discomfort, irritation and a variety of physiological disturbances.

1. Loss of hearing:

The most common ill effect of noise pollution is impairment of hearing ability of an individual. Prolonged exposures to loud noise can cause temporary or permanent loss of hearing. People working in noisy places such as industrial establishments, factories etc. often suffer from temporary loss of hearing. The ciliary cells in the inner ear are inactivated or numbed and the threshold of hearing of the subject is raised. If the loudness of noise is moderate or the duration of exposure is short, the damage is only temporary. The auditory system recovers itself when the exposure ceases. In Audiometric tests, the phenomenon is referred to as **Temporary Threshold Shifts** or TTS. Longer exposures to louder noises may cause permanent shift in the threshold of hearing of an individual. The individual in such cases suffers from partial but permanent loss of hearing. He is no longer able to hear low sounds, which are audible to normal persons. This is caused by slow and chronic damage to ciliary cells in the inner ear. Still, medical science is of little help in such cases (3;4).

Very loud, sudden and impulsive noises, such as a bomb blast, are capable of causing acute damage to auditory system and an abrupt loss of hearing. With or without involvement of inner ear, it is the middle ear, which is affected in most of the cases. High intensity sound waves damage the eardrums and may disrupt the delicate bony chain which carries sensation from ear drums to the inner ear. Very fine surgical techniques have been developed to restore the hearing ability where only middle ear is involved (5;7).

2. Ill-health effects of high levels of noise pollution:

High levels of unwanted sound or noise can adversely affect both our physiological and psychological health. It can cause annoyance and aggression, tinnitus, sleep disturbances, and other harmful effects. Work, which needs a high degree of skill and precision, is considerably affected. It may cause headache, irritability, and fatigue. It is interesting to note that our Optical system is considerably affected by noise pollution. Dilation of pupils, impairment of night vision and decrease in ability to perceive colours are some of the effects caused by exposure to loud noise for long durations (5;6).

Tinnitus may be caused by chronic exposure to high noises. The perception of sound within the human ear in absence of corresponding external sound is known as Tinnitus. The affected person hears buzzing, humming or ringing sounds while there is actually no such sound. Tinnitus can in turn lead to forgetfulness, severe depression and at times panic attacks (8).

Stress and hypertension induced by higher levels of noise are the leading causes which result into most of the health problems. Higher noise levels have been shown to result in cardiovascular effects and exposure to moderately high levels during a single eight hour period causes a statistical rise in blood pressure of five to ten points and a rise in stress and vasoconstriction (2). This leads to an increased incidence of diseases of coronary artery. Loud noises tend to decrease the output of blood from heart, cause arterial blood pressure to fluctuate and smaller blood vessels of the body constrict reducing the flow of blood to the organs concerned. Heart beat rate is affected. Changes in breathing amplitude have been reported due to sudden and impulsive noises. Eosinophilia, hyperglyaemia, hypokalaemia and hypoglycaemia may also be caused by changes in blood circulation and other body fluids due to noise pollution (5;6;9).

3. Environmental effects of noise pollution:

Loud noise has detrimental effect on animals which causes stress and can lead to temporary or permanent loss of hearing. It usually makes an individual communicate louder placing additional strain on their energy resource. Disturbed communication among members of the same species or between different species increases the risk of mortality by depressing the ability of individuals to spot prey or predators or the mate to reproduce. The unheard voices might be warnings, finding of prey, or preparations to reproduce. The delicately balanced prey-predator relationships and reproductive processes are affected. Failure to communicate could result in reproductive failures and raised mortality risks which could in turn lead to reduced populations, inbreeding depression and finally extinction of the species. When one species begins speaking louder, it will mask other species' voice, causing the whole ecosystem to eventually speak louder. The entire ecosystem is, therefore, affected in one way or the other. The overall affect of noise pollution is reduction of usable habitats for a number of species (4;6;10) .

(V) PREVENTION AND CONTROL OF NOISE POLLUTION

Loud noise is the form of pollution, which often causes much public concern. Therefore, necessary steps have to be taken to control the nuisance. Some of these are:

1. Reduction of noise at the source of its origin: Often a little precaution can reduce much of the nuisance caused by loud noise. This can be achieved by replacement of noisy rattling devices or machines with quieter ones. Noise level can be reduced effectively by replacement of noisy and rattling parts, providing better cushioning to check the vibrations, proper oiling and greasing to ensure smooth running and using effective silencers etc.

2. Application of sound proofing techniques to muffle down loud noises: Sound waves are absorbed by porous material such as perforated sheets and other objects. Just as putting cotton plugs in the ears reduces noise level for the individual concerned, sound barriers placed around the source of origin of loud noises drastically reduce the intensity of sound on the other side of the obstacle. For example, little of loud noise produced in picture halls and auditoria escapes out because of effective soundproofing and acoustic techniques applied for the purpose. The same can be done for industrial units also.

3. Keeping residential localities free of noisy industries, busy highways, aerodromes etc: Residential localities should be established away from noisy industries, busy highways, and aerodromes or else these noisy establishments should be developed away from quiet residential areas. Industrial units can be displaced to some industrial area whereas bypasses may be developed to divert busy railway tracks and highways from domestic establishments. Only that part of traffic should be allowed to get into a residential area, which is barely necessary. This shall curb much of the nuisance caused by noise pollution.

4. Enactment of strict legislation and its effective compliance: In most of the countries, including our own, legal framework against noise pollution has been developed. However, in most of the cases little efforts are made to enforce these rules and regulations effectively. If we ensure only effective compliance of these rules much of the nuisance of noise pollution shall automatically be curtailed.

(B) THERMAL POLLUTION

In an aquatic system temperature is an important factor which affects physico-chemical parameters of water quality as well as plant, animal and microbial life considerably. Aquatic life is usually well adjusted to diurnal and seasonal changes in temperatures, which occur naturally. However, man-made changes in temperature of water often cause adverse changes which damage its utility and productivity (11).

(I) IMPORTANT SOURCES OF SURPLUS HEAT:

Important sources of surplus heat which is ultimately discharged into the environment are:

1. Thermal or nuclear power plants.
2. Industrial effluents.
3. Sewage effluents and other waters.
4. Biochemical activity.

The most important source of surplus heat are thermal or nuclear power generation plants which are responsible for contributing almost 70-80% of discharge of heated waters to aquatic bodies. For example, of the total amount of heat received by river Thames, England, 75% comes from Power plants, about 6% from industrial effluents, about 15% from sewage effluents and other heated waters and only 4% is contributed by biochemical activity within the system.

About 70% of the total heat produced in a thermal power station is discharged as waste heat. This becomes more significant when we look at the total amount of coal consumed for power generation, which is almost 60% or more of the total global coal production of about 2730 million metric tons. Nuclear power plants discharge a little more heat than those of coal-fired power plants, producing an equivalent amount of electricity. A large fraction of this heat goes to the cooling waters drawn from some aquatic body and returned to the system as heated waters. In India, about 262.7 million metric tons of coal was produced in the year 1994, out of which about 170 million metric tons were used to produce electricity (12).

Most of the sewage effluents and waste water, on an average, have temperatures 4°–8°C higher than the aquatic bodies in which they are discharged. Heated waters from thermal or nuclear power plants are, however, more injurious as they have temperatures about 10°–14°C higher than the temperatures of the receiving waters. The enormous volume of these heated waters causes significant changes in temperatures and temperature-induced changes in the aquatic system.

(II) DAMAGES CAUSED BY HEATED DISCHARGES ON AQUATIC LIFE

It is the living organisms within the aquatic body which are adversely affected by the additional heat (12). Most of the biota in aquatic system consists of poikilotherms whose body temperature is regulated by the temperatures of the outside environment. The adverse effects of heated discharges in an aquatic system may be summarized as follows:

(1) During their passage through the cooling systems, which consist of a series of channels and pipes, small organisms, fish larvae, and plankton have to face abrupt changes in temperatures. Sensitive ones are killed due to the temperature shock.

(2) Mechanical impingement on the sidewalls and screens placed to check entry of large organisms and plants, which could choke the pipes, may cause mechanical injury and damage to the living organisms. Small fishes drawn in by pumps along with water to be passed through the condenser cooling system may get suffocated to death by pressure exerted on their gill chambers.

(3) The chemicals introduced in these waters to remove slime from the cooling systems can cause toxicity and death to the living organisms.

(4) At higher temperatures cellular fats may melt, coagulation of cell-proteins may take place and permeability of cell membrane may be affected. The activity of most of the enzymes within the cells of the aquatic organisms is speeded up till a certain optimum level is reached. Any further rise in temperatures causes disruption of the enzymatic machinery.

(5) Metabolic reactions in a biological system are, therefore, speeded up at higher temperatures, which cause an increased demand of oxygen. Increased irritability and swimming speed is observed in fishes. The temperatures in which most of the fishes can survive ranges below 35°C. Temperatures beyond 35°C may be lethal to many fish-species. Higher temperatures also cause increased activity, which exhausts many aquatic organisms and shortens their life span. Reproductive processes may be triggered earlier causing pre-mature deposition of eggs in some aquatic animals.

(6) The solubility of oxygen dissolved in water decreases with rise in temperatures. At 5°C about 12.8 mg per litre of oxygen is needed to saturate the aquatic system. However, at 35°C only 7.1 mg per litre saturates the water, whereas at 40°C only 6.6 mg per litre of oxygen is sufficient to saturate the system. So waters at 40°C cannot hold more than 6.6 mg per litre of oxygen. In an aquatic system saturated with oxygen at lower temperatures, rise in water temperature causes the excess oxygen to form bubbles and leave the system. Gas bubbles may be formed within the bodies of aquatic organisms. For most of the fishes, these bubbles are injurious as they cling to the gills and block oxygen intake. This has virtually the effect of forcing the fishes to breathe in air, which they cannot. The phenomenon is known as gas-bubble disease, which chokes the fishes to death.

(7) At higher temperatures pathogenic forms become more active while plants and animals become an easy victim to various types of infections.

(III) ECOLOGICAL EFFECTS OF HEATED WATERS ON AQUATIC SYSTEM

Heated waters are less dense and hence are lighter. They float over the water surface and cover it causing stratification. This often insulates, effectively, the lower layers of water from the atmosphere above. Exchange of gases from water surface stops. At high temperatures, the solubility of oxygen in water also decreases while rapid metabolic activity requires more oxygen. Species intolerant to higher temperatures migrate to lower layers, which are cooler and so also are many planktonic forms, which settle down from lighter water above to denser water below. All this cumulatively results in creation of oxygen deficit in both upper and lower layers. If the heated waters contain some organic matter, microbial activity is stepped up for its degradation. Various toxic chemicals such as organic acids, hydrogen sulphide, ammonia etc. accumulate and damage the plant and animal life, which is already under duress due to the depletion of oxygen (12;13).

Solubilities of many organic and inorganic chemicals are higher in heated waters than in cooler waters, while the solubilities of gases decrease with rise in temperatures. This results in dissolution of more salts while the amount of dissolved gases diminishes. This causes an altogether changed chemical picture, which either kills or forces some organisms to migrate to habitats that are more suitable.

The elimination of some species or their migration to other places leaves the tolerant forms to grow and multiply without much competition. The diversity in species composition is lost. Number of species present in the system decreases while the number of individuals of the few tolerant species, which are present, rises abruptly. This often causes disappearance of economically desirable forms. The entire biological spectrum is, therefore, changes as a consequence of discharge of heated waters or effluents from various thermal power plants and industries (13).

(IV) CONTROL OF THERMAL POLLUTION

There is little one can do about this type of pollution except to wait and allow the waters to cool down to biologically harmless temperatures before being discharged into some aquatic system. If the volume of heated waters is manageable, they can be passed over a system of cascades or through fountains, which causes rapid cooling. To handle large quantities of heated effluents large tanks or reservoirs should be constructed to retain the water for a little longer time. After the waters have cooled down to a tolerable temperature, they may be released in natural water bodies. It is advisable to discharge the waters into some lotic system, wherein active churning occurs due to flow of water, which causes rapid cooling, instead of a lentic system.

(C) RADIO-ACTIVE POLLUTANTS

Radiation is a phenomenon, which involves movement of energy through space. A number of atoms possess the ability to emit radiations and, thereby, cause radioactive pollution. Radiations originate from instability of the nuclei of an atom, which loses sub-nuclear particles and energy to acquire a stable state. The phenomenon is termed as radioactivity. No physical, chemical or biological process can influence, reduce or terminate these emissions from a radioactive element, as it is the state of atomic nuclei, which is responsible for the phenomenon.

(I) ATOM OF AN ELEMENT AND ITS ISOTOPES

The core of an atom of any element is constituted by a super dense object, the nucleus that is composed of neutrons and protons. The number of protons equals the number of electrons, which revolve round the nucleus. The chemical properties of an element depend upon the number of electrons, which revolve round the nucleus particularly in its outer orbits. As far as the chemical properties are concerned, an element shall remain the same as long as the number of electrons (or protons) is the same. An element will change or transmutate into another if the number of protons is altered. However, a change in number of neutrons in the nucleus shall not affect the chemical properties of the element, instead it will cause the formation of different isotope of the same element. Most of the elements occur in nature as a mixture of two or more than two atomic forms or isotopes.

(II) RADIOACTIVITY AND RADIATIONS

There are some combinations of neutrons and protons in nuclei of an element, which are more stable than others are. Stable nuclei do not emit radiations or lose nuclear particles and energy but unstable ones lose electrons, protons, or neutrons and emit electromagnetic radiations. It is the emission of these particles and energy as electromagnetic radiations, which causes the nuclei of atom to acquire a stable state. The acquisition of stable state is usually accompanied by change in the number of protons as well and the element transmutates or decays into another element. The following types of radiations are given out when an element transmutates or decays into another element.

1. **Emission of alpha-particles:** Emission of alpha-particles takes place in radio-active isotopes of heavy elements such as $_{92}U^{238}$, $_{88}Ra^{226}$ etc. α-particles are nothing but Helium nuclei and their emission causes the number of protons and neutrons to change. The element will thus move a little lower down the period table, i.e., it will change into another element with lower atomic number.
2. **Emission of β-particles:** It is electrons only, which constitute β-particles. Emission of β-particles involves reduction in the neutrons but that of protons goes up (a neutron splits to produce an electron and a proton). Thus, the element involved moves a little higher up in the periodic table, i.e., changes into another element with a higher atomic number.
3. **Emission of high-energy electromagnetic radiations:** These high-energy electromagnetic radiations almost always appear when either alpha or beta particles are given out and transmutation of an atom is in progress. These rays, such as x-rays and γ-rays have a very high penetrating power.

(III) DAMAGES CAUSED TO A BIOLOGICAL SYSTEM BY RADIATIONS:

Most of the damages caused by radioactive pollutants stem from their capacity to produce high-energy radiations, which are very harmful to a living system. There are two main modes in which radioactive pollution can be dangerous to a biological system:

1. Damages caused by radiations from outside sources.
2. Damages caused by radiations from sources inside the body.

Radiation is a phenomenon in which energy or particles carrying high energy traverse the environment. Those radiations, which cause ionization, are high-energy radiations capable of removing electrons from an atom and attaching them to others there by producing positive and negatively charged ions. These are known as ionizing radiations and include x-rays and γ-rays, emission of α and β particles etc. In contrast to these, there are some radiations, which do not cause any ionization. They have a shorter wavelength. Due to their high-energy content, they are capable of causing harms to microorganisms and damage only the surface tissue of man, plant and animal bodies. A familiar example of such type of radiations is that of Ultra-violet rays.

As the chemical properties of radioactive isotopes of an element are similar to those of its non-radioactive isotopes a biological, system is unable to distinguish between the two. Many radioactive isotopes are taken up by biological systems used, accumulated and passed on to the higher trophic levels in the food chain. Thus, the radioactive source is lodged within the body, often in concentration many times higher than the original medium in which the organism concerned lives. An internal source of radioactivity is much more dangerous than an external source and may cause galls, tumours, degenerative changes and other malfunctions in the tissues of the body. Transmutation of the radioactive isotopes incorporated in the structural component of the body of the organisms causes complete chaos in the system as the very basis of structure may crumble down. Moreover, such an individual who is loaded with radioactive material is a moving source of radioactivity and a health hazard for others who come in contact with him (14). Damages caused by radiations at different levels in a biological organization may be summed up as follows:

1. Damages at molecular level: Damages to macromolecules such as enzymes. DNA, RNA etc. through ionization, cross-linkages within and between the two affected molecules.
2. Damages at sub-cellular level: Damages to cell membranes, nuclei, chromosomes such as fragmentation, mitochondria, lysozymes etc.
3. Damages at cellular level : Inhibition of cell division, death, decay and transformation to malignant state etc.
4. Damages to tissues and organs: Disruption of such systems as Central Nervous System, loss of sight, inactivation of bone marrow activity resulting into blood cancer, malignancy and ulceration of intestinal tract etc.
5. Damages to an individual and whole population: Death or shortening of life due to radiations. Changes in characteristics due to mutations etc.

Table 18.3 *Natural and Man-Made Sources of Radiations to Mankind*

	Source	Exposure*
1	Natural sources	130.00
2	Diognostic and therapeutic uses.	72.00
3	Atomic tests	4.00
4	Nuclear power industr	0.02
5	Industrial and miscellaneous.	3.04

* As mRem per year. (Source: Meril Eisenbud 1973)

In human beings exposure to radiations results in little visible effects in early stages. However, after 12-24 hours injury symptoms manifest themselves. These include erythema of skin (reddening of skin), anaemia, anorexia (loss of appetite), vomiting diarrhoea and with high heavy doses, blister formation, pigmentation of skin, burning sensation all over the body, loss of sight, paraplasia caused by disruption of central nervous system etc. A single whole body dose of over 500 rems could result in death within four or five days, Delayed effects, which may take as much as twenty years to appear, are blood cancer or leukemia, carcinogenesis, foetal developmental anomalies, radiation shortening of life etc. It must be pointed out at this stage that for all this there is no remedy available – once a person is exposed to radiations he has to endure its consequences. Medical aid can do little for it. There are some sulphur compounds such as Cystaemines, Amino-ethylisothiouronium, Betamercaptoethylguanidine etc. that are able to perform some reparative treatment but these too are not effective at higher doses of radiations (14;15).

(IV) SOURCES OF RADIATION IN THE ENVIRONMENT

Important sources of radiation exposure to plants, animals and humans can be grouped into the following two categories.

1. Natural sources of radiations.
2. Man-made sources of radiations.

Nuclear activity involving sub-atomic particles (neutrons, protons and electrons) and a huge amount of energy have caused the formation of this planet. These reactions have also produced a large number of unstable nuclei or isotopes, which lose sub-atomic particles and emit high-energy radiations to acquire a stable state. Though much of this radioactivity has already been spent, the process still continues and constitutes an important source of radiations for the biosphere. In addition to these, earth's surface is regularly irradiated by X-rays, cosmic and ultra-violet rays from outer space, which also constitutes another natural source of radiations. Natural sources may constitute as much as 130 mRem of radiations, which an average human being receives annually (Table 18.3).

Man-made sources of radio-activity include nuclear reactors, diagnostic and therapeutic application of radio-activity, atomic tests and weapons and miscellaneous industrial uses of radio-active material. Controlled nuclear fission within a reactor gives rise to a large number of isotopes, which are confined within a reactor and are separated, refined and used for a variety of purposes. The structural material of the nuclear reactor also becomes radioactive due to constant irradiation by particulate and high-energy rays. The corrosion products of reactor are, therefore, strongly radio-active. The spent fuel and the corrosion products, both of which may include solids, liquids and gases are an important source of radio-activity.

Nuclear tests and weapons disseminate radioactivity and radio-active material far and wide as they are usually detonated in open. The fissile material is usually a heavy element, which splits to produce a number of radioisotopes with lower atomic mass number. However, these tests affect only localized areas and by the time wind, water or other agencies carry the radioactive material to distant places their harmful constituents are diluted considerably. But they still cause a little rise in the overall radio-activity. Medical and diagnostic uses of radioactivity are by far the most important source of radiations to general public. There are numerous X-ray clinics and hospitals located within thickly populated areas all over the world. Radioactive isotopes of Cobalt, Iodine etc. are frequently used for diagnostic or curative purposes in the treatment of cancers, hyperthyroidism etc. in these clinics. A large number of people frequent these places and receive radiations daily. Colour TVs computer screens, video games etc. are also mild sources of radiations to general public. Important radioactive isotopes which are hazardous to the biosphere, and their half lives are listed in table 18.4.

(V) HAZARDS ASSOCIATED WITH RADIOACTIVE POLLUTION

Radioactive pollutants are not like other pollutants, which are sooner or later converted into harmless material and degraded into simpler constituents to be recycled into the ecosystem and used

	Radioisotope	Half life		Radioisotope	Half life
1	Tritium	15 hrs	16	Hafnium-181	40 days
2	Carbon-14	5000 yrs	17	Tantalum-182	111 days
3	Sodium-22	26 yrs	18	Polonium-210	138.4 days
4	Sodium-24	16 hrs	19	Radon-222	3.83 days
5	Manganese-54	300 days	20	Radium-226	1622 yrs
6	Iron-55	2.9 yrs	21	Thorium-230	77,000 yrs
7	Cobalt-58	71 days	22	Thorium-232	14 million yrs
8	Cobalt-60	5.2 yrs	23	Uranium-233	1,62,000 yrs
9	Strontium-90	27 yrs	24	Uranium-235	710,000 yrs
10	Zirconium-95	65 days	25	Uranium-238	4,500,000 yrs
11	Iodine-129	15,800,000 yrs	26	Uranium-239	23.5 min
12	Iodine-131	8 days	27	Neptunium-237	2,100,000 yrs
13	Copper-64	12.8 hrs	28	Plutonium-239	24,400 yrs
14	Caesium-135	2,300,000 yrs	29	Plutonium-240	6,540 yrs
15	Caesium-137	30 yrs	30	Plutonium-244	380,000 yrs

again. It is not the element itself, but the instability of its nuclei, which is responsible for damages caused by these pollutants. As long as radiations continue, these wastes are dangerous for the living beings. After the emission of radiations, nuclei attain stable state and behave like any other element in the environment or the biosphere. Life on earth's crust could evolve only when the nuclear activity ceased, atoms acquired stable state and radiations were reduced to the level, which could be tolerated by living beings. Major hazards (15;16) associated with radioactive pollution can be summed up as follows:

1. No physical, chemical or biological process can influence the process of radioactive emissions. The unstable nuclei have to decay and acquire a stable state.
2. A number of radioactive isotopes have a very long half-life. Thorium-232 ($_{90}Th^{232}$) takes 14,000,000,000 years to lose half of its radioactivity. Half of Uranium-235 ($_{92}U^{235}$) takes 710,000 years to disintegrate. Half of Neptunium-237 ($_{93}Np^{237}$) decays in 2,100,000 years. This makes these radioactive wastes almost a permanent hazard for the biosphere.
3. Most of the radiations have a high penetrating power. Thick sheets of steel, cement concrete walls etc. cannot contain them. They can easily penetrate to deep-seated organs and cause injury.
4. Nucleic acids (DNA & RNA) effectively absorb these radiations. Even low-level radiations, which do not cause any visible damage, are completely absorbed by nuclear material, which causes carcinogenic, mutagenic and teratogenic effects.
5. A biological system is unable to distinguish between a radioactive and a normal isotope of an element as their physical and chemical properties are similar. Radioactive isotopes are therefore, absorbed and incorporated within the bodies of living organisms as normal isotopes are. This lodges a radioactive source within the body of the organism itself.
6. Like any other element, radioactive isotopes are also absorbed, accumulated and bio-magnified thousands of times. Thus, the entire food chain becomes contaminated. Organisms at higher trophic levels may, therefore, receive a highly concentrated source of radioactive material through their food supply.
7. There is no other way to dispose of these hazardous wastes except to store them for thousands or millions of years away from living beings. This is too long a period on human scale of time. Even the safest burial places for radioactive wastes, which represent the best of human efforts, have shown signs of leakage. At present it appears very difficult, though not impossible to store radio-active wastes away from the biosphere for such long periods.
8. In spite of all these hazards, nuclear reactors and tests are continuing and an increasingly large amount of radioactive wastes is accumulating every day while no solution to the problem of their safe disposal is in sight until date.

(VI) GLOBAL ATCCUMULATION OF NUCLEAR WASTES

Indeed, the longest lasting legacy of the nuclear age, which Fermi inaugurated in December 1942, will be the packages or irradiated wastes, which shall persist in a state hazardous to the biosphere for hundreds of thousand years. These wastes contain radioactive isotopes which have a very long life. For example Plutonium-239, which has a half life of 24,400 years, is dangerous for about quarter million years or 12,000 human generations. It decays to produce uranium-235, which has a half-life of 710,000 years, a span of time longer than the parent material. Uranium-235 may decay to produce another radioactive isotope with a long half-life. Thus, the problem of nuclear waste disposal acquires almost a permanent character. Radioactive wastes or irradiated wastes may be grouped into the following two categories:

1. High level irradiated wastes.
2. Low level irradiated wastes.

(1) High Level Irradiated Wastes: High-level irradiated wastes consist of spent fuel from nuclear reactors and left over material from a nuclear fuel processing units. They contain radioactive isotopes of a large number of elements whose half-lives range from fraction of a second to hundreds of thousand years. Each metric ton of spent fuel produces about 180,000,000 curies of radioactivity, which gradually declines with time, but the rate is very slow. Even after 10,000 years, each metric ton of this fuel shall emit 470 curies (Table 18.5). Some countries like U.K. and France reprocess their spent fuel to be used again. The overall impact of which is the production of more high-level radio-active wastes. About four hundred and twelve or more commercial reactors all over the world generate about 5% of the world's power supply but they also produce about 10,000 metric tons of high-level nuclear wastes – solids, liquids and gases. Fig. 18.1 shows the approximate trend in accumulation of high-level wastes since the year 1965. At present about 100,000 metric tons of these wastes have already accumulated (17,18).

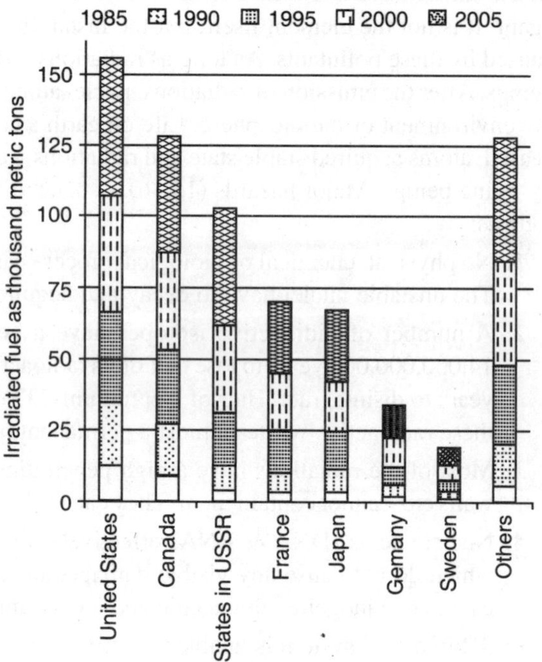

Fig 18.1 *Global accumulation of irradiated waters (Estimatss)*

(2) Low Level Irradiated Wastes: Low-level irradiated wastes are those wastes, which have a smaller amount of radioactive material with half-lives about thirty years or less. The material, which is regularly irradiated by radiations from nuclear reactors, also becomes radioactive. Though there is little radioactivity, these wastes may contain radioactive isotopes with very long half-life – such as those of technitium-94 (half-life 210,000 years) and iodine-129 (half-life 15.8 million years). The largest volume of radioactive wastes, however, comes from uranium mines, milling and processing establishments. The uranium ore which also contains a number of radio-active disintegration products of uranium as well as other isotopes, is finely powdered, most of the uranium is extracted and tailings are left behind which possess about 85% of the total radio-activity originally present in the ore.

Low-level wastes having little radioactivity are almost as dangerous as high level wastes. Indeed, they do not cause high-level injury, which occurs in cases of acute exposures. However, both low as well as high level wastes are capable of causing delayed carcinogenic, teratogenic and mutagenic responses. There is no 'threshold dose' or 'dose of no consequences' for radiation exposures (17;18).

(VII) DISPOSAL OF RADIO-ACTIVE WASTES

Only the process of natural decay, which requires hundreds of thousands or even millions of years, reduces the radioactivity of nuclear wastes. There is no physical, chemical or biological process which can reduce this span of time. All we can do is to store the radioactive material at some remote place away from the biosphere for a million years – task which is too difficult for the mankind. High level irradiated wastes also produce a lot of heat and hence have to cooled to avoid explosion of the cask or the containers in which they are kept. Ever since the beginning of nuclear age a number of options have been put forth for the isolation of radio-active wastes from the biosphere some of which may be summed up as follows:

1. Burial under Antarctica Ice cap: A leakage could be catastrophic. With temperature changes, ice could melt disturbing the burials. Abandoned.
2. Burial under seabed: The radioactivity may leak out in the surrounding waters and from there to other places. Abandoned.
3. Disposal in space outside earth's gravitational field: A single launch failure could introduce the radioactivity in the atmosphere from where it can contaminate the entire planet. Abandoned.
4. Disposal in deep underground surface: This involves difficulties arising due to ground water flows and uncertain geology, which may cause tectonic disturbances, volcanic activity etc. Moreover, human intrusion over such long periods may not be ruled out. However, this proposal has been followed till date.
5. Burial in specially constructed buildings under human care: The proposal involves too much dependence on human care, which may not be possible for such long periods as required.
6. Reprocessing of nuclear wastes: Chemical separation of uranium and plutonium from nuclear wastes reduces the radioactivity by about 3% but increases the volume of wastes almost 150 times. Hence, abandoned.
7. Transmutation to isotopes with shorter half-lives: Bombardment by neutrons may convert isotopes with long half-lives into isotopes with shorter half-lives. However, there are many technical uncertainties in the process, which can be dangerous as well. The method is, anyway, enormously expensive. It is already under active study.

At present, much of the scientific opinion is in favour of burying the high-level irradiated wastes deep inside earth's crust. However, an important factor in such burials is underground water table, which feeds a number of fresh water deposits and rivers on earth's surface. Tectonic movements of rocks and volcanic activity may also disturb these burials (18).

The United States Government's Deptt of Energy has proposed two such deep burial sites – the projects which may be considered as representing the best of human efforts. One at Nevada's Yucca Mountains and another, called Waste Isolation Pilot Plant (WIPP), in Southeastern New Mexico. The Pilot Plant at New Mexico has burial rooms 600 metres deep amidst salt formations under which there is a large reservoir of saline water (the brine). Under the brine runs a ground water aquifer, which feeds a branch of river, Rio Grande. The brine constantly seeps through walls of repository. It can easily damage the steel containers and create radioactive slurry. The heat produced by wastes could cause an explosion and leakage will follow which could contaminate the under -ground fresh water aquifer. At Yucca Mountain Nevada, whose rock formations are criss-crossed by at least 30 seismic faults deep burial is proposed at a depth of about 300 metres, immediately below which there is a fresh water aquifer. If, due to seismic activity or tectonic movements or volcanic activity, this water comes in contact with the nuclear wastes the heat generated could produce enough steam to blow off the plug at the top of the tunnel. Or else the radioactive material shall contaminate the ground water aquifer, which feeds many hot springs in the Death Valley of Nevada, USA. Thus mankind's enormous knowledge bows low before the challenge of isolating nuclear wastes from the biosphere for periods long enough to convert them into a harmless state. It is note-worthy here to remember that there was no English Channel only 7000 years ago, that volcanoes abounded in what is now Central France barely 10,000 years ago and that Sahara desert was a fertile land only 5000 years ago. In a world of such a rapid change it appears almost impossible to find a safe, inviolable and permanent resting place for the nuclear wastes for periods as long as hundreds of thousands or millions of years (18).

QUESTIONS

1. What is noise pollution? Discuss its causes and consequences.
2. How sound is measured? Discuss the ill effects associated with high level of noise on humans. What measures you would suggest for prevention and control of noise pollution?
3. What is thermal pollution? Discuss damages caused by heated effluents, on aquatic systems. Suggest control measures.
4. What are the major sources of heated waters, which are discharged in our water bodies? Enumerate ecological effects caused by such waters on our aquatic bodies.
5. What is radioactive pollution? Discuss the damages caused to a biological system by radiations.
6. What are major sources of radiations in the environment? Discuss the hazards associated with radioactive pollution.
7. Give a brief account of global accumulation of nuclear wastes. Suggest methods to minimize it.
8. Is there any foolproof and safe way to dispose of radioactive wastes? What are the options suggested from time to time for their disposal. Discuss their merits and demerits.

19
Chapter

Air Pollution

The mantle of gases, which surrounds our planet, is referred to as "atmosphere". It is a complex mixture of gases, water vapours and a variety of fine particulate material. It consists of about 5.137 x 10^{15} metric ton of gases, which exert a pressure of about 1 kg per sq cm. on earth' surface (1). Most of these gases are compressed in the lowermost layer of the atmosphere. Pressure decreases as we move upward. Living beings inhale a considerable quantity of air. About 90% of man's total daily intake of materials – food, water and air – is contributed by air. An average human being breathes about 22,000 times a day and inhales about 16 kg of oxygen. The pollution of air, therefore, may have a profound influence on living organisms. Unfortunately, an alarming quantity of gases, particulate material, fumes, vapours and smoke is discharged daily into the atmosphere. Dilution, dissemination, transformation and clearance of the pollutants are strongly influenced by the structure and prevailing meteorological conditions.

(I) MAJOR POLLUTANTS OF THE ATMOSPHERE

The gaseous cover over the surface of our planet essentially consists of 78.0% nitrogen, 20.95% oxygen, 0.93% argon, 0.03% carbon dioxide, a number of other gases and water vapours. In addition to these gases, air also contains a variety of particulate matter, which includes dust, pollen grains, algae, bacteria, fungal spores etc. so also are various odours, fumes and vapours, which occur in small traces. In this huge and diverse admixture, a number of substances harmful to a biological system also occur. But their concentration is usually, so low that they hardly pose any problems. Human activity has, however, been introducing a number of undesirable and harmful materials in large quantities which pollute the atmosphere. Major pollutants of the atmosphere are:

(1) Particulate matter: Most of particulate material in the environment is contributed by dust particles from various sources, fibrous material of plant and animal origin, finer particles or fly-ash derived from combustion of coal, oil and organic matter or from mining and processing operations carried on to obtain metals and other elements. To these may be added pollen grains, fungal spores, algae, bacteria, viruses etc, which are capable of causing various diseases, allergy etc.

Particles larger than 10 μm are introduced into the air primarily by physical processes. These particles tend to settle down rather rapidly. It is these particles, which are gathered in settling jars, weighed and reported as gm per sq m. per day. Dust fall or Settleable particulate material (SPM), as these particles are often called, create problems locally, usually in the vicinity of their source of origin as they tend to settle down before winds can transport them to far off places.

Particles smaller than 10 μm in size usually remain suspended in the air for long durations. Being smaller and lighter the rate of settlement of these particles is very slow and the turbulence of air keeps them afloat in the atmosphere almost permanently. They constitute the most important and largest part of particulate material in air by weight. Since these particles settle down with difficulty, they have to be separated from air with the help of high volume air sampler for study and observations. The finest particles, e.g., those below 0.1 μm in size are so small that they cannot be filtered out. The cleanest air usually contains at least several hundreds of such particles per cubic millimetre while in a polluted urban environment as many as 100,000 of these particles may be found in a single cubic millimetre. They do not amount much by weight and are so small that it is difficult to study them. In a scientific study, they are often ignored.

Air passage in higher animals, including man, is lined internally with small ciliated cells, which are covered by a layer of watery, jelly-like substance (the mucus). Particles above 5 μm in diameter are trapped in the mucous lining, propelled up by coordinated movement of cilia to the throat from where they are coughed out or swallowed. Within about twenty minutes of inhalation, they are cleared from the respiratory track. Besides thin-walled cells, which line the alveolar chambers, there are also present larger cells called alveolar macrophages. Fine particles or aerosols, which do reach the alveolar lumen, are engulfed by these macrophages, which are then passively transferred to the regional lymph nodes. From nodes, they are disposed of via upper respiratory track along with coarser particles, filtered earlier. Inert particles may be retained in the lymph node or in the macrophage cytoplasm throughout the life of an individual without causing any harm (2).

However, toxic particulate material like coal, mica, talc and some of crystalline forms of silica may injure or kill alveolar macrophages (of lungs). Elongate asbestos fibres, which are less than 5 μm in cross-section, may also reach the alveolar lumen. These fibres are many times longer than the actual diameter of the macrophage and hence are incompletely covered. In such cases a cytoplasmic sleeve forms covering the entire fibre and a number of macrophages may have to fuse together to engulf these fibres. Particles released from dead macrophages are engulfed again by newly formed macrophages. As these macrophages tend to adhere to alveolar wall, they are covered by newly formed epithelial cells. The cyclic engulfment of the particle by macrophages, their release and epithelialization results in the formation of nodules, which may grow large enough to block the alveolar lumen, making it completely defunct. With aging of nodules, collagen fibres are laid down and a fully developed nodule resembles a ball of wool. The phenomenon is known as fibrosis. Since a large number of such nodules are formed simultaneously, a considerably large respiratory area is lost. In order to compensate for the lost area the alveoli of other parts of lung get dilated – a feature which is known as compensatory emphysema. Thus, aerosols, which have a diameter less than 5 μm are dangerous while fibrous material with a cross-section less than 5 μm, but a length many times greater is much more risky to breathe in. This result in diseases like silicosis, asbestosis etc. and the individual is rendered susceptible to many types of infectious diseases. Tuberculosis is one of such diseases, which is quite frequent in dirty, dusty and crowded cities in India (2).

(2) Oxides of Carbon: Combustion of organic matter and fossil fuels in presence of oxygen results in the formation of carbon dioxide while carbon monoxide is formed due to incomplete combustion of these materials under oxygen deficient conditions.

Carbon monoxide is the most important atmospheric pollutant as it accounts for about 52% of the total air pollution. The commonest source of carbon monoxide pollution in cities is automobile exhausts. In USA alone, as far back as 1960, it was estimated that about 66 million tons of carbon monoxide was injected into the environment by automobile exhausts (3). Annual global emission of this gas was 841.08 million metric tons in 1990 which shot up to 1.076 billion metric tons in the year 2000 AD (4). Large cosmopolitan cities of India may have carbon monoxide concentration almost 250-300 times higher than the unpolluted rural areas.

Carbon monoxide is a colourless, odourless gas. The presence of this gas cannot be sensed and unknowingly an individual may inhale large quantities with air. The toxic action of this gas is due to its property to form a complex, carboxy-haemoglobin, with haemoglobin. Carboxy-haemoglobin cannot carry oxygen. The oxygen-carrying capacity of blood is, therefore, lowered. An inhalation of 1000 ppm of carbon monoxide may convert as much as 50-60% of haemoglobin of human blood to carboxy-haemoglobin, which could be serious. Green plants, soils and oceans are the natural sinks of carbon monoxide (5). It has been shown that temperate forests including the forest soils are capable of removing this gas with extraordinary efficiency (6). In atmosphere, carbon monoxide is oxidized to carbon dioxide and hence this gas does not cause any long-range problems.

Carbon dioxide is produced in large amounts when organisms respire but almost a similar quantity is assimilated by green plants. Combustion of fossil fuels, organic matter, incineration of limestone for the production of cement and lime, various metallurgical operations and a number of industrial practices, however, produce a large amount of this gas. In addition oxidation of carbon monoxide, methane etc. also contributes some carbon dioxide to the atmosphere. Extensive deforestation further adds to our miseries, as it tends to depress the global carbon assimilation. An increased input and a reduced output, results in accumulation of carbon dioxide in the system. The figures, published by the US National Oceanic and Atmospheric Administration on its website, reports that CO_2 is accumulating in the atmosphere faster than expected. From 1970 to 2000, the concentration rose by about 1.5 ppm each year, but since 2000 the annual rise has leapt to an average 2.1 ppm. Global atmospheric CO_2 levels during June-July 2007 were found to be 383.9 ppm a level which was never attained during the preceding 650,000 years (7). Using 5.137×10^{18} kg as the mass of the atmosphere, 1 ppmv of CO_2 = 2.13 Gt of carbon.

Carbon dioxide is an important resource for plants. It is raw material for photosynthesis – a process that is essential for sustenance of life on this planet. A higher concentration of this gas shall have little effect on plants, which could be benefited due to stepped up photosynthetic rate as suggested by some workers. Animals and humans may experience breathlessness and anoxia under higher concentration of carbon dioxide. However, a rise in carbon dioxide content on global scale shall further strengthen the gaseous insulation around the globe resulting in global warming and a chain of catastrophic events such as rise in mean sea level, changed climatic conditions, precipitation patterns etc.

(3) Oxides of Sulphur:

Oxides of sulphur, primarily sulphur dioxide is the second most important pollutant of the atmosphere, next only to oxides of carbon in magnitude and severity of effects on a biological system. Most of the atmospheric sulphur dioxide comes from burning of organic matter, fossil fuels and from industrial units processing sulphur-bearing ores. Global emission of Sulphur dioxide amounts to about 150 million metric tons per year (4). Usually oxides of sulphur dissolve in the film of water, which covers fine particulate material, or in droplets of water and are carried down with rains or snow. They are also cleared by dry precipitation. Usually oxides of sulphur are cleared from the atmosphere within a period of about 40-42 days.

Sulphur dioxide is capable of causing a wide variety of harmful effects both on plants and on animals. The oxidizing power of dissolved sulphur dioxide affects electron transport system adversely. It competes with carbon dioxide, retards photosynthesis and thereby carbon assimilation. Hydrogen-ion concentration is shifted towards acidic side when this gas dissolves in water, which affects the activity of various enzyme systems drastically and this in turn disturbs the entire metabolism of the living beings. The gas is capable of causing thickening of mucous layer of trachea, hypertrophy of goblet cells & mucous glands and bronchial constriction that results in a marked degree of resistance to flow of gases – features resembling the pathology of chronic bronchitis. The thickening of mucous membrane and the thicker and viscous layer of mucus, which develops over it adversely, affects the function of ciliary cells because of which the ability of the respiratory track to clear out foreign particulate material is badly damaged. Plants exposed to this gas for long durations suffer from necrosis, interveinal chlorosis and damaged chloroplasts (2).

(4) Oxides of Nitrogen:

The three oxides of nitrogen, nitrous oxide (N_2O), nitric oxide (N0) and nitrogen dioxide (NO_2) are important pollutants of the atmosphere. Most of these gaseous oxides are introduced in the atmosphere in considerable amounts by burning fossil fuels, organic matter as well as by microbial action on nitrogenous organic matter and on chemical fertilizers applied in agricultural fields. It has been estimated that about 70-80% of the total input of nitrous oxide comes from natural sources and from the breakdown of nitrogenous fertilizers. A considerable amount of these gases are also generated by natural events like solar flares and volcanic emissions (8). Nitrous oxide is reacted upon very slowly in the atmosphere, slower than the rate of its input and hence it has been accumulating in the atmosphere at a rate of 0.3% per year. Currently concentration is about 321 ppb – an enhancement of about 51 ppb over the its pre-industrial level (4;9). To plants and animals, nitrous oxide is not very harmful but it gives rise to nitric oxide and nitrogen dioxide, which are more injurious. These gases are responsible for the formation of Peroxyacetyl nitrate (**PAN**) and Peroxybenzonyl nitrate (**PBN**). Nitric oxide is responsible for depletion of the stratospheric ozone layer catalytically while nitrogen dioxide is an important constituent of the photochemical smog or oxidizing type of air pollution and is at least partially responsible for causing acid rains. Nitrogen dioxide irritates lungs and is capable of causing pulmonary oedema at 5 ppm and pulmonary emphysema at concentration between 10 to 40 ppm (2).

(5) Ozone:

Ozone is generated in the lower atmosphere during the formation of photochemical smog when nitrogen dioxide splits up to produce reactive oxygen atoms, which combine with molecular oxygen to give rise to ozone. High up in the stratosphere oxygen molecules split up under ultraviolet radiations to produce oxygen atoms which recombine with molecular oxygen to form ozone. It is this gas, which forms the protective ozone umbrella in the stratosphere and shields life from biocidal high-energy radiations. However, this gas is a serious pollutant lower down in the troposphere (10).

Ozone is highly reactive gas, which causes oxidation of a number of macromolecules within a biological system and produces free radicals, which have been implicated in a number of ozone, induced pulmonary or extra-pulmonary damages. It has been suspected that these free radicals can damage DNA molecules and cause carcinogenesis. Ozone can crack and deteriorate rubber, damage textiles and other materials due to its strong oxidizing action. This gas is a deep lung irritant and when inhaled in large concentrations may cause pulmonary oedema leading to death of individuals (10;11).

(6) Hydrocarbons:

Biological decomposition of organic matter, natural gas, volatile emissions, incomplete combustion of fossil fuels and biomass, automobile exhausts etc. are some of the sources of hydrocarbons in the atmosphere. Harmful effects caused by these substances vary from compound to compound. However, some of these in general cause necrosis of leaves, chlorosis of flower buds, growth inhibition etc. in plants. In animals, hydrocarbons may cause irritation of mucous membrane and bronchial constriction. Many hydrocarbons are known to have carcinogenic properties.

Methane: Methane is a gaseous hydrocarbon and is a greenhouse gas. Like carbon dioxide, it is capable of absorbing infrared and heat waves, radiated back to space from earth's surface, thereby causing a heating effect. Human activity has caused an increase in the production of this gas, the main sources of which are decomposition of organic matter under anaerobic conditions, burning of biomass and excessive use of fossil fuels. About 360 million hectares of rice fields and about 1.2 billion odd heads of ruminating livestock, which release methane when they belch, or break wind, collectively inject an enormous quantity of methane into the atmosphere. About 400-765 × 10^{12} gms per year of methane have been estimated to be produced on global scale (8). Methane is oxidized to carbon dioxide and water, which adds to carbon dioxide and water content of atmospheric air. Recent studies suggest that methane reacts with stratospheric chlorine to form hydrochloric acid and thus exerts a protective influence on ozone concentration by eliminating the ozone depleting chlorine content from the medium.

Chlorofluorocarbons (CFCs): Another group of hydrocarbons, which is causing great concern these days, is that of Chlorofluorocarbons (CFCs). These are non-toxic, colourless, odourless and inert chemicals, which persist in the atmosphere for long durations. They can easily be liquefied. These inert chemicals find widespread use in refrigeration, air-conditioners, foam blowing, spray cans and as solvents etc. Though highly persistent in the lower atmosphere, high up in the stratosphere they undergo dissociation under the influence of ultraviolet radiations to yield chlorine atoms, which catalytically decompose ozone molecules. The total world production of these halogenated hydrocarbons has increased from 41 million kg in 1950 to 700 million kg in the year 1980. The use of CFCs in refrigerators and air conditioners involves circulation in a close circuit from where their release is only accidental. However, use of these chemicals in spray cans or aerosol sprays, plastic foam blowing, instant foam shaving creams and as cleaning solvents etc causes the release of substantial amount of CFCs in the atmosphere on global scale (12).

Aldehydes: Exhausts from automobiles and incomplete combustion of fossil fuels, biomass, and the glue used in plywood etc generate a good amount of aldehydes most of which are toxic chemicals. Formaldehyde, acetaldehyde and acrolein, an unsaturated aldehyde, probably contribute much to the odour, eye and lung irritation produced by the photochemical smog. Formaldehyde usually accounts for about 50% while acrolein accounts for about 5% of the total estimated aldehydes present in polluted air.

Two important aspects of formaldehyde have rather recently attracted much public attention. Firstly, their presence in indoor atmosphere where it comes from incomplete combustion of wood and fuels, from plywood, from foam insulators and a number of objects in which formaldehyde containing adhesives are used. Many pesticides also contain formaldehyde. Secondly, the finding of nasal cancers in rodents chronically exposed to 3-15 ppm of formaldehyde has caused much concern. However, this aspect has not fully been investigated and many workers argue that formaldehyde has been in use since very long time. If it has any such properties there should have been complaints of increased nasal carcinoma in at least occupationally involved workers. Being highly reactive chemicals, these are rather rapidly degraded in the atmosphere (13).

Peroxy-alkyl-Nitrates: Reactions involving hydrocarbons, particularly aldehydes and oxides of nitrogen in the atmosphere at times result in the formation of compounds like **Peroxyacetyl nitrate** (PAN) and **Peroxybenzonyl nitrate** (PBN). These compounds cause acute irritation of eyes – a feature often met with in photochemical or oxidizing type of air pollution. These chemicals have also been implicated in causing various respiratory disorders, in plants, they are capable of suppressing photosynthesis, and thereby the plant yield.

Though formaldehyde, the simplest member of aldehyde group also forms peroxyformyl nitrate, it is the presence of peroxyacetyl nitrate, formed by acetaldehyde which causes more problems. Peroxybenzonyl nitrate derived from aromatic aldehydes is also of considerable importance in many cases. A number of workers have recorded glazed appearance on the lower side of bean and lettuce plants sensitive to these chemicals. PAN and PBN are also capable of damaging chloroplasts and thus causing drastic reduction in photosynthetic efficiency, inhibition of electron transport system and interference with a number of enzyme systems, which play an important role in cellular metabolism (13).

Aromatic Hydrocarbons: In general, aromatic hydrocarbons are considered more toxic than simple aliphatic compounds. Much of aromatic hydrocarbons present in the atmosphere are derived from combustion of biomass, fossil fuels such as coal, oil, tar etc Edible material such as fish and meat smoked for preservation may contain appreciable quantities of these aromatic compounds. Benzene (C_6H_6) is the simplest aromatic hydrocarbon used in various industries as solvent and in fuel oils because of its antiknock properties. Of the total amount of hydrocarbons emitted with exhausts of internal combustion engines, benzene may constitute almost 2.0 – 2.5% by volume. About 470,000 tons of benzene was estimated to have escaped into the environment in 1977 alone. Total daily intake of benzene by humans on an average basis is 1.1–1.3 mg of which about 80% comes from air, which we breathe (2). Benzene toxicity is characterized by reduction in white cell counts in laboratory animals and leukemia in humans. Benzene has also been shown to be carcinogenic.

Poly-Nuclear Aromatic Compounds: Polynuclear aromatic hydrocarbons (PAHs) are organic chemicals, which possess two or more benzene rings fused together. More than a hundred of these compounds have been identified in the environment of which 11 have been shown to be carcinogenic. Three important carcinogenic polynuclear hydrocarbons are: benzopyrene, benzoanthracine and dibenzine. Many of these compounds are produced naturally. However, much of PAHs present in the atmosphere is derived from pyrolysis of organic matter. Polynuclear aromatic hydrocarbons are lipophilic compounds, which have a low solubility in water. Metabolism of these compounds involves the formation of epoxies, which are finally converted to phenols. The carcinogenic action of PAHs is due to the formation of strongly electrophilic epoxides, which bind to various macromolecules such as DNA and RNA within the cell. Lung cancer risk multiplies if plenty of PAHs is present in the air, which we breathe (2). Ophthalmic preparation like Kajal and Surma common in many Indian households, contains a significant amount of PAHs (230-250 mg/gm) of which Pyrene is most abundant!

(II) ACUTE VERSUS CHRONIC EFFECTS OF AIR POLLUTION

Air pollution usually causes slow and gradual effects, which are of chronic type. As the main route of entry of pollutants is through respiratory track various chronic, non-specific respiratory diseases such as bronchitis, chronic obstructive ventilatory diseases, pulmonary emphysema, bronchial asthma etc. may be caused by air pollution.

Very large doses of most of the common pollutants of air are required to produce acute toxicities. Such concentrations are very rare. Most of the instances of acute toxicity so far recorded involve industrial emission of toxic gases and vapours, which are not usually present in the environment, or if present, they occur in traces only. The famous Bhopal tragedy of 1984, in which about 40 tons of Methyl isocyanate was accidentally released into the atmosphere, left about 2800 people dead within 15 minutes and many more crippled for life. The release of chlorine gas in a subway tunnel in Brooklyn is another such example. Both these cases were caused by toxic agents not commonly found in toxic concentration in the atmosphere. It was only the accidental release of these gases from industries which resulted in the disasters.

(III) FOG, SMOG AND REDUCING TYPE OF POLLUTION

The pollution, which is characterized by presence of large quantities of oxides of sulphur, nitrogen, smoke and fumes, is known as reducing type of pollution. Burning of fossil fuels, smelting of metallic ores and various industrial operations yield a large amount of particulate materials as well as oxides of sulphur, carbon and nitrogen. The gaseous mixture and particulate material, under normal conditions are carried up and dispersed in the atmosphere causing little harm to the living beings. However, in cases when there is an abrupt cooling of the lower layers of air while upper layers remain warm, the temperature profile of the atmosphere becomes inverted. Lower layers become heavier than the upper layers. Pollutants discharged into the atmosphere tend to sink down and stay near the surface. Consequently, a high concentration of pollutants gradually builds up in the lower layers of the atmospheric air.

As temperature drops water vapours condense on the surface of fine particles and form tiny liquid droplets. An accumulation of such small droplets results in formation of fog or smog, which due to still atmosphere stays near the surface of earth and forms a blanket enveloping the locality.

$$SO_2 + H_2O \rightarrow H_2SO_3 + O \rightarrow H_2SO_4$$

$$2\,NO + O_2 \rightarrow 2\,NO_2 + H_2O + O \rightarrow 2\,HNO_3$$

$$NH_3 + H_2O \rightarrow NH_4OH$$

$$NH_4OH + H_2SO_4 \rightarrow (NH_4)_2\,SO_4 + H_2O$$

$$NH_4OH + NHO_3 \rightarrow NH_4NO_3 + H_2O$$

Various gases, fumes and vapours dissolve or adsorb on to the fine film, which is thus formed around the tiny particles. Of these gases sulphur dioxide is the most important, though other gases like oxides of nitrogen, carbon, ammonia, various hydrocarbons etc. are also there. The tiny solid material inside the aerosol may consist of oxides of various metals and other elements, which act as catalytic agents for various atmospheric reactions occurring on the aerosol surface. Sulphur dioxide is first converted to the corresponding acid and then to sulphates upon reacting with ammonia or ammonium hydroxide bearing aerosols, ammonium sulphates and bisulphates may be formed. Likewise nitrogen dioxide is changed to corresponding acid and then to nitrates. Thus secondary aerosols are formed which bear sulphates, nitrates and acids like sulphuric and nitric acids (13).

The occurrence and extent of these reactions is dependent upon the quantities of the reactants and the presence of aerosols capable of providing suitable locus and the catalyst. The first documented case of this type of pollution occurred in Meuse Valley, Belgium, in 1930. The second took place in Donora, Pennsylvania in 1948, while the third and the most disastrous one took place in London in 1952. These instances had much in common. Rapid cooling of surface layers of air prevented the normal mixing of pollutants which occurs due to the warmer layers rising up to be replaced by the upper cooler layers which sink down. Innumerable sources kept on adding various pollutants such as sulphur dioxide smoke, fumes and oxides of carbon, nitrogen etc. as the main fuel for domestic heating and power generation was coal. These conditions lasted for many days at a stretch during which time sufficiently high concentration of these pollutants was built up. Various atmospheric reactions occurring on the surface of aerosols caused the formation of more toxic secondary aerosols. The concentration of acids, sulphates, nitrates, oxides of carbon, hydrocarbons etc. became sufficiently high to cause adverse effects at least on the weak and infirm. Though in Meuse valley only 65 and in Donora 20 deaths were recorded, in London disaster about 95,000 people were affected and by the time fog lifted about 4000 deaths had occurred. As hydrothermal and nuclear power has replaced coal to a varying extent in large crowded industrial cities of the developed world, the menace of reducing type of pollution has been on the decline. However, in many developing countries like India, coal, mineral oils and organic matter continues as an important source of energy and thermal power plants still use about half of the total national output of coal. The chances of occurrence of episodes of London type are far greater in rapidly expanding Indian cities as compared to developed western countries (13).

(IV) ACID RAINS

The phenomenon of acid rains is a consequence of accumulation of huge amounts of oxides of sulphur, nitrogen and fine particulate materials or aerosols. However, unlike the reducing type of pollution, which involves fog or smog, these gases and aerosols accumulate high up in the atmosphere. Water vapours condense on aerosol surface and form a fine film providing suitable loci as well as catalysts for the oxides of sulphur and nitrogen to dissolve in water and react to form to corresponding acids. These acids form salts, such as sulphates and nitrates, when they come in contact with aerosols of basic nature. The secondary aerosols are thus formed which bear acids and salts of these acids. Large and heavier aerosols are developed when smaller ones coalesce together and these tend to drift downwards. As water vapours continue to condense on the aerosols large droplets of water are produced which rain down as acid rains. Hydrochloric acid derived from the oxidation of organochlorine compounds and polyvinyl chloride in plastics may also be present in these droplets. Rain also traps much of windblown particles dust and other gaseous constituents, which may also be acidic in nature as it falls down (14;15).

$$SO_2 + H_2O \rightarrow H_2SO_3 + O \rightarrow H_2SO_4$$
$$2NO + O_2 \rightarrow 2NO_2 + H_2O + O \rightarrow HNO_3$$
$$NH_3 + H_2O \rightarrow NH_4OH$$
$$NH_4OH + H_2SO_4 \rightarrow (NH_4)_2\,SO_4 + H_2O$$
$$NH_4OH + NHO_3 \rightarrow NH_4NO_3 + H_2O$$

Thus, all chemical reactions, which result in the formation of fog or smog and reducing type of pollution, occur higher up in the atmosphere and give rise to rains of acidic nature, which may have a pH in the range of 3.0 – 5.8. Sulphuric acid, nitric acid, various sulphates and nitrates are the main chemical constituents of these rains, though small amount of other acids like hydrochloric acid and their salts may also be present. The damages caused by acid precipitation are of a very diverse nature (16). The harmful effect of acid precipitation can be summed up as follows:

(1) The corrosive action of the acid content in these rains damages buildings, wood, steel and cement concrete structures. It also damages fabric and other fibrous materials. Many of our monuments, which are built up of steel, stone of cement concrete, are threatened by the corrosive action of acid rains (like the Tajmahal at Agra).

(2) Low pH causes release of many toxic metals and trace elements in excessive amounts as the rate of decay and decomposition of parent rocks is enhanced. However, in traces, these elements do not pose many problems but in higher concentration, many are highly toxic. Moreover, these elements are bio-accumulated by living organisms and are bio-magnified to concentration many thousand times higher than their concentration in the surrounding medium as they move up along the food chain. So even a small rise in concentration of many of these elements in the environment could be dangerous to the living beings.

(3) Acid precipitation affects aquatic system drastically as due to low mineral content, the capacity of natural waters to assimilate the introduced acid is limited. There is a reduction in the alkalinity. Solubilities of various chemicals and gases are affected with decline in pH and the entire chemical picture of the water concerned undergoes a change. Soft waters are the worst sufferers because of their low salt content. Changes in pH also affect microbial community, which are responsible for decomposition of organic matter, maintenance of plant nutrients and properly balanced mineral cycles.

(4) Soils in general have a greater buffering capacity than aquatic systems. However, excessive amount of acids introduced by acid rains may disturb the entire soil chemistry. The concentrations of sulphates and nitrates rise while more phosphates are released in soluble form, so also are many toxic heavy metals and trace elements, which are converted to their available state. Acidity, as well as harmful action of toxic elements damages vegetation while susceptible microbial species are eliminated. The soil loses its normal functions such as decay and decomposition of organic debris and the capacity of a balanced regulation of nutrients. Heavy downpours leach away much of the precious mineral nutrients and desertification follows.

(5) Much of the diverse effects of acid rains on plant and animal life stem from lowering of pH which acid produces when introduced in a biological system. Excessive amount of hydrogen ions adversely affects biological membrane, the electron transport system and a number of pH specific biochemical reactions. In aquatic environment, the effect of lowering of pH is usually more pronounced. Small planktons, flagellates, algae, crustacea etc. may disappear and so also are fishes like trout and salmon which are intolerant to pH below 5. Molluscs suffer because low pH inhibits formation of calcareous shell, which protects them in addition to the diverse effects caused by disruption of enzymes and electron transport system. In higher plants chlorophyll is degraded and leaves lose their green pigmentation which results in drastic reduction in productivity. In higher animals including man which have got a sufficiently thick epidermis, sulphuric acid bearing aerosols and waters irritate eyes and when inhaled drastically reduce the tracheo-bronchial muco-ciliary clearance activity. They also cause a marked degree of bronchial constriction, which results in significant resistance to the flow of gases into and out of lungs (2). This makes an individual susceptible to other respiratory ailments.

(V) PHOTOCHEMICAL SMOG OR OXIDIZING TYPE OF POLLUTION

In contrast to reducing type of pollution, oxidizing type of pollution is characterized by the presence of large concentrations of ozone, oxides of nitrogen and various hydrocarbons. In fact, it is a chain of reactions initiated by oxides of nitrogen, which is responsible for the accumulation of large quantities of ozone, nitrogen dioxide and formation of other harmful substances.

These reactions are extremely complex in nature. The oxidant formed in largest concentration is ozone, which is a serious pollutant in lower atmosphere. Higher up in the atmosphere there are sufficient short wavelength ultraviolet radiations (1700Å–2400Å) available to convert oxygen to ozone. However, these wavelengths do not reach earth's surface. The excessive build up of ozone in lower atmosphere has been thought to be due to the presence of nitrogen dioxide which absorbs ultra violet radiations of a little higher wavelengths (3000Å–4000Å) and produces an atom of oxygen. This atom reacts with molecular oxygen to produce ozone. In unpolluted atmosphere, ozone thus produced reacts with nitric oxide, also a product of above reaction and results in the formation of nitrogen dioxide and molecular oxygen again (Fig 19.1)

A cyclic reaction is, therefore, established and as a consequence most of the ozone produced is broken down. The system tends to stay in a steady state condition. The net result of this cyclic reaction is a little warming up of the atmosphere due to the absorption of energy of solar radiations. (Fig 19.1). The presence of hydrocarbons, especially the unsaturated ones (olifins), however, upsets the natural balance. The oxygen atoms produced by nitrogen dioxide react with hydrocarbons to convert them to their highly reactive oxidized state, which starts reacting with nitric oxide to form nitrogen dioxide again.

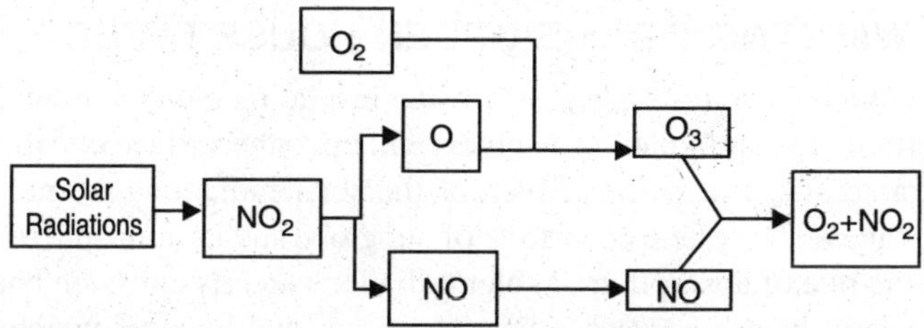

Fig. 19.1 *Formation and breakdown of Ozone in atmosphere.*

Due to the greater affinity, nitric oxide present in the atmosphere reacts preferentially with reactive hydrocarbons to produce Nitrogen NO_2, CO_2 and H_2O. Ozone is left out. Consequently, a gradual build up of the concentrations of ozone and nitrogen dioxide occurs while that of nitric oxide declines. The brownish colour of the smog is due to the colour of nitrogen dioxide, which is orangish brown.

The entire chain of these chemical reactions is initiated by absorption of solar radiations. Early in the morning the concentration of oxidants, nitrogen dioxide and ozone are almost negligible while that of nitric oxide and hydrocarbons are a little higher. With the rise of life-activity and morning traffic, the concentration of nitric oxide and hydrocarbons begins to rise. As the sun comes up nitrogen dioxide absorbs solar radiations, splits up to yield oxygen atoms, which produce ozone on reacting with molecular oxygen. The nitric oxide produced reacts with hydrocarbons to produce hydrocarbons in highly reactive oxidized state. The concentration of ozone begins to rise more rapidly as the sun comes up, while hydrocarbons are converted to their reactive state, which triggers the reaction in which nitric oxide is rapidly changed to nitrogen dioxide. Within two or three hours of sunrise, the concentration of ozone and nitrogen dioxide reaches the threshold where problems start. In the morning as earth's surface is cooler, the air near it is cooler than the layers above. This brings ozone and nitrogen dioxide right down to the surface of the earth (Fig 19.2). As the day advances, by 1.00 – 2.00 PM due to solar radiations earth's crust warms up which gradually displaces the brown misty haze upwards (13).

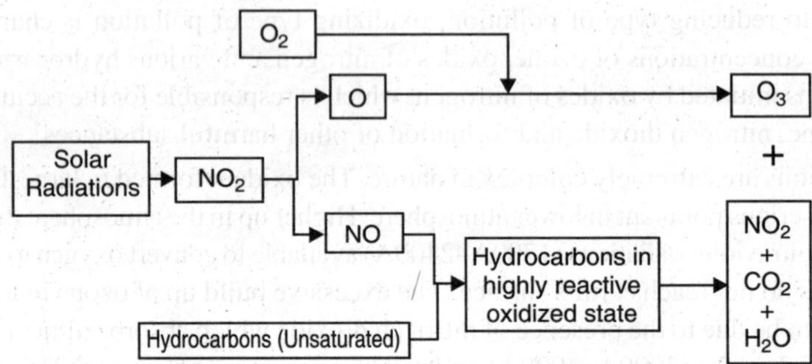

Fig. 19.2 *Accumulation of Ozone and Nitrogen dioxide in atmosphere*

The brown misty haze is a hazardous mixture to breathe as it contains ozone and nitrogen dioxide both of which adversely affect respiratory system and general health of living organisms. It also contains aldehydes, such as formaldehyde, acetaldehyde acrolein, peroxyacetyl nitrate (PAN), peroxybenzonyl nitrate (PBN) which contribute much to odour, eye and lung irritation and lacrimations caused by the photochemical smog. In plants, prolonged exposures damage chloroplasts, disturb electron transport system and adversely affect a number of enzymes, which play a vital role in photosynthesis. In addition, the strong oxidizing action of ozone and the harmful effects caused by nitrogen dioxide are injurious to a number of objects and materials made up of wood cotton, rubber and other fibres and cause many troubles (13).

(VI) GLOBAL WARMING OR THE GREEN HOUSE EFFECT

The gaseous mantle, the atmosphere, forms an insulating blanket around our globe and allows a considerable portion of solar radiations to enter right up to the surface of earth, which absorbs it and radiates back infrared and heat waves. Much of the solar radiations are radiated back to space as infrared and heat waves. The system consisting of our globe and its atmosphere is in a state of dynamic equilibrium with the rate of absorption of solar radiations and its emission back to space as infrared and heat waves, nearly balancing each other. The gases and vapours which allow free passage to radiations of relatively shorter wavelengths (2900 Å – 7000 Å) while absorbing effectively infra-red and heat waves (700 Å onwards) play a very important role in maintaining surface temperatures within a range in which life can exist. The phenomenon is similar to that of greenhouse in which the glass-enclosed atmosphere gets heated up due to its insulation from the rest of the environment. Hence, global warming is also known as **Green House effect** and the gases responsible for it are called **green house gases** (9).

(1) Causes of Global Warming

Very slow, almost imperceptible rise of about 4–5ºC in the global temperatures has occurred in the past 20,000 years. However, a rise of about 0.3º–0.8ºC was recorded during the last century alone, which is remarkably faster as compared to change that occurred in the past (Fig 19.3). This acceleration in the pace of global warming coincides with a rise in the concentration of green house gases in the atmosphere (Fig. 19.4-19.7). The insulation of earth's surface from the outer space caused by green house gases tends to become more and more effective as the concentration of these gases rises. More heat and infrared radiations are trapped by the gaseous mantle around the globe, which accelerates the pace of global warming (17).

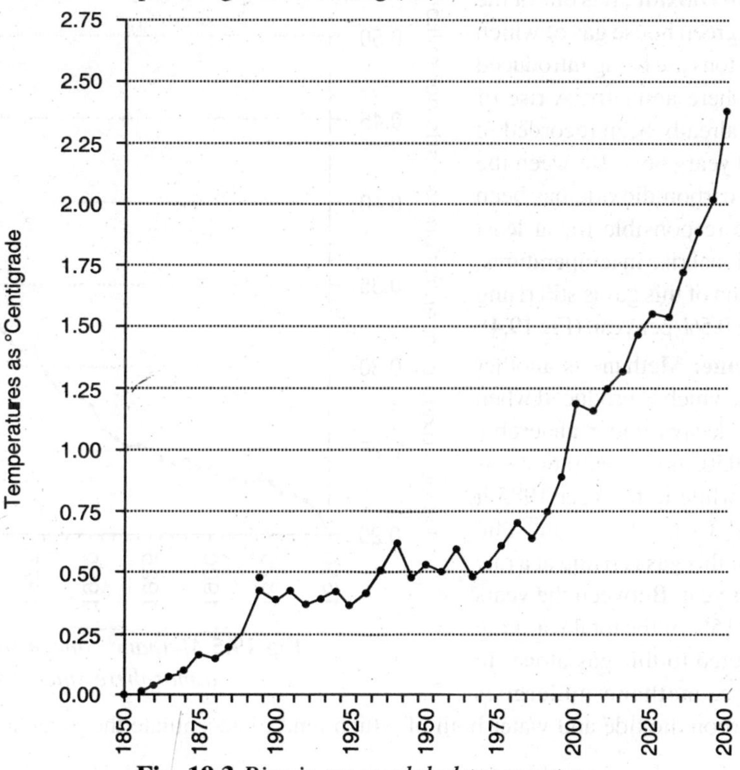

Fig. 19.3 *Rise in mean global temperature*

(2) Gases Responsible for Global Warming

There are a number of gases present in the atmosphere, which are capable of absorbing effectively heat waves and infrared rays while being transparent to radiations of lower wavelengths. Carbon dioxide, methane, oxides of nitrogen, sulphur dioxide, ozone, chlorofluorocarbons and water vapours are some of the gaseous constituents of troposphere, which come in this category. From the point of global warming, however, only those gases are important which maintain an effective concentration in the troposphere, *i.e.*, the region of atmosphere immediately covering earth's

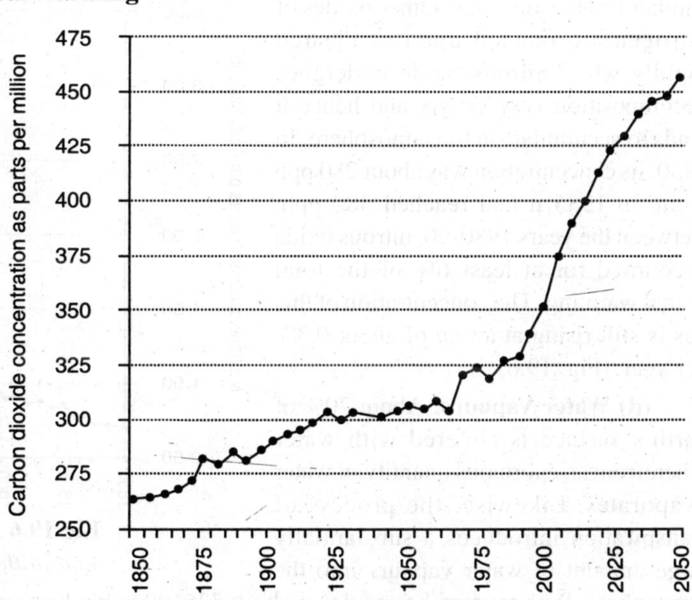

Fig 19.4 *Carbon dioxide concentration in the atmosphere since 1850*

surface. These gases are carbon dioxide, methane, chlorofluorocarbons, nitrogen oxide and water vapours.

(a) Carbon Dioxide: It is one of the most important green house gas of which about 18 billion tons are being introduced into the atmosphere annually. A rise of about 26% has already been recorded in a period of 200 years only. Between the years 1980-90, carbon dioxide has been estimated to be responsible for at least about 55% of global rise in temperatures. The concentration of this gas is still rising at a rate of about 0.5% per year. (Fig 19.4).

(b) Methane: Methane is another greenhouse gas, which is produced when organic matter decays under anaerobic conditions. In 1950, its concentration was about 1.1 ppm while in the year 1985 it was estimated to be 1.7 ppm. The concentration of this gas is rising at a rate of about 1% per year. Between the years 1980-90, about 15% of the total warming has been attributed to this gas alone. In the atmosphere, methane undergoes oxidation to carbon dioxide and water both of which tend to accentuate the greenhouse effect (Fig. 19.5).

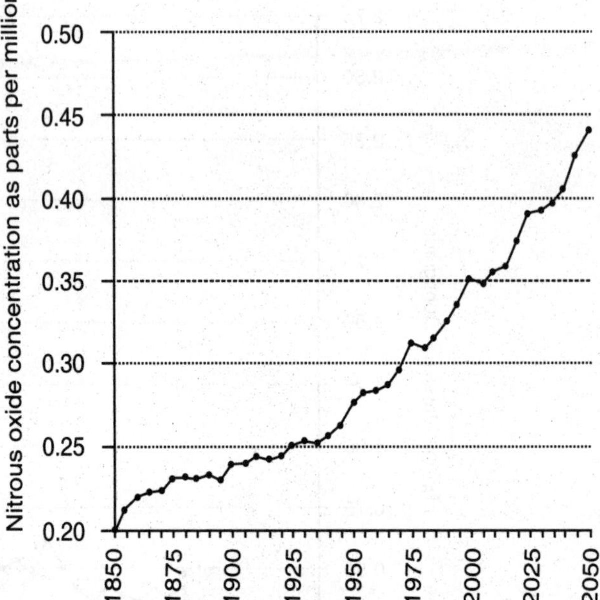

Fig 19.5 *Methane concentration in the atmosphere since 1850*

(c) Nitrous Oxide: Nitrous oxide is another troublesome gas. Other oxides of nitrogen are reacted upon or cleared rapidly while nitrous oxide undergoes decomposition very slowly and hence it tends to accumulate in the atmosphere. In 1950, its concentration was about 280 ppb while in 1985 it had reached 380 ppb. Between the years 1980-90, nitrous oxide accounted for at least 6% of the total global warming. The concentration of this gas is still rising at a rate of about 0.3% per year. (Fig. 19.6)

(d) Water Vapours: About 70% of earth's surface is covered with water wherefrom an enormous quantity of water evaporates. Likewise, the process of transpiration introduces a substantially large amount of water vapours into the atmosphere. Precipitation brings down about 525100 cubic kms of water to earth's surface whereas vapours equivalent to 14,000 cubic kms of water stay back permanently in the atmosphere (18). Water vapour like any other greenhouse gases contributes significantly to the global warming. With an

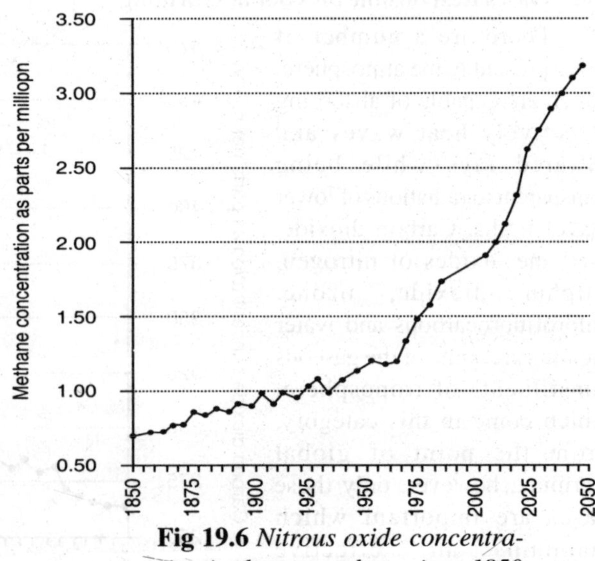

Fig 19.6 *Nitrous oxide concentration in the atmosphere since 1850*

overall rise in temperatures the rate of global transpiration and evaporation shall also go up which shall introduce more water vapours into the atmosphere and could in turn influence the pattern of global warming.

(e) Chlorofluorocarbons: Chlorofluorocarbons represent a group of man-made, colourless, odourless, easily liquefiable chemicals, which have more potential for global warming than any other greenhouse molecules. They are very stable compounds, which may persist in the atmosphere for periods as long as 80-100 years. Though first introduced only in the fifties, chlorofluorocarbons have rapidly attained such levels that between the years 1980-90, they were responsible for 24% of global warming. Till 1985, about 15 million tons of these compounds had been released in the atmosphere. In spite of much international efforts to check the use of these chemicals, CFCs are still rising at a rate of about 5% per year (Fig 19.7). Apart from contributing substantially to global warming, the persistent nature of chlorofluorocarbons enables them to accumulate and rise at random to reach the stratosphere. Strong ultra-violet radiations present in the stratosphere decompose these compounds to yield chlorine atoms, which catalytically destroy the vital ozone shield.

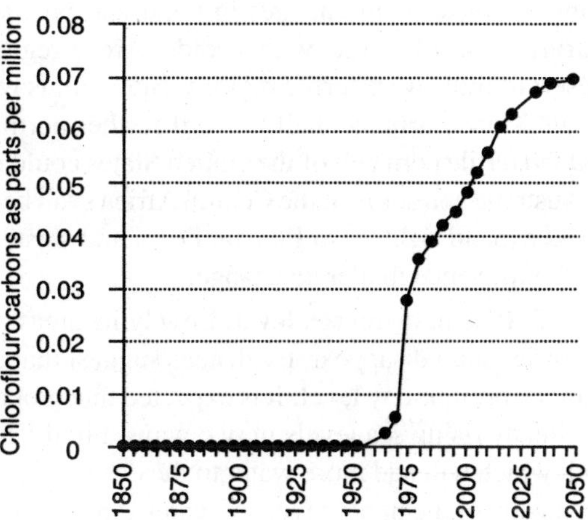

Fig 19.7 *Electromagnetic spectrum of solar radiations*

It is human activity, which is mainly responsible for the accumulation of green house gases in the atmosphere. Energy production and its use, intensive agriculture, maintenance of huge livestock population, use of chlorofluorocarbons, land use modification and industrial production are some of the aspects of human activity, which are responsible for this accumulation.

(3) Consequences of Global Warming

Global records of earth's surface temperatures indicate that a warming of about 0.5° (0.3° - 0.7°C) has occurred during the last century alone. Results from recent climatic models suggest that mean global temperatures shall rise by 2° - 6°C during the next century if we assume that the carbon dioxide concentration in the troposphere increases to 600 ppm. (Fig. 19.3). The projected change in mean surface temperatures may appear insignificant because variations of this magnitude are experienced in course of seasonal or even daily weather. In fact it is not so. During the last great Ice Age, about 12,000 years ago, when much of Northern America and Europe was covered with a sheet of ice, the mean surface temperatures were only about 5°C lower than today. The world climate was very much different from what it is now. The transition from the great Ice Age to present-day climate during which average surface temperature rose by 5°C took almost twelve thousand years. However, the variations of almost similar magnitude shall be experienced within the course of a single century if greenhouse gases continue to rise at the present rate.

There are considerable uncertainties, regarding the precise consequences of global warming. However, some of the obvious consequences of the general heating up of earth's surface as suggested by most of the climatic models shall be:

1. Global climatic change: While tropics shall expand, world's climatic belts shall shift away from equator while polar areas shall shrink. The rise in global temperatures shall not be uniform all over the surface area of the world. Most of the workers agree that Polar Regions of the world would undergo larger increase in temperature, about ten to twelve times as much as the tropics. This shall bring unprecedented changes in wind and precipitation patterns within a span of a single century. During the last Ice Age, with a colder Arctic region, North Africa, the deserts of Arabia and the Thar Desert of India were fertile regions. But with its warming the precipitation belt has shifted northward today. More warming shall move it further north. Therefore, North Africa, Europe, parts of Russia and the fertile corn belt of the United States could become drier, while much of peninsular India, parts of Australia and some of the Central Africa shall become more humid. In India, the deserts of Rajasthan could expand right up to Punjab, Delhi and western part of Uttar Pradesh while eastern part of India shall experience moderate changes.

2. Rise in mean sea level: Low lying areas of the world shall be submerged. Many small island countries shall disappear. Evidences suggest that about 12,000 years back it was nearly 100 m lower than the present-day level. It is expected that global rise in temperatures shall further enhance the rate of already rising sea levels in two ways. Firstly, large deposits of ice present on earth's surface shall melt which will add more water to the oceans. Secondly, rise in temperatures shall also cause thermal expansion of the upper layers of water. An increase of 4° – 5°C could cause enough expansion of this enormous mass of water so as to raise the mean sea level 5 - 6 cms. If all ice present on earth's crust was to melt, sea level could rise by about 60 m. Large stretches of low-lying areas shall be submerged. As much of the world population live near shore, this could be a total catastrophe. About sixty odd island countries shall face deep encroachments by seawater and some like Maldives may disappear altogether.

3. Raised intensity and destructive power of weather events like rains, thunderstorms and cyclones: The intensity and destructing power of whether events like rains, thunderstorms, cyclones etc shall increase. Rains for example shall take on the shape of heavy downpours and shall deposit large volumes of water in a brief spell of time which could cause flash floods, a raised rate of nutrient-leaching and soil erosion and a highly reduced rate of recharge of ground water table. Higher temperature shall cause a rise in evapo-transpiration and ground water table may be affected resulting water stress in many localities of the world. Salt waters encroaching upon the low-lying areas may result in wastage of many of our fresh water reservoirs and could spoil much of our underground water resources.

4. Drastic reduction in biodiversity: As the climatic belts shift away from equator towards poles, vegetation shall have to shift in the same direction to stay in favourable climatic conditions. Those species, which are unable to do so, shall die. There will be losses of genetic resources on large scale. Hardy and resistant forms shall come up and survive. An altogether changed biotic spectrum shall replace the earlier ones and almost all important biomes shall be affected. As temperature, changes will affect wind and precipitation patterns also water could play an important part in altering the biotic communities. It has been suggested that some rise in precipitation, however, shall be balanced by an enhanced evapo-transpiration and this could lead to water deficit and moisture stress in many regions of the world (19). The effect of global warming on agriculture will be of a varied type in different parts of the world. Wheat and maize crops may suffer from moisture stress. More fertilizers shall have to be used to sustain productivity. In places where temperature conditions are already near the upper tolerance levels, even a rise of 1 - 2°C may be quite harmful. Alterations in cropping pattern shall occur and pest resistant varieties more suitable to warmer conditions shall have to be developed. In short, green house warming shall bring with it an entirely new environment in which life though not impossible yet its existence shall be tougher to maintain.

5. Increased losses due to insects, pests and pathogens and changes in disease patterns: Insects and pests and pathogens may increase as warmer conditions could be more favourable to their growth and coupled with higher humidity pathogenic diseases shall multiply. The small size, an active metabolic rate, prolific reproduction, a life cycle which can be completed within a few hours or days provide an enormous evolutionary advantage to insects, pests and pathogens over man. New unforeseen diseases of plants, animals and man shall appear and could acquire the dimensions of an epidemic. Under the changing climatic conditions these organisms could inflict heavy damages on human health and those of his dependents.

(VII)OCEAN ACIDIFICATION

An unfortunate consequence of rising concentration of CO_2 in the atmosphere is the danger of ocean acidification. In the past 200 years the oceans have absorbed nearly half of the CO_2 produced by fossil fuel burning and cement production. CO_2 dissolves in water to produce Carbonic acid, H_2CO_3, a weak acid which lowers the pH of oceans waters. Observations on measurements of pH of the surface waters of oceans and our understanding of ocean chemistry indicate that the uptake of CO_2 by oceans has led to a reduction of the pH of surface waters by about 0.1 units. The pH of sea-surface waters around the year 1700 AD was 8.18. Today it is 8.07. This value may seem small because of the manner in which pH is measured, however, it is equivalent to about 30% increase in the concentration of hydrogen ions (20). Increasing atmospheric concentration of CO_2 will lead to further acidification of the oceans. By the years 2050 and 2100 the pH could decline to about 7.95 and 7.82 respectively. (21).

The chemical changes in the oceans caused by increases in the concentration of CO_2 in the atmosphere will include a lowering of the pH, an increase in dissolved CO_2, a reduction in the concentration of carbonate ions and an increase in bicarbonate ions. It is believed that the resulting decrease in pH will have negative consequences, primarily for calcifying organisms of the oceans which deposit $CaCO_3$ on their cell coverings and in skeletons. Calcifiers include organisms such as coccolithophores, corals, foraminifera, echinoderms, crustaceans and molluscs. Under normal conditions, $CaCO_3$ are stable in surface waters since the carbonate ions are at saturated concentrations. But as ocean pH falls, the concentration of carbonate ions shall decline resulting in dissolution of structures made of calcium carbonate. Researches have already found that corals, coccolithophore algae, coralline algae, foraminifera, shellfish and pteropods experience reduced calcification or enhanced dissolution when exposed to elevated CO_2 concentrations (22). Besides the effect of calcification the process of reproduction in many species could be disrupted and a number of species could suffer from CO_2 induced acidification of body fluids (hypercapnia). Changes in nutrient levels may adversely affect the overall productivity and the food resources of a number of species higher up the food chain. These changes may change the entire biotic spectrum of our oceans leading to unpredictable consequences.

Although full ecological consequences of these changes in calcification are still uncertain, it appears likely that a decline in the coccolithophores may have secondary effects on climate change, by decreasing the earth's albedo, which is the proportion of solar radiation reflected back into space, via their effects on oceanic cloud cover. A lowered albedo shall raise earth's temperature while higher albedo shall cool the planet (23). The capacity of the world's oceans to take up and sequester CO_2 present in atmosphere may decline which could also enhance global warming. On the contrary it has also been suggested that ocean acidification in the future will lead to a significant decrease in the burial of carbonate sediments for several centuries, and even the dissolution of existing carbonate sediments (24). This will cause an elevation of ocean alkalinity, leading to the enhancement of the ocean as a reservoir for CO_2 with potentially beneficial implications for climate change as more CO_2 leaves the atmosphere for the ocean (25).

The report of the Royal Society, London, concludes on the basis of current evidence that ocean acidification is an inevitable consequence of continued emissions of CO_2 into the atmosphere, and the magnitude of this acidification can be predicted with a high level of confidence. However, its impacts are much less certain and require substantial research efforts (22).

(VIII) THE PROBLEM OF STRATOSPHERIC OZONE DEPLETION

Sun emits radiations at all wavelengths. Of these, radiations important to our planet are : ultraviolet radiations, light rays, and infrared rays or heat waves. Radiations of visible spectrum and infrared rays carry little energy, which does not harm living beings. The energy content of ultraviolet radiations is however, larger than the limits of tolerance of a living cell and hence is harmful or even lethal to a living system. Ultra-violet radiations are capable of causing damages to DNA and RNA and other macromolecules and could result in carcinogenic, teratogenic and mutagenic effects. Life therefore, had to await the development of an effective shield of atmospheric gases, which could check the biocidal radiations high up in the atmosphere before it could come out on land.

(1) Ultraviolet Radiations and Ozone

Ultraviolet radiations (Fig. 19.8) are usually grouped, rather arbitrarily, into the following three categories:

1. **ULTRAVIOLET-A (UV-A):** With a wavelength range between 3150-4000 Å which do not cause much harm to a living system. Only a part of these radiations reaches earth's surface, which are tolerated by the living beings.
2. **ULTRAVIOLET-B (UV-B):** With a wavelength from 2800 Å to 3150 Å which are more damaging than UV-A. Absorption maxima of DNA and RNA fall within this range.
3. **ULTRAVIOLET-C (UV-C):** With a wavelength range from 2000 Å to 2800 Å, which carry larger amount of energy. These are the most damaging radiations for the biosphere. However, they are almost completely absorbed by atmospheric gases.

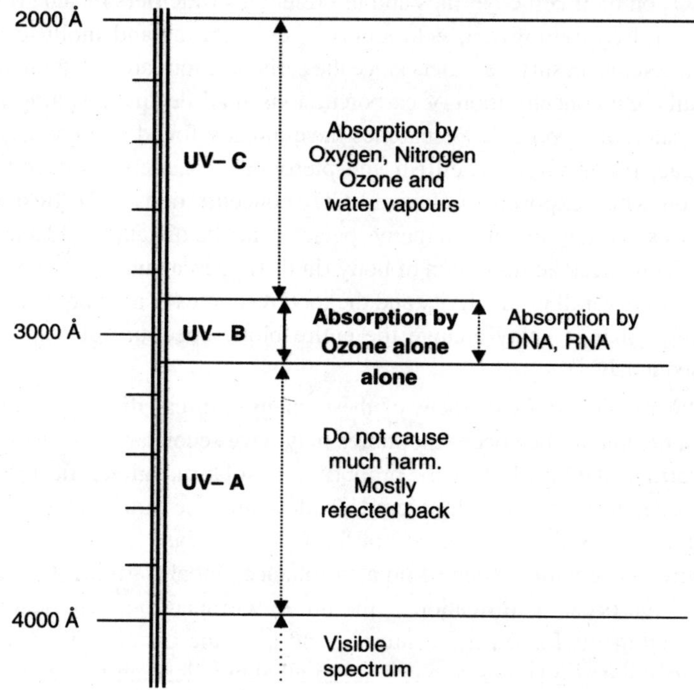

Fig. 19.8 *Electromagnetic spectrum of solar radiations*

Much of the harmful high-energy solar radiations are absorbed at various heights by the gaseous mantle which **surrounds** our planet. Though oxygen, nitrogen, ozone, water vapour and a number of other constituents of the atmosphere absorb short wavelength ultraviolet radiations, none of these gases can absorb effectively wavelengths greater than 2800 Å. This leaves a gap, which is filled by ozone alone. It absorbs all radiations between 2800 to 3150 Å. Radiations above 3150 Å are not completely absorbed by this gas, which however are considerably diluted by the ozone layer. A depletion of ozone content of the atmosphere shall result in increased penetration of 2900 Å to 3150 Å radiations (UV-B), which are very injurious to the biosphere. In extreme cases, radiations with wavelengths lower than 2800Å may also reach earth's surface.

Laboratory studies have shown that absorption of ultraviolet radiations within the range of UV-B by a living cell is largely due to the absorption of energy by nucleic acids, the DNA and RNA molecules. The energy thus absorbed breaks up these macromolecules and affects adversely such vital processes as protein synthesis, growth and reproduction. Alterations in DNA molecules could cause long-range genetic effects, which could change the very shape of life on this planet.

(2) The Ozone Layer

High up in the stratosphere, about 15-40 kms above earth's surface, short wavelength ultraviolet radiations in the range of 1800 Å to 2200 Å are absorbed by molecular oxygen which splits up into its constituent atoms. Oxygen atoms combine with oxygen molecules to form ozone.

Therefore, ozone is a result of photochemical reactions in which the starting molecule is oxygen. Along with this reaction another photochemical reaction which causes breakdown of ozone molecules due to absorption of 2000-2900 Å radiations also occurs. The two reactions, i.e., the formation and destruction of ozone molecule normally balance each other and ultimately result in effective absorption of short wavelength ultraviolet radiations in the stratospheric region. Life underneath is thus protected from the biocidal solar radiations.

Ozone occurs in the atmosphere because the atmosphere contains oxygen. Without oxygen there would have been no ozone umbrella to protect the living systems on earth's surface. Evolution of atmospheric oxygen and subsequently the ozone layer has, therefore, been intimately connected with the evolution of life on this planet. Primitive atmosphere contained very little oxygen. The only process, which could have contributed some oxygen, is photo-dissociation of water molecules by 1950 Å - 2000 Å ultraviolet radiations, which are soon cut off as some oxygen accumulates and this stops the photo-dissociation reaction. It was only after the evolution of photosynthesis that a gradual accumulation of oxygen occurred. With the accumulation of this vital gas more and more ozone was produced which provided means to check UV-C and UV-B radiations effectively.

Most of the ozone present in the atmosphere, almost 90% is concentrated in the stratosphere at an altitude of about 15-40 kms. If the entire ozone present in the stratosphere is condensed into a single sheet covering the globe, it will form only 2.4 – 2.6 mm thick layer at equator and 3.1 – 4.5 mm thick layer at latitudes about 70°N and S. At ground level, ozone occurs in traces only. Though maximum amount of ozone production occurs over equator, its concentration is lowest here as most of the ozone formed is displaced towards the poles with massive movement of atmospheric air.

(3) Stratospheric Ozone Depletion

The concern over stratospheric ozone depletion was first aroused by the U.S. chemist Harold Johnson in 1971, when he pointed out that Supersonic Aircrafts then under fabrication would introduce large quantities of nitric oxide at precisely those altitudes where ozone content is maximum. Nitric oxide shall catalytically attack ozone molecules and convert them to oxygen. Large-scale use of supersonic aircrafts shall weaken the protective ozone layer on global scale and could result in a variety of adverse consequences such as massive increase in number of skin cancers, modification of flora, fauna and climatic conditions.

In May 1985, Farmen and co-workers reported the ***Ozone hole*** and created a sensation. They showed that the total average quantity of ozone measured over the South Pole in the month of October (the beginning of southern spring) was gradually diminishing. From 1979 to 1985, a reduction of about 40% in ozone content over Antarctica was recorded (26). It was soon realized that it was not only nitric oxide but there were other hazards as well which could bring about catalytic destruction of ozone content. Depletion of stratospheric ozone has been found to occur mainly due to the three following constituents of the stratosphere.

1. **Nitric Oxide Molecules:** Nitric oxide usually present in stratosphere reacts with ozone to form nitrogen dioxide which in turn reacts with atomic oxygen to produce nitric oxide again.

$$NO + O_3 \rightarrow NO_2 + O_2$$
$$NO_2 + O \rightarrow NO + O_2$$

2. **Chlorine Atoms:** Chlorine atoms react with ozone to yield chlorine monoxide, which reacts with atomic oxygen to regenerate chlorine atoms again.

$$Cl + O_3 \rightarrow ClO + O_2$$
$$ClO + O_3 \rightarrow Cl + 2O_2$$

3. **Hydroxyl Ions:** Hydroxyl ions produced by photo-dissociation of water molecules in the stratosphere, react with ozone molecules to produce HO_2, which reacts with another ozone molecule to yield hydroxyl ions again.

$$H2O \rightarrow H^+ + OH$$
$$OH + O_3 \rightarrow HO_2 + O_2$$
$$HO_2 + O_3 \rightarrow OH + 2O_2$$

One of the most important features to note in the three set of reactions above is that the principal constituent, which starts the reactions, is regenerated at the end of cycle and is available again to cause further ozone breakdown. Under normal conditions, nitrogen dioxide combines with chlorine monoxide to form chlorine nitrate, which is further converted to nitric, and hydrochloric acids – products that are harmless. Similarly methane normally present in the system in the stratosphere has a protective role as it reacts with chlorine atoms to produce the harmless hydrochloric acid. These reactions are very important as they serve as a means of elimination of both oxides of nitrogen as well as chlorine atoms from the system. Recent data collected by NASA's Upper Atmospheric Research Satellite indicates that fine particulate material present in the atmosphere plays a vital role in destruction of stratospheric ozone. Reactions catalyzed by sulphuric acid aerosols immobilize nitrogen dioxide and block its protective functions. These aerosols also retard the reaction between methane and chlorine, thus damaging another important defensive reaction, which protects the ozone umbrella by locking of chlorine atoms in harmless hydrochloric acid molecules.

$$ClO + NO_2 \rightarrow ClNO_3 + H \rightarrow HCl + HNO_3$$
$$CH_4 + 2Cl + 3O \rightarrow CO_2 + 2HCl + H_2O$$

(4) The Destruction of Ozone Molecules over Poles

Much of the stratospheric ozone is formed at equator and is displaced over to Polar Regions so that the concentration of this gas is maximum at the poles. Incidentally, it is at the poles only, the Antarctic and Arctic regions that the destruction of ozone appears to be fastest. Strong westerly winds around the poles, which often reach a speed of 90-100 metres per second isolate a column of cold air inside a fast moving vortex. Cut off from the warmer air around the temperatures in this column of air drop forming clouds up to an altitude of 25-30 kms. Sub-zero temperatures cause the formation of ice crystals, which provide suitable loci for atmospheric reactions to occur.

Oxides of nitrogen are locked in these fine crystals before they can immobilize free chlorine atoms, which are held over the long polar winters on fine aerosols as chlorine nitrate and hydrochloric acid. With the return of sun and its high-energy radiations, hydrochloric acid and chlorine nitrate react to release free chlorine atoms, which destroy ozone molecules catalytically. Thus instead of dissociation into harmless acids, weather conditions promote the reaction of chlorine nitrate with hydrochloric acid causing the regeneration of chlorine atoms. By the beginning of summers, when the polar vortex begins to break up the ozone concentration over the Polar Regions has been reduced appreciably.

Ozone hole was first discovered over Antarctica and soon similar ozone depletion was observed over Arctic also. The process of thinning out of this protective ozone umbrella is not confined to Polar Regions alone. Observations conducted from highflying research planes have shown that blobs of chlorine monoxide rich air slip off the poles and hang over Northern Europe from London to Moscow, high up in the stratosphere. Global ozone monitoring porgramme have pointed out a reduction of about 3.0% in mean annual ozone content between the latitudes 53° and 40° North, which covers much of Europe, Great Lake areas, the Black Sea, the Caspian, part of Mongolia, Manchuria and Japan.

(5) Causes of Stratospheric Ozone Depletion

The three main reactants, namely nitric oxide, chlorine atoms and hydroxyl ions are normally present in the environment. Though much of these are derived from natural sources, human activity is also equally responsible for contributing many of such constituents in the atmosphere, which give rise to these reactants. Temperatures decrease with height in the troposphere, the zone of atmosphere immediately covering the surface of our planet while in stratosphere there is a rise in temperatures. In between the two lies a narrower region, called the tropopause, which has almost uniform temperature. It is this temperature profile, which prevents largely the exchange of materials between the stratosphere and the troposphere. However, some exchange does occur and many constituents of the troposphere do reach the stratosphere. In addition to this exchange, many pollutants are injected straight into the stratosphere by supersonic transport planes, rockets and space shuttles, surface nuclear tests, volcanic emissions etc. Human activities responsible for polluting the stratosphere may be summed up as follows:

(1) **Combustion of Fossil Fuels and Organic Matter :** Huge quantities of various oxides such as oxides of carbon, nitrogen and sulphur, hydrocarbons and fine particulate material is introduced daily into the troposphere some of which goes high enough to reach the stratosphere. To this may be added microbial decomposition products of organic matter, which discharge plenty of hydrocarbons into the atmosphere of which methane is one of the most important constituents which reaches high enough to pollute the stratosphere.

(2) **Excessive Use of Nitrogenous Fertilizers:** Microbial action on nitrogenous fertilizers produces nitrous oxide, which escapes into the atmosphere. As this gas is decomposed with difficulty, it tends to accumulate in the atmosphere. The estimated tropospheric load of nitrous oxide is 1.7×10^{18} gms while the stratospheric content is about 8.0×10^{15} gms. Action of high-energy radiations converts it to nitric oxide, which is an important ozone-destroying constituent of the stratosphere.

(3) **Excessive Use of Chlorofluorocarbons:** Chlorofluorocarbons are inert, highly stable, colourless, odourless chemicals, which can be easily liquefied. These are mostly halogenated hydrocarbons, which under the influence of high-energy radiations break up to yield a chlorine atom. Chlorofluorocarbons are considered the most damaging pollutants of the stratosphere, as they persist in the atmosphere for periods as long as 150 years. A single chlorine atom can destroy 1,00,000 ozone molecules.

(4) **Supersonic Transports, Rockets and Space Shuttles:** Supersonic jet-liners, which fly in lower regions of the stratosphere, discharge various oxides of nitrogen, carbon, sulphur, hydrocarbons and particulate material. Space flights, which are powered by powerful rockets, also release large quantities of these pollutants right into the stratosphere. Ammonium perchlorate used in many of these rockets as oxidant releases plenty of chlorides.

(5) **Nuclear Tests:** Surface nuclear explosions produce huge quantities of various gases, dust shoot and debris with enormous force which carries much of material straight into the stratosphere. Much of this material damages the ozone layer.

The three reactants, which play a major role in destruction of ozone, namely nitric oxide, chlorine atoms and hydroxyl ions, may also be derived from natural sources. The quantities of ozone destroying constituents introduced into the stratosphere by natural means are also considerable and some workers (27) believe that natural sources exceed the anthropogenic additions of pollutants in the stratosphere. Major natural sources of pollutants, which are responsible for stratospheric ozone depletion, are solar flairs and volcanic eruptions.

1. Volcanic Eruptions: If we look at the list of volcanic eruptions since 1500 A.D., it becomes apparent that these events are not so infrequent. One of the biggest eruptions in the recent history was that of Krakota in 1882 and a more recent one of almost similar magnitude was that of Mt. Agung in 1963. Though events of such magnitude are rare, smaller ones are more frequent which occur almost every alternate year at one place or another on our globe. Elchichon in Mexico, Mount Pinatubo in Philippines are quite recent examples of such small eruptions. The cumulative effect of these smaller or larger eruptions is, however, responsible for adding enormous quantities of waste gases and other pollutants, which are often thrown out with force to reach high up in the atmosphere. Mt. Erebus on Antarctica alone ejects about 1000 tons of chlorine a day annually venting out 50 times chlorine than an entire year's CFC equivalent (27).

2. Solar Flares: Violent eruption from the sun or the solar flares force out plenty of energetic particles at velocities almost close to that of light which enter earth's atmosphere right down to the stratospheric ozone layer. High energy content of these particles causes stratospheric nitrogen to combine with oxygen or ozone molecules and produce nitric oxide. Solar flares are also not infrequent and some may inject catastrophically large amounts of nitric oxide into the stratosphere. The effect of high-energy particles on other components and reactions occurring normally in the stratosphere should also be significant. The solar flare of August 4, 1972, alone has been estimated to have caused about 13% reduction in ozone content.

(6) Consequences of Stratospheric Ozone Depletion

Ozone layer shields us from biocidal ultraviolet radiations. Depletion of stratospheric ozone concentration shall result in an increased penetration of Ultraviolet-B radiations. A higher loss of stratospheric ozone could cause entry of not only UV-B radiations but also Ultraviolet-C radiations. This could turn the geological clock back to the era when the ozone umbrella was very thin. Terrestrial life shall be drastically affected so also will be aquatic life, which occurs in shallower waters. In brief, consequences of ozone depletion may be summed up as follows:

1. As absorption of ultraviolet radiations in the range of 2800-3150 Å is largely due to the absorption by DNA and RNA molecules, which therefore break down. Increased UV-B penetration shall cause a rapid rise in ailments and damages associated with DNA and RNA disruption. Cancers of the parts exposed to solar radiations shall become more frequent. Crude estimates suggest that there could be 100% rise in incidence of skin cancers for a 25% reduction in stratospheric ozone content.

2. A direct correlation has been observed between cataract formation in eyes and ultraviolet radiations. Our skin and eye-lens contains chromatophores, which absorb ultraviolet radiations and generate highly reactive oxygen, hydroxyl radicals and hydrogen peroxide, which damage cell's structural and functional components.

3. Higher animals possess thick skin, hairs etc. as protective devices. However, plants and lower animals, which are often at a lower level in the trophic structure, shall be the worst sufferers, as they have nothing but a simple cell wall for their protection against high-energy radiations. With

the primary trophic levels drastically impaired, the entire ecosystems could collapse. This is more likely in marine environment where the tiny phytoplanktons are the sole producers.

4. With rise in high-energy ultraviolet radiations a number of complex photochemical reactions are likely to occur which shall produce a variety of toxic highly irritating chemicals – the photochemical smog – from seemingly harmless constituents. These chemicals could be highly injurious to living systems. In plants, adverse effects on the process of photosynthesis could drastically affect primary production, which in turn could disturb the entire ecosystem.

QUESTIONS

1. What are the major pollutants of atmosphere? Enumerate the adverse effects caused by these pollutants on a biological system.
2. What are aerosols? Discuss the problems associated with aerosol accumulation in the atmosphere.
3. What do you understand by smog or reducing type of pollution? Describe the conditions under which it is developed. What can you do to control the problem?
4. What are acid rains? Discuss their causes, consequences and control measures.
5. What do you understand by photochemical smog or oxidizing type of pollution? Under what conditions it is developed. Enumerate the adverse effects caused by photochemical smog.
6. What is green house effect or global warming? Discuss its causes, consequences and control measures.
7. Discuss the roles which Carbon dioxide, Methane, Water vapour, chlorofluorocarbons and oxides of nitrogen play in Global warming.
8. What do you know about stratospheric ozone depletion? Discuss its causes and consequences. What you will do to control the problem?
9. What roles does ozone play in the stratosphere? What will happen if all ozone in the stratosphere is exhausted?
10. Discuss the role played by chloroflourocarbons, oxides of nitrogen, water vapours and methane in depletion of stratospheric ozone layer.
11. What is ozone hole? Where it was first discovered? Discuss the role of meteorological conditions in its formation.

PART – V

Social Issues and the Environment

20
Chapter

Environmental Ethics and Value

The term ethics is derived from Greek word "ethos" which means character. It refers to one's ability to distinguish the right from the wrong, the values, beliefs and actions, which shape the character of a person and the society. The discipline concerned is also referred to as moral philosophy.

The term is also applied to any system or theory of moral values or principles. How should we live? Shall we aim at happiness or at knowledge, virtue, or the creation of beautiful objects? If we choose happiness, will it be our own or the happiness of all? Is it right to be dishonest in a good cause? Can we justify living in opulence while elsewhere in the world people are starving? If conscripted to fight in a war we do not support, should we disobey the orders? What are our obligations to other creatures with whom we share this planet and to the generations of humans who will come after us? Ethics deals with such questions at all levels. Its subject consists of the fundamental issues of practical decision making and its major concerns include the nature of ultimate value and the standards by which human actions can be judged right or wrong.

The notion of "right" and "wrong" has varied from time to time, so have the "believes" and "values" of people. However, there are certain ethical principles, which have been universally accepted and have remained unchanged throughout the entire course of human history. Some of the widely accepted ethical principles are honesty, integrity, righteousness, being caring and compassionate, having respect for human dignity and a fair and open mind, which is willing to admit mistakes etc. It is these universally accepted ethical believes which help us to formulate a relatively new discipline of philosophy, which has been referred to ***Environmental Ethics***.

Environmental ethic, therefore, can be defined as a system of ethical values, human reasoning and knowledge of nature which endeavours to forge a pattern of right conduct towards environment so that needs of living beings of the present generation are fulfilled without compromising the ability of the future generations to meet their own needs. Environmental ethics concerns itself with the moral relationship between humans and the world around us in contrast to the traditional ethics which deals primarily with relationship among peoples only (1).

(I) THE NEED OF ENVIRONMENTAL ETHICS:

As humanity enters the third millennium, news reports are peppered with reference to our "fragile and endangered planet." The use of such an adjective for our planet is certainly an exaggeration. In its five billion years of existence, our planet has endured bombardments by meteors, abrupt changes in its magnetic fields, dramatic realignment of its land mass and the advance and retreat of massive glaciers, the blocks of ice that reshaped earth's surface. It is in this turmoil that life appearing nearly three and a half billion years ago has proved resilient. Species have come and gone but life has persisted without interruption. In fact, no matter what we do it is unlikely that we could disrupt the powerful physical and chemical forces, which drive earth systems (2).

Although we cannot completely suppress earth systems, we do affect it significantly, as we use energy and emit pollution in our quest to provide food, shelter and other products for world's ever-growing population. Scientific evidence indicates that because of our activities, the global environment is undergoing profound changes. Man has since long affected his local environment but it is only in twentieth century, specifically in the last fifty years, that consequences of his actions have expanded to a global scale. Now it is apparent that with regular repetition, even seemingly innocuous actions such as driving a car or cutting down a tree can influence the physical and chemical systems, which govern our planet. Over the geologic past, conditions in the atmosphere, ocean and biosphere have largely followed natural cycles. Today, cumulative effect of human activities is a significant force driving the changes in the global environment. The changes brought by human activity tend to make life more and more difficult every day. To some species, it is a total catastrophe (3).

We have to stem the tide. We have to stop this damaging trend. We have to device the right code of conduct to regulate our behaviour towards the environment and natural resources so that the quality of life we live is not compromised and the ability of our future generations to meet their own ends is not affected. As an appeal to human conscience, his ethical convictions about the good and bad, the right or wrong is often an effective way to control unmanageable problems, Environmental Ethics assumes a pivotal role in our efforts to slacken or halt the damages we are causing to the environment, wildlife and natural resources (4;5;6).

(II) GUIDING PRINCIPLES OF ENVIRONMENTAL ETHICS:

Although there could be disagreements, many of the writers such as Milbreth (1), Delors (9), Schrader-Frechte (7) on the subject agree that the guiding principles of environmental ethics should be:

1. The distribution of the resources of the world should be egalitarian as far as possible. All men are equal. For all there should be equal opportunities to compete for the comforts and riches of the world.
2. The 'rights' of the environment and natural resources should take precedence over the right of individuals as they are linked to the welfare of the entire biosphere.

(III) BASIC DIRECTIVES OF ENVIRONMENTAL ETHICS

Environmental issues raise many difficult ethical questions. We are living in a transient world. Every living being is born to die some day. No one has come here to live forever. In this cycle of life, which has been operating since the first representatives of the biosphere appeared on our planet, species and the individuals of the species have been changing. Older generations are replaced by newer ones. We have inherited the riches of the world and the environs, in which we live from our predecessors. Today they belong to us. Tomorrow they shall belong to our future generations. Man is simply a temporary resident custodian of the living and non-living resources of this planet. His unique mental capabilities, erect posture and hands with thumb opposed to fingers, which are free to handle the spear, spade, or tools have placed him in a commanding position. He shall be failing in his duties if he does not hand over the riches of the world in a usable state to his successors or spoils the environment with the wastes he discharges. In doing so his 'rights' may be compromised but that is a small price to pay for carrying out the obligations he has to the society, the nature and the residents which shall occupy this planet in future (1;7;8;9).

From a study of the current perspectives of Environmental ethics certain directives crystallize out to regulate our conduct towards the human society, the nature and our future generations which are yet to come. These are:

1. Individual's obligations to the community and the society: As a sentient, knowledgeable creature, everyone has certain obligations to the society. If an individual has rights, he has duties also.

He should refrain from such actions, which create difficulties or damage life and property of fellow citizen or the community or the country. If his seemingly innocuous actions, repeated several times and by so many people add up to sizeable problems, environmental ethics directs him to place a check on such practices so that a little restrain on the part of every one may reduce the overall impact. Many global environmental problems are there simply because so many people at so many places contribute small bits and pieces to it. The affect of Individual life style decisions such as using a bike instead of a car, taken by a large number of people may finally accumulate to cause significant reduction in the amount of green house gases we discharge in the atmosphere. Avoiding the use of polythene and plastic bags, or using wastes from kitchen to manure kitchen garden may reduce the burden of solid garbage which municipalities have to dispose of. Using domestic wastewaters to irrigate our kitchen gardens and saving a bucket of water or two every day by everyone could save enough fresh water to irrigate many additional hectares of crops. Enough electricity can be saved if everyone uses a fan instead of air-conditioner or puts on woolen pullovers instead of heating the entire house, to light hundreds of additional homes. Population growth rate can be significantly reduced if most of the couples decide to have only two children (8).

2. The rights of nature: Does nature has rights? Do other species have right also? The view that the environment and its resources are exclusively created for man to use or abuse has been replaced by a more realistic concept that man is just a single species among millions of species, which live on our plant. He cannot live alone. The realization that the biosphere on earth forms a life support system changes in which shall endanger his life also compels man to expand his moral concern and ethical values to cover other life forms as well. If the pleasure and pains of humans have some intrinsic significance, why not those of non-humans? What if a species becomes extinct? Or a rain forest is cut down? Are these things to be regretted only because of the loss to humans or other sentient creatures? Human conscience provokes man to respect the pleasure and pains of non-humans also, simply not because they are useful to us but because they are also a part of biosphere as man is (10;11;12).

3. Our obligations to the future generations: Our concern for the environment also raises the question of our obligations to future generations. How much do we owe to the future? From a 'social contract' view of ethics, the answer would seem to be nothing. Because we can benefit them, but they are unable to reciprocate. Most other ethical theories, however, do give weight to the interests of coming generations. The fact that members of future generations do not yet exist is no reason for giving less consideration to their interests than we give to our own. That it is certain that they will come and shall have interests that will be affected by what we do today, should be a reason enough for us to be more considerate. For example, in the case of the storage of radioactive wastes, it seems clear that what we do will indeed affect the interests of generations to come. The question becomes much more complex, however, when we consider that we can affect the size of future generations by the population policies we adopt. Most environmentalists believe that the world is already dangerously overcrowded. We are responsible for the existence of the future generations. We can manage their size so that the resources of the world are more than enough for their needs and they live a life of affluence and plenty. Or else we can allow the maximum numbers to come into existence so that per capita availability of resources or rough and harsh environment compels them to live a life of want, misery and starvation. The choice is difficult to make (7).

Man is a temporary resident custodian of this planet, simply a link in the cycle of events, which has been operating since times immemorial. We do not have the right to spoil the resources, pollute the environment and disrupt the very cycle of life of which we form a simple insignificant link. It will affect the future generations whether small or large. Just as we have inherited the resources of the world from our predecessors, should not we pass it on to our future generations in a clean and usable state? In 1949, Aldo Leopold put forward the concept of "Land ethics" which affirms the right of all resources, including microbes, plants, animals and earth materials as well to continued existence in a natural state. Humans have the ethical responsibility for the preservation and maintenance of the biosphere, the atmosphere, the hydrosphere and the lithosphere (13).

(IV) ETHICS AND ENVIRONMENTAL ISSUES:

Environmental issues are the most pressing national and international issues of the present century. We have already committed our planet to a major environmental change in the years ahead. Elevated concentrations of greenhouse gases already emitted will persist for many centuries to come no matter what we do. The chlorofluorocarbons in the atmosphere today will continue to deplete the ozone shield for centuries. Complex tropical forest ecosystems once cleared shall take many decades to regenerate. We cannot halt these changes altogether but we can reduce the pace of these changes. Environmental problems like fog smog and acid rains, thermal and noise pollution have paled out of existence before the global issues, which now threaten man with drastic consequences (14;15). Major environmental issues which humanity faces today are:

1. Global climatic change: global warming and stratospheric ozone depletion.
2. Diminishing biological capital: disappearing wild life and forests.
3. Diminishing fertile land and freshwater resources: the threat to future food security.

Today, the changes in global environment are driven by ever-greater number of people, increasing economic development and its attendant increase in industrial activity and consumption of energy by humans. Our numbers are increasing today at the rate of about 90 million per year and by the end of the current century, it is projected to reach nearly 10 billion. About 90% of these people shall be borne in poor countries of today's developing world. While population has shot up, so have the standards of living for many of world's people, the consumption of fossil fuels and expansion of world economy. Yes, these changes have caused astonishing improvements in human welfare but unfortunately, many of the processes that produce gains degrade the environment and deplete the ecological capital – the soils, forests, biodiversity and freshwater resources – on which humanity subsists. People are richer in developed countries – with nearly one quarter of global population consuming most of the world's energy. They have at their command about 80% of world riches, use most of world's natural resources and generate most of the total wastes produced on global scale. People in developing countries, with three quarters of world's population have less than one quarter of world's riches. They are also contributing to resource depletion and environmental degradation. They are often compelled to destroy their environment simply to survive. Forests are cut down, biodiversity, which is a very important biological capital in these countries, is degenerating at a fast rate and their agricultural practices deplete more and more fertile soils every year.

The three major global environmental issues of today arise from our daily activities, the cumulative effects of which have reached a scale and pace sufficient to disrupt the natural systems that took millions of years to evolve (3;16). Each of us contributes a little to these global problems!

1. When we drive a car, when we cut down a tree, put an over-dose of chemical fertilizer to raise our crops or use a hand-full of insecticide to kill insects and pests, we are adding our share, howsoever small, to the global problems.
2. The thriving 'markets' and the 'black-markets' of wild life products exist because we decorate our houses with ivory, fur, animal trophies and use other animal products. Our effort to conserve wildlife and biodiversity is often thwarted by numerous people who are engaged in collection of firewood and other plant products. Poachers ply on hapless creatures with their traps and arms. In absence of the demand, trade in wildlife products shall come to a grinding halt and a number species shall be saved from wasteful elimination.
3. The smoke, fumes and acidic fallouts from our domestic and industrial establishments adversely affect plant and animal life.

4. When we thoughtlessly cut down a tree it adds to the global problem of deforestation. Fertile land covered with vegetation, crops, grasses, bushes or trees acts as sponge to soak up enough water deposited by rains to serve the man as well as plants and animals during drier seasons. The absence of plant-cover causes rainwater to flow down to the sea quickly without having much time to recharge the ground water table. Withdrawal from an already reduced ground water stock causes the ground water table to go deeper down the soil strata. When it recedes beyond the reach of plant roots, greenery disappears and desertification follows.
5. We use the land bared of plant cover over large areas to grow crops. To raise productivity we provide chemical fertilizers and river or ground water for irrigation. Salts accumulate and problems of salinity and alkalinity appear in poorly drained soils. If we grow our crops amidst land covered with trees and other vegetation more water will be retained in the soil while environment around shall remain more humid and congenial. Soil drainage shall improve. We shall have to use less water for irrigation. This shall improve appreciably the problems of salinity and alkalinity.

Much can be achieved simply by an appeal to individual's conscience. If most of the people realize and believe that by their actions they are adding something to worldwide problems, many shall try to limit such activities. A little restrain on the part of most of the individuals shall help the cause of environment tremendously. Global environmental policies, if they are to be successful require long-term changes in believes and perception of the people at large. To save our physical environment from degradation, we shall have to alter our conceptual environment and in altering our conceptual environment, ethics or moral philosophy plays a major role. This is because ethics helps people to determine the decisions we make. The way people answer ethical questions affects the way they live their lives. To a great degree, ethics determines, for example, whether to have an abortion, invest in weapon producing company, or become a vegetarian. It is for this reason that the worst form of environmental pollution is the 'pollution of mind' and rarest global resources are well-thought-out ethical principles (17).

It is not force or authority but gentle persuasion and an appeal to one's conscience, which can make people, realize that just as development is detrimental, lack of development also contributes much to the degradation of natural resources and the environment. A little restrain and a shift to more environment friendly technologies is the need of the time. The pattern of development so far followed by the developed countries is no longer acceptable as it has never been environment friendly. We may have to make some painful choices. We shall have to move beyond compartmentalization and out-moded patterns to draw the very best of our intellectual and moral resources from every field of endeavour. If all men are equal, they deserve at least a reasonable share of the resources of the world, opportunities and a legitimate quality of life. A resort to ones moral conscience and ethics only shall alter the public opinion at large. In due course of time, the altered public opinion shall translate in to policies, which shall be followed by action in technologically advanced democracies of the world. The rapid rate at which we are degrading the environment, wild life and natural resources forces a huge re-learning task on the humankind. It has to make a choice thoughtfully and foresightedly (18).

(V) VALUE EDUCATION

Education is process of learning which continues throughout one's life. It begins right in the cradle. It enables us to walk, know about things around us, communicate with others and finally acquire means of living and lead a life. It is education only, which forges the pattern of right conduct and behaviour, like truthfulness, honesty, sincerity, integrity, righteousness, etc. towards the family, the society and the country at large. No education can be considered adequate if it fails to inculcate these values among its subjects.

Though there have been many sporadic instances of benevolent rule during the two millennium which have preceded our times, the humanity at large has reeled under the whims of autocratic rule in which "values" which shape the character of a person and the society have been changing. It has largely been during the last two hundred years, which have witness three great revolutions – the American Revolution, the French Revolution and the Russian Revolution that the concepts of liberty, equality and fraternity have emerged. Today, the Constitutions of most of the countries ensure justices liberty, equality and fraternity to all of their citizens. In India we have struggled for freedom, liberty, equality and social justice for a long period and have finally accepted and enshrined these values in our Constitution.

However, as we look around us, the national scenario is both disturbing and distressing. The tragedy, today, seems to be that we are not committed to any value system at all and have little regard to any 'values' beyond our crude self-interests. Money and power have become the highest values regardless of the well-being of fellow citizen and the interests of the nation as a whole. In such a situation, the only hope of building up a unified and integrated country in which the rights of individuals are respected lies in the growth of civil institutions, the people of India. If the people rise above petty self interests, recognizing the benevolent spirit of our Constitution and make a determined effort to protect and promote the inherent 'values', nothing can prevent us from making a just society and a country of our dreams.

1. What is Value education?

Value Education refers to wide range of learning processes, which builds the character of its subject by inculcating values like truthfulness, honesty, sincerity, integrity, righteousness, civic sense, a respect for the principles of equality, liberty, fraternity and compassion for all living creatures. Values are intrinsic principles, which direct our actions. They are virtues of an individual.

2. Erosion of values:

There have been great advancements in the field of science and technology. However, wisdom languishes and there has been a gradual erosion of values, which is, reflected in day-today life of a large section of our society. Erosion of values causes rapid breakdown of families, the society and the nation as a whole. Today the integrative moral imperatives of social and cultural values are being replaced by disintegrative forces in almost all parts of our country. The present political and developmental processes are themselves acting as a catalyst to the collective degeneration of our society. Social stresses tend to pitch individuals against individuals, groups against groups, communities against communities. Such horizontal and vertical cleavages in our society are further enhanced by the so-called 'elites' who show a tendency to amass wealth and power at the cost of ethics and morality (19).

Since independence, a number of high-powered commissions and committees have been set up to advice the Government on matters connected with Education. Right from the beginning of post-independent era, almost all commissions and committees, namely, Radhakrishnan Commission (1948-49), Kothari Commission (1964-66), Ramamurti Committee (1990), Central Advisory Board of Education committee on policy (1992), National Policy on Education (1986), have stressed the necessity of making our education system 'value oriented'. It is disappointing that such well-concerted efforts of nearly five decades have failed to achieve the desired results. The reports of these bodies have been gathering dust in government almirahs. The proposed plans and strategies remain largely on paper.

It was due to the extensive erosion of values in the nineteen eighties that Value Education received a shot in the arms. Earlier the National policy on Education had already highlighted the need while Planning Commission Core Group on Value Orientation of Education (1992) emphasized the urgent need of making our education system value oriented. A number of public bodies were advocating the cause of value education by nineteen nineties (20).

3. The strategy of value education

Value education touches every aspect of life, personality and learning. It includes cultivation of appropriate sensibilities – physical, moral, cultural – in the personality of its subjects and develops the capability to take just and right decisions in times of need. Value education has to be so developed as to achieve the objectives of a just society. The National Policy on Education (1986) emphasizes equity and social justice in education, which shall endorse our unique cultural identity, contribute to national integration and promote the concerns enshrined in the Indian Constitution. It was therefore considered appropriate to introduce value education on the lines as embodied in the Constitution of India. Justice J. S. Verma in 1999 identified the following basic values from the text of Preamble of the Constitution of India.

The right to equality

- Sovereignty
- Socialism
- Secularism
- Democracy
- Republican character
- Justice
- Liberty
- Equality
- Fraternity
- Dignity of individual
- Unity & integrity of the Nation

The Constitution of India guarantees six fundamental rights to all Indian citizens, which are:

- The right to equality as presented in Articles 14,15,16,17.
- The right to freedom as presented in Articles 19(1-6), 20, 21, 22,
- The right against exploitation as presented in Articles 23, 24.
- The right to freedom of religion as presented in Article 25, 26, 27, 28.
- The Cultural and Educational rights as stated in Articles 29, 30.
- The right to constitutional remedies as stated in Articles 32, 33, 34, 35.

In addition to fundamental right, the Constitution also prescribes a set of fundamental duties to every citizen of India. Article 51A, Part IV A states that it shall be the duty of every citizen of India. To abide by the constitution and respect its ideals and institution, the National Flag and the National anthem.

- To cherish and follow the noble ideals which inspired our national struggle for freedom.
- To uphold and protect the sovereignty, unity and integrity of India.
- To defend the country and render national service when called to do so.
- To promote harmony and spirit of common brotherhood amongst all the people of India transcending religious, linguistic and regional or sectional diversities and to renounce practices derogatory to the dignity of women.
- To value and preserve the rich heritage of our composite culture.
- To protect and improve the natural environment including forests, lakes, rivers and wildlife and to have compassion for creatures.
- To develop scientific temper, humanism and the spirit of inquiry and reform.
- To safeguard public property and abjure violence.
- To strive towards excellence in all spheres of individual and collective activity so that nation constantly rises to higher levels of endeavours and achievement.

Justic Ranganath Mishra, former Chief Justice of India, wrote a letter to then Chief Justice of India pointing out that acquainting the students about fundamental duties in every educational institution was necessary as a nation-building mission. It was treated as Writ petition (Civil) No. 239/1998 by the Honourable Court and was supported by the Government of India as well. A Committee was set up under Sri S.B. Chavan to Report on Value Based education, which tabled its report in Rajya Sabha on February 26, 1999. The report emphasized:

- That Truth, Righteous conduct, Peace, Love and Nonviolence are the core universal values which can be co-related with five major objectives of education, namely knowledge, skill, balance, vision and identity. Primary school stage is the period in a child's life when seeds of value education can be implanted in their impressionable minds in a very subtle way. Commencing from primary school level value-based education should be extended to the school, college and university levels while its purview should be enlarged to provide an understanding of the constitutional, fundamental rights and duties as enshrined in the Constitution of IndiaThat in view of the diverse character of our country, it is necessary that students should be acquainted with the history of India's freedom struggle, cultural heritages and features comprising our national identity.
- That due thought should be given to the Religion which is the most misused and misunderstood concept. Students should be acquainted with the basics of all religions, the values inherent therein and a comparative study of philosophy of all religions should begin at the middle stage in schools and continue up to university level. Students have to be made aware of that the basic concept behind every religion is common only practices differ. Even if there are differences of opinions in certain areas, people have to learn to co-exist and carry no hatred or feeling of ill will against any religion.

The National Council for Education Research and Training (NCERT) was asked to implement the suggestions made in the report. NCERT published a book "Education in Values - A source book in 1992. The council also publishes a periodical, "The Journal of Value Education'. The third volume, 3(2), of the journal, published in 2003 was exclusively devoted to environmental values. The council has been issuing directives and taking steps to introduce Value Education in Schools and College at various stages in accordance to the policy introduced by the Sri S. B. Chavan Committee Report.

QUESTIONS

1. What do you understand by "Environmental ethics"? Why a 'pattern of right conduct towards environment is so much important today? Explain.
2. What are the basic directives of environmental ethics? Why the needs of environment should take precedence over the rights of man?
3. What are the basic principles of environmental ethics which help us to forge a pattern of right conduct towards the environment? Explain.
4. What is value education? How the current erosion of values can be stopped by promoting value education at different levels in the education system?
5. What is the strategy of value education? Give salient features of the Report tabled in Rajyasabha on Value education.

21 Chapter

The Population Explosion

Historians of the future, if they are able to look back at the events of the preceding century shall certainly comment upon the population factor as one of most remarkable features of twentieth century. Humanity took several thousand years to attain the one billion mark in the year 1800 AD, but another billion was added in just 130 years and this population was tripled in just 70 years. (Please see Fig 3.1). The dramatic rise in human population has caused much concern about the strains it places on the world resources and the quality of our lives. If the world population continues to grow at the current rate for a few more generations, it is likely that the carrying capacity of our planet shall be exceeded.

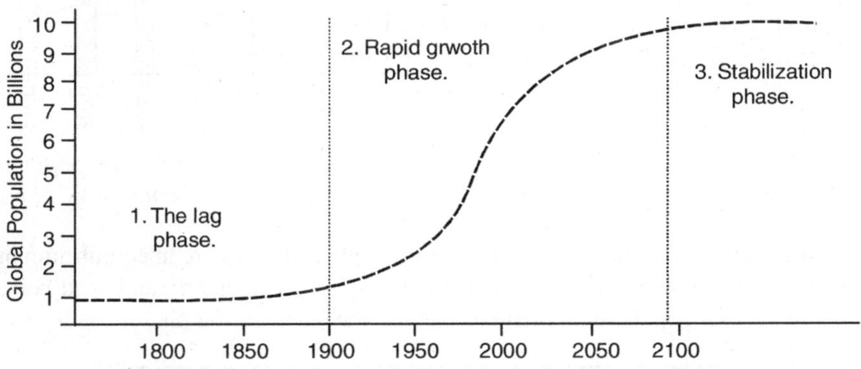

Fig. 21.1 *The three phases of growth of human population.*

(I) THE THREE PHASES OF POPULATION GROWTH AND THE DEMO GRAPHIC TRANSITION

The term 'demos' means people and a study of human populations, birth rates, death rates and human numbers is referred to as 'demography'. As early as 1945, the eminent demographer Frank Notestein outlined the theory of demographic changes, which stems from the effect of economic and social development on the growth of population. His theory, which drew heavily from the European experience, has provided a conceptual basis for demographic research since then (1). He classifies all human societies as being in any of the three phases of population growth, which are:

1. The lag phase or the phase of slow growth:This phase is characteristic of rather primitive societies in which population grows slowly. As the society tries to establish itself in its surroundings living conditions are difficult and life expectancy is low. This phase is usually a long phase in which growth and development is very slow. Birth rates are high and so are death rates. They balance each other, *i.e.*, they are in equilibrium and the net result is a very slow or nominal growth. The relative percentage of different age groups, the age-structure is pyramidal with little height as life expectancy is low. The successive age groups are smaller and smaller with the oldest being at the top.

2. Rapid growth phase: With improvement in living conditions, food production expands. Growing health awareness, education, better medical facilities like mass immunization and use of antibiotics etc reduce the burden of heavy mortality, which had depressed life expectancy earlier. Great strides in mass communication and transportation enables man to carry food supplies and medicines where required in case of drought, floods and other natural calamities. This results in a highly reduced death rate and raised life expectancy while birth rates remain high. The population, therefore, enters a phase of rapid growth. The age structure assumes a shape like a bell with successive tiers from base upward nearly similar in size. One or two uppermost tiers are smaller as they represent the oldest age groups.

3. Stabilization phase: The population enters its third phase of growth with rising per capita income, higher economic and social gains along with a highly reduced infant mortality rate which reduce the desire for larger families. Late marriages and a reduced desire to have a larger family is generally an outcome of the desire to live a life of comfort, convenience and plenty. In smaller families, incomes are shared by few, which results in a higher per capita availability of resources. With better medical facilities and a reduced mortality at all stages of life, the uncertainties that surround small families, is gone. Birth rates come down and populations move to the stabilization phase with little or no growth at all. Like the first stage, birth and death rates are in equilibrium but are at a much lower level whereas life expectancy is high. There are more elderly and aged persons in the society than youngs in a population which has entered the stabilization phase.

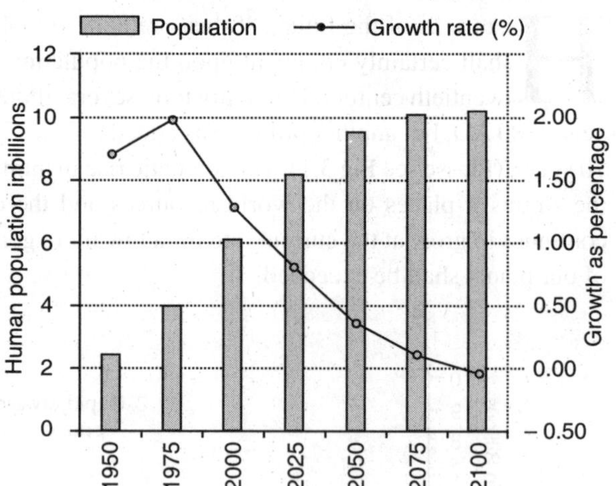

Fig. 21.2 *Growth of human population since 1950 and future projections*

(II) ANALYZING THE PATTERN OF POPULATION GROWTH

During the two preceding millennia, the fortunes of humanity have largely been subject to the physical and chemical forces, which drive earth systems. It has largely followed natural cycles and has ebbed and flowed at the mercy of the natural forces, which govern our planet. Today, scientific and technological advancements have enabled man to govern his own numbers, destiny and the well-being of all-life forms, which inhabit this planet. Population Growth rate nearly touched 2% following the baby boom which ensued after the Second World War. Since 1975 growth rates are steadily declining and are expected touch the zero mark by the year 2100 AD when the world population is expected to stabilize at about 10 billion people (Fig 21.2). As the third millennium unfolds an analysis of the pattern of population growth and its distribution in different regions of the world reveals many welcome signs as well as disturbing trends which can be summed up as follows:

(1) Uneven growth and distribution of global population in different regions of the world:

Today nearly 6.12 billion people live on our planet, which are distributed very unevenly in different regions of the world. It has been only after the Second World War that the population growth rates in poor countries began to outpace that of rich countries. About 3.97 billion or nearly 66.2% of the total world population now lives in Central America, Central Africa, the Far East and South and South East Asia, which are referred to as poor countries or developing countries (Fig 21.3).

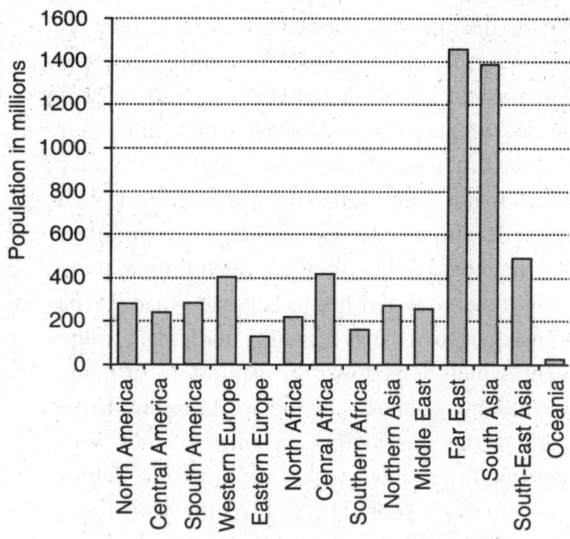

(For conutries included in each region see the table at the end of this chapter)

Fig. 21.3 *Uneven distribution of human population*

Demographically the world was more uniform earlier than it is today. There are countries in which the annual population growth rate is as high as 3.93%. In some countries the population is not growing at all, it is rather declining. If we observe regional average, we shall find that Central Africa is the region of the world, which has the highest population growth rate followed by North Africa and the Middle East, all of which have a growth rates above 2% per year. Differences in the growth rates among different sections of our society tend to change in the long-run ratios between the populations of different cultural, ethnic or religious groups, which in turn may have serious socio-economic and political implications (2)

(2) The number of poors in the world is rising:

The third quarter of the twentieth was characterized by a baby boom all the world over. Global population grew to four billion people by 1975 AD from about 2.5 billion in the year 1950 AD. In 1950, nearly 1.00 billion lived in the countries, which we refer to as developed (North America, Europe, Japan, Australia and New Zee Land) and about 1.5 billion in developing countries – a ratio of 1: 1.5. This ratio changed to 1: 2.57 by 1975 AD and was about 1: 3.81 by the close of the millennium.

The population ratio between the developed and the developing countries is expected to rise more and more each year. The population in developed countries is now growing at an average rate of about 0.40-0.65% per year while those in developing countries is growing at a rate of 1.5% per year or more. Global population is expected to cross ten billions by the year 2100 AD and the most worrying feature of the population growth is that more than 90% of this growth shall occur in developing countries. Each year nearly 70-80 million people are added to the overall population of developing countries. The ratio between the populations of developed and the developing countries is expected to rise to 1 : 6.17 by 2100 AD (Figs 21.4). To some these projections appear to be unrealistic as the life support system in developing countries shall be unable to support the enormous number and something drastic may happen (3;4).

(3) The rich are getting richer and the poors poorer:

With the rising demands of a rapidly growing human number, the life support system is strained to the limits, which it cannot sustain. This is happening in the developing countries of the world or is likely to happen in near future. As the demands persist the very resource base, on which the life support system depends to provide us the food we eat, the water we drink or use to irrigate our crops, the biotic wealth we use to produce food etc slowly begins to degenerate. The society starts consuming the very resource base on which all life depends. And with degenerating resource base agricultural yields decline, sub-surface water table recedes and desertification follows. More and more people are pushed below the poverty line as market prices of grains, vegetables, fruits and other commodities rise (5;6). For a developing country there is no option left but to import food from other developed countries, which have surpluses. As the means to buy food are often limited there is no other way left but to beg and borrow. The consequence of this is the rising burden of foreign debt, which bogs down the aspirations of many developing countries of the world today. Huge sums of national output, in terms of money or commodities, have to be given to the debtor country each year. The rich of the world are, therefore, getting richer and the poors poorer (7;8).

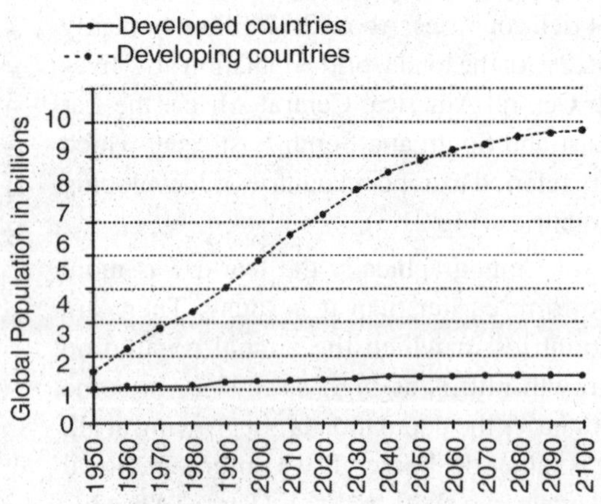

Fig. 21.4 *Population in developing and developed countries from 1950 to 2000 and futrue estimates*

(4) The developed world has become more uniform while the developing world has become more diverse:

Both birth rates and death rates are low and somewhat similar throughout the developed world. A high per capita income, the desire to live a life of comfort, convenience and plenty, small households and lowering fertility has reduced population growth to the point where populations are declining in some countries of the world. In many countries the current generation of parents shall not have enough children even to replace themselves. Differences among nations, regions, different sections of the society or the casts and creeds are generally small or almost non-existent (9).

In developing countries, however, enormous differences in population growth rates has come into being – Africa, the Middle East, much of Central America continue to have a high birth rates while death rates are declining and life expectancy rises. In Somalia population growth rate is as high as 3.93 % per year today, in Eritrea it is 3.89 % per year, in Mali it is 3.1% per year. Social and economic conditions in these high growth regions vary considerably. On one hand are the wealthy oil producing countries whose money can buy anything whereas on the other hand is the hunger, misery and poverty of Sub-Saharan Africa. Age-old traditions, socio-economic and cultural factors highlight the differences between different sections, castes and religious groups. It is the horizontal and vertical segmentation of the society which sustains a high population growth rate. The belt of high fertility, low death rates and rising life expectancy sweeps eastward through areas of Muslim culture and other Christian countries to the edges of China where nearly one quarter of the world's people live.

In china, which is a low-income country population-growth-rates have declined miraculously to rates comparable to the developed countries. In Central and South America, South and South East Asia many countries have entered the third phase of demographic transition but at speeds which vary considerably from region to region. To a certain extent, cultural differences which have emerged among different countries explain the variations, but economic, socio-political differences and policy matters also play a major role. Most of the countries in which population growth rates have slowed down appreciably, have made sincere efforts to bring modern means of birth control within the reach of the entire population and have pursued policies which benefit the majority of people at large (10;11;12).

(5) Growing rural landlessness and expanding cities:

Throughout the developing world, the cities are growing rapidly while in those of developed world growth of cities has been rather slow or nominal. The population in third world cities is growing twice as fast as the total population of the country. Most of these increases are due to migration from rural areas. High birth rates have crowded many of the domestic household to the point where per capita availability of land is no longer sufficient to cater to the demands of all. Many, therefore, are forced to migrate to the cities. Higher incomes, industrial development that generates job opportunities and glitter of the city life absorbs them (13;14). In 1950, only 18% of the total population of third world countries lived in cities. By 1980, this number grew to 35%. By 2000 AD it is expected to rise to about 55%. Today, London, world's second largest urban area in 1975 is now smaller than Jakarta or Cairo (15).

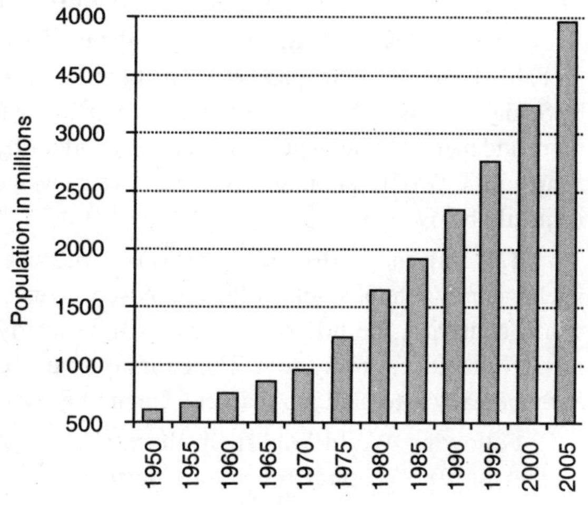

Fig. 21.5 *Growth of world's Urban population*

Large cities present a multitude of problems. The huge numbers concentrated in a small area have to be fed, provided freshwater, houses to live, medical facilities, means of transport and communication on one hand. On the other hand, the filth and waste they create has to be disposed off. As the cities of the developing world attain enormous proportions, they tend to grow beyond the administrative capacity of local governments. Many of these are struggling to provide the basic civic services. The sewage system built in Alexandria, Egypt, designed to cater to the needs of one million people now serves more than five million people. Everywhere in third world cities basic civic amenities, such as housing, roads, water supply systems, sanitation etc are overwhelmed by the sheer number of people using them (13;14;16).

The growth of cities in industrial nations has been an integral part of national economic strategy and a locomotive for country's development. Today they have turned in to a financial drain and liability. In most of the developing world, cities account for 40-60% of gross domestic production. For example, nearly 50% of Mexico's gross domestic production in 1990 came from a single metropolis, the Mexico City. Nearly 60% of Philippine's manufacturing establishments in 1980 were located in or near Manila. Similar figures can be quoted from Jakarta, Cairo, Calcutta, Bombay, Khartoum, Lagos and others (8). The preferential treatment to the cities has caused the rural area to be a victim of benign neglect. The growing landlessness and poverty of rural area has been forcing people to migrate to the cities. Often they are too many to manage. However, the very working force which fuels to the engines of development in cities and thereby that of the country as whole has also been a victim of neglect and apathy. Many third world cities have gained notoriety for large number of people living in makeshift burrow, tin and tarpaper shacks, hillside garbage dumps (13;14;16).

In the past urban development has been an outcome of agricultural success. However, today, expanding cities reflect agricultural bankruptcy. Large-scale migration from rural areas is symbolic of the severe imbalance, which marks national economic strategies in most of the developing countries of the world. High rural population growth rate, skewed land distribution, poverty and nominal or non-existent government investment in agricultural sector have made even the urban slums to appear more appealing than the life in rural households

(6) Population growth, crimes, conflicts and controversies:

The relationship between the population growth and crimes, conflicts and controversies has probably been lost in the gap between demography and political science. It has largely been ignored by Social scientists. When the growth of a population accelerates sharply, the age structure is dominated more and more by young people. As the second millennium closes, much of the developing world has nearly 35-45% of its population below 15 years in age whereas in developed countries this percentage is mostly below 20%.

This shift in age structure is the source of trouble. Educational institutions are flooded, initially at elementary school levels, followed by secondary and finally at college levels. In a number of third world countries, the tide of youngsters has already overwhelmed schools and colleges making a mockery of education system. The enormous number has to be provided jobs, which adds pressures to the already saturated job markets. Unemployment or under-employment grows (17;18).

Following World War II, in the third quarter of 20th century, the economic tide which swept the world, raised incomes everywhere, in every country regardless of its economic system or the stage of development. Between 1950 AD and 1975 AD, the world economy expanded at a healthy rate of 5% per year following which a decline has set in. Now these figures range between 2-3% per year for many countries of the world and a little or substantially lower for many developing countries. In order to maintain steady growth rates many third world countries had to resort to foreign aid. Heavy borrowing to sustain the per capita gains eased the pressure of local demands for a while but soaring interest rates and sagging global economy left a number of third world countries heavily indebted and unable even to make interest payment. The third world countries now spend a lion's share of export earnings for the payment of interests on foreign debts. With a sagging economy, it is difficult to create jobs for so many people. This show down has come just at the time when huge number of restless youths are entering the job markets in third world countries adding enormous pressure on the concerned government's ability to manage the situation and provide the needed civic services (19;20;21).

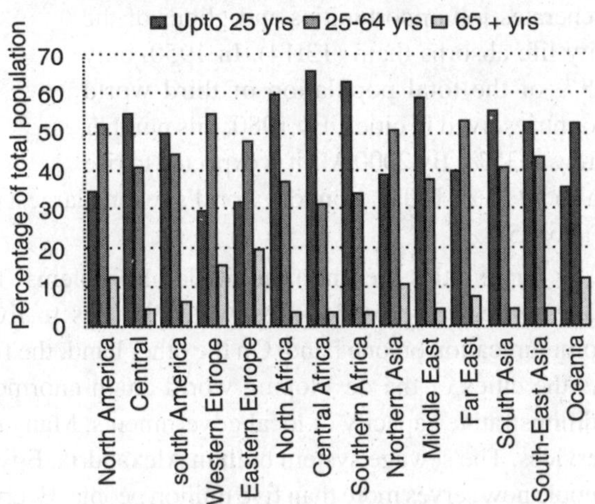

Fig. 21.6 *Age structure in diffeerent regions of the worlds*

The sight of growing numbers of restless young people, full of energy and seeking employment in an already saturated job market does not convey an image of social tranquility. Restless hungry people are forced to beg, borrow or steel. The crime graph rises. It has been observed that a crowding does provide a set of conditions, a context in which tension, violence and various forms of aberrant behaviour are more likely to occur. The effect of crowding cannot be separated from those of poverty with which it is usually closely associated. Within societies, crowding may exacerbate long-standing religious, tribal, regional or ethnic differences. People are more likely to be swayed by promises of unscrupulous leaders who may lead them to new yet unpredictable movements and bloodshed. No doubt, many mass uprisings and political movements are probably in their budding stages in poverty-stricken developing countries (22).

When deteriorating environmental support system can no longer support the growing population, conflicts arise as people are forced to migrate in search of livelihood and encroach upon the jurisdiction of other countries. Such conflicts are common in Africa where these ecological refugees with their herds and flocks come into conflict with farmers in the region they are trying to enter. Intensification of competition for fresh water resources can be observed along the Nile River. The countries, which depend on its flow, Egypt, Sudan and Ethiopia, have rapidly growing populations. The competing claims of these countries, where the industrial and agricultural growth is keyed to the waters of river Nile could become a matter of life and death and contentious political issue (23).

(III) THE BEGINNING OF DECLINING TREND IN POPULATION GROWTH

A look at the statistics of population growth trends reveals that we are now witnessing the receding floods. Global rate of population growth touched 2 percent per year sometimes in the years 1970-71 but has subsequently declined (Fig 21.2). This is a welcome sign and has been considered as a significant change by demographers. However, this does not mean that the pressure of growing population on our resources and environment shall be relieved in times to come, as nearly 80 to 90 million peoples will still be added to the world population every year for two or three decades. It is interesting to note here that the flood of population growth has not been checked by resource limitation – poverty, famine, war or pestilence. It is rather late marriages, the desire to have smaller families and falling fertility in marriages, which have reduced the growth rate even when death rates are declining and life expectancy is increasing (12). Repeatedly it has been demonstrated that with affluence and prosperity, birthrate declines, life expectancy rises and population growth slows down. This is apparent from statistical co-relation between affluence, prosperity and major parameters of population growth in nearly two hundred diverse countries of the world. Gross domestic production (GDP) in terms of US Dollars has been found to be co-related at highly significant levels with declining population growth rates, birth rates and rising life expectancy as shown in Fig 21.7-9. Since long control over fertility has been exercised in traditional communities through restrictions on marriages and sexual activity, frequent abstinence, prolonged breastfeeding, contraception, abortions etc. As the societies modernize, these controls over fertility are weakened. They are replaced by individual decisions in which both men and women take an equal part. With better living standards higher literacy and better status and education for women, which goes with affluence and prosperity, the desire to live a life of convenience comfort and plenty develops. Couples tend to avoid prolonged childbearing and produce many children. They prefer to have one or two children and a small household the needs of which can

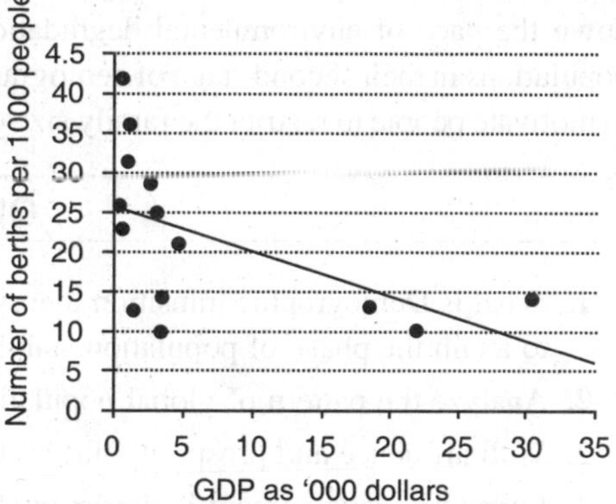

Fig. 21.7 *With afluence and prosperity both rate declines*

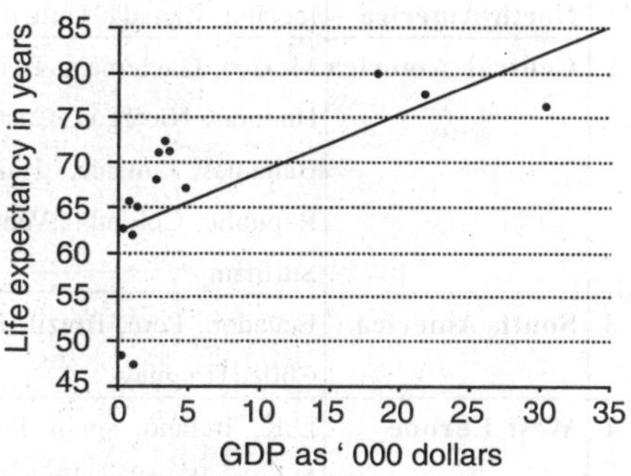

Fig. 21.8 *With affluence and prosperity life expectancyrises*

be catered with efficiency. In large households per capita availability of resources and incomes declines as the total assets have to be divided into many portions. This leads to smaller establishments where the available resources are shared by few and the rate of population growth slows down. This is apparent from a significant relationship between the gross domestic production and population growth rates as depicted in Fig 21.9.

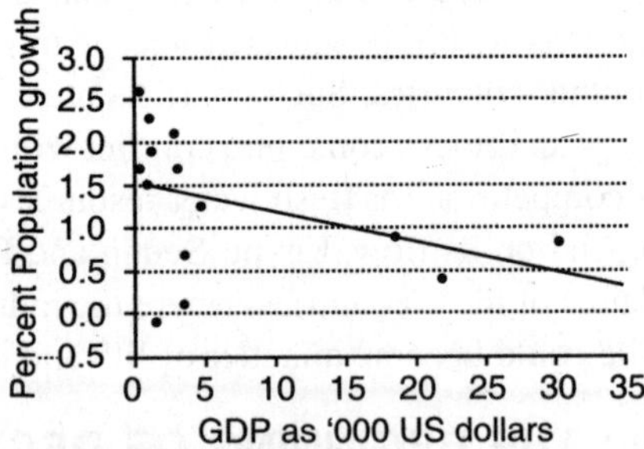

Fig. 21.9 *With affluence and prosperity population grwoth rate declines.*

The relationship between affluence, prosperity as depicted by gross domestic production in US Dollar and the current birth rates, death rates, life expectancy and population growth rates in most of the countries of the world carries an important message. This message is that development is necessary to check population growth, reduce the pressure of the growing demands of humanity on natural resources, and slow down the pace of environmental degradation. It is the development only, which shall enable the populations in their second stage of demographic transition to achieve socio-economic targets needed to motivate people to restrict the family size (24).

QUESTIONS

1. What is Demographic transition theory? Discuss the conditions necessary for human societies to attain the phase of population stabilization.
2. Analyze the pattern of global population growth and the consequences associated with it.
3. With affluence and prosperity birth rates decline and population growth slows down. Justify.
4. Economic development is detrimental to the environment yet it is necessary to enhance our capacity to conserve the enivornment. Discuss.

Table 21.1 World Population in different regions of the world and populations projections for 2025 AD & 2050 AD

			Estimates of World Population		
		Countries included	**2000 AD**	**2025 AD**	**2050 AD**
1	**North America**	Iceland, Canada, United States of America.	278,350,662	373,645,203	435,010,091
2	**Central America**	Mexico, Gautemala, Belize, El Salvador, Hnduras, Nicargua, Costa Rica, Panama, Cuba, Bahamas, Jamaica, Haiti, Dominican Republic, Columbia,Venezuela, Guyana, Surinam.	232,528,144	327,877,800	397,793,838
3	**South America**	Ecuador., Peru, Brazil, Bolivia, Paraguay, Chile, Uruguay.	283,330,977	359,442,161	392,326,463
4	**West Europe**	U.K., Ireland, Spain, Portugal, France, Norway, Sweden, Finland, Belgium, Denmark, Netherland, Luxemberg, Germany, Switzerland, Czech Republic. Slovakia, Austria, Hungary, Italy.	401,627,510	382,866,025	312,880,922

5	**East Europe**	Estonia, Latvia, Lithuania, Belarus, Poland, Slovania, Croatia, Bosnia-Herzegovinia, Yugoslavia, Romania, Bulgaria, Albania, Macedonia,Moldova, Greece, Malta, Cyprus.	130,022,684	118,418,274	104,487,114
6	**North Africa**	Egypt, Lybia, Tunisia, Algeria, Morocco, Western Sahara, Mauritania, Mali,Niger, Sudan, Eritria.	216,351,406	345,463,290	466,695,137
7	**Central Afrca**	Ethiopia, Somalia, Senegal, Gambia, GuineaB, Guinea, SiereLeone, Liberia, CoteD'Voire, B.Fasso, Ghana, Togo, Benin, Nigeria, Cameroon, C.A.R. , Eq.Guinea, Gabon,.Congo, Demo.Rep. Congo, Uganda, Rwanda, Burundi, Kenya.	413,726,383	714,041,691	1,198,915,167
8	**South Africa**	Angola, Zambia, Tanzania, Malavi, Namibia,Botswana, Zimbabwe, Mozambique, South Africa,Lesotho, Swaziland, Madagasker.	159,962,202	233,978,804	290,201,300
9	**Northern Asia**	Russia, Ukrain, Georgia, Armenia, Azerbaijan,Uzbekistan, Kyrgystan, Tajikistan,Turkmenistan, Kazakistan.	268,250,217	276,646,057	271,516,448
10	**Middle East**	Turkey, Syria, Iraq, Iran, Jordan, Lebanon, Israil, Soudi Arabia, Kuwait, Qatar, United Arab Emirate,Oman, Yemen, Afganistan.	257,043,243	430,663,885	616,213,720
11	**Far East**	China, Mongolia, North Korea, South Korea, Japan.	1,454,294,822	1,610,900,481	1,506,381,268
12	**South Asia**	Pakistan, India, Sri Lanka, Nepal,Bhutan, Bangladesh,Myanmar.	1,383,060,785	1,944,189,546	2,357,737,501
13	**S E Asia**	Laos, Thailand, Combodia, Vietnam, Malaysia,Indonesia, Philippines, Papua-New Guinea, Fiji Islands, Solomon-Islands, Maldives, Mauritius.	1,383,060,785	1,944,189,546	2,357,737,501
14	**Oceania**	Australia, New Zealand.	22,647,958	26,635,924	27,407,636
Source: See No.25.					

22

Chapter

Accelerated Urbanization

Although big cities arose from time to time earlier also, urbanization is largely a 20th century phenomenon. The first permanent human settlements evolved when nomadic people began to cultivate crops. Successive agricultural advances such as harnessing of draft animals and the development of irrigation enabled farmers to produce enough food to support the nascent villages and towns of the pre-historic times. With the diversification of trade and the production of a much wider variety of goods larger number of people were encouraged to settle in places, which later turned into towns. Neolithic towns began to enlarge into cities about 5000 years ago in fertile regions of world where water was available in abundance. Flourishing urban centres soon became the seedbeds of culture and commerce. The relative density of humans in cities appears to have speeded up the exchange of ideas and innovations on which depended the advances in science, arts and technology of the time. However, these cities could never grow beyond certain limits as they depended largely on the productivity of cropland in immediate neighbourhood. As transportation largely depended on draft animals, their supply lines were short. The world population has been overwhelmingly rural since the Neolithic times. As recently as 1900 AD, only 12-14% of world's people lived in cities (1).

(I) URBANIZATION DURING PRECEDING TWO CENTURY:

The Industrial Revolution, which began in nineteen century in Europe, fostered the development of large cities on modern lines. Coal replaced firewood as the major source of energy in Europe and spurred the growth of early industrial cities. Development of railways and steamships enabled cities to obtain food supplies and raw materials from long distances. Soon mineral coal was replaced by mineral oil, which in turn hastened massive urbanization providing a cheaper, cleaner and convenient fuel for transportation and consolidation of the industrial process. The population of London grew from 1.1 million in 1800 AD to about 7.3 million within span of hundred years. The population of Paris rose from 0.55 million to nearly three million within century. The spate of urbanization, which enlarged European cities, however, is dwarfed by the rapid expansion that marks the cities in Third World countries now (2). At the current rate of growth, the number of people living in cities of the world doubles within

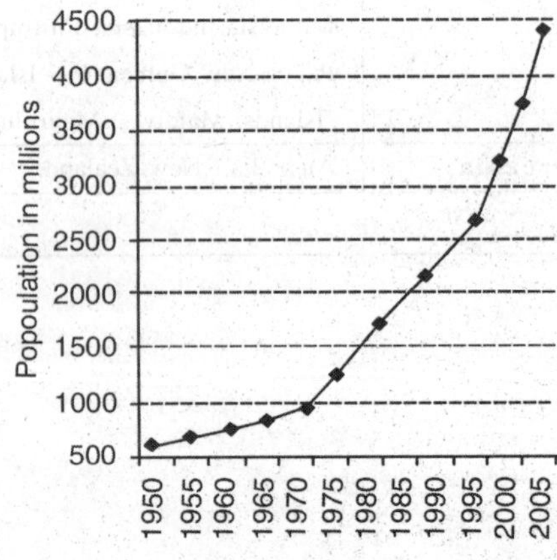

Fig. 22.1 *Growth of world's urban population*

a period of 28 years (Fig 22.1). Nearly 90% of this growth occurs in third world countries where the annual urban growth rate approaches 3.6%, more than three times that of industrial world (3:4).

(II) ACCELERATED URBANIZATION IS MAKING COUNTRYSIDE POORER:

It was an affluent and prosperous countryside, which prompted the development of ancient cities. Today, the cities flourish at the cost of rural areas, which are a victim of benign neglect. Cities are enlarging so fast because of the poverty of the countryside. Accelerated urbanization tends to concentrate political power within the cities. This leads to policies that favour urban over rural areas. Subsidies and price structure which are tilted more and more in favour of urban growth, often makes food and other basic commodities cheaper in cities. This discourages investment in agricultural sector causing the countryside to suffer. In many developing countries, governments encourage growth of large cities and industries, which usually control domestic markets as well as international trade. The benign neglect and apathy accorded to the countryside impoverishes rural populations. Rising populations, growing rural landlessness, a highly skewed land ownership, rising costs of fertilizers, seeds and other agricultural inputs and poor returns from the agricultural produce makes life difficult for small farmers. They are forced to migrate to the cities in search of jobs, livelihood and prosperity. (5)

(III) CITIES DEVELOPED AS ASSETS ARE TURNING INTO LIABILITIES:

The growth of cities in industrialized nations has been an important constituent of the national economic development earlier. The produce from these industries enriches the country. This has been so in the developing countries also as long as there is an affluent and prosperous countryside. Productivities from rural and urban sectors complement each other to make the nation affluent and wealthy. However, rising populations, growing landlessness and poverty in the rural sector force people to migrate to the cities. The urban sector starts expanding rapidly. The uncontrolled expansion of urban sector, which draws heavily upon the rural sectors as its very supply-lines originate in the countryside, places additional strain on the rural sector and forces more people to move to the cities. As the pace of urbanization accelerates the supply-lines snap under the strain. Basic necessities of life have to be imported from far off places. Unbridled urban expansion, today, is a product of agricultural failure!

As the cities acquire huge proportions, they tend surpass the administrative capacity of local governments. Most of the Third World cities are now struggling hard to provide the most basic services of day-to-day life to its inhabitants. The demand for electricity has gone up so much while generation lags behind that power cuts as long as 7-10 hours have to be imposed in many cities of North India. Majority of people in large African cities, like Lagos, Nairobi, Kinshasa, Addis Ababa and Lusaka lack piped water and sanitation. The cost of supporting huge cities has become so high as to parallel or at places exceed their contribution in terms of goods and services. In 1981-82, Sao Paulo, a Brazilian city had less than 10% of the country's population but accounted for 44% of country' electricity consumption, 39% of telephones and well over half of the entire industrial output and services (6). In 1983 the cost of maintaining Mexico city and providing necessities of daily life took nearly half of the total national income (7).

(IV) THE PROBLEMS OF A LARGE CITY

Huge cities present a number of management problems. The large numbers of people living in a small area require food supplies, freshwater, houses to live, commodities of daily use, power supply, mean of transport and communication, health care units, educational institutions etc. The wastes they produce have to be disposed off. Providing the basic necessities of day to day life, becomes a burden which assumes enormous magnitude. With the countryside already reeling under malnutrition, hunger and poverty many countries have to resort to food imports. When means of the country are limited, they burden themselves with heavy foreign debts.

1. Food supplies: The cities which developed before the Industrial revolution were largely fed with grain surpluses produced in fields in the immediately neighbourhood. Lack of efficient transportation prevented movement of supplies to long distances. With Industrial Revolution, however, the picture changed. As the populations of cities grew so large as to exceed the capacity of local crop-fields to provide grains and other edibles, the supplies had to be brought in from distant places. Great Britain, the first country to industrialize and develop large cities, like London, began to import food and other raw material from far of places in exchange of finished goods from its industries. The practice was soon followed by other industrializing European countries also. Use of mineral coal, the development of railways and steam ships, enabled these countries to carry their finished products to far off places and bring back supplies of food and other raw material from there. Much of imported food brought in this way was used to feed the growing cities of the time when the supplies from countryside were inadequate.

After the end of the Second World War, advances in agricultural technology in North America created large quantities of exportable grain surpluses. By 1950 AD, the continent was already exporting 23 million metric tons of grains, which in 1980 AD had grown to about 131 million metric tons. It is mainly North America, which has sustained much of the urban growth of the later half of 20th century. Australia and New Zealand also started contributing its share to the global food grain trade as an exporter by 1950 AD. Before mid-twentieth century, most of the Asian and African countries were food exporters earlier (1). By 1960 rapid population growth, urbanization, the neglected countryside and agricultural sectors had created substantial grain deficits in African and Asian countries. Asia had to import nearly 17 million metric ton of grains while Africa also entered the world market buying about 2 million metric ton of cereals. Within a period of twenty years, by 1980 AD, the food grain imports of African countries shot up 7.5 times and those of Asian countries about four times. Eastern Europe and Soviet Union emerged as a major grain importer while Central America, which was an exporter in 1950, had to import nearly 10 million metric ton of grains largely to feed its rapidly growing urban population (5).

As the Green Revolution opened up opportunities of achieving self-sufficiency in food grain production, an increased emphasis to develop agricultural sector along with the urban sector had been raising agricultural productivity all over the world since 1975 AD. The cumulative effect of these efforts could be witnessed in 1990 AD when the Western Europe, a food grain importer since 1800 AD had enough surplus to export it to other countries. India and China large importers of cereals nearly acquired self-sufficiency in food grain production. The food imports of Latin America declined by nearly 50%. By 1990 AD Africa, Eastern Europe, Soviet Russia and Asia (excluding India and China) and some Latin American countries were the only regions of the world, which depended on grain imports from other countries largely to feed their cities (USDA, ERS, World Indices). Countries, which have attained self-sufficiency in food grain production, have done so by intensifying their agricultural production and increasing the use of fertilizers, chemical and energy inputs, which have their own drawbacks. They shall continue to do so as long as they are able to provide these inputs to their fields and manage soil fertility, productivity and freshwater resource in a sustainable manner – tasks which appear to be too difficult for many third world countries (8).

2. Supply of Freshwater: While feeding the growing population in cities of today is a problem, providing freshwater in sufficient quantities is also not easy. Many cities the world over rely on groundwater sources for much of the water they use. Nearly 50% of Hong Kong's freshwater supply of 133 million cubic meters is drawn from underground sources. The rest comes from surface catchments across its borders from China (9). Withdrawals of water from ground water deposits have their own limitations. Extraction of ground water beyond the quantity, which enters the subsurface strata, is ecologically detrimental. Like Hong Kong, a large number of cities in the developing world rely on their underground water table for the supply of freshwater. Surface waters are either insufficient or are grossly polluted. Heavy withdrawals of freshwater from ground water deposits in many third world cities are pushing the water table down at a rapid rate. The ill effects of ground water overdraft are slowly becoming apparent in many third world cities. Desertification, wind and soil erosions are the usual consequences of the heavy withdrawal of ground water, which is slowly eroding soils and soil productivity in and around many third world cities (10).

In a number of third world countries, the water supplied to the city has to undergo costly treatments to get rid of suspended material, impurities and pathogenic organisms. In many third world cities, we have been treating our surface water deposits as convenient bodies to dispose of untreated sewage and wastewaters. The absence of proper organic waste collection and treatment in many third world cities results in serious health problems and epidemics. Raw night soil disposed off in cities drains and wastewater conduits is a blue print for enteric diseases prevalent in the community in many developing cities. Pathogens present in human wastes include hookworm, tapeworm, viruses and organisms, which cause cholera and typhoid. Use of inadequately treated water from our surface deposits ensures the spread of these diseases.

3. Heavy energy requirements of the expanding cities: Rapid growth of cities since the onset of Industrialization has been closely associated with the use of fossil fuels. Coal burnt to run steam engines powered both factories and rail transport developing early industrial societies and cities. Coal dominated the era of fossil fuel use until a few decades ago after which it is the mineral oil which has made massive urbanization possible.

The nature and the quantity of urban energy needs differ with the degree of development of a city. In developing Third World Cities, cooking dominates energy use, much of which is provided by firewood whereas in developed countries, it is the transportation system, which takes up a large share of energy used to provide food, water and other amenities to the people. The rising demand of firewood has placed severe strain on the forests surrounding the cities in developing countries. All forests have been razed within 70 kilometers of Niamey the capital of Niger and Ouagadougou the capital of Burkina Faso (11). Although the decline in forest cover around Coimbatore has been modest, about 15% only, around Delhi, a staggering 60% of the forests have been lost within a decade (1970-1980) (12).

As world's oil production declines, cities shall have to turn to renewable sources of energy. If harvested evenly throughout a country's forests, fuel wood a renewable source of energy could provide enough for the growing demand. However, because the demand is heavily confined to places around the cities only nearby forests are affected. The distant ones are left untouched. As more and more people congregate in cities, the inability to manage the forests resources for optimum yields could become a costly and ecologically disastrous failure. Other important sources of renewable energy, which are being used today, include wind energy, hydroelectricity, waste fueled electric generation, solar collectors and geothermal energy. Many cities are well under way in the process of transition from oil to these renewable energy sources. Just as the cities were shaped by the shift from wood to coal and then from coal to oil, so also the future of our cities shall be shaped by the necessary shift from fossil fuels to renewable sources of energy.

4. Nutrient recycling: Every day thousand tons of basic plant nutrients, such as nitrogen phosphorus and potassium, enter into the cities from countryside in the flow of foodstuff, which sustains population growth. Worldwide enormous organic wastes, society's most ubiquitous disposable material, are created. Nearly two third of the nutrients present in human wastes are released into the environment as unreclaimed sewage which pollute the streams, rivers and lakes. For example, in 1985 AD sewage systems served only about 5% of people in India, which generated nearly 3.6 million cubic meters of wastewater. If all of it were recycled, it could yield about 82000 tons of nitrogen and

24000 tons of phosphorus annually (13). Naturally as cost of materials and energy needed to produce fertilizers rises, the viability of intensive agriculture, which rests heavily on fertilizer and energy inputs, shall be endangered. The viability of agriculture and of urban sector today largely hinges on how successfully the cities can recycle this immense volume of nutrients, which enter every day in our cities in the form of food supplies and finds its way to streams, rivers and lakes where it creates other problems. Closing the nutrient cycles is one of the most important aspects of an ecologically sustainable society (13;14;15).

In European cities, the practice of fertilizing crops with human organic wastes was started in the mid-nineteenth century with purpose of using organic wastes and minimizing pollution. By 1875, nearly 50 sewage farms came into existence in Britain. However, early efforts in this direction failed because the volume of wastes grew so much as to overwhelm the capacity of the sewage farms. As cities grew, sewage farms had to be displaced away from the cities, while the untreated sewage was recognized as a major source of health problems. The practice was terminated (13).

During the past two or three decades, however, the attitude towards nutrient recycling has changed. Higher fertilizer prices, a better understanding of natural fertility of soils, ecological limitations and better waste management technologies have renewed interest in nutrient recycling both in developed and in the developing countries alike. Recycling organic wastes saves money improves soil-structure, fertility, water retention capacity that increases food production and thereby enhances urban self-sufficiency. It also enhances the ability of crops to draw on both natural and synthetic nutrients. In other words, they repair the damages caused to soils by repeated use of fertilizers and chemicals, which we apply to intensify agricultural productivity. Wheat yields in India increased from 28% to 45% with application of about 5 tons of compost per hectare (16). As many as half a dozen Chinese cities were already producing more or less 85% of their vegetables within the city boundaries reclaiming nutrients from human organic wastes and garbage by 1985 AD (17). According to a World Bank estimate as recently as 1980, nearly one third of fertilizer needs of China were met with by night soil (13). Although widely adopted in developed countries, it is however, a little discouraging that appropriate technologies for reducing sewage related health risks, have not been fully implemented in developing countries.

5. The problem of environmental deterioration: Most of the problems of urban sector stem from the concentration of humanity in a localized area. The large numbers of people, which live in a small area, require food, water, shelter and other commodities, which have to be imported from long distances. Just as Nature cannot concentrate the resources required by so many people, it can neither disperse nor dispose of the wastes which so many people generate. Fuel wood or fossil fuels burning in so many homes collectively produce so much smoke, fumes and other waste gases, which add up to form huge quantities. Trains running on coal, enumerable vehicles plying on petrol or diesel together make up so much amount that adds significantly to the atmospheric burden of green house gases. To supply fuel wood to the cities an increasingly larger area is deforested annually. Nutrient rich organic wastes, which are an asset to rural sectors, are in fact an economic burden in our cities. Money, labour and energy is required to dispose the huge quantities of wastes which are produced in a city every day. Much of these wastes are disposed off in rivers, streams, and lakes as untreated sewage a practice that results in health problems and causes an accelerated eutrophication of our aquatic systems. Bared and denuded soils are subjected to a high rate of erosion by the battering action of wind and rains. Waters percolating through the heaps of garbage, persistent and toxic wastes renders the soil useless for cultivation. In the long run, the toxic leachates may spoil the subsurface water table. The list of damaging effects of busy and bustling city life is a long one. It should suffice here to add that many of global problems environmental degradation have their genesis in the huge cities of the world.

Of all the attention, efforts and investment needed to improve living conditions in cities the most important is the treatment and disposal of domestic and industrial wastes – solid, liquid or gaseous wastes. Reports on the adverse consequences of air pollution in Third World cities are numerous. An estimated 60% of Calcutta's people are believed to suffer from respiratory diseases related with air pollution. Deaths due to lung cancer in cities of China are reported to be four to seven times higher than the country as a whole. The difference has been explained to be due to air pollution (5). Water pollution in Third World cities may even be worse than air pollution. Only 210 of India's 3120 towns and cities had even partial sewage system and treatment facilities till 1980s. Nearly 114 towns and cities dump raw sewage in the Ganges, our holy river. At Kanpur, U. P., India, the average faecal coliform count is about 6.5 millions per ml a level, which as compared to safe drinking water limit of 100 is astronomically high. Heavy metal contamination of water supplies is also common around many Third World Cities. Both concentrating the resources needed to sustain a city as well as disposal of the wastes that threaten to make the cities inhabitable have become stupendous tasks. The deteriorating environment and ecology of cities is a matter of great concern today.

(V) STRIKING A RURAL-URBAN BALANCE

The relationship between rural and urban sectors has however, been so obvious that it has ignored by most of the people and policy makers. Early Greek cities have been described as both small and relatively self-contained largely dependent on local countryside for food supplies and other materials (1). A stable and productive farm economy ensures long-term food security. Improvement in living standards is usually associated with urbanization, which capitalizes on the economic growth associated with industrialization. It is the productivity of a affluent and prosperous rural sector which sustains both industrialization and urbanization. Countries that underwent urbanization early in nineteenth century built their cities on successful agragarian economy.

The uncontrolled urban growth in Third World Cities is an outcome of failed economic and population policies. Today, urban growth is driven more and more by rural poverty rather than urban prosperity and promises of a comfortable, convenient and secure lifestyle. Most of the third world governments, with a few exceptions, have favoured cities at the expense of countryside. It is not uncommon for these countries to allocate only 20% of their annual budget to more than 70% of their population living in rural areas. A policy, which results in stagnating or inadequate growth of income in rural areas, and has caused misery and poverty to grow at a rapid pace. Landlessness and a high rate of growth of rural population dampen the prospects of the agragarian future for majority of young men in many of the developing countries. Subsistence farmers migrate to cities on seasonal basis looking forward to supplementing their incomes but many move to the cities permanently. Massive unemployment results in the cities where there are few jobs and too many people chasing them while declining food surpluses in the countryside forces the urban areas to become increasingly dependent on imported food. Shanty dwellings, slums and extremes of misery and poverty occur along with the other extreme of affluence and prosperity in high-rise skyscrapers in most of the Third World Cities.

The need of the time is a set of foresighted national policies, which encourage national equity, develop a prosperous and affluent countryside so that productivity and incomes in rural sector could be at least comparable to those of urban sectors. It is apparent that the most effective way to ameliorate the problems facing cities may well be those to increase investment and hence employment and productivity in the countryside. Only this could check the flow of migrant to cities and provide a sound basis for the development of the urban sectors. The World Bank and International Monetary Fund are encouraging heavily indebted governments of third world countries seeking new loans to abandon the urban-biased policies, which have contributed so much to their dilemma and adopt agricultural policies that stimulate domestic food production and reduce dependence on imported food.

QUESTIONS

1. What do understand by accelerated urbanization? Discuss its consequences.
2. Discuss the problems large cities have to face in developing countries.
3. Discuss the environmental consequences of accelerated urbanization.
4. Rapid growth of most of the third world cities is based on rural poverty. Discuss.

23

Chapter

Human Health and the Environment

Human health and well-being is affected appreciably by the environment around us. Life in mild, peaceful calm and quite environment encourages health and well-being while life in unhealthy and stressful environment causes ill health, morbidity and a general shortening of one's life span. An unhealthy and severe environment taxes our mental abilities, strains our physical capabilities directly or indirectly, induces anxieties and diseases of various kinds.

Since times immemorial there has been a tug of war between humans and the microbes conditioned by generations of contact, exposures to immune system and human behaviour. As scientific knowledge developed better and more sophisticated measures were adapted to control the diseases. Many infectious diseases were gradually brought under control. However, microbial organisms appear to be better-equipped to invade new places, take on to new hosts or ecological niches, change their virulence or means of transmission and develop resistance to drugs. Certainly, an organism capable of replicating itself a million times within a day has an evolutionary advantage over its host, with chance and surprise on its side. Therefore, no matter how much sophisticated disease control measures become, there is always the possibility of another outbreak of an epidemic at anytime and anywhere in the world. Today enormous mobility and speed has been added to this battle. In 2006 international airlines carried nearly 2.1 billion passengers which mean that the diseases have the capability to spread much faster than at any time in the known history of mankind. An epidemic in one part of the world could become a health threat to people across the world within a few hours only. To add to our miseries, these diseases appear to be emerging much faster than they did earlier. Since 1970, WHO has identified 40 new diseases which were unknown a few generations ago and recorded 1100 new epidemic outbreaks (1;2).

While infectious diseases has troubled humanity since times immemorial and have caused health problems and deaths on unprecedented scales, the advancing frontiers of science and technology have added a new threat to the human health. This threat comes from the variety of chemicals and radioactive material which we manufacture, use or discard. During their manufacture, storage, and transportation, some of the toxic material may escape and contaminate the environment. Twentieth century has witness numerous incidences of accidental or deliberate discharge of toxic chemicals which have caused adverse health effects and deaths on large scales.

(I) INTERNATIONAL CO-OPERATION:

The realization that diseases do not recognize national boundaries drawn by man necessitated the international collaboration to combat infectious disease. By nineteenth century the outbreak of cholera epidemic, the threat of plague and ineffectiveness of quarantine measures etc made it apparent that to prevent the outbreak of infectious diseases international efforts and co-operation was required. From 1851 to 1900, 10 International Sanitary Conferences were convened, which mostly comprised of a group of about 12 European countries or states, These conferences focused exclusively on the containment of epidemics within their territories. The inaugural 1851 conference in Paris was a long one which covered a period of six months and established the vital principle that health protection was a proper subject for international consultations (3).

During the 1880s, a small group of South American nations signed the first set of international public health agreements in America. In addition to cholera and plague, often carried among the huge number of immigrants arriving from Europe, these agreements covered yellow fever, which was endemic in much of the region. In 1892, the first International Sanitary Convention dealing only with cholera was signed. Five years later, at the 10th International Sanitary Conference, a similar convention focusing on plague was also signed. The ***Pan American Sanitary Bureau*** (now called the Pan American Health Organization) came into existence in 1902. Its counterpart in Europe, the ***Office International d'Hygiène Publique*** (OIHP), was established in 1907 with its headquarters in Paris (4)

Apart from its immediate toll on human lives, the First World War brought in its wake many epidemics resulting from the destruction of public health infrastructure, from typhus in Russia that threatened to spread to Western Europe, to cholera, smallpox, dysentery and typhoid in the Ottoman Empire. These epidemics were the basis for the formation of the League of Nation's health organization. In 1923 AD the Health Organization of the League of Nations was came into being with its headquarters at Paris which established communication among the public health officials of diverse countries of the world. Later it served as a back-ground for establishment of ***World Health Organization*** which is United Nation's specialized agency set up under Article 57 of the of UN Charter on July 22, 1946. In 1951, a few years after its creation, WHO adopted a revised version of the International Sanitary Regulations first approved in 1892 AD. They focused on the control of cholera, plague, smallpox, typhoid fever and yellow fever and were succeeded by International Health Regulations (1969) which required the Member States to report the outbreak of certain diseases (5).

Recent events have demonstrated the urgent need for a revised set of regulations with broader disease coverage, and measures to stop their spread across borders. In 1986 a massive reactor explosion occurred at Chernobyl, former USSR, which left hundreds of thousands exposed to radioactive gases across the European continent. Such public health threats from accidental or deliberate release into the environment of chemicals, biological or radioactive materials, agents causing adverse health effects were not included in the regulations adopted earlier. The International Health Regulations-2005 respond to this need and have now come into force. The broad definition of "public health emergency of international concern" and "disease" allows for the inclusion in International Health Regulations (2005) of threats beyond infectious diseases. The inclusion of public health emergencies other than infectious diseases extends the scope of the Regulations to protect global public health security in a comprehensive way (6).

(II) HUMAN EFFORTS TO CONTROL INFECTIOUS DISEASES:

Only a hundred years ago, infectious diseases such as cholera, plague and yellow fever and many more such as diarrhoeal diseases other than cholera, influenza, malaria, pneumonias and tuberculosis ravaged most of the civilizations and threatened public health security. If we trace the history of man's fight against diseases we shall find that the human success, although partial, came through three major stages which helped the mankind to save millions of human lives:

1. Quarantine:

The term "quarantine" refers to the 14th century practice of the isolation of people for about forty days arriving from plague-infected areas to the port of Ragusa under the control of the Venetian Republic at the time. The practice of separating people with disease from the healthy population is an ancient one. Both biblical and koranic scriptures refer to the isolation of lepers to prevent the disease from spreading to healthy people. By the 7th century, China practiced the well-established policy of preventing sailors and foreign travelers suffering from plague from coming in contact with locals. Such public health measures became widespread and international over the following centuries (7). In the 17th century, an attempt to keep plague, which was spreading through continental Europe, from reaching England obliged all London-bound ships to wait at the mouth of the River Thames for at least 40 days. The attempt failed and plague caused devastation in England in 1665 and 1666. The continuing devastation regularly wrought by plague and other epidemic diseases demonstrated that crude quarantine measures alone were largely ineffective.

2. Sanitation and hygiene:

John Snow's famous work on cholera highlighted the importance of sanitation and hygiene. His observation that most of the reported cases of cholera were clustered around a particular water-source with foul smelling water led him to associate the epidemic with insanitary conditions and infested waters. His work eventually led to improvements in sanitation in the United Kingdom that reduced the threat of cholera – though not to the same extent as endemic diarrhoeal disease from other causes (8;9). Latin America which had been free of Cholera epidemic for more than a century was ravaged by the epidemic which struck in 1991 and by 1995 there were more than 1 million cases and just over 10000 deaths reported in the region. Cholera continues to be a major health risk all over the world. The need to provide sanitation both for drinking-water and hygiene remains a huge challenge today in developing countries. Currently 1.1 billion people lack access to safe water and 2.6 billion people lack access to proper sanitation. As a result, more than 4500 children under five years of age die every day from easily preventable diseases (10).

3. Immunization

Smallpox is the oldest known human disease. The mummified head of Ramses V, dating back to 1157 BC shows postular eruptions caused by smallpox. It was the casual observation of an English Physician, Edward Jenner that people exposed to cowpox infection, a milder but related disease, were immune to smallpox. In the year 1796 he inoculated a ten year old boy with cowpox virus and after observing the reaction the boy was inoculated with smallpox virus. The boy did not develop the disease. Jenner's experiment demonstrated that smallpox could be controlled by cowpox inoculation. Jenner's procedure was soon widely accepted, resulting in sharp falls in smallpox death rates. At the beginning of the 20th century, smallpox was still endemic in almost every country in the world. In the early 1950s, an estimated 50 million cases occurred globally each year with an estimated 15 million deaths, figures which fell to around 10–15 million cases and 3 million deaths by 1967 as access to immunizations increased. A global smallpox eradication campaign was begun in 1967 which was successfully completed by 1979 (11). Today deliberate release of smallpox virus by terrorist groups who may be storing the virus has been causing much concern in a number of industrialized countries of the world. Work is already under way on a new and safer vaccine against smallpox, which would need to be produced in huge quantities if immunization against a deliberate release were to be undertaken (12).

(III) HEALTH PROBLEMS ASSOCIATED WITH DEGRADED ENVIRONMENT:

The environment can affect our life drastically by bringing about changes in conditions of life around us. An unhealthy and degraded environment may reduce productivity of agriculture, animal husbandry, and fisheries etc causing shortages of all types. Scarcity, poverty and the consequent malnutrition which is rampant in many of the African, South American and Asian countries of the world is a blue print for ill health, morbidity and outbreak of infectious diseases. Malnutrition in particular, predisposes people to a variety of deficiency and infectious diseases which could be avoided, otherwise, on a healthy diet. A degraded environment may bring about adverse qualitative or quantitative changes in many of the resources we use which in turn may affect human health. The pollution of the stratosphere, for example, today threatens to destroy the protective ozone shield. Damage to the stratospheric ozone layer shall result in higher amounts of solar ultraviolet radiations to reach down to earth's surface causing increased incidences of cataract of eyes, skin diseases and cancers etc. Major human-health problems associated with a degraded environment may be of the following types:

1. The threat of infectious diseases.
2. The threat of exposure to toxic chemicals, radioactive materials, heavy metals and toxic trace elements.
3. Hidden health effects of contaminated environment on human lives.

1. The threat of infectious diseases:

A degraded and unhealthy environment may promote the growth of other life forms (insects & pests) that may become a potential threat to human health and well-being. Organic wastes, sewage effluents, excreta, exudates and faecal matter etc. support a rich population of microbes. Numerous bacteria, algae fungi, protozoans, helminthes, annelids, larval stages of various insects, pests etc thrive on organic matter. Some of these are responsible for causing dangerous diseases of man, animals and plants.

Waterborne diseases are caused by pathogenic microorganisms which are directly transmitted when contaminated drinking water is consumed. According to the World Health Organization, diarrheal diseases account for an estimated 4.1% of the total global burden of disease and is responsible for the deaths of 1.8 million people every year. It was estimated that 88% of that burden is attributable to unsafe water supply, sanitation and hygiene, and is mostly concentrated in children in developing countries. Chance contamination of water supplies, food and other edible material with sewage effluents and organic wastes is often responsible for the outbreak of a number of diseases. Typhoid caused by *Salmonella typhii* and cholera caused by *Vibrio cholerae* for example mainly spread through contaminated waters.

Air borne diseases are diseases which are transmitted through air. Acute respiratory infections kill more than 4 million people each year and are the leading cause of death among children under age 5. This range of infections, which includes pneumonia in its most serious form, accounts for more than 8 percent of the global burden of disease. Overcrowding and unsanitary household conditions favour the transmission of the disease, which is spread by droplets from a cough or a sneeze or unwashed hands. Although cause-and-effect relationship between indoor air pollution and acute respiratory infections is difficult to prove in part because people who use biomass fuels tend to be poor and exposed to multiple risks such as overcrowding, tobacco smoke, and malnutrition. Even so, the World Bank estimated in 1992 that switching to better fuels could reduce the number of pneumonia deaths drastically (2;13).

Other airborne diseases also thrive in conditions of poverty, crowding, and poor hygienic conditions. Tuberculosis (TB), to name just one, killed an estimated 3 million people in 1996, and nearly 7.5 million others developed the disease. Roughly 95 percent of all TB sufferers are in the developing world, mostly in Southeast Asia, Western Pacific, and Africa. TB has resurfaced in developed countries, where it is concentrated among poor populations. Measles and diphtheria, vaccine-preventable childhood diseases of crowding and poverty, have been all but eliminated in the developed world since the advent of successful vaccines. In the developing world, however, measles still affects 42 million children per year who lack access to the vaccine; roughly 1 million of these children die. Since 1990, diphtheria has resurfaced in the former Soviet Union, triggered by social disruption and a drop in immunization rates. Other familiar diseases in this group are neonatal tetanus, poliomyelitis, and pertussis. This group, all linked with environmental conditions, accounts for nearly 15 percent of the total disease burden globally for children under age of 5 years. Despite widespread immunization programs, these diseases claimed the lives of 1,985,000 children in 1990 (2;13).

Dead and decaying organic matter is the starting point of the detritus food chain. A series of microbes, insects, pests and rodents feed on organic wastes or organisms, which live on these wastes. Abundance of biodegradable wastes, therefore, causes a substantial rise in the population of aphids, mosquitoes, houseflies, rodents and such other animals. These become a nuisance for day-to-day human activity. Many of these organisms are carriers or vectors of serious disease of man, animals and plants. A number of pathogenic organisms are physically carried from one place to another by these insects and pests. Some virus, bacteria and protozoans may enter the body of the insect or pest feed, breed, multiply and persist therein for long periods of time. The pathogen is transferred to other animal or man with the saliva when the insect bites or feeds upon them. Or else such carriers simply continue to excrete the pathogen for long duration. Malarial parasite is transferred by the bite of the female mosquito of genus *Anopheles*. African sleeping sickness, caused by *Trypanosoma gambiens* and *T.rhodesiencse* is spread through the bite of Tse-tse flies (*Glossina palpalis* and *G.morsitans*). Like-wise sand fly, (*Phlebotomonas* sp.) disseminates Cutaneous Leishmaniasis and Kala-azar both caused by *Leishmania* sp. The mosquitoes of genus *Aedes* (*A.aegippti, A.allopicus* and *A.acutellaris* etc.) transmit viruses causing Dangue fever, Yellow fever and Japanese Encephalitis (14;15;16).

2. **Threat of exposure to toxic chemicals, radioactive materials, heavy metals and toxic trace elements:**

A large number of chemicals are in common use these days. Insecticides, pesticides, toxic trace elements and heavy metals contaminate our soils and water supplies. Many of these toxic agents are accumulated and biomagnified in plants and animals which provide us the food we eat. Man is, therefore, exposed to a much larger concentrations of these toxic agents. In big industrial towns the air is often contaminated with fine particulates, harmful gases, vapours and products of atmospheric reactions etc which is a hazardous mixture to breath. The toxic exposures in a contaminated environment could be of the following types:

a. **Acute Exposures:** Usually, intentional or accidental release of toxic material in the environment causes this type of toxicity. Such exposures cause drastic symptoms of toxicity, which appear suddenly and result in emergencies. The concentration of toxic agent within the body of the organism rises sharply well beyond the threshold limit and this is attended by rapid development of symptoms of toxicity, often with a possible lethal outcome.

b. **Chronic Exposures:** Repeated or continuous exposure to small concentrations of toxic agent: An organism may be exposed to small concentration of toxic agents being absorbed repeatedly or continuously by the system. Our body has to work persistently to detoxify and/or eliminate the toxic material, which strains the metabolism and affects our vitality. Chronic toxicity is characterized by a slow and gradual development of malfunctions in the normal activity of the subject when the effects of the prolonged exposures gradually accumulate.

In today's technologically advanced society a large number of chemicals are used in our daily life. The likelihood of the chemical contamination of water we drink, the food we eat and the air we breathe is always there. The toxic material absorbed in smaller quantities is eventually detoxified and/or excreted from the biological system. In many cases, the toxic response does not appear. The toxic response, however, appears only when the concentration of the toxic material exceeds the threshold concentration for toxic action.

There are situations when harmful and toxic chemicals have to be used and toxic exposures are inevitable. In such cases a number of standards are used to regulate the use of chemicals in the society and determination of ***acceptable risks***. However, environmentally sound management of toxic chemicals is still a serious challenge for many nations and the international community. For this reason, almost every nation in the world has heavily regulated the manufacturing, distribution, usage, transportation and storage of various types of chemicals, in one way or another. However, different substances are often regulated according to different standards. The same substances may be subject to one regulatory standard when used in agriculture and another when used in industries and still other when used for domestic purposes. The statutory risk standards are commonly vague and often confusing (17). The criteria usually adopted for regulating the use of chemicals in various spheres of our daily life are:

a. **No-risk criteria:** For the chemicals, which cause delayed toxic effects such as cancers or mutations, are to be avoided completely. No chances have to taken with such chemical whatever may be the cost of imposing such restrictions (18). The criteria are oblivious of the cost of regulation as the effects produced by such exposures are of a very serious nature. The chemical could, if allowed to circulate in the environment, change the very course of evolution.

b. **Margin of Safety criteria:** If a substance causes reversible toxic effects, which are manageable within a comfortable margin of safety, its use may be permitted. Smaller concentrations of such chemicals may be allowed as they cause little obvious toxicity and even if the toxic response is stronger remedial measures could be adopted, as there is no danger of late and latent effect. Regulatory agencies permit the use of these chemicals with little additional margin of safety (19).

c. **Criteria using Risk-benefit analyses:** At times, it becomes necessary to use a chemical even if it causes some toxic effects. Regulatory agencies are required to weigh the damages caused by the exposures to such chemicals and the benefits, which accrue from its use. If the benefits derived are substantial as compared to the costs involved in restricting the use of the chemical or the cost involved in repairing the damages caused by the exposures the use of the chemical is allowed (18).

3. Hidden health effects on life in environment contaminated with pathogenic organisms and toxic chemicals

We remain healthy and free of diseases in a grossly polluted & degraded environment because of the miraculous defense mechanism within our bodies against pathogenic organisms and the versatile machinery of detoxification and elimination of toxic chemicals which are regularly absorbed by our systems. Often we are not even aware of the work which our bodies do to save us from these exposures. However, unknown to us these exposures drain our energy and material resources and shorten the period of our healthy existence.

Exposure to pathogenic organisms activates the natural defense mechanism within human body which may be external or internal. When the pathogen has penetrated the external defenses of the human body, such as skin, mucosal linings and other mechanical barriers internal defense mechanism is activated. The internal defense mechanism may be *Non-specific* defense which consists of complement protein system in blood serum, phagocytosis, interferons and natural killer cells and *Specific* immune system which consists of an integrated cellular system producing serum antibodies and sensitized cells which together eliminate the pathogen. The defense mechanism of human bodies requires a number of proteins, enzymes, co-enzymes and co-factors and an integrated cellular system to do its work in a concerted manner. The assembly and maintenance of the defense machinery, therefore, places a demand on energy and material resources of our bodies. In an environment contaminated with a variety of pathogenic organisms our bodies are subjected to a persistent onslaught of invading micro-organisms for which the defense mechanism has to work incessantly. As these onslaughts intensify, more and more strain is placed on the defense system which in turn places more and more demands on energy and material resources of our bodies. So, although an individual may remain healthy, life in a contaminated environment is more taxing and tiresome. With time the defense system may wear out and finally diseases set in. General nonspecific symptoms, such as weakness, debility and morbidity appear. There is a general shortening of one's life span (20).

The mammalian system possesses a versatile machinery of biotransformation, detoxification and elimination of the toxic chemicals which regularly enter our systems. Like the defense mechanism against pathogens this system also uses many proteins, enzymes, co-enzymes, cofactors and energy to carry out its work. In a healthy uncontaminated environment these exposures are few or rare. In a degraded and contaminated our bodies absorb almost a regular supply of these toxic materials usually in small doses. Our bodies have to work persistently to detoxify and eliminate them from our systems. Naturally this places demands on the energy and material resources of our body " which multiplies with the level of toxic chemicals absorbed. The toxic symptoms do not appear because of the incessant activity of this versatile cleansing machinery within our bodies. But with time or age this machinery may wear out resulting in accumulation of a variety of chemical toxicants within our bodies beyond the threshold of toxic action. And there is a general shortening of one's life span

(IV) THE IMPACT OF CLIMATE CHANGE ON HUMAN HEALTH

Climatic change is a grim reality now. The scientific evidence continues to mount. The effects of the changing climate are already being felt and human activities are the principal cause of this change. Although climate change is a global phenomenon, its consequences will not be evenly distributed. Scientists agree that developing countries and small island nations will be the first and hardest hit. The warming of the planet will be gradual, but the effects of extreme weather events – more storms, floods, droughts and heat waves – will be abrupt and acutely felt. Both trends can affect some of the most fundamental determinants of health: air, water, food, shelter, and freedom from disease. Major health consequences of climate change can be summarized as follows:

a. As the agricultural sector is extremely sensitive to climate variability rising temperatures and more frequent droughts and floods shall affect our food security adversely. A significant rise in malnutrition is expected in a number of countries of the world which shall be especially severe in countries where large populations depend on rain-fed subsistence farming. Malnutrition, much of which is caused by periodic droughts, is already responsible for an estimated 3.5 million deaths each year.

b. Both scarcity and excess of water due to frequent torrential rains will increase the burden of diarrhoeal disease which is spread through contaminated food and water. Freshwater is an essential ingredient of hygiene, a cleansing agent which removes infections but is also instrumental in carrying infections from one place to another. Diarrhoeal disease is already the second leading cause of childhood mortality and accounts for a total of approximately 1.8 million deaths each year.

c. More frequent extreme weather events such as storms, floods and droughts shall result in more destruction, injuries and deaths. In addition, flooding can be followed by outbreaks of infectious diseases, such as cholera, especially when water and sanitation services are paralyzed. storms and floods are already among the most frequent and deadly forms of natural disasters.

d. Heat waves, especially in urban "heat islands", can directly increase morbidity and mortality, mainly in elderly people with cardiovascular or respiratory disease. Apart from heat waves, higher temperatures can increase ground-level ozone and hasten the onset of the pollen season, contributing to asthma attacks.

e. Changing temperatures and patterns of rainfall are expected to alter the geographical distribution of insect vectors that spread infectious diseases. Of these diseases, malaria and dengue are of greatest public health concern.

In short, climate change can affect problems that are already huge, largely concentrated in the developing world, and are difficult to combat. The World Health Organization estimates that the warming and precipitation trends due to anthropogenic climate change of the past 30 years already claim over 150,000 lives annually. Many prevalent human diseases are linked to climate fluctuations, from cardiovascular mortality and respiratory illnesses due to heat waves, to altered transmission of infectious diseases and malnutrition from crop failures. The experience of the 2003 heat wave in Europe shows that high-income countries might also be adversely affected. Uncertainty remains in attributing the expansion or resurgence of diseases to climate change, owing to lack of long-term, high-quality data sets as well as the large influence of socio-economic factors and changes in immunity and drug resistance. (22;23).

(V) HUMAN HEALTH: THE FUTURE CHALLENGES.

Although health management in developed countries has acquired a remarkable degree of success, in developing countries of the world crowding, scarcity and malnutrition still prevails which causes deficiency diseases and provides an ideal setting for outbreak of infectious diseases. *A look at the future trends in population growth in different regions of the world reveals that nearly 3.5 to 4.0 billion people are still to be added to the world population. Most of this growth, nearly 95% shall occur in developing parts of the world which shall add significantly to the growing rural landlessness and expanding cities in the third world countries.* The urban slums and poverty infested regions of our cities shall multiply providing an environment just fit for the proliferation of infectious diseases. Accelerated economic activity and expanding industries shall intensify chemical pollution and subject people to ever rising doses of toxic chemicals and radioactivity. In such a setting deliberate or accidental discharge of radioactive material, toxic gases and other harmful agents in water ways or air is distinct possibility.

Today we face a worldwide threat of terrorist attacks which have largely been using conventional weapons of mass destruction such as bullets, granades and bombs so far. However the possibility of terrorists using chemical or biological agents of mass destruction is not a remote one. These agents are too easy to acquire or manufacture and too easy to disperse. A terrorist attack with an agent of low concentration that is ineffectively delivered shall still cause mass casualties which will threaten civil order and overwhelm available medical facilities. The creation of this chaos is as much within the capability of a lone, skilled, and determined person with his or her own warped agenda. Beyond the immediate crisis, any chemical or biological incident may produce delayed and chronic physical and psychiatric disorders.

Health management in times to come shall face problems which would have assumed much larger proportions and intensity. The task has already been a huge one and we have had only a partial success. Controlling malnutrition, epidemic diseases, the threat of radioactive and chemical exposures and management of natural disasters as well as terrorist strikes, shall require major investment and co-operation of the diverse countries of the world. We shall have to be in a state of perpetual preparedness with medicines, antibiotics, vaccines and other provisions with mobile highly efficient machinery of disaster management to face emergencies at any time and in any part of the world.

QUESTIONS

1. Trace the efforts made by International community to control the outbreak of epidemics like Cholera, Plague and Smallpox.
2. Discuss the role of Quarantine, Immunization, Sanitation and hygiene in preventing the outbreak of epidemics.
3. What are the health-problems associated with degraded environment?
4. Use of toxic chemical and radioactive material has added a new Public health challenge, Discuss.
5. Discuss the criteria used regulate the use of toxic chemicals in public life agriculture and industry.
6. Discuss the likely impact of Climate change on human health.

24 Chapter

Human Rights and the Environment

Human rights are those rights, which are inherent to human beings. Humans enjoy these rights simply because they are humans. Every human being is entitled to enjoy these rights without any distinction of race, colour, sex, language, religion, political or other opinion, national or social origin, property, birth or any other status. 'Inherent human rights' find their expression through International Human Rights law which are a series of international human rights treaties and other instruments which have emerged since 1945 AD and provide a legal status to the rights of humans. The full body of International human rights law consists of more than 100 treaties, declarations, guidelines, recommendations and principles, which together set out international human rights standards (1).

The Universal Declaration of Human Rights, adopted by Resolution 217(III) of the United Nations' General Assembly consists of a preamble and 30 articles which sets out the human rights and fundamental freedoms to which all men and women are entitled without distinction of any kind (2;3). It recognizes that the inherent dignity of all members of human family. It's provisions involve:

- The right to life, liberty and security of persons.
- The right to adequate standard of living
- The right to seek and enjoy asylum from persecution in other countries.
- The right to freedom of opinion and expression.
- The right to education, freedom of thought, conscience and religion.
- The right to freedom from torture and degrading treatment.

Shocking, as it may appear now, slavery was generally legal under the national law in some countries until twentieth century. In England it was made illegal by 1772 AD, in the United States it was abolished in 1863, and in Brazil it was abolished in 1880 only, whereas in many African States and New World Countries under the colonial rule it enjoyed the state-protection right up to early years of twentieth century. The earliest texts and treaties related to the problems of human rights were, therefore, concerned with abolition of slavery, the ill treatment meted out to the prisoners of war, the defeated people and the protection of minorities. The atrocities, which accompanied wars in Europe in the eighteenth and nineteenth centuries, were notorious and so has been the tale of many communities defeated in conflicts and wars, which called for international efforts. This necessitated agreements and treaties to protect the interests and wellbeing of minorities. The job was well handled by the League of Nations, which came into existence after the First World War. The political turmoil, which preceded the Second World War, overwhelmed these efforts. These developments, however, were instrumental in establishing the liberal traditions of Western Democracies, principles of equality and fraternity and respect for inherent human rights and dignity (3).

It is England, which took the lead in this connection. As early as 1215 AD Magna Carta guaranteed English citizens freedom from imprisonment or from dispossession of their property and freedom from prosecution and exile unless by a lawful judgment of its peers or by law of the land. It also included the primitive formulation of the right of fair trial in the famous words *'to non will we sell deny or delay the right of justice.'* The Civil War and the Peaceful revolution of 17th century later resulted in the Habeas Corpus Acts and the Bill of Rights of 1689 AD.

The political philosophy behind these acts was inherited by the colonists in Northern America when Thomas Jefferson asserted that Americans were a 'free people' claiming their rights as derived from the laws of nature and not as the gift of their 'Chief Magistrate'. The American Declaration of Independence states that all men are created equal, that they are endowed by their Creator with certain unalienable rights, that among these are the rights to life, liberty and the pursuit of happiness (4). Nearly a century after the Habeas Corpus acts and the Bill of Rights of 1689, the philosophy behind the liberal democratic traditions found expression in the famous French Revolution. The French declaration of the rights of man proclaimed a number of entitlements, which are now generally called Civil and Political Rights. It conferred on its subjects equality, liberty, freedom of opinion, freedom of expression and the well-known definition of liberty as freedom to do anything, which is not harmful to others. (3;5)

(I) HUMAN RIGHT TO HEALTH AND ENVIRONMENT

Although right to life, equality, liberty, adequate standard of living, freedom of expression, freedom from oppression and degrading treatment etc found due expression in the Charter on Human rights, adopted in 1948, there was no mention of the "human right to healthy environment" (6). International Covenant on Economic, Social, and Cultural Rights, adopted by the UN in 1966, recognizes "*the right of everyone to the enjoyment of the highest attainable standard of physical and mental health*" (7). The constitution of the World Health Organization (WHO) also affirmed that "*the enjoyment of the highest attainable standard of health is one of the fundamental rights of every human being*" (8).

The declaration of the human right to health does not have any meaning if we do not recognize the intimate link between ecosystem quality and our health and well-being? No life can be healthy without a healthy environment. Without the right to a healthy environment, how can we fulfill our promise of a healthy life? This uncoupling of the right to health and the absence of a right to a healthy environment demonstrates the inadequacies of the current system of rights and the necessity to craft out an integrated approach. (9).

As early as 1972, the international community considered the links between human rights and environmental quality. The Stockholm Declaration 1972 included one of the earliest references to the links between human rights and the environment. The first principle of the Stockholm Declaration states that: "*Man has the fundamental right to freedom, equality and adequate conditions of life, in an environment of a quality that permits a life of dignity and well being, and he bears a solemn responsibility to protect and improve the environment for present and future generations.*" (10). For reasons better known to the contracting parties the initial emphasis on human rights *vis-à-vis* environment remained in a state of suspension for almost two decades. Even the UN conference on Environment and Development held in Rio de Janeiro in 1992 tactfully avoided the terminology of "rights" altogether. The Rio declaration proclaimed that human beings are at the centre of concerns for sustainable development and are entitled to a healthy and productive life in harmony with nature. Thus the link between human rights and the environmental quality remained ambiguous.

(II) POST STOCKHOLM DEVELOPMENTS

In a world with a steadily rising population, deteriorating environment and shrinking resource base it is very difficult or rather impossible to defend human rights as enshrined in the United Nations Charter. It is obvious that many economic social and cultural rights, including freedom from hunger, the right to an adequate standard of living and the right to the enjoyment of physical and mental health cannot be secured in localities where much of the population is living on or below the poverty line. The realization of the necessity of economic development prompted the declaration on the Right to Development, which was adopted by the U. N. General Assembly in the year 1986 AD. The 'Development' was recognized as a comprehensive economic, social, cultural and political process, which aims at continuously improving the well-being of the entire population and of each individual. The Right to development includes individual's right to the benefits of development, equality of access to basic resources, education, health services, food, housing, employment, and a fair distribution of income. (11).

The 'African Charter on Human and People's rights' tabled at the 18th Summit meeting of the Organization of African Unity (OAU) in Nairobi In June 1981 contained provisions relating to right to a *general satisfactory environment favourable to development*. It came into force in October 1986. The Convention on the Rights of the Child (1989) in article 24(2) (c) also requires State parties in the matter of combating disease and malnutrition to take into consideration, '*the damage and risks of environmental pollution*'. The Final Report of the Special Rapporteur on Prevention of Discrimination and Protection of Minorities listed over 15 rights relative to environmental quality. Some of these include: the right to freedom from pollution, environmental degradation and activities which threaten life, health or livelihood; protection and preservation of the air, soil, water, flora and fauna; healthy food and water; a safe and healthy working environment.

The Draft Declaration on the Rights of Indigenous Peoples also places environmental protection of peoples' traditional territories at the center of protection of their human rights. It recognizes their "distinctive and profound relationship with their lands" and provides that they have the "collective and individual right to be protected from cultural genocide, including the prevention and redress for . . . dispossession of their lands, territories, or resources." Article 11 of the Additional Protocol to the Inter-American Convention on Human Rights (1994) popularly known as the San Salvador Protocol, states that: everyone shall have the right to live in a healthy environment and to have access to basic public services; the state parties shall promote the protection, preservation and improvement of the environment.

(III) THE RIO DECLARATION

Although avoiding the terminology of "rights" Chapter 6 of Agenda 21 adopted at UN Conference on Environment and Development 1992 at Rio de Janeiro, in is entirely devoted to "protecting and promoting human health condition", while the Rio Declaration itself (Principle 1) proclaims that human beings are entitled to a healthy and productive life in harmony with nature and provides that states should effectively cooperate to discourage or prevent the relocation and transfer to other states of any activities and substances that, *inter alia*, are found to be harmful to human health (Principle 14).

(IV) DRAFT DECLARATION OF PRINCIPLES ON HUMAN RIGHTS AND THE ENVIRONMENT (1994)

It was under the auspices of United Nations that the first ever 'Declaration of Human Rights and Environment' was drafted in Geneva on May 16, 1994. The Draft Declaration is the first international instrument that comprehensively addresses the linkage between human rights and the environment. It demonstrates that accepted environmental and human rights principles embody the right of everyone to a secure, healthy and ecologically sound environment. The Draft Declaration describes the environmental dimension of established human rights, such as the rights to life, health and culture. It also describes the procedural rights, such as the right to participation necessary for realization of the substantive rights. It was appended to the Report of the UN Special Rapporteur on Human Rights and the Environment. The Report was presented to the Sub-Commission on Prevention of Discrimination and Protection of Minorities at its 46th Session (UN Doc, E/CN.4/Sub.2/1994/9). The declaration affirms:

- The human rights to healthy safe and secure environment free from pollution and degradation.
- The right to enjoyment of natural ecosystems and their rich biodiversity in just and equitable manner
- The right of humans to lead a dignified life and fulfill their legitimate needs.
- The rights of future generations to fulfill their own needs
- The right to environmental information, awareness, education and participation in environmental decision-making.

The declaration covers the duties also which apply to everyone. It makes protection, preservation of the environment and prevention environmental harms the duty of every one including individuals, governments, international organizations and multinational corporations. The States are directed to avoid using environment as means of war and shall respect international law for the protection of environment.

In spirit, the draft declaration tends to link the right to 'development' to the right to 'safe and clean environment', which should be considered not only at the level of individuals, but also at community, national and global levels. The development has to be so regulated that it does not deteriorate the environmental quality beyond limits, which have not been defined! Much scientific uncertainty prevail today regarding the 'thresholds' or limits below which the environmental quality shall be considered safe or healthy. Therefore, the levels of environmental quality beyond which it must degenerate to amount to the breach of human rights have not been precisely defined in the draft declaration. Although this instrument is non-binding legally, many national courts have used the Draft Declaration as a basis for decisions on environment matters and have found legal support in the Draft Declaration in deciding in favour for the protection of the fundamental right to a healthy environment.

QUESTIONS

1. Trace the emergence of worldwide concern for human right to a healthy environment.
2. What do you understand by human rights? How do human rights come in fonflict with environmental imperatives in third world countries?
3. Why it is difficult to defend the UN covenant on economic, social and cultural right.
4. Give salient features of the draft declaration on the principles of human rights and environment of 1994.

25

Chapter

Acquired Immunodeficiency Syndrome (AIDS)

Acquired immunodeficiency syndrome (AIDS) is a disease caused by Human Immunodeficiency Virus (HIV). The disease is spreading at an alarming rate. The stigma, prejudice, fears and silence that surrounds AIDS makes it a difficult problem to address. Today, AIDS confronts the humanity as a serious public health hazard and a challenge to the mankind (1).

We are surrounded by a variety of life which live as parasites and cause diseases of various types like virus, bacteria, protozoans, fungi etc. These organisms constantly invade our bodies through respiratory tract, the gastrointestinal track or through cuts and wounds on our bodies. Our bodies have to work and fight these invaders constantly to keep us healthy, free of diseases. A complicated system of defense responses, the **Immune system**, helps us to repel disease-causing organisms (pathogens). Immunity from disease is actually conferred by two cooperative defense systems, called nonspecific, innate immunity and specific, acquired immunity. Nonspecific protective mechanisms repel all microorganisms equally, while the specific immune responses are tailored to particular types of invaders. Both systems work together to prevent organisms from entering and/or proliferating within the body. These immune mechanisms also help eliminate abnormal cells of the body that can develop into cancers. In absence of this versatile defense mechanism, our body becomes an easy prey to the disease causing organisms. AIDS virus weakens and destroys the immune system in our bodies. In fact, human immunodeficiency virus does not kill, it simply leaves the subject defenseless against the army of invading pathogens, which finally take their toll (2).

(I) IMPORTANT FEATURES OF AIDS

Acquired Immune Deficiency Syndrome (AIDS) has been defined as any sign or symptom of acquired cellular immunodeficiency in persons having confirmatory serological and/or viral evidence of infection with Human Immunodeficiency virus.

1. An HIV positive person, carries the HIV in his body. The virus damages the immune system, the part of the body, which fights against infections of various types.
2. With time, the immune system becomes very weak. Actually, it is this stage, which is referred to as AIDS. It is the last stage of HIV infection.
3. Symptoms of AIDS include Persistent cough, Generalized pruritic dermatitis, Oropharyngeal candidiasis, Chronic progressive and disseminated Herpes Simplex and Generalized lymphadenopathy, Recurrent Herpes Zoster, Weight loss, Chronic diarrhoea and Prolonged fever.
4. Nobody knows exactly when a person with HIV will become sick of AIDS. Many people who are HIV positive have stayed healthy for years together. Generally, AIDS can take 7-12 years to develop after the initial infection with HIV.

5. Human immunodeficiency virus is transmitted through semen and vaginal fluids, infected blood and blood products and from the infected mother to her baby before birth, during birth, or through breast milk.

(II) THE VIRUS AND COURSE OF INFECTION

The main cellular target of HIV is a special group of white blood cells critical to the immune system known as Helper T-lymphocytes, or helper T-cells. Helper T-cells play a key role in normal immune responses by producing factors that activate almost all other immune system cells. These include B-lymphocytes, which produce antibodies needed to fight infection; cytotoxic T-lymphocytes, which kill cells infected with a virus, and macrophages and other effector cells, which attack invading pathogens. AIDS results from the loss of most of the helper T-cells in the body (3).

HIV is a retrovirus, the genetic material of which consist of RNA (instead of DNA) surrounded by a lipoprotein envelope. HIV cannot multiply on its own and instead relies on the machinery of the host cell to produce new viral particles. Once the virus has infected a T cell, HIV copies its RNA into a double-stranded DNA copy by means of the enzyme, reverse transcriptase. Because reverse transcriptase lacks the 'proof-reading' function that most DNA synthesizing enzymes have, many mutations may arise as the virus replicates, further hindering the ability of the immune system to combat the virus. Because of the high rate at which the genetic material of HIV mutates, the virus in each infected individual is slightly different. Genetic variants of HIV have been categorized into several major subtypes, which have different geographical distribution. These mutations allow the virus to evolve very rapidly, approximately one million times faster than the human genome evolves. This rapid evolution allows the virus to escape from antiviral immune responses and antiretroviral drugs. By constantly changing the structure of its predominant surface proteins, the virus can avoid recognition by antibodies produced by the immune system (4;5).

This is followed by integration of the viral genome into the host cell DNA. Integration may occur at any accessible site in the host genome and results in the permanent acquisition of viral genes by the host cell. Under appropriate conditions, these genes are transcribed into viral RNA molecules. Some viral RNA molecules are incorporated into new virus particles, while others are used as messenger RNA for the production of new viral proteins. Most infected cells die quickly (in about one day). The number of helper T-cells that are lost through direct infection or other mechanisms exceeds the number of new cells produced by the immune system, eventually resulting in a decline in the number of helper T-cells. Physicians follow the course of the disease by determining the number of helper T-cells in the blood of the infected persons. This measurement, called the CD-4 count, provides a good indication of the status of the immune system. Physicians also measure the amount of virus in the bloodstream – i.e. the viral load – which provides an indication of how fast the virus is replicating and destroying helper T cells (4;5).

The course of HIV infection involves three stages: Primary HIV infection, Asymptomatic phase, and AIDS. During the first stage, the transmitted HIV replicates rapidly, and some persons may experience an acute flu-like illness that usually persists for one to two weeks. During this time a variety of symptoms may occur, such as fever, enlarged lymph nodes, sore throat, muscle and joint pain, rashes, and malaise. Standard HIV tests, which measure antibodies to the virus, are initially negative because HIV antibodies generally do not reach detectable levels in the blood until a few weeks after the onset of the acute illness. As the immune response to the virus develops the level of HIV in the blood decreases.

The second phase of HIV infection, the Asymptomatic period, lasts generally for 8-12 years. During this period, which is a symptom-less phase, the virus continues to replicate, there is a slow decrease in the number of helper T cells. When the helper T-cells-count falls to about 200 cells per microlitre of blood (in an uninfected adult it is typically about 1,000 cells per microlitre), patients begin to experience opportunistic infections, i.e., infections that arise only in individuals with a defective immune system. This is AIDS, the final stage of HIV infection. The most common opportunistic infections are Pneumocystis carinii pneumonia, tuberculosis, Mycobacterium avium infection, herpes simplex infection, bacterial pneumonia, toxoplasmosis, and cytomegalovirus infection. In addition, patients can develop dementia and certain cancers, including Kaposi sarcoma and lymphomas. Death ultimately results from the relentless attack of opportunistic pathogens or from the body's inability to fight off malignancies (6;7).

(III) THE MAGNITUDE OF THE PROBLEM

The pandemic nature of and the magnitude of health problems associated with HIV infections have been well recognized. The serological evidence of the virus infections first became available from North America and Europe in 1970. Since then the virus has spread very rapidly. By July 1992, well over 5,00,000 cases of AIDS had been reported to the World Health Organization in 168 countries. By 2002 AD about 42,000,000 people in world were estimated to be either HIV positive or had AIDS. The number could be higher as the veil of secrecy, silence and fear, which envelops the disease, hinders a true appraisal of the problem (1).

The first HIV positive case in India was detected in the year 1986 among the prostitutes of Chennai. The first case of AIDS was recorded in May 1986. The patient had received infection by the blood transfusion given during coronary by-pass surgery. According to the National Aids Control Organization (NACO), the estimates of HIV positive cases for the year 2001 AD and 2002 AD have been worked out to be 3.97 million and 4.58 million respectively, in the adult population (15-49 yrs age group). This means that there has been a rise of about 6.6 lakh infections in a period of about a year, which is an alarming figure (8;9).

(IV) HOW HIV TRANSMISSIONS CAN BE PREVENTED

The disease spreads through infected blood and blood products, infected needles, from infected mother to the babies, before, during or after the birth and injectable drug abuse. Global statistics point out that unprotected sexual intercourse with HIV positive persons accounts for nearly 75-80% of transmission of the disease. Injections with contaminated blood and use of non-sterilized equipments about for 3-5% of the infections and 5-10% of worldwide infections come from the reuse of contaminated needles by intravenous drug users (2;8).

A number of base-less believes about the transmission of Human Immunodeficiency Virus are prevalent in the society which often make the life of an HIV infected person difficult, desperate and unbearable. It must be borne in mind that HIV is NOT transmitted through:

- Through air – the infection cannot spread through sneezing coughing or breathing.
- Through casual physical contact – like touching, caressing, hugging or kissing.
- Through water – using common bath tubs or swimming pools
- Through toilets
- Through mosquitoes
- Through using the same utensils, phones etc.

An HIV positive individual is a perfectly safe person to live with as long as you avoid contact with semen/vaginal fluids, needles and other injectable instruments used by the infected person and use of the infected person's blood for transfusion. The transmission of HIV can be effectively prevented:

- By practicing safer sex – confining to a single partner, use of condoms etc.
- By using clean, sterilized needles and avoiding unnecessary skin perforation.
- By avoiding the use injectable drug equipment used by another person.
- By avoiding blood transfusion, which should be resorted to only when necessary while only properly screened blood should be used.

(V) HIV/AIDS AND HUMAN RIGHTS

Because of the dreaded nature of the disease caused by the HIV people suffering with this disease are treated with indifference, indignation and apathy. People tend to avoid them for the fear of contracting chance infection. They are treated as a 'gone case' and are often forced to live in isolation. However, such a treatment amounts to violation of Human Rights. The provisions of human rights usually violated are:

- Denial of health care and treatment.
- Denial of employment.
- Denial of various services including insurance, medical benefits etc.
- Denial of access to information
- Denial of access to legal remedies.
- Discrimination of against children of HIV positive parents.
- Ostracisation of HIV positive persons from community and family.
- Prevention of children from playing and interacting with other children.

The provisions International Human Rights Instruments can play an important role in issues related with HIV positive persons and people suffering with AIDS. These norms can be used to erect procedural, institutional and social mechanisms to counter the epidemic. In June 2001, the Heads and Representatives of 189 States met at the United Nations General Assembly, Special Session on HIV/AIDS and a Declaration of Commitment was adopted by the delegates. The Declaration provided a framework for an extended response to the global HIV/AIDs epidemic. In September 1996, the second International Consultation on HIV/AIDS and Human Rights was convened by the United Nations group and the Office of the U. N. High Commissioner for Human Rights, which lead to the formulation of International Guidelines on HIV/AIDS and Human Rights. The Guidelines addressed multisectoral responsibilities and accountability, including improving the role of the governments and private sector, placing more stress on the duty of the States to engage in law reforms and identifying legal obstacles so as to form an effective strategy for the care of the diseased and prevention of the epidemic. The experts who gathered at the meeting recognized that:

- The protection of human rights is essential to safeguard human dignity in the context of the disease and to ensure an effective, right based response HIV/AIDS. An effective response requires the implementation of all human rights in accordance with existing international human rights standards.
- A right based effective response to the disease epidemic involves establishing appropriate governmental institutional responsibilities, implementing law reforms and support services and promoting a supporting environment for groups vulnerable to the disease and those living with disease.

(VI) THE INDIAN EFFORTS TO CONTROL HIV/AIDS

In India, a National Human Rights Commission has been established and a National AIDS Prevention and Control Policy have been formulated. The Government recognizes the fact that without the protection of human rights people who are vulnerable and afflicted with HIV/AIDS, the response to HIV/AIDS epidemic will remain incomplete. In order to implement an effective response and control measure the Government adopted following right based strategy:

1. The government will review periodically and reform criminal laws and correctional system to ensure that they are consistent with international human right obligation and are not misused in the context of HIV/AIDS targeted against the vulnerable groups.

2. The Government will strengthen anti-discrimination and other protective laws that protect vulnerable groups, people living with HIV/AIDS and people with disabilities from discrimination in the public and private sectors, ensuring privacy, confidentiality and ethics in research involving human subjects, emphasize education and conciliation and provide for speedy and effective administration and civil remedies.
3. Government will ensure widespread availability of qualitative prevention measures and services, adequate HIV prevention and care information and services.
4. The Government will ensure support services that will educate people affected by HIV/AIDS about their rights, provide legal services to enforce these rights and develop expertise on HIV related legal issues.
5. Government will promote distribution of creative, educational, training and media programs explicitly designed to change the attitude of community towards discrimination and stigmatization associated with HIV/AIDS.
6. The government will in collaboration with and through the community, promote supportive and enabling environment for women, children and other vulnerable groups by addressing underlying prejudices and inequalities through community dialogues, specially designed social and health services and support to community groups.
7. The Government will co-operate with all relevant programs and agencies of the United Nations System, to share knowledge and experience concerning HIV related human rights issues and would ensure effective mechanisms to protect human rights in the context of HIV/AIDS at international levels.

QUESTIONS:

1. What is AIDS? Why it is a serious disease?
2. Give important features of AIDS .
3. What is the primary cellular target of human immunodeficiency virus? Discus the course of infection of the disease.
4. How the transmission of AIDS can be prevented?
5. How the treatment meted out to AIDS/HIV positive patients violates Human Rights?
6. What are the efforts made by Government of India to control HIV/AIDS

26

Chapter

Disaster Management, Resettlement and Rehabilitation

Even the enormous scientific and technological progress of our society has not been able to reduce the loss of lives and property due to natural or manmade disasters. In fact, the human toll and economic losses have mounted. There has been an increase in the number of natural disasters over the past years, and with it, increasing losses on account of crowding, population growth and urbanization. Devastations in the aftermath of powerful earthquakes that struck Gujarat, El Salvador and Peru; floods that ravaged many countries in Africa, Asia and elsewhere; droughts that plagued Central Asia including Afghanistan, Africa and Central America; the cyclone in Madagascar and Orissa; and floods in Bolivia are global events in recent memory. Although many of these disasters represent natural phenomena, human activities also contribute to enhance their intensity and frequency up to some extent. These disasters whether man-made or natural cause extensive damage and require proper management so that the loss of property, buildings and other structures, human and animal lives could be minimized. However, what is disturbing is the knowledge that these trends of destruction and devastation are on the rise instead of our efforts to keep them check.

(I) MAJOR PHASES OF DISASTER MANAGEMENT:

Disaster management (or emergency management) is the discipline of dealing with and avoiding disaster-risks (1). It involves preparations for disaster before it occurs, disaster response like emergency evacuation, quarantine, mass decontamination etc as well as supporting, and rebuilding the society after natural or man-made disasters have occurred. Usually any disaster management is the continuous process which prepares individuals, groups or communities to manage hazards and minimize the damages to life and property caused by natural or manmade disaster.

The nature of management depends on local, economic and social conditions. It has long been observed that the process of emergency management must include long-term work on infrastructure, public awareness, and even human justice issues (2). Activities at each level of social organization, individual, groups, community have to be co-ordinated to rescue people affected by the disaster and provide effective relief to the people. Effective emergency management relies on thorough integration of emergency plans at all levels of government and non-government machinery. The process of emergency management involves four phases:

1. Risk assessment and vulnerability analysis
2. Mitigation,
3. Preparedness,
4. Response,
5. Recovery.

1. Risk assessment and vulnerability analysis:

A very important aspect of the today's disaster management strategies is to identify hotspots where a particular type of disaster is likely to occur. Today we have gathered plenty of knowledge of earth's crusts and it's interior to identify areas of earth's surface where volcanoes may erupt or earthquakes may occur. We have enough knowledge of earth's topography, the cover of vegetation and human habitations to predict with adequate precision the magnitude and timing of the flow of floodwaters deposited at higher altitude. Meteorologists who keep a close watch over pressure and temperature pattern all round the clock and all over the world can predict the likelihood of torrential rains, hailstorms and cyclones. They can issue warnings well in advance about the impending hailstorm or cyclone with a fair degree of precision. Technologies like Remote sensing, Satellite communication, Global-positioning system (GPS), Geographical Information System (GIS), Global meteorological Surveys etc play a vital role in our efforts to anticipate, pinpoint and estimate the extent of the likely damage. These institutions should be strengthened and made more efficient.

2. Mitigation

Mitigation efforts involve prevention of the incidence from developing into disasters altogether, or to reduce the damages to life and property due to disasters when they develop. This phase differs from others because it places more emphasis on long term strategies to reduce or eliminate the risk of hazards from developing into disasters. When applied after the disaster has already occurred the implementation of mitigation measures is considered as a part of recovery process (1). Strategies adopted to reduce or eliminate the damages caused by a disaster can be two types:

1. **Structural measures** which involve technological solution like erection of flood levees, spurs or embankments to contain flood water, provision of reinforcement and other structural adaptation to prevent buildings from collapsing in case of earthquakes, installation of fire fighting instruments and provision of escape routes in case of fires etc. Some structural mitigation measures may have adverse effects on the ecosystem.
2. **Non-structural measures** include legislation, land use planning, insurance etc. such as imposing restriction on construction of high-rise building in earth-quake prone areas, provision of buffer zone to accommodate flood waters etc. An effective personal insurance or house-hold insurance, crop-insurance can provide much relief to people affected by the disaster.

An activity which should be undertaken before the mitigation measures are adopted is the identification of risks. Physical risk assessment refers to the process of identifying and evaluating hazards and should combine both the probability and the level of impact of a specific hazard (1). It is only after identification and evaluation of the hazard that proper measures may be adopted to minimize the damages caused or to eliminate the chances of occurrence of the disaster. Although not always suitable, mitigation is the most cost effective method for reducing the impact of the disaster. It also includes regulations regarding evacuation, sanctions against those who refuse to obey and communication system to communicate the potential risks to the public (3)

3. Preparedness

As disaster can strike anywhere and emergency could arise at any time, the machinery of disaster management has to be in a state of preparedness to deal with it at any time or at any place. Simple well thought-out plans of action have to developed and well rehearsed before the occurrence of the disaster so that in emergency there is little confusion and people are aware as to what has to be done. In general most of the preparedness measures include:

- Plans for communication with easily understandable terminology and methods.
- Proper maintenance and training of emergency services, including mass human resources such as community emergency response teams.

- Installation and exercise of emergency population warning methods combined with emergency shelters and evacuation plans.
- Stockpiling, inventory, and maintenance disaster supplies and equipment.
- As professional emergency workers are often overwhelmed, a battery of trained organized and responsible volunteers drawn from civilian population are often an asset to the rescue efforts. Such organizations should be developed in this phase.

A very important aspect of the phase of preparedness is casualty prediction. A rough estimate of injuries or deaths to expect for given type of disaster gives rescue and relief workers an idea of the resources required for a particular type of event. Emergency plans in the planning phase should be flexible, and cover all possible contingencies. Depending on the region, nature of disaster, severity of episode - municipal or private sector emergency services can rapidly be depleted and heavily taxed. Non-governmental organizations that offer desired resources i.e. transportation of displaced home owners to be conducted by local school buses, evacuation of flood victims to be performed by mutual aide agreements between fire departments and rescue squads etc, should be identified early in planning stages, and practiced with regularity (4).

4. Response

The response phase includes the mobilization of the necessary emergency services at the first intimation of the occurrence of the disaster. This is likely to include a first wave of core emergency services, such as police, fire-fighters, medical help and ambulance crews etc. These should reach as soon as possible since in many cases delays or time lost could mean lives lost. They may be supported by a number of secondary emergency services, such as specialist rescue teams. A well rehearsed emergency plan developed as part of the preparedness phase enables efficient coordination of disaster management efforts which enables an effective search, rescue and relief operation to commence at an early stage (3;4).

5. Recovery

The recovery phase aims at restoration of the affected area to its normal state. It differs from the response phase in that it is concerned with issues and decisions after immediate needs of the people affected by the disaster has been addressed. Recovery efforts are primarily concerned with activities such as rebuilding the destroyed property, re-employment, and the repair of other essential infrastructure such as water & power-lines, restoration of rail and road links etc. During the recovery phase many mitigative measures may also be implemented which would be otherwise unpopular. People are more likely to accept these when a disaster is fresh in memory (4).

(II) A NEW APPROACH TO DISASTER MANAGEMENT:

The traditional practice of disaster management in most of the cases has been limited to "calamity relief" which involved provision of rescue and relief operation only when the disaster has occurred. However, the impact of major disasters cannot be mitigated by the provision of immediate relief alone, which is the primary focus of calamity relief efforts. With the kind of economic losses and developmental setbacks which many countries have been suffering year after year, it is apparent that development process has to incorporate disaster prevention and mitigation aspects in the development planning (2;4;5).

Over the past couple of years, the Government of India has brought about a paradigm shift in the approach to disaster management. The new approach stems from the conviction that development cannot be sustainable unless disaster mitigation is built-in into the development process. Another aspect of the approach is that mitigation has to be multi-disciplinary spanning all sectors of development. The new policy also emanates from the belief that investments in mitigation are much more cost effective than expenditure on relief and rehabilitation (3;5).

(III) FLOODS: MANAGEMENT STRATEGIES:

Places away from equator are characterized by a marked seasonality in the pattern of precipitation. Rains are confined to three or four month of the year which often deposit large amounts of water at higher altitudes. This water trickles down, forms streams and rivers and constitutes the flood flow, which rushes down to the sea. In valleys the flow spills over banks causing enormous floods and devastating large areas of agricultural land and human habitations.

An important reason for raised intensity and frequency of floods is extensive deforestation. On bared land, there is nothing to impede the flow of water, retain it a little longer and ensure its gradual discharge so that the channels down streams are not overwhelmed. Flood flow carry large quantities of silt and debris, which is, deposited down-streams raising the bottom of the river channels, which become shallow. It is largely human activity, which is responsible for recurrence, and raised intensity of floods, which have assumed the proportions of a disaster today (6). Flood management strategies today, largely involve efforts:

1. To retain the water deposited by precipitation at higher altitudes in dams and reservoirs from where it could be released gradually and flow of the rivers could be regulated. However, instance of a breach in the dam or rainwater overwhelming the storage capacity of the reservoirs are also many. Reforestation and proper watershed management have been suggested as a better alternative, which could withhold large quantities of water, recharge underground water table, check soil erosion, inhibit landslides and prevent the river channels downstream from turning shallower every year. And at the same it will help in maintaining a sustained flow of water in the rivers during the drier months of the year. At many places, reforestation programs in the catchment area of rivers have already been undertaken.
2. To improve flow conditions in the channel, dredging and cleaning operations and construction of embankments and walls to keep human establishment protected while the flood-flow passes quickly away to the sea. This may include straightening the meandering course of the river channels, bypassing curves and bends to improve the gradient of the flow and maintenance of adequate buffer zones and wetlands, which could moderate the flood fury.
3. To ensure proper management of flood plains it's zoning and effective preparedness for fighting floods, adequate disaster relief measures and improved public health facilities.
4. To install flood forecasting and warning system as well as flood insurance.

(IV) CYCLONES: MANAGEMENT STRATEGIES

Cyclones are low-pressure, calm oval or circular, moving area, the 'eye of the storm' in the centre surrounded by a high-pressure zone outside. It forms a powerful swirling column that may measure up to 6 to12 kms in height and 250-1000 kms in diameter, which could move 250-2500 kms per day devastating everything on the surface. The cyclones originate at the front separating two air masses of different temperature, density and direction. While moving over oceans they collect moisture and energy from warm waters, which form intense clouds as warm air, rises over the cold air causing torrential rains and high-speed winds, which may attain velocities over 120 kms per hour. They are referred to by different names in different parts of the world. In Pacific ocean and China they are called 'typhoons', in North Atlantic and Indian ocean they are called 'cyclones', in America they are known as 'hurricanes'. The main danger from cyclone stems from the very strong, high-speed winds, torrential rains and high storm tides. Much of the casualties are caused by inundation of coasts from high storm tides, which are often followed by heavy rains and floods (7).

Science and technology has enabled us to know broad range or the length of coastline likely to be affected, about 20-24 hours before the cyclone strikes the coastline. There should be an efficient public notification system to take the warning to the residents of the locality as early as possible so that people may get some time to prepare or leave for safer place. Often this warning fails to reach the people in distant localities who are suddenly trapped in difficult situations. During the storm or a cyclone, communication system is usually the first casualty. Many distant areas are cut off completely from the centre administering rescue and relief operations. With no one aware of their plight, chances of relief or help reaching these localities also become meagre. Mobile wireless communication system is a great asset in such situations which should be made available to all communities dotted along the extensive coastline in India.

As it is well known that a cyclone may strike anywhere along the long coastline of our country during pre-monsoon or post monsoon months, we should maintain adequate buffer zones, develop wetlands, shelter belts and construct appropriately designed houses which can withstand the fury of strong winds, torrential rains and high storm tides. Proper disaster management strategies such as development of cyclone shelter, training and education to fight the ravages of the cyclone, medical and other infra-structural facilities and adequate rescue and relief operations etc should be adopted when the disaster strikes.

(V) EARTHQUAKES: MANAGEMENT STRATEGIES

An earthquake is a the movement of earth's surface caused by endogenic disturbances deep inside earth's crust such as movement of molten rock material, faulting, folding, warping, and shuffling of smaller or larger tectonic plates etc. The movement of earth's surface varies from very faint tremours detected by instruments only or very sensitive people to wild vibratory motion capable of shattering buildings apart and causing gaping fissures in earth's surface.

The place of origin of earthquakes, called '**focus**' usually lies deep inside the earth with the depth varying from 20 kms to 650 kms. The point, which lies immediately above the focus on earth's surface, is referred to as '**epicentre**'. An enormous amount of energy is released during an earthquake as seismic waves, which may be of three types: Primary or pressures waves (P-waves), secondary or transverse waves (S-waves) and Long waves or surface waves (L-waves). These waves are recorded and studied with the help of an instrument called **Seismograph**. It is the epicentre which is the first to record the seismic waves. The magnitude of energy released by an earthquake is usually measured on Richter scale, which ranges between 0 to 9. The damage caused by an earthquake begins at 5.0 on the Richter scale, which cause slight damage to buildings, and rises to magnitudes as high as 8.9 at which nearly total destruction could occur (8).

An analysis of the situation of major earthquakes all over the world shows that there are wide belts or zones, corresponding to the margins of major tectonic plates along which most of the earthquakes occur. These broad belts are:

- Circum-Pacific Belt which is a wide zone skirting the Pacific ocean covering the Western region of North and South America and the eastern margin of Asia, Australia and Northern parts of Antarctica.
- Mid-Atlantic belt covering the mid-Atlantic ridge and its extensions in the Atlantic Ocean.
- Mid-continental belt representing epicentres located along the Alpine-Himalayas chains of Eurasia and North Africa extending as a wide belt along the course of river Nile deep into the fault zones of East Africa.
- A wide belt extending from the Southern tip of Africa right up to Southern coast of Australia with extensions in the Indian oceans right up to the Arabian Sea and the bay of Bengal.

Indian earthquakes, which occurred in Gujarat and along the Himalayas and its foothill zones, can be explained on the basis of plate tectonics. The Asiatic plate is moving southward whereas the Indian plate is moving northward and hence the northern margins of the Indian plate, which run along the Ran of Kutch, Gujarat, and northern region of the Aravalis, Vindhyachal and parts of Bihar is being subducted below the Asiatic plate. This collision of Asiatic plate and Indian plates has resulted in folding and faulting in region of Himalayan foothills and is responsible for the gradual rise of Himalayas at a rate of about 50 mm per year. On the other hand, the process of expansion of the floor of Indian Ocean has caused deformity, fold and faults in the Indian plate, and is probably responsible for earthquakes in the peninsular India particularly along its western margins. A list of earthquakes that have occurred in India since eighteenth century reveals that most of the earthquakes had their epicentres along the interface of Asiatic and Indian plates and along the western region of the peninsular India

Much of the damage caused by an earthquake when it strikes a populated area stem from slope failures, landslides in hilly areas and damage to buildings as well as other structures due to violent tremours. Man can do nothing to reduce the intensity and frequency of earthquakes. However, resort to appropriate earthquake mitigation measures such as properly designed earthquake resistant buildings and other structures can prevent many buildings from tumbling down resulting in reduced number of casualties. A quick and efficient rescue and relief service, adequate medical facilities and provision of other necessities required by the affected people can do much to reduce the damages caused to human life and property (9).

(VI) LANDSLIDES: MANAGEMENT STRATEGIES

Landslides are often recurrent phenomena in mountainous regions of the world. Gravitational forces constantly exert a downward pull over debris and the mass of soil and rocks, which may slide downward the moment their grip on the surface of the slope is loosened. Heavy rains, which may wipe away soil particles and small rock fragment or slight tremours are the usual cause landslides in most of the hilly regions. Rocks and debris sliding down a slope may gather other loose material, which comes in their way all of which are finally deposited when level ground is reached. Villages or towns, which happen come in the way of the avalanche, are buried under the debris and rock fragments. Although weathering, changes in geology and climate may also cause landslides it is human activity, which has raised the frequency and in many cases the destructive potential of landslides. Extensive deforestation, clearance of hill slopes in stepwise manner to grow food crops, intensive building activity and road construction etc have added considerably to the destabilization forces.

To minimize the threat of landslides, long-term land use planning is important. Landslide mitigation practices today involve prevention of landslide by modification of topography, construction of ridges and props, provision of safer drainage, control of soil erosion by making check dams terracing, jute and coir nettings, promotion of plant growth and reforestation. One of the most effective, though time taking, landslide mitigation measure is the development plant cover on loosened rock fragments and hill-soils. Plants roots although capable of tearing rock fragments apart also bind the loosened rock material. Plant-roots weave a network that prevents the rock fragments from sliding down under the influence of gravity. The organic matter derived from plant parts provides humus to the soil, improves its texture and holds the soil particles from being washed away by rapidly flowing waters while improving its fertility at the same time (10).

It is important that most of the establishments, which are under the threat of landslides, should be equipped with adequate system of rescue and relief measures and medical facilities as far as possible. In the event of a landslide that is caused by earthquake tremours or heavy rains the lines of communication are usually disrupted. Communication may be possible through wireless but the roads are often blocked by fallen soil and rock materials at many places, which takes time to be cleared. Like that of earthquakes, in a landslide there are often a large number of physical injuries. People are crushed under buildings or flowing boulders and rock fragments. First aid and physicians who could take care of fractured bones and bleeding victims of the disaster is the primary need in the case of landslides.

(VII)RESETTLEMENT AND REHABILITATION

In a democratic country, which is governed by the people for the welfare of the people resettlement and rehabilitation assumes an added importance. People who have suffered damages or have substantially lost their assets due to a natural disaster, the responsibilities of a Welfare State could be limited to provision of adequate or rather legitimate help to enable them to re-establish in their lives and begin earning their own bread. Victims of a cyclone or of floods may return to their homes and to their lands and continue their vocation with a little help, financial or otherwise. However, the people who have to sacrifice their assets for the benefit of the entire region or the country as a whole deserve a just and adequate compensation as well as resettlement and rehabilitation package. Thus, resettlement and rehabilitation activities may be grouped into two categories:

(1) Resettlement and rehabilitation of the victims of the disaster:

Disasters like floods, cyclones, earthquakes and landslides leave a large number people bereft of their homes, robbed off their agricultural crops, domestic animals and other assets on which their lives depend. The immediate need is to feed these people, provide shelter, medicines and other necessities. It is usually not very difficult to cater to their demands since help pours in from all corners of the country and even from all over the world. The primary requirements of the area devastated by floods, a cyclone, an earthquake or a landslide are:

- Restoration of means of communication and infrastructural facilities like power supply, water supply system, roads, railways etc. This ensures overall development of the area and its even distribution.
- Helping the people through loans, grants, subsidies etc to arrange for their necessities themselves and also to arrange for employment of the people in development works and project so that they could earn their own living.
- Seek co-operation of the affected people and provide for their social and cultural needs while ensuring that the 'help' is properly utilized or utilized for the purpose for which it was awarded.

Many, social institution, government and non-government organization come forward to help the affected communities for resettlement and rehabilitation. Though well intentioned, they tend to work in isolated packets helping people here and there. Therefore, the most important aspect of the resettlement and rehabilitation are the identification of priority areas, just and equitable allocation of resources, a proper planning and/execution of the program.

(2) Resettlement and rehabilitation of people displaced by developmental projects:

A number of development activities such as major dams, irrigation projects, big industries, mining and processing operations force people to move to other places leaving their traditional houses, land holdings and other property behind. Their life support system is dismantled and productive assets are lost. Often they have to be relocated to environments where their capabilities fetch diminished returns and their skills are no longer useful. The competition for resources at these places may be greater and to make a living a little harder. The community structure and social circles may be weakened and the cultural identity, the traditional authority could be lost. Involuntary resettlement may cause severe long-term hardships to the people, impoverishment and environmental damage unless appropriate measures are taken and executed. The people who are uprooted from their traditional holdings, their environment, means of livelihood and the resources they have been using since ancient times have a right to adequate compensation and resettlement in such a way as to ensure that at least their former living standards and earning capacity is maintained.

The World Bank was first multilateral agency to formulate resettlement and rehabilitation policy that is enshrined in its document entitled "Involuntary Resettlement" adopted in June 1990. The resettlement policies followed by the Government of India as well as State Governments are nearly similar to World Bank's policy in intent and purpose while making appropriate adjustments to suite the local conditions. The policy makes resettlement and rehabilitation an integral part of all development project which should be taken up right from earliest stages of project preparation and its execution. It ensures that:

- Involuntary resettlement should be avoided or minimized as far as possible by examining all feasible project designs and alternatives.
- Where there is no option, resettlement plans should be developed and executed just as the development project. The displaced person should be provided with sufficient resources to invest for their well-being and given opportunity to share in the benefits from the project.
- Compensation for their losses at full replacement cost may be awarded prior to the movement of the people. They should be assisted and supported during the transition period.
- They should be helped in their effort to improve their living standards, earning capacity and productivity so that they may attain at least their former living standards.
- Proper opportunity of participation in planning and implementation of resettlement activity should be given to the people being displaced and attempts should be made to preserve their social organization and cultural institutions, as they were earlier.
- Efforts should be made to integrate the settlers economically and socially into the host communities, which should also be consulted before the planning of resettlement activities.

In India a large number of development project which mostly involve construction of huge water reservoirs have been undertaken which have displaced thousands of people or shall do so in near future. Till date, however, the resettlement and rehabilitation package offered to the people displaced by the project has been found to be grossly inadequate. In fact, little attention was paid to the affected people, apart from allotting a little often barren land in the case of projects started and completed three or four decades back of which Damodar River valley project is just an example. The plight of these people failed to make much impression on the powerful lobby of interests, which executed the project. Public awareness and strong opposition by people, scientists and environmentalists met with by a number of such development schemes have forced the policy makers to provide for a better resettlement and rehabilitation package to the displaced persons. Sardar Sarovar Project over Narmada, planned relocation of 33014 people in the first phase for which resettlement sites have been selected in Madhyapradesh and Gujarat. All of 30739 people affected by Indira Sagar project are to be resettled in Madhyapradesh. Similarly, nearly 4000 people affected in 61 villages by the Maheshwar Project have to be resettled elsewhere. The Tehri Dam in the outer Himalayas when completed will drown Tehri town and about 100 villages. There has been a very strong opposition to its construction. Resettlement and rehabilitation package offers to relocate the people to higher places and some have already been resettled. However, the problem is more difficult here as no cultivable land is available in the area. Resettlement and rehabilitations involves more than just allotting land to the affected people, which have been outlined in the Resettlement and Rehabilitation Policy above. In many cases, even today the basic aspects of the policy are being ignored.

QUESTIONS

1. What are the main phases of disaster management? Discuss.
2. Why more emphasis is being laid on the mitigation phase of disaster relief operations?
3. Discuss the strategies adopted to control recurrent floods in Indian plains.
4. What are strategies adopted to minimize the damages cause by earthquakes?
5. How landslides can be prevented and the damages caused minimized?

27

Chapter

Sustainable Development

Sustainability, in a broad sense, is the capacity to endure. For humans it is the potential for long-term improvements in wellbeing, which in turn depend on the wellbeing of the natural world and the responsible use of natural resources. Long-lived and healthy ecosystems like wetlands and forests are examples of sustainable biological systems. The chemical cycles circulate water, oxygen, nitrogen and carbon through the world's living and non-living systems, and have sustained life for millions of years. With rising population the demands on natural resources has multiplied. Natural ecosystems have declined and changes in the balance of natural cycles have had a negative impact on both humans and other living systems. In 1960 almost all countries in the world had more capacity than enough to meet their own demand. However, by 2005 the situation had changed completely. A number of countries were able to meet their needs only by importing resources from other nations (1). Rapid population growth and rising individual consumption has more than doubled humanity's demand on resources of the planet. There is now plenty of scientific evidence that humanity is living unsustainably. Returning human use of natural resources to level of sustainable limits will require a major collective effort.

In 1983, United Nations' General assembly established the World Commission on Environment and Development (WCED) under the Chairmanship of Cro Harlem Brundtland. A team of Scientists from all walks of life, from all over the world were pressed into action to study all aspects of environmental degradation and economic development. The Commission submitted its report in 1987 which was entitled "Our Common future". It was in this report that emphasis was laid on "**Sustainable Development**". The Commission emphasized that it is the pattern of development which humanity has been following, the greedy ways of our society, are responsible for causing massive degeneration of natural resources and pollution of environment which threaten to destroy the vital life support system on our planet.

(I) SUSTAINABLE DEVELOPMENT – DEFINITION:

The world commission on environment defined sustainable development as "the *development which provides for the needs of the present generation without compromising the ability of the future generations to meet their own needs*" (2). Although this definition is frequently quoted it is not universally accepted and has undergone various interpretations (3;4). It has often been criticized as a vague, ambiguous and a feel-good buzzword with little meaning or substance (5;6). The idea of sustainable development is sometimes also viewed as a synthesis of mutually contrasting concepts because development inevitably depletes and degrades the environment (7). Consequently some definitions either avoid the word development and use the term sustainability exclusively, or emphasize the environmental component, as in "environmentally sustainable development". The concept of living within the limits defined by environmental capabilities forms the basis of the IUCN, UNEP and

WWF definition of sustainability which is: "improving the quality of human life while living within the carrying capacity of supporting eco-systems" (8). The Earth Charter seeks to establish the values and direction needed to achieve sustainability. "We must join together to bring forth a sustainable global society founded on respect for nature, universal human rights, economic justice, and a culture of peace. Towards this end, it is imperative that we, the peoples of Earth, declare our responsibility to one another, to the greater community of life, and to future generations" (9).

(II) CHARTING A SUSTAINABLE COURSE

A commitment to meet the needs of present and future generations has various implications. It is by no means easy to carry out the obligation of the commitment in a divided world. "Meeting the needs of the present generation," means satisfying:

1. Human needs economically – including access to an adequate livelihood or productive assets, also economic security when unemployed, ill, disabled, or otherwise unable to secure a livelihood. It should provide for the legitimate needs of all people of the present generation and bring them to a reasonable standard of life.
2. Social, cultural, and health requirements of mankind – including a shelter which is healthy, safe, affordable, and secure, within a neighbourhood with provision for piped water, drainage, transport, health care, education, child development, and protection from environmental hazards. Services must meet the specific needs of children and of adults responsible for children (mostly women). Achieving this implies a more just distribution of income between nations and, in most cases, within nations.
3. Political aspirations of peoples – including freedom to participate in national and local politics and in decisions regarding the management and development of the home and neighbourhood

This has to be done in such a way that the vital life support system on our planet is not adversely affected so that its potential to serve our future generations is not compromised. Meeting such needs of our generation "without undermining the ability of our future generations to meet their own needs" means:

1. Reducing the use or wastage of non-renewable resources – including minimizing the consumption of fossil fuels and substituting it with renewable resources where feasible. Also, minimizing the wastage of scarce mineral resources (by reducing use, reusing, recycling, and reclaiming).
2. Sustainable use of renewable resources – including use of fresh water, soils, and forests in a manner, which ensures a natural rate of replacement.
3. Keeping within the absorptive capacity of local and global sinks for wastes – including the capacity of rivers to break down biodegradable wastes as well as the capacity of global environmental systems, such as climate, to absorb greenhouse gases.

We have already committed the world to a grave environmental change. The green house gases which we have emitted shall persist in the atmosphere for many centuries to come. The ozone hole which we have created shall take hundreds of years to repair. The lush green forests we have destroyed shall take many decades to regenerate. It has been the cumulative action, spread over several decades of so many people all over the world, which has caused the degeneration of natural resources and adverse changes in the environment of today. A versatile, multi-directional strategy is required to redress environmental problems both at regional and global levels. Agenda 21 highlights the need of such a strategy (10;11). It calls for concerted efforts and co-operation of individuals, small or large communities, governments and international agencies. However, there is by no means a general agreement among experts regarding the precise action or the ways, which should be adopted for the purpose (11;12). There are ample reasons for us to move ahead on at least three or four major issues, which confront the humanity today (13;14). These are:

1. Reducing waste generation, stabilizing chemical cycles and moderating the pace of global climatic change.
2. Eliminating the threat to the global food security.
3. Controlling the degeneration of biodiversity.
4. Stabilizing the world population.

(III) REDUCING WASTE GENERATION, STABILIZING THE CHEMICAL CYCLES, AND MODERATING THE PACE OF GLOBAL CLIMATIC CHANGE

There is little we can do about changes, which have already occurred in chemistry of our environment – the nature shall take care of these abuses in its own way. In order to avoid further changes we shall have to put restrictions on human activities, which cause these changes. These activities are, however, our day-to-day activities, which cannot be stopped altogether (14). In-order to moderate the adverse impact of human activity the following strategies have to be adopted:

1. Setting standards for effluents discharged in the environment.
2. Adopting more environment friendly technology and switching over to renewable & least pollutive sources of energy.
3. Avoiding concentration of polluting industries and raising assimilative capacity of natural systems.
4. Controlling Consumerism and minimizing waste production.

1. Setting standards for effluent to be discharged in the environment: This approach provides statuary standards for each polluting unit – be it an automobile or an industry. These standards specify permissible limits in terms of different parameters of effluent quality and the polluting unit is required to ensure that the effluent it discharges are within the specified limits. For example, wastewaters from Tannery should not have a BOD higher than 30 mg/l over 5-day period at 20°C. Tanneries are required to treat their liquid effluents in such a way that its BOD is reduced to levels below 30 mg/l over 5-day period at 20°C. Thus, management based on standards distributes the responsibility of reducing the load of pollution to the polluting units.

2. Adopting more environment friendly technology and switching over to renewable & least pollutive sources of energy: A much better way to restrict the pollution of environment is adoption of more environment friendly technology. For example, of the three kinds of fossil fuels, (coal, oil and natural gas) in common use these days, natural gas is the least polluting source of energy. Its use should be promoted. This is true for a number of industries. Alternative process which cause much less pollution are available for a number of industries and where they are not available further research and studies should be conducted to develop such methods and technology.

Global energy requirement which is largely met with conventional sources of energy, such as fossil fuels, nuclear fuels and fuel wood, creates massive pollution of environment. Alternatives sources of energy exist which could replace the use of these energy sources if not completely, at least partially. We are already using wind energy (wind-mills) and potential energy of running water to generate electricity which is a pollution free source of energy at many places. Wind energy, geothermal energy, energy of ocean-waves, energy from decomposing organic matter (gobar-gas plants) etc are newer sources of pollution free energy which have not been fully tapped. A huge potential to provide for energy needs of man lies unutilized, waiting to be harnessed. We have to develop skill and technology to use them. Switching over to these, shall drastically reduce the burden of polluting emissions which humans are currently adding to environment.

3. Avoiding concentration of polluting industries and raising assimilative capacity of natural systems: Natural systems such as green cover, forests soils, and aquatic systems have a remarkable capacity to absorb, accumulate, assimilate and transform various pollutants of the environment. The adverse effects caused by smoke and waste gases from local industry can significantly be moderated by green cover surrounding the industry. We can enhance this capacity of the natural system by promoting green cover in areas surrounding the industry. Barren land should be treated properly, provided fertilizers and water so that it turns into a eutrophic system – capable of supporting suitable plants and a variety of microbes so as to serve as an effective agent for cleaning up of the environment. Avoiding concentration of polluting industry at a particular place is a much wise policy. With little efforts and may be a little more cost these industrial units may be located away from each other so that their effluents do not collect and over-burden the surrounding environment.

4. Controlling Consumerism and minimizing waste products: Consumerism is a practice, which involves chronic acquisition of new products and services without any regard to their true worth, utility, durability, origin or the environmental effects of their manufacture and disposal. It is usually driven by a desire to follow the 'trend' or 'fashion' and the huge sums which are put in advertising by profit-motives of the manufacturers or the marketing agencies. Life styles today are characterized by huge consumption of resources, materials as well as energy and generation of large amounts of wastes. Just for a little convenience, self-satisfaction or to follow the trends, we spend rather unreasonably large sums of money to acquire things or services without which we could have done equally well.

A. Adverse consequences of consumerism: When we replace the 'old things' with 'new ones', our waste dumps grow. The energy and materials used in its manufacture is wasted. If we add up the environmental cost, the investment needed to repair the environmental damages caused for the production of the energy and material used in its manufacture, the overall cost of rejected commodity is staggering. What was the need of such a replacement when the 'old thing' was giving equally cheap and efficient service as the 'new ones'? We often replace things simply because we are fed up with them or simply because the new ones are in vogue. We could have saved a lot of money individually by avoiding such a replacement and if this practice were adopted by a large number of people, it would reduce its demand in the market, which could, in turn, reduce its production. In this way we would have saved the energy and material already used in the production of 'old things' while a decline in the production of 'new things' would have resulted in substantial savings of virgin material and energy and thereby a little less damage to the environment.

Equally wasteful is the 'use and throw' practices of our society. Today many things are made simply to be used once or for a definite period of time and then discarded. These products are so designed and manufactured that they serve us only once or for a definite period or definite number of times. It is the manufacturer and retailers in the market who are largely benefited. Huge sums are spent to promote this practice in advertisements. May be, such products are cheaper to produce but they add enormous burden of wastes to the environment. Advances in packaging technology have enabled manufacturers of consumer goods to pack little quantities in small pouches which can be easily afforded by poors and which have become very popular today. A person with a modest earning will not buy a bottle of shampoo, which costs nearly a hundred rupees but shall readily buy a pouch of shampoo for two or three rupees, which he can carry in his pocket and use it once or twice, although, he has to pay much more for the unit weight of the product as compared to a large pack. The unfortunate consequence of this is the growing accumulation packaging material, the plastic pouches, small bottles, cartons and wrappers in our waste dumps. The energy and material used for the same is simply wasted. The use of these pouches has grown very rapidly. Even edibles, like cooking oils, curd, tea, coffee, sweets and syrups may be carried in pouches, which are discarded once the product is consumed. Such practices, therefore, involve deliberate waste of energy and virgin materials. We are generating huge quantities of these wastes, which has been taxing our waste management capacity and which threatens to overwhelm the society with heaps of garbage all around.

B. Restricting consumerism and wasteful use of materials and energy: Today over-exploitation of natural resources threatens the ever-growing number of people with shortages of all types. The pollution of environment has begun to effect the environment adversely causing global warming, stratospheric ozone depletion and many unpredictable changes, which shall make life more difficult to live. A more sensible approach to the problem should be economy, reuse and recycling of resources. We shall have to control consumerism which wastes materials and energy. The profit motives of our society should be replaced by environment-friendly approach and restorative economy.

a. **Economy in use of finite resources:** The simplest way to conserve our finite resources is to practice economy. Careful use or use only where it is necessary can reduce much of consumption of metals and minerals. Most of the metals and minerals are cheap because Governments provide plenty of subsidies in the form of land at a throwaway prices, cheap power, subsidized fossil fuels, large loans and generous tax exemptions. Many industrialized nations also try to ensure access to cheap mineral supplies through their international trade and policies. The lure of large revenues, generous financial assistance, finished products, arms and other necessities force poor countries to sell their mineral wealth at a cheap price. It is mainly due to their low prices that many commodities are freely used, even at places where they are hardly required. A little restriction and taxation could cause these commodities to become more expensive which in turn shall curtail unnecessary over-consumption.

b. **Making finished products long lasting:** The 'use and throw away practice' of our society is a wasteful practice. The metal components in the product are also discarded and wasted. A study by the United States Office of Technology (Washington D.C.) in 1979 pointed out that repairs and re-use is a very promising method of conservation of many our finite resources. Metal containing products, for example automobiles, should be so manufactured as to last longer, and be repairable to be used again or their components if in working condition may be used again and again. Some materials are irrecoverably lost to the environment due to mishandling, corrosion, wear and tear. Of about 600 kg of per capita iron and steel used in USA, for example, more than one-third is lost never to be recovered. The average life span of all steel in use varies between 25-30 years while that of other metals is somewhat shorter. To make a commodity last longer we shall have to prevent its corrosion, wear and tear and the irrecoverable losses.

c. **Re-use and re-cycling of finite resources:** As rapid consumption of virgin minerals depletes our resources – why not use mineral products more efficiently or again and again. A machine having brass components can be pulled apart, its components used to make another object of utility. In India, most of the copper, brass, bronze and aluminium objects are regularly recycled. The residual waste can be converted to another useable resource. In many developed countries, wastes are used to produce energy. Biodegradable wastes can be composted to produce organic fertilizers. Wastes from one industry are often valuable resource for another industry such as bagasse, the waste of sugar industry is the raw material for paper and ply-board industries. The waste of textile industries or the cloth-rags discarded by us may be shredded to produce fibre again to be woven into another cloth. All this shall serve the useful purpose of reducing pressures of demand on our virgin resources and enhance their life spans.

d. **Shifting to more environment friendly commodities:** Today we are using a large number of products which when discarded do not undergo degradation for long periods. They add to the heaps of garbage, which collects in our waste dumps or are littered all around us. Instead of using plastic or polyethylene packing material why not use paper and for plastic or polyurethane cups, earthen pots may serve the purpose. Why not use permanent glass syringes, which had to be sterilized after each use instead of using disposable plastic syringes. A little care and efforts are needed but such practices shall help enormously in waste management and disposal.

(IV) ELIMINATING THE THREAT TO THE GLOBAL FOOD SECURITY

Agriculture today lacks an element of permanence. Nearly four decade have passed since new high yielding varieties of crops were introduced to the farmers all over the world. The new varieties being more responsive to chemical fertilizers and irrigation boosted agricultural production. However, the new 'Green Revolution Technology' rests heavily on chemical and energy inputs which tends to ignore the natural nutrient regeneration capacity of soils and has made agriculture costly.

Scientists the world over now realize that to introduce an element of permanence, modern agriculture shall have to go back to the traditional farming systems. Until recently, few agricultural scientists recognized the ecological and agronomic wisdom of the traditional agriculture, which has allowed farmers to maintain soil fertility for centuries together. In a desperate attempt to raise productivity, scientists have ignored the need for long-term sustainability. The traditional farming systems, which have persisted at many places since times immemorial, involved careful management of soil, water and nutrients to grow nearly enough year after years, i.e., sustainably. Traditional farming system may break down under growing population pressures, as working under severe agronomic limitations, it cannot provide large yields, which modern high input methods do. Heavy reliance on chemical and energy inputs, irrigation by river and tube-well waters coupled with rapid removal of tree cover have been degrading our soils. No doubt, we are growing more food than we actually require. But this we are doing at the cost of deteriorating resource base, which forms the bedrock of all agricultural productivity. Agriculture rests heavily on: fertile soils, supply of freshwater and biotic resources.

(1) Conserving fertile soils and wasteland reclamation:

Of about 167.5 million sq.kms of world's land surface, nearly 105.5 million sq.kms consist barren deserts, inhospitable ice-covered terrains, or rocky surface, which cannot be used for any productive purposes. On nearly 44.1 million sq.kms stand, our forests with more than 40% of canopy cover, which cannot be reduced much due to the obvious importance of forests. This leaves about 18.25 million sq. kms of arable land on which humanity can carry on his agricultural activity. We have to mark out some of our land as pastures. A large area has to be devoted for the cultivation of non-edible crops like tobacco, jute, cotton etc. Most of our cities stand on the most fertile agricultural land. This leaves a limited area for food grain production. The global area devoted to food grain production attained its peak value in the year 1976 when nearly 7.20 million sq.kms of arable land was put under cereal cultivation. Since then it has fluctuated between 7.10 to 7.15 million sq.kms. It is not expected to rise any more in near future.

Thus, land is a scarce commodity. The demands of an expanding humanity shall certainly put more and more strain on our land resources in times to come. Agricultural practices should be so modified that there is little damage to soils' nutrient regeneration and retention capacity and consequent soil degradation. Along with it, we should make a sincere effort to reclaim all **culturable wasteland** that is capable of or possesses the potential for agricultural use or for use as pastures or can be reforested. A **culturable** wasteland is the land, which has the potentials of productive use but is laying waste due to limitations such as high salinity or alkalinity, lack of water, soil erosion, water logging or simply human apathy in contrast to the **unculturable** wasteland, which is too barren, rocky or covered with snow to be of any productive use.

Today wastelands form a significant part of land of every country. A **wasteland** can be defined as a land, which is bereft of forest cover and is neither being used for agricultural purposes nor for any other purpose such as a sanctuary or a national park, power project or for industries. Wastelands are usually ecologically unstable lands, incapable of sustaining plant growth. They are badly eroded and degraded portion of land, which is not capable of producing materials or services of any value. It is mainly human activity, which is responsible for turning most of the arable land into wasteland. Extensive deforestation, indiscriminate over-exploitation of forest resources, over grazing, shifting cultivation, the adverse effects of developmental schemes and unscientific management of land resources are the primary causes of degradation of arable land.

The strategy of wasteland reclamation revolves around methods to control soil erosion, retain more water on land surface, enhance its downward percolation and provide more organic matter so as to improve the soil structure, water retention capacity and fertility of the soils. These methods can be summed up as follows:

1. Management of the topography: Water running down over the soil erodes it. Little of it percolates down to the subsurface water table. The speed of the flow depends on the gradient of the slope and the presence of obstacles in the passage. By creating obstructions in the passage of water as it moves down a gradient we can check the speed of the flow, which shall allow more water to percolate down through the surface to recharge ground water deposits. This is done by:

Strip farming: Planting different crops in strips across the slope in alternating strips may prevent the loss of soil as when one crop is harvested the other is still is present to protect the soils and impede the flood flow down the slope.

Contour ploughing: Fields may be ploughed across the slope rather than up and down. The tiny ridges thus created act like little dams, which checks the water from moving quickly down the slope, and more of it percolates down the soil surface.

Creating tied ridges: A series of earthen ridges running at right angles to each other are referred to as tied ridges. These are usually helpful in localities, which have heavy rainfalls. The movement of water is in quadrangles thus created which slackens its speed and hastens its downward percolation.

Terracing: The soil along the slopes may be shaped to create steps, shelves or wide strips of land, which holds more water. The edges of steps are protected by placing bamboo rafts stones or plant whose roots bind the soil particles effectively. The procedure is costly, as it requires much labour. However, it enables us to grow food crops even on the steepest hill slopes.

Gully control: The formation of widening gullies due to rapid flood flow may be controlled by leveling the fields and erecting check bunds, dams or drains, which divert the flow of water. This prevents widening and deepening of gullies, checks the rapid flood flow and consequently erosion of the soil.

Protection of the banks of the streams and rivers: Much of our soils on stream banks are carried away by rapid water flow in streams and channels. To prevent the loss of soil along the banks of rivers and streams physical obstructions such as stone may be placed or brick lining may be laid. The streambeds should be dredged regularly to ensure rapid clearance of the flood flow from the area.

2. Providing cover to the fields: The ground is covered with leaves, twigs and crop residues, to minimize the drying or erosive effect of wind and rains. It also moderates soil temperatures and provides protection to soil microbial life, which enhances fertility, water retention capacity and the structure of the soil. If the crop residues are insufficient or the other organic matter is difficult to be arranged, a **cover-crop** such as rye or alfalfa may be planted after the harvest which provides cover and protects the soil. Later this crop may be ploughed with the soil to add to the organic matter content of the soil.

3. Promotion of cover of vegetation over the denuded surfaces: Plant cover is extremely helpful in wasteland reclamation and prevention of soil erosion. It impedes the rapid flow of water, check drying and erosive action of wind binds the soil particles together, provides organic matter, improves soil structure and water retention capacity etc. A dense plant cover serves as a useful habitat for wildlife. Much of our land has been turned into wasteland by overgrazing, shifting cultivation, excessive felling of trees, forest fires or due to implementation of developmental projects etc. We should promote reforestation wastelands, which are not suitable to be put to agricultural uses, thereby raising its productivity. We shall at least get something, such as firewood, rather than getting nothing from our wastelands.

Cultivation of perennial crops, crops that grow for more than two years, on land, which can support their growth, could be better option for using some of our wastelands. Planting grasses, crops like tea, coffee or other plants that provide fruits, which could require a little more care in the beginning. However, such a practice shall make the investment and labour put in for cultivation every year unnecessary.

4. Handling saline and alkaline soils: Soils made useless due to accumulation of salts and alkaline material or disposal of chemical wastes should be put to extensive leaching to wash away the salts or chemicals concerned. Salinity and alkalinity usually develops due to poor drainage of water through the soils. Many such soils have a bed of hardened soil (Kankar) underneath which prevents water to percolate down and remove the salt to deeper layers. If this hard bed underneath is shaped like a pan or trough water collects over it. During dry season this water rises and comes up by capillary action and evaporates leaving the salt behind.

The only way to reclaim such soils is to add some cheap material, which reacts with the salts and neutralizes it or to cause holes or cracks in the subterranean pan of hardened soil mechanically so that waters may percolate down through these passages to deeper strata. Planting trees, like Babool (*Acacia* sp.) which can be grown on saline and usar soils in very helpful as its roots go deep inside the soil often penetrating the hardened strata. With time, the hard pan is fragmented and soil drainage improves while the fallen leaves of the tree provide organic matter to the soil.

(2) FRESHWATER CONSERVATION, RAINWATER HARVESTING AND WATER SHED MANAGEMENT:

Earth has an enormous quantity of water but most of it is useless for us as it contains too much of salts in dissolved state. Our freshwater supplies are limited to the amount water, which precipitates as rains or snow minus the amount, which is lost through evapo-transpiration. Land surface receives about 113,500 cubic kms of freshwater as rains or snow every year of which about 72,500 cubic kms are lost due to evaporation or transpired by plants. The balance 41,000 cubic kms is the amount which mankind can safely use. Of this enormous quantity, nearly 27,000 cubic kms per year are lost as flood flow through streams and rivers and about 5,000 cubic kms happens to be placed in barren inhabitable localities. The actual amount of water available to mankind is about 9,000 cubic kms per year. The pressure of demand on our freshwater resources is expected to rise as human numbers multiply. The problem of water scarcity can be tackled only by reducing wastage, using water economically and augmenting our freshwater supplies.

(A) FRESHWATER CONSERVATION:

Water is the basic need of a living organism. No one can live without water. The demand for freshwater is likely to exceed its supply by the second or third decade of the current century. An acute crisis is expected to follow in some regions of the world. The shortage of water shall make many localities barren, devoid of life. Fertile land will become desert. Conservation of fresh water is, therefore, an absolute necessity of today. Otherwise, the tomorrow will be grim, drier and barren to live through.

It is somewhat consoling that though the actual quantity of water drawn has already reached the level of total water availability, the irrecoverable consumption of water is well below this level. Needless to say, this is due to a number of steps undertaken to minimize wastage of fresh water resources and to make more efficient use of the available water. Some of these may be summarized as follows:

1. **Water Economy, Re-Use and Recycling:** For almost all spheres of human activity till date, more water is drawn than the amount actually needed. Much of the surplus water is returned to surface flow in an impure state. A little care can reduce the over-consumption. We waste water because of its easy availability. If a water meter is installed and money charged for every bucket of water we use, water consumption in domestic establishments, livestock management and industries shall drastically decline. The resultant surplus may be diverted to regions of water scarcity. This shall also reduce production of wastewaters, which pollute our aquatic systems.

   Power generation is another sphere of human activity wherein a large amount of water is needed. Most of it, however, is used as coolant (about 90-95%). Irrecoverable consumption is only 5-10%. The heated waters from thermal power plants may be utilized elsewhere after proper cooling. The same is true for many industries. Water used once may be used again for another purpose. Not all processes require good-quality water. Agricultural runoffs from fields can, likewise, be used to irrigate cropland down the stream while an efficient use of water with conditions of proper drainage can significantly reduce the agricultural runoffs.

2. **Development of an Efficient Distribution System:** Water resources are not distributed evenly. Some localities have plenty of water. Others have little of it. Therefore, transport of water from one place to another becomes an essential part of water conservation efforts. Many river basins have plenty of water, which flows down unused to the sea. This surplus can be diverted to drier regions through a system of canals and pipes. Water drawn out from underground sources can also be transported to zones where underground water cannot be tapped. The surplus of one basin can be used to make up the deficit at another.

3. **Reduction of Pollution and Recycling of Water:** Pollution spoils huge quantities of our surface water. All possible efforts should be undertaken to divert wastewaters to some treatment plant instead of releasing them into our surface waters. While treated water can be safely discharged in our aquatic systems, it may also be recycled where there is more pressing need. Ordinarily the biodegradable impurities of wastewaters make them most undesirable. These can be conveniently decomposed by some biological treatment followed by treatment with a suitable disinfectant. This makes the water concerned almost as good as, it was earlier. If not for direct consumption, these waters can be used satisfactorily for other purposes such as washing and cleaning, as coolant in industries, for irrigation etc.

4. **Conserving freshwater in agriculture:** Of the total amount of water used by mankind, nearly 75-80% is used to grow food crops. Improved irrigation practices such as drip-irrigation, use of sprinklers, and devices to control excessive evaporation can conserve nearly 20-30% of freshwater. Losses due to leakage from canals should be curtailed by immediate repairs and the channels should be lined with brickwork to prevent the loss of water due to seepage.

(B) AUGMENTATION OF EXISTING SUPPLIES OF FRESH WATER:

We can augment our freshwater supplies by retaining some of the 27,000 cubic kms of freshwater, which is lost annually as flood flow. Soils covered with rich vegetation act as a sponge, which soaks up large quantities of fresh water, which serves the humanity during drier periods. We have bared a large area of earth's surface of plant cover. The exposed land surface dries out quickly and with time, compaction occurs. The water which rains down cannot percolate through the hardened soil and flows away quickly to streams, rivers and to the sea without recharging the underground water tables. As withdrawal of huge quantities of fresh water continues, underground water table is receding deeper and deeper everywhere in the world. It is apparent from the shortages of water recorded from so many places all over the world that there is an appreciable reduction each year in the amount of precipitation, which is retained on earth's surface or in its underground deposits. The problem of shortage of freshwater can be solved if we adopt the following:

1. Rainwater harvesting: Rainwater harvesting involves collection, storage and subsequent use of water deposited by rains. In water stressed, dry regions of the world, rainwater harvesting is an ancient practice. In parts of Rajasthan and Gujarat, people collect, whatever they can, the meagre quantity of water that is deposited by rains in large storage tanks and vessels. This water serves them during most of the year. Today, the development of a centralized water distribution system and the depletion of underground water table have forced people to look back to the ancient practices of rainwater harvesting. In many states in India such as Haryana, Kerala and Gujarat rainwater harvesting and ground water recharge projects have been initiated.

Roof tops properly cleaned are used to collect rain water which is directed through pipes to large underground pucca, storage tank which are sealed off from all sides except for a small opening used for cleaning and withdrawal purposes. In cities like Dwarika, Gujarat, people use this water for drinking and cooking purpose for most of the year.

2. Enhancement of surface storage Capacity: The flood flow, which results after heavy rains rushes down to the oceans through streams, and rivers of the world and is of no use to the humanity. We can store this water in tanks and reservoirs for use during drier seasons. This can be done by erecting embankments and dams, which check the flood-flows and detain water for longer duration on land surface. Construction of 'talabs,' 'talaiya,' 'pokhras' is an ancient practice in many parts of India which functioned to store water of monsoon rains to be used for agriculture and a variety purposes. Through a system of pipes and canals, the water can be supplied wherever needed.

In many parts of the world, huge dams and reservoirs have been constructed to store water deposited over the catchment area of large rivers. The potential energy, the energy of water flow as it moves from a higher place to lower may be used in hydroelectric power generation, while the reservoirs, which develop behind the dam, may be used for fisheries and other purposes.

However, surface storage of water, in huge quantities, is a risky and costly venture. Water losses through evaporation and seepage are enormous both from the reservoir and from the distribution system. Much of the water infiltrates the soil and moves to the ground water table. Large areas of fertile land are submerged which may include human settlements as well. Natural ecosystems are destroyed. The pressure of standing water table enhances seismic activity, which may precipitate an earthquake. A crack in the dam, sabotage or bombardment during wars is catastrophic to the people living down streams. Entire localities may be washed away by rapidly gushing waters. The multi-billion rupees projects are gradually made defunct by silt and debris carried by the river water if little is done to reduce soil erosion upstreams in the catchment area of the river.

3. Improvement of underground storage capacity: An enormous amount of fresh water is stored in underground deposits. It represents accumulations over a long period of time. Every year, about 10-15% of total precipitation enters the ground water table. These deposits regularly feed streams and rivers during drier periods. Ground water deposits are cheap and easily obtainable source of freshwater – except for the cost involved in its withdrawal. We can improve the ground water storage capacity of earth's crust by:

- Collection of water deposited over roof tops open fields and slopes by erecting impediments to check the flow.
- Directing this water to underground water table through wells, bore wells and deep shafts drilled in the soil for the purpose so that the collected water goes to the underground

deposits quickly.

- Providing an effective plant cover over the soil surface. Plants obtain most of their water from soil moisture and keep the surroundings cool and humid, thereby, preventing excessive loss of water through evaporation. They impede air currents which prevents drying and check the flow of water retaining it for longer durationin the area because of which more water percolates down the soil surface to add to the ground deposits.

(C) WATERSHED MANAGEMENT:

Rainwater as it collects at any place flows under the influence of gravity to form small stream many of which join each other to develop larger flows or tributaries that converge in a valley to constitute a river. The entire drainage system is referred to as a 'watershed' or the 'catchment' of the river concerned. The management of a watershed with objectives to conserve the freshwater, the fertility of its soils and enhancement of the plant cover for the benefit of the communities, which inhabit the locality, is referred to as 'watershed management'.

All over the world, human communities inhabiting the watersheds of most of the river have been tempering with drainage systems, removing the plant cover, enhancing water losses and soil erosion. Water flowing over the bared soil surface collect large amount of silt and rock debris, which suffocates aquatic life downstream and chokes up river channels. Large quantities of precious plant nutrients are lost. The rapid flood flow during rains allows little time to the water to percolate down the soil surface to recharge underground water table while during dry seasons soil surface hardens and compaction occurs. The watershed becomes a stretch of bared, denuded and degraded land.

The basic techniques of watershed management involve promotion of plant cover in the watershed and delaying the flood flow for longer duration in the area so that more of it moves down to recharge sub-surface water table. This becomes more complicated when the watershed is colonized by numerous human settlements. Some of the measures adopted for watershed management include:

- Minimizing forest clearance and removal of plant cover
- Promotion of plant cover in general, particularly over the slopes of the hills.
- Adoption of sound farming and forestry practices to reduce surface runoff.
- Enhancing organic matter content of soils by leaving crop residues in the fields and adding more of it to prevent desiccation and erosion.
- Erection of props, check dams and digging of pits, trenches and small check dams so that more water is retained in the area.
- Conservation of wetland in the way of river channels to enhance the natural surface storage capacity of river channels.

(V) CONTROLLING THE DEGENERATION OF BIODIVERSITY

Nobody knows exactly how many species occur on this planet. Roughly, 1.4 million species have so far been identified. Scientists believe that the total number of species of all kind with whom man shares this planet is somewhere between 10 to 80 millions. Most of these are small creatures, which inhabit the unexplored, or little explored environments all round the world. Nature however, retains its mystery even in the most familiar places also. A few grams of soil from your kitchen garden may contain some species, which are unknown to science. This simply illustrates the vast gap in our knowledge of nature around us (Wilson 1988).

However, even with such a fragmentary state of our knowledge it has become apparent that the biological diversity or the total variety of life on our planet – the genes, species, and ecosystems – is collapsing at a fast rate. Human activity is dismantling bit by bit the vital life support system which took nearly three and a half billion years' to develop. The basic reason behind such a rapid degeneration of biodiversity, among others, is the expansion of human enterprises at a fast rate. Expanding agriculture, enlarging cities, sprawling industries encroach upon natural ecosystems splitting them into small fragments and destroying many life forms. Smoke, shoot, dust, harmful gases, domestic and industrial effluents that the cities and its industries export to natural system, affect them adversely. The stressful environment of a highly restricted habitat causes a large number of species to leave or be doomed (Diamond 1989, Kenton et al 1991).

Both ex-situ and in-situ conservation efforts are already underway for the conservation of biodiversity which occurs on our planet. Although **ex-situ** conservation efforts have been able to save many species, its scope is limited. We cannot attend to the large number of species, which are now believed to be threatened with prospects of extinction due to lack or resources. There may also be a large number of species unknown to us, which may be under the threat of extinction. A better way to handle the problem is **in-situ** conservation, which aims at conserving the entire ecosystem as a whole. A large number of species whether known or unknown to us may thus be conserved together in the system which we refer to as 'sanctuaries.' 'protected areas,' 'wild life reserves' or 'biosphere reserves.' World wide, there are nearly 7200 such areas which have been set aside exclusively for wild life to continue their existence. They cover an impressive 655 million hectares, nearly 5% of the total land surface of the world. Although there are many inadequacies in the system, these efforts have contributed much to conserve a large number of species (Jenkins 1992, Olney & Ellis 1992, Raven 1981).

These efforts, good as they are, are not enough. The world is dominated by a single species, the *Homo sapiens*, and it shall continue to be so in future also. Although the pace of growth has slowed down, the world population shall continue to grow until the close of 21st century when, as many population projections predict, it shall stabilize at about 10 billion people. This means that another four billion people shall be added to the world population in the current century. The growing numbers shall definitely encroach upon natural habitats and place severe stresses on the natural systems pushing wild life to recede to small isolated patches. A large number of species, known or unknown to us shall disappear in the process. Unless the human society can learn to tolerate and permit wilderness in their civilized landscape biodiversity appears to have a bleak future. Urban areas are antithesis of natural diversity. Nevertheless, even these concrete jungles can support much diversity. Unused land in the cities, corridors along highways, spaces around large buildings, waterways running through the cities etc. may be used to support a large number of plants and infuse biodiversity in urban settings. Man is also a creation of biosphere why cannot he be a part of it. Why ecological decline is considered inevitable result of economic development? Why can't we orient economic development in such a way as to preserve environmental health and quality? Why economic development and environmental well-being cannot go together within a natural system in which man plays the part of a caring and intelligent custodian looking for the needs of all? Wild life and man have existed together ever since man appeared on this planet and so they will in future also. Man with his superior intellectual capacities should assume stewardship of this planet manage things in such a way that both man and wild life may live together for all times to come. After all, man cannot exist alone on this planet (Wilson 1988, Mc Neelay et al 1990, Todd 1988)

(VI) STABILIZING WORLD POPULATION

Growing populations pass through three stages of demographic transition. Initially the growth is slow because both the birth rates and death rates are high. As living conditions improve death rates decline, life expectancy rises and the population enters a phase of rapid growth. Rising incomes, affluence and prosperity, education and better health care facilities etc. usually results in a desire to have smaller households, late marriages, decline in fertility, which reduces the birth rates. The population reaches its optimum size at which birth rates and death rates are at a much lower level and are nearly equal. The population growth stabilizes, i.e., there is no growth in numbers.

China has miraculously controlled her population. The key to Chinese success lies in the nation wide effort to foster public understanding of the consequences of the rapid population growth. Using long-term projections, the people were informed about the highly reduced per capita supplies of food, water, cropland, energy supplies and jobs. The consequence of these efforts was a shift in the focus of childbearing decisions. Traditionally the concerns of couples centred on how many children they would need to support them in their old age. They began to think in terms of how the size of their family shall affect the world in which their children would live. This subtle shift from a narrow self-centred thinking to the concern for the well-being of future generations has been the key to the Chinese success in controlling her population (24) (Brown & Wolf 1987). If the teaming millions in China can do this, why other countries of the world cannot achieve these targets?

Of course, trying to slow population growth rapidly when living standards are declining is one of most difficult, politically hazardous issue any government can face. Many countries now face a demographic emergency where failure to check population growth will lead to continued environmental degradation, economic degeneration and eventually social disintegration.

1. Launching an intensive Family Planning Drive: An intensive family planning drive is essential to control the rapidly growing populations in many of the third world countries. Economic development, higher literacy and adequate health care facilities are instrumental in reducing the population growth rates. With better health care facilities the uncertainties associated with small family size shall disappear. The realization that lower per capita resource availability with growing family size shall make people to adopt family planning measures and control family size. Better education and prosperity shall naturally force people to realize that rapid population growth shall bring about conditions of life in which no one would like their children to live. It will naturally curb the desire to have larger families which so far has been interpreted as providing better financial returns and an adequate old-age security.

2. Introducing an Effective Women and Child Welfare Programs: Through out the world, it has been demonstrated that with rising literacy among women, affluence and prosperity of the household improves, children get better health care and education while life expectancy rises and the population growth rates come down. This is because of the silent, sincere and painstaking industry of our women folk, who represent half of the total world population and play a pivotal role in the structure and organization of a family. Probably the most effective way to control the rapidly expanding populations in third world countries is to tackle it from grass root levels. Literacy of women, better education facilities at par with male members of the society and adequate knowledge of family planning and health measures shall confer more power and a better status to women. This will enable the women to improve in subtle and silent way her household, the health of her children as well as her own. Many problems for which we run costly schemes and programs shall be automatically solved.

Each year about eleven million children die because of diseases and inadequate nutrition. Nearly 20% children die before the age of five year. Most of these deaths occur in developing countries. Pneumonia, diarrhea, measles, malaria and malnutrition claim about 70-75% of these lives. Contaminated water, pollution of air, infected food material and unsanitary conditions all around enhance the problem. In many developing countries of the world, the condition of women is equally deplorable. They suffer from various infections, malnutrition and unregulated child bearing. Although their duty is to cook and feed the male members of the family, they are often denied a healthy diet. They are often denied opportunities of proper education at par with the male members of the family. The three major factors, which deprive women and children of a cheerful, healthy and content life, are:

a. **Malnutrition:** Most of the women and children in developing countries have to bear the burden of inadequate nutrition on their health. They become susceptible to many deficiency diseases and infections, which they could avoid on healthy diets. Pregnant women, nursing mothers are particularly vulnerable to the effects of malnutrition. Maternal depletion, low weights at birth, anaemia, toxemias of various types during pregnancy, post-partum hemorrhage etc. lead to high morbidity and mortality of both women and children.

b. **Infectious diseases:** Maternal infections are often responsible or causing a variety of adverse health effects such as abortions, foetal growth retardation, low birth weights and embryopathy. Infection in children may begin with labour and delivery and increase as the child grows older. Major diseases of children are pneumonia, diarrhoea, measles, malaria, tuberculosis, diphtheria etc. In many villages in developing countries, the usual vaccination programs, which confer to the children immunity against many contagious diseases, are not readily available.

c. **Uncontrolled Reproduction:** In many developing countries ill health and morbidity in women and children is a result of uncontrolled and unregulated reproduction. Conception after conception in series year after year for a dozen times result in shattering effect on women's health. This either results in repeated abortions or in a row of children which are difficult to manage and feed. Malnutrition and ailments of all types follow.

3. The Indian effort to control population growth and National family welfare programme :

In India a national family planning programme was launched in the year 1951 with the objective of reducing the brith rate to the extent necessary to stabilize the population at a level consistent with the requirement of the National economy. The programme was accorded a high priority and was centrally sponsored by the Government of India. In the beginning, during the First and Second Five year plans the Famil Welfare Programme endeavoured to provide basic infrastructure needed to reduce the birth rate, i.e. its approach was largely 'Clinical'.

The census of 1961 revealed the necessity of educating the eligible couples to plan their families and adopt small family norms. During the Third and Fourth Five Year Plans the emphasis of the programme was shifted to Extension and Education approach which involved extension of clinical facilities along with spreading of the message of family planning. In Fourth Five Year Plan (1969-74) about 16.5 million couples of reproductive age group were protected against conception. A phenomenal increase in performance of steri lization was observed during the years 1975-1977. The programme received set-backs during the years 1977-78 due to coercive means adopted by some field workers in implementation of the rigid targets set forth by the project. Consequently the Government had to make it clear that there was no place for coercion or compulsion or pressure of any type and that the programme had to be enforced on purely voluntary basis with the sole purpose of 'family welfare' and has to rely only on mass-education and motivation for its implementation. The name of the programme was also changed to Family Welfare Programme. Earlier the programme was called Family Planning Programme.

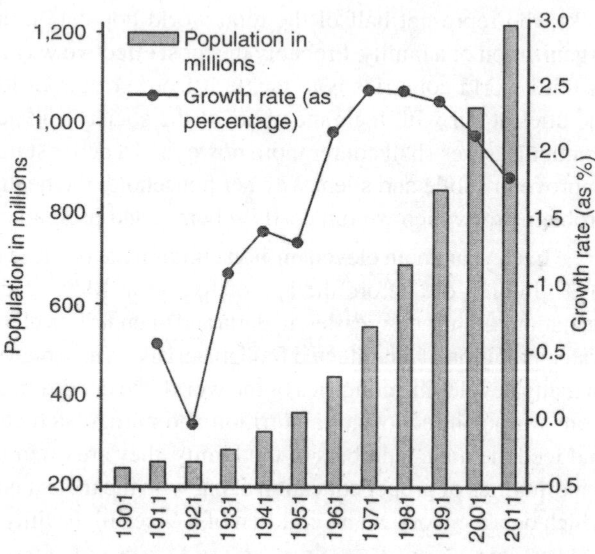

Fig. 27.1. *Indian Population and its growth rate since 1901*

During 1980-1985, the Sixth Five Year Plan, long term goal of achieving net reproductive rate of unity was envisaged whereas the programme was continued on purely voluntary basis during the Seventh Five Year plan (1985-1990) also. Emphasis was laid on promotion of spacing between consecutive child births, securing maximum community participation and promotion of maternal and child health care. It was in the year 1985 that **Universal Immunization programme** was launched to provide safety cover to infants and pregnant mothers against identified preventable diseases. The project was enlarged to cover all districts of the country.

The approach adopted during the preceding decade was continued during 1990-1995. To ensure effective community participation **Mahila Swasthya Sanghs** (MSS) were brought into being in the year 1991. These organizations consisted of people from varied socia-economical segments of the society, the workers involved in welfare activities, adult education instructors, primary school teachers and Anganwari workers with Auxiliary Nurse Midwives as its member-convener. From 1993 onwards, Universal Immunization programme has been strengthened and enlarge into child survival and safe motherhood project. The project involves sustaining the high immunization coverage under UIP, and augmenting activities under Oral Rehydration Therapy, prophylaxis for control of blindness in children and control of acute respiratory infections. Under the Safe Motherhood component, training of traditional birth attendants, provision of aseptic delivery kits and strengthening of first referral units to deal with high risk and obstetric emergencies were being taken up. To impart new dynamism to the Family Welfare Programme, several new initiatives were introduced and ongoing schemes were revamped in the Eighth Plan (1992-97). Realizing that Government effOlts alone in propagating and motivating the people for adaptation of small family norm would not be sufficient, greater stress has been laid on the involvement of NGOs to supplement and complement the Government efforts. Reduction in the population growth rate has been recognized as one of the priority objectives during the Ninth & Tenth Plan period.

Indeed these efforts have yielded results which are encouraging. Figure 27.1 presents the over-all population and the rate of growth of Indian population since 1901. To begin with the gowth of Indian population was slow. It accelerated after the year 1951 and during 1961-1971 a high growth rate above 2% per year was maintained for about three decades .. However, thanks to the ational Family Welfare programme that a slow decline has set in after 1971 and the census of 2011 has recorded a growth rate of 1.76% per year (25).

(VI) THE CRUCIAL ROLE OF ENVIRONMENTAL EDUCATION

Environmental education has a fundamental role to play in motivating people to adopt environmental friendly practices. We cannot expect people to act in appropriate way without awareness of the problems, its causes, the impact on our daily life and the long-range consequences. Many problems concerning environment and biosphere are simply there because so many people contribute little bits and pieces to it, all of which put together assume enormous dimensions. A little effort, a little care exercised by each individual in the society could eliminate the entire problem. Starting from grass-root levels environmental education should involve all sections of our society and should be able to:

1. Create a general awareness concerning environmental problems.
2. Motivate people to conserve resources, protect environment and avoid extravagance.
3. Promote understanding and co-operation among people to face ecological issues.
4. Conserve indigenous knowledge, traditions and culture friendly to the environment.

Environmental education should foster a positive pattern of conduct towards the use of natural resources. The resources of the world are enough for man's needs but not for his greed. With proper management, just and equitable distribution we can feed every mouth on earth's surface. However, the world today is divided into nations, nations in races, races into communities, communities into the rich and the poor, the privileged and the deprived. The poor and deprived, unfortunately, constitute a majority in today's world. Little can be expected from people whose main concern is acquisition of food just enough to survive.

Issues related to the environmental degradation are and shall continue to be ignored in favour of economic considerations as long as gross injustices in the pattern of distribution of natural resources prevail. The establishment of new economic order both at national and at international levels, which provides for the deprived and the needy, is necessary if we are to expect something concrete from the millions of inhabitants of the developing countries of the world. There should be a free transfer of technology and means between the developed and developing countries of the world. The teaming millions fighting for life or for better life in developing countries of the world could serve as an effective instrument to clean up habitat, re-instate and protect the wild life.

(VIII) ROLE OF INFORMATION TECHNOLOGY IN ENVIRONMENT AND HEALTH:

Information technology (IT) involves the retrieval, processing, storage and dissemination of vocal, pictorial, textual or numerical information by a microelectronics-based combination of computing and telecommunications. It has played a key role in the development of human society and has changed the human lifestyle. Development of sophisticated instruments like computers, satellites, telecommunication instruments etc have resulted in total revolution in almost all spheres of life.

The new technology has enabled man to store huge volume of vocal pictorial textual or numerical information in exceedingly small chips from where any information can be retrieved at random whenever required within milliseconds. The development of satellite communication and inter! inkage of computers through Internet has made communication and dissemination of information exceedingly easy. Sitting anywhere in the world one can access a wealth of information through internet within a few milliseconds. In almost every field of human activity this has revolutionized our lives. Information technology has tremendous potential in the field of environment education and health as in any other field like business, sciences, economics, politics or culture, environmental management and conservation of wild life and natural resources. It has reduced the need of keeping records on paper charts, files data-sheets and map which are bulky to maintain while combing out the desired information is often very difficult and time consuming (26).

World Wide Web (WWW), Geographical Information System (GIS), Remote Sensing, and informatio'l through satellites have developed a wealth of up to date information on various aspects of environment and health. A number of user friendly soft-wares have been developed for studies on environment and health. The comprehensive database includes infonnation on wild life, conservation forest cover, water logged areas, spread of desertification, deforestation, urban sprawl, networks of rivers and canals, mineral and energy reserves, extent and occurrence of diseases like HIV, AIDs, Malaria, Flourosis, Cholera etc.

Geographical Information System (GIS) has proved to be a very effective tool in environmental management. GIS is a technique of superimposing various thematic maps using digital data on a large number of inter-related or interdependent aspects. Different thematic maps containing information on a number of aspects like water resources industrial growth, human settlements, road network, soil types, forest lands, cropland, grassland, degraded land and polluted localities can be superimposed on a layered form in computer using soft-wares and has become an effective tool for decision making, evaluating the impact of changes, determining priorities, and directing conservation action. The prospects of keeping entire lifetime record of drugs taken, pathological tests conducted, X-ray taken, diseases suffered etc in short the complete medical history of all people in the locality which can be accessed by a health care worker from anywhere in the world is poised to make medical care exceeding efficient, prompt and effective (27).

(IX) THE ROLE OF AN INDIVIDUAL IN CONSERVATION OF WILD LIFE, NATURAL RESOURCES AND ENVIRONMENT.

Most of the problems of wildlife degeneration, wastage of natural resources and the pollution of the environment are simply there because so many people around the world, knowingly or unknowing contribute small bits and pieces to it which adds up to make a huge quantity. A little care and thoughtful efforts by everyone in the society shall add up to make a big difference.

Avoiding or minimizing the use of resources so that they last longer. Resources which are being over-exploited and are now threatened with the prospects of exhaustion can be made to last longer if their use is avoided on minimized. For example there are many metals the use of which can be avoided altogether. Why use a brass bucket for keeping and carrying water when plastic bucket can serve the purpose? Cheaper and at times better substitutes are available for a number of metal products, use of which by majority of individuals can reduce the burden of demand on our resources and make them last longer.

Avoiding or minimizing practices which pollute the environment and damage wild life and natural resources. In cold winters we often burn a lot of firewood or coal or consume electricity to keep us warm. Why heat up the entire house or room when we can stay comfortable in a woolen pullover or a quilt? When conserving our body's heat can serve the purpose the use of an external source of heat may be avoided. If a large number of people act in this way there could be substantial saving of energy which may be use for any other useful work or the electricity thus saved may light a number of additional homes. A little less shoot, smoke and carbon dioxide shall go into the atmosphere. Large market exist in many countries which sell animal skin, crown, fur, ivory and other wild life products on which poachers thrive. If majority of people avoid the use of these products there would be no demand. These markets shall collapse and a number of animals would be saved from merciless killing. Here, it is the decision of individuals which matters.

Our surface waters are often used as conduits for the disposal of waste waters and sewage which as a consequence require costly treatment and disinfection if they have to be used again or else we have to depend on our ground waters which are now being over-exploited. As a result our ground water tables are receding deeper and deeper. If we use the available water economically, we can reduce the volume of wastes waters which we create and by using pits and bore-holes we can dispose of the sewage where biological machinery of decay and decomposition shall take care of the filth and dirt we create. The practice shall also help to recharge our ground water resource. Individuals in the society can bring about these changes with little effort.

Individual life style decisions by majority can greatly help conservation of wild life natural resources and the environment. If majority of couples decide to have only two children there will be substantial reduction in population growth rates. If each of us may decide to plant a tree and look after the sapling shall greatly help in promotion the green cover. Majority of individual may decide to use a bike instead of a four-wheeler. This shall save a little petrol which shall finally add up to form a huge quantity and a little less carbon dioxide shall go up into the atmosphere. Policies which originate at grass-root level, from general public at large reflect the basic need of the time are effectively implemented rather than policies which are introduced from top downwards. But for such movements to start an effective drive to mould the public opinion is needed which makes environmetal education very important. Individuals may greatly help the cause conservation of will life, natural resources and environment by spreading the awareness of the problems we face today or shall face tomorrow or our coming generations are likely face in future.

QUESTIONS

1. Why the ways of the today's world are unsustainable? Discuss.
2. What do you understand by sustainable development? Trace the emergence of global concern for sustainable development.
3. Enumerate the crucial issues faced by humanity which must be tackles and strategy adopted for sustainable development.
4. Traditional economic development has to be re-oriented to more environment friendly development to transform the world into a sustaible system. Discuss.
5. Discuss the crucial role of Environmetal education for the development of sustainable society.
6. What role can an individual play our efforts to conserve wildlife, natural resources and environment ?

28

Chapter

Environment and Law

Evolution of the legal Framework for the protection of wildlife and Environment

The environmental movement is an expression of fundamental change in man's perceptionf life on earth. The evidence of this change has appeared at every level of social organization local, regional, national and international. The shift from the view of earth with unlimited resources in abundance and exclusively created for man's use to the concept of earth as a domain of life or biosphere in which man is just one of the species among millions or simply a temporary resident custodian has been remarkable. It has occurred within a span of a century or so and is a development of great significance for the future of all life on this planet.

The global concern for environment, wild life and natural resources caused the development of legal framework both at national and international levels. Rules were established by authority, mutual consensus or traditions to regulate the behaviour of individuals, a community or concerned countries for the protection of environment and natural resources.

(A) INTERNATIONAL LAW:THE UNIQUE SYSTEM OF GLOBAL GOVERNANCE WITHOUT A GLOBAL GOVERNMENT

International law is a collective body of agreement regarding mutual rights and obligations of the countries of the world. It is a framework of conventions and treaties between contracting parties, international declarations, collective principles, opinions and generally accepted practices among the states. The enforcement of the provisions of the international law is brought about through negotiations rather than by adjudication.

The conventions and treaties between different countries, international declarations, common principles, opinions and generally accepted practices concerned with the Environment constitute what we know as the International Environmental Law. It is a system of regulating the practices and behaviour of concerned countries in environmental matters, violation of which may cause a surge of adverse public opinion and indignant ire of the world community which in turn could be followed by troubling consequences and even war. Thus a unique system of *global governance without a global government* has come into being. **United Nations Organization** is the leading institution which monitors activity of this "global governance". It is an organization consisting of 185 states which offers a platform for discussion and resolution of all sort of problems of its member states. The charter for the establishment of this world body was approved and signed by 51 countries representing 80% of the world's population on June 26, 1945, at San Francisco (1).

(I) THE BASIS OF INTERNATIONAL CO-OPERATION

Concern for deteriorating conditions of environment and degeneration of wild life arose simultaneously in many countries of the world, especially the leading industrialized democracies, like U.S.A., U.K., France, Germany etc. People in these countries had been pressurizing their Governments for legislative and administrative action to protect and restore the quality of environment. Many groups of these people and organizations communicated with each other across their national borders to establish international bodies. Rachel Carson's "The Silent Spring" (1962), Stewart Udall's "The Quiet Crisis" (1963), Jean Dorst's "Before Nature Dies" (1965), Rolf Edberg's "On the Shred of Cloud" (1966) and a number of publications which appeared in popular press warned the people of adverse consequences of the misused technologies, over-stressed ecosystems, wasted resources and contaminated environment.

Soon for a number of issues it became apparent that national action alone cannot successfully redress the environmental problems. For example, no country alone could protect itself from nuclear fall outs from surface atomic tests. How could geographically concentrated industrial states of Western Europe protect themselves from trans-boundary air and water pollution? How could African States protect their wild life from merciless destruction when lucrative markets of skins and trophies existed in Europe and America? The chemical contamination of oceans or depletion of marine fishes in International waters could not be prevented by the action of a single country alone. A cumulative effort was needed to resolve these issues (1).

(II) THE ROLE OF NON-GOVERNMENTAL ORGANIZATIONS (NGOS)

Private bodies and non-governmental organizations have played a very important role in the development of environmental movements and international co-operation for the conservation of wild life and their natural habitats. They were instrumental in initiating dialogues between concerned parties for most of the treaties and agreements or conventions signed during the early part of twentieth century. Scientific congresses also are non-governmental organizations intended chiefly for the advancement of science and technology many of which have acquired an international status. After mid-nineteenth century these international scientific meets became more frequent. International Botanical, Zoological and Ornithological congresses were held regularly. Although their basic purpose was exchange of information and views between people working in different countries, early in the twentieth century, their agenda were molded to include environmental issues and conservation of wild life and natural resources (2).

Following World War II, the responsibilities of many Governments expanded to cover a number of aspects of public life which had hitherto been accorded much less attention. Many Governments and inter-governmental organizations became increasingly involved in conservation of natural resources and subsequently in environmental conferences. During 1960–1970, international net-works of non-governmental organizations matured and began to effect the transition from the traditional narrow and limited conservation policies to a more comprehensive open ended concern for the quality of life and environment. The most important non-government organization which emerged during the period was **International Union for Conservation of Nature and Natural Resources (IUCN)**, the draft constitution of which was adopted by a conference sponsored by UNESCO at Fontainebleau in 1948 (2).

(III) U.N. CONFERENCES FROM 1946 TO 1971

A number of conferences were convened during 1946 to 1971, however, non these adopted a truly holistic view and need of integrating ecological aspects for conservation action. The **Biosphere Conference of 1968** represented a landmark in the recognition of the importance of man's relations with the natural world. The conference named "Inter Governmental Conference of Experts on Scientific Basis for Rational Use and Conservation of Resources of the Biosphere" was held in Paris in September 1968. It was sponsored by a number of agencies under the leadership of UNESCO. The Conference examined a broad spectrum of issues with emphasis on ecological consideration.

Unlike earlier conferences, this conference adopted a draft containing twenty recommendations for the future action by participating governments, the U.N. system and the UNESCO (3).

(IV) THE STOCKHOLM CONFERENCE AND ITS LEGACY

The United Nations Conference on Human Environment for the Preservation and Enhancement of its quality met at Stockholm on June 6–16, 1972. The conference marked the culmination of efforts to place the protection of the environment & biosphere on the official agenda of international policy and law. Apart from increasing awareness of environmental issues among public and governments, the Stockholm Conference laid framework for future environmental cooperation. A number of governments subsequently created Ministries for the Environment and/or national agencies for environmental monitoring and regulation. The meeting agreed upon a Declaration containing 26 principles concerning the environment and development, an Action Plan with 109 recommendations, and a Resolution.

Probably one of the most important recommendation of the Stockholm Conference, when we compare it with so many which were convened in the past, was creation of institutional arrangement to implement its recommendations and setting up of an environmental fund to carry out the same. It was on the recommendation of the Conference that United Nations General Assembly resolved on Dec. 15, 1972, through its Resolution-2997(XXVII) to establish the necessary machinery and make financial arrangements required thereof and the **United Nations Environment Programme (UNEP)** was borne. At the insistence of developing countries the size of governing body of UN Environment Programme was enlarged to 58 nations to accommodate greater representation from Asia and its Headquarters were shifted to Nairobi, Kenya (4;5;6)..

(V) POST STOCKHOLM DEVELOPMENTS

The Stockholm Declaration and the persistent efforts of United Nations Environment Programme (UNEP) have been the prime driving force behind the cumulative efforts of nations to protect their environment, wild life and natural resources which constitute a vital life support system for all life on this planet. Separate roles have been played by other international organizations such as OECD, ICSU, IUCN etc. without which it would have been very difficult to compel many reluctant countries to adopt the necessary national or international measures for the protection of environment and natural resources. The United Nations Environment Programme (UNEP) assisted by many non-governmental organizations (NGOs) was able to obtain ratification of the following set of treaties which were negotiated during or after the Stockholm Conference and which have now come into effect:

1. Convention on the Protection of the World Cultural and Natural Heritage, Paris, 1972.
2. Convention on the Prevention of Marine Pollution by Dumping of wastes and other matter, London, 1972.
3. Convention on International Trade in Endangered species of wild Fauna and Flora, Washington D.C., 1973.
4. Convention on the Prevention of Pollution from Ships, London, 1973 (MARPOL).

The Convention on **Trade in Endangered Species of Wild Flora and Fauna** (CITES), signed in Washington D.C. in 1973, is a very significant treaty as far as wild life conservation is concerned. The lucrative profits and the great demand of products of wild life, such as musk pods, horns of rhinos, skin of crocodiles, and larger mammals, elephant tusks etc., have attracted poachers and smugglers to engage in clandestine trade to earn huge profits. CITES has been designed to establish a system by which States may strictly control international trade in specimen of species which are on the verge of extinction. India became a party to this convention in 1976 and has adopted the international system of licensing and legal procurement of certificates to control the trade of all the species listed in Appendices to the CITES Convention. Under the Convention countries which are signatories to the Convention cannot import or export the species listed in the CITES appendix without proper licenses (1;6;7).

It was during the 1980-1990 that UNEP assisted by other international agencies, was more successful in getting a number of treaties negotiated and signed.

1. Convention on Long range Transboundary Air Pollution of 1979
2. Protocol on the Reduction of Sulphur Emission or their Transboundary Fluxes by at least 30 per cent in 1985.
3. Protocol Concerning the Control of Emission of Nitrogen oxides or their transboundary fluxes in 1988.
4. Vienna Convention for the Protection of Ozone layer, 1985.
5. Montreal Protocol on Substances that Deplete Ozone layer, 1987.
6. Convention on Early Notification of a Nuclear Accident, 1986.
7. Convention on Assistance in the Case of Nuclear Accidents or Radiological Emergency, 1986.
8. Convention on the Control of Transboundary Movements of Hazardous Wastes and Their Disposal, 1989.

In Additions to these a number of older treaties and conventions were gradually receiving ratification. Though the credit of all these achievements cannot be attributed to UNEP alone, yet this specialized institution of the United Nations has served well as an effective instrument of the international community for the implementation of the Stockholm Declaration (6;7).

(VI) WORLD CONSERVATION STRATEGY AND THE WORLD CHARTER FOR NATURE

The World Conservation Strategy, the strategy to conserve and sustainable use of living resources of the world was developed by International Union for Conservation of Nature and Natural Resources (IUCN) in collaboration with UNEP, World Wild life Fund (WWF), Food and Agricultural Organization (FAO) and United Nations Educational Scientific and Cultural Organization (UNESCO). The strategy defined living resource conservation, explained its objectives and its contribution to the human survival and development. It also identified main obstacles in achieving these objectives

In retrospect the World Conservation Strategy 1980, appears to be a fore-runner of the World Charter for Nature which was adopted by the United Nations General Assembly in 1982, on the insistence of a majority of Third World Countries. It was on October 28, 1982, that the World Charter for Nature came before the Assembly for final action and was adopted by a majority of 111 votes with 18 abstentions and a single negative vote cast by USA. The Charter was an important expression of the intent among the member states of United Nations, to achieve a more harmonious and sustainable relationship between the humanity and rest of the biosphere. Point 14 of the Charter stated that principles set forth in the charter should be reflected in the law and practice of each state as well as at the international levels. Thus an effort was made to launch a cumulative action by all countries of the world for the protection of nature and natural resources (1;7).

(VII)UNITED NATIONS CONFERENCE ON ENVIRONMENT AND DEVELOPMENT (UNCED) 1992

On December 22, 1989, the General Assembly of United Nations by its Resolution No. 228 voted to accept an invitation from Brazil Government to hold a major Conference on Environment and development at Rio de Janeiro. It was in 1992 that the United Nations Conference on Environment and Development, also known as the Rio Summit or Earth Summit was convened in Rio de Janeiro from June 3 to June 14, 1992. One hundred eight heads of States or Governments were present at the conference in which 172 Governments took part. About 2,400 representatives of non-governmental organizations (NGOs) also attended the conference.

None of the earlier UN Conferences held so far can be compared with the Earth Summit as far as the size, participation and the enormous scope of its deliberations was concerned. Meeting twenty years after the first global conference on human environment in Stockholm, the summit forced rulers of the world to ponder on the course of economic development the world had been following and device ways to check the destruction of irreplaceable natural resources and pollution of the planet. The fact that both the poverty in third world countries and affluence, prosperity and over-consumption of resources in the developed countries of the world places damaging stress on the environment was recognized by the world community. The rulers of the world perceived the need to re-orient national and international plans and policies to ensure that their economic decisions have none or the least adverse impact on the environment. And the message has produced results, making eco-efficiency a guiding principle for business and governments alike. The Earth Summit has influenced all subsequent UN conferences, which have examined the relationship between human rights, population, social development, women's welfare and human settlements — and the need for environmentally sustainable development (1;7). Some of the major achievements of the Rio conference were:

1. Rio Declaration on Environment and Development
2. Agenda 21
3. Convention on Biological Diversity.
4. UN Framework Convention on Climate Change.
5. Forest Principles

1. THE RIO DECLARATION ON ENVIRONMENT AND DEVELOPMENT:

It was a short document produced at the 1992 United Nations Conference on Environment and Development (UNCED). The Rio Declaration consisted of 27 principles intended to guide future sustainable development around the world. It covered multiple aspects of the problem of human environment and development and the conservation of biodiversity. It was another important land mark in our efforts to protect the environment and natural resources (8). The salient features of the Convention can be summed up as follows:

1. That the contracting parties were conscious of the intrinsic value of biological diversity and its components and its importance as a life support system.
2. That the conservation of biological diversity is common concern of all mankind.
3. That the states are responsible for conserving their biological resources and use them in a sustainable manner.
4. That access to biological resources for which the states have sovereign rights, will be determined by the national governments on mutually agreed terms subject to the national legislation.
5. That there shall be free transfer of scientific and technical information, technical and scientific co-operation in handling of biotechnology patents and distribution of its benefits on mutually agreed terms
6. That the states should integrate considerations of conservation and sustainable use of biological resources into their national decision-making and adopt measures relating to the use of biological resources and avoid or minimize adverse impact on biological diversity.
7. That the states should protect and encourage customary use of biological resources in accordance with the traditional cultural practices that are compatible with the conservation or sustainable use requirements.
8. That the states should support the local population to develop and implement remedial action in degraded areas where the biological diversity has been reduced and encourage co-operation between its governmental authority and its private sector in developing methods for the sustainable use of biological resources.

2. AGENDA-21:

Agenda-21 is a programme run by the United Nations (UN) related to sustainable development. It is a comprehensive plan of action to be taken globally, nationally and locally by organizations of the UN, governments, and major groups in every area in which humans impact the environment. The number 21 refers to the 21st century. The full text of Agenda-21 was submitted at the United Nations Conference on Environment and Development (Earth Summit), held in Rio de Janeiro on June 14, 1992, where 179 governments voted to adopt the programme (9). Agenda 21 contains well over 900 pages, divided into four sections and 40 chapters:

Section I: Includes Social and Economic aspect of development such as combating poverty, changing consumption patterns, promoting health, promoting sustainable settlement patterns and population demographic dynamics and integrating environment and development into decision-making.

Section II: Includes conservation and management of resources for development such as atmospheric protection, control of pollution, checking deforestation, protecting fragile environments and conservation of biological diversity.

Section III: Aims at strengthening the role of major groups such as youths, particularly women and children, indigenous tribes, local authorities, workers in business and industries, NGOs etc.

Section IV: Includes the mechanism of implementation, education, science, technology transfer, international institutions and financial mechanisms

The Agenda focuses on the problems of today while aiming at preparing the world for the challenges of the next century. It reflects a global unanimity of opinion and political consensus at the highest level on development and environmental cooperation. The dynamism of Agenda-21 makes it a befitting plan of action for different situations, capacities and priorities of diverse countries and regions of the world while fully respecting the principles contained in the Rio Declaration and is expected to evolve over time in the light of changing needs and circumstances of the people concerned.

3. CONVENTION ON BIOLOGICAL DIVERSITY:

The Convention on Biological Diversity, known informally as the Biodiversity Convention, is an international treaty that was adopted by the Earth Summit at Rio de Janeiro (10). The Convention has three main goals:

- Conservation of biological diversity (or biodiversity);
- Sustainable use of its components; and
- Fair and equitable sharing of benefits arising from genetic resources.

In other words, its objective is to develop national strategies for the conservation and sustainable use of biological diversity. It is often seen as the key document regarding sustainable development. The Convention was opened for signature at the Earth Summit in Rio de Janeiro on 5 June 1992 and entered into force on 29 December 1993. The agreement covers all ecosystems, species, and genetic resources. It links traditional conservation efforts to the economic goal of using biological resources sustainably. It sets principles for the fair and equitable sharing of the benefits arising from the use of genetic resources, notably those destined for commercial use. It also covers the rapidly expanding field of biotechnology through its Cartagena Protocol on Biosafety, addressing technology development and transfer, benefit-sharing and biosafety issues. Importantly, the Convention is legally binding, countries that join it ('Parties') are obliged to implement its provisions.

4. UNITED NATIONS FRAMEWORK CONVENTION ON CLIMATE CHANGE.

Another significant outcome of the conference on Environment and Development (UNCED) held in Rio de Janeiro was the international environmental treaty referred to as United Nations Framework Convention on Climate Change (UNFCCC or FCCC). The treaty is aimed at stabilizing greenhouse gas concentrations in the atmosphere at a level that would prevent dangerous anthropogenic interference with the climate system. On June 12, 1992, 154 nations signed the convention which upon ratification committed signatory governments to a voluntary 'non-binding aim' to reduce atmospheric concentrations of greenhouse gases with the goal of preventing dangerous anthropogenic interference with Earth's climate system. On October 13, 1992, US president George Bush signed the instrument of ratification with which the United States of America also became a party to the convention.

The treaty as framed originally did not define the mandatory levels of greenhouse gas emissions for individual nations nor did it contain any enforcement provisions. It was, therefore, a legally non-binding instrument. However, the treaty did contain provision for updates (referred to as 'protocols') that would set mandatory emission limits. Therefore, in accordance to the provisions of the treaty Signatories to the Convention have been meeting every year at the **Conference of Parties (COP)** since the UNFCCC came into force on March 21, 1994. As a matter of fact it has been the annual conferences of parties signatory to UNFCCC which have played a very significant role in the global efforts to reduce green house gas emission. There have been about a dozen such conferences many of which are better known than UNFCCC itself (Such as the Kyoto Protocol).

The **Kyoto Protocol** was adopted at the third session of the Conference of Parties to the UNFCCC (COP-3) in 1997 in Kyoto, Japan. It is an agreement under which industrialized countries will reduce their collective emissions of greenhouse gases by 5.2% compared to the year 1990 (compared to the emissions levels that would be expected by 2010 without the Protocol, this limitation represents a 29% cut). The goal is to lower overall emissions of six greenhouse gases - carbon dioxide, methane, nitrous oxide, sulfur hexafluoride, hydrofluorocarbons and perfluorocarbons - averaged over the period of 2008-2012. National limitations range from 8% reductions for the European Union and some others to 7% for the US, 6% for Japan, 0% for Russia, and permitted increases of 8% for Australia and 10% for Iceland (11).

United States of America provoked worldwide criticism by rejecting the Kyoto protocol in 2001. The Kyoto protocol of 1997, which was revised at the Bonn climate change conference in 2001, binds industrialized countries to cut greenhouse gas emissions. United States contains 4% of the world's population but produces about 25% of all carbon dioxide emissions. By comparison, Great Britain emits 3% - about the same as India which is the second most populous country in the world. An average American produces six tonnes of carbon dioxide, average Briton three tonnes, Chinese 0.7 tonnes and an Indian 0.25 tonnes annually. Reducing environmentally harmful emission to the levels required by Kyoto protocol and the Bonn conference would mean substantial curbs on US industries and the cost to US business could be heavy. It would mean some fundamental changes in American lifestyle. US citizens tend to drive larger cars and make more frequent trips. The argument in the US is that the Kyoto protocol would unfairly penalize Americans and the American economy (12;13). Without US which are world's biggest polluter, little success could be achieved in moderating global warming. It has been at the insistence of the United States of America that many compromises had to be made which were negotiated at COP-6 held at Bonn Germany in July 2001 whereas COP-7 held at Marrakech, Morocco in November 2001 finalized most of the operational details and prepared the stage for nations to ratify the Protocol. The changes include:

a. Flexible Mechanisms: The flexibility mechanisms were incorporated, which the United States had strongly argued for including **Emissions trading**, **Joint Implementation** (JI), and the **Clean Development Mechanism** (CDM) which allow industrialized countries to *fund emission reduction activities in developing countries as an alternative to domestic emission* reductions. An important feature of this agreement was that there would be no quantitative limit on the credit a country could claim from use of these mechanisms, but that domestic action should constitute a significant part of the efforts of each country to meet their emission reduction targets.

b. Carbon sinks: Credit was granted for broad activities which absorb carbon from the atmosphere or store it, including forest and cropland management and re-vegetation, with no over-all cap on the amount of credit that a country could claim for sinks activities. For cropland management, countries could receive credit only for carbon sequestration increases above 1990 levels.

c. Compliance: Final action on compliance procedures and mechanisms that would take care of non-compliance of Protocol's provisions was deferred to COP-7, but included broad outlines of consequences for failing to meet emissions targets that would include a requirement to make up for the shortfall, suspension of the right to sell credits for surplus emissions reductions and a required compliance action plans for those not meeting their targets.

d. Financing: Three new funds were brought into being to provide assistance for needs associated with climate change, a least-developed-country fund to support National Adaptation Programs of Action, and a Kyoto Protocol adaptation fund supported by a CDM levy and voluntary contributions.

Conference of Parties (**COP-11**) which was held November – December 2005 at Montreal, Canada, hosting more than 10,000 delegates, was one of Canada's largest international events ever. This meeting happened to be the first meeting of parties since their meeting at Kyoto in 1997 and has also been referred to as the First Meeting of Parties or **MOP-1**. While the United States of America continued to evade inclusion as a member of the conference of parties, it was at this meeting that Kyoto protocol entered into force. An agreement to extend the life of the Kyoto Protocol beyond 2012 the date at which it expires and to negotiate deeper cuts in greenhouse gas emission was reached at this conference which has been referred to as **Montreal action plan** (14;15).

5. FOREST PRINCIPLES

The Forest Principles is the informal name given to the 'Non-Legally Binding Authoritative Statement of Principles for a Global Consensus on the Management, Conservation and Sustainable Development of All Types of Forests,' a document produced at the 1992 United Nations Conference on Environment and Development (UNCED). While recognizing the sovereign rights of countries to utilize and manage their forest resources on a sustainable basis the document makes several recommendations for forestry which include the development and application of sustainable forest management to all types of forests and timbers, enhancement of efforts toward the greening of the world and raising the world's forest cover, a commitment to provide financial resources as well as environmentally sound technology on favourable terms to developing countries to enhance their capacity to sustainably manage, conserve and develop their forest resources and promotion of a supportive international economic climate and trade in forest products. At the Earth Summit, the negotiation of the document was complicated by demands of developing nations in the Group of 77 to enhance foreign aid in order to pay for the setting aside of forest reserves. Developed nations resisted those demands, and the final document was a compromise.

(VIII) WORLD SUMMIT ON SUSTAINABLE DEVELOPMENT 2002

On the 30th anniversary of the United Nations Conference on the Human Environment (UNCHE), Stockholm, in 1972, and the 10th anniversary of the United Nations Conference on Environment and Development (UNCED) held at Rio de Janeiro, United Nations (UN) organized the World Summit on Sustainable Development (WSSD) from August 26 to September 4, 2002, at the Sandston Convention Centre in Johannesburg, South Africa (16). The 2002 summit is also informally known as "Rio+10".

It was apparent that the world had changed. The new era of global environmental cooperation that had been promised at Rio proved false. The developed countries of the world continued to be reluctant to provide the support and resources they were expected to provide to the developing countries while the United States in particular and some other industrialized countries were hesitant to co-operate on the key issues such as climatic change. It was in an environment of hopelessness that the World summit on Sustainable development started its deliberations. However the two magic words "sustainable development" were there in the title of the World Summit Johannesburg was testimony to the fact that the term 'sustainable development' had gained worldwide policy acceptance. For those who believed in the concept, this was a chance to put meaning into it (16;17).

The Johannesburg Declaration was a main outcome of the Summit. The declaration is a collection of general political statements, reaffirming a commitment to agreements made at the Rio de Janeiro and Stockholm Summits. International cooperation, decreasing world poverty, special attention for developing nations, empowering women, and maintaining biodiversity, among other things, are outlined as key points to building a sustainable future. The document is meant to serve as a contract for the participants of the summit, binding them to the outlined agreements. The Plan of Implementation laid down more specific goals for the nations and organizations that participated in the summit(16;17). Some of these goals include:

- The establishment of a solidarity fund to wipe out poverty. This fund would be sustained by voluntary contributions; however, developed nations are urged to dedicate 0.7% of their national income to this cause.
- Cutting in half by 2015 the proportion of the world's population living on less than a dollar a day. This is a reaffirmation of a UN Millennium Summit goal.
- Cutting in half by 2015 the number of people who lack clean drinking water and basic sanitation
- Substantially increase the global share of renewable energy
- Cut significantly by 2010 the rate at which rare plants and animals are becoming extinct
- Restore (where possible) depleted fish stocks by 2015, and
- Halving the number of people suffering from hunger.

(B) NATIONAL LEGISLATIONS

Principles, policies, directives, and regulations enacted and enforced by local or national authorities to regulate human treatment of the nonhuman world come under the purview of Environmental law. During the second half of 20th century, environmental law beginning humbly as a set of clauses within the laws regulating public health matured into a globally recognized independent field protecting both human health and nonhuman world. It covers a wide area of concerns in diverse legal settings, like the Indian Air (Prevention and Control of Pollution) Act, 1981, Regulatory standards for emissions from coal-fired power plants in Germany, initiatives in China to create a "Green Great Wall"– a shelter belt of trees – to protect Beijing from sandstorms.

As early as early 1960s the Japanese government had begun to consider a comprehensive pollution-control policy, and in 1967 Japan enacted the world's first such overarching law, the Basic Law for Environmental Pollution Control. Many countries have included the right to environmental quality in their national constitutions. Since 1994, for example, environmental protection has been placed in the German Grundgesetz (Basic Law), which now states that the government must protect for "future generations the natural foundations of life." The Chinese constitution guarantees to each citizen a "right to life and health" and requires the state to ensure "the rational use of natural resources and protect rare animals and plants". The South African constitution recognizes the right to "an environment that is not harmful to health or well-being and to have the environment protected, for the benefit of present and future generations". The Bulgarian constitution provides for a "right to a healthy and favourable environment, consistent with stipulated standards and regulations" and the Chilean constitution contains a "right to live in an environment free from contamination."

(I) TYPES OF ENVIRONMENTAL LAW

An examination of the law formulated by different countries of the world reveals that most of Environmental Laws fall into any of the following four categories:

1. Command-and-control legislations: Many environmental laws fall into the general category of laws referred to as "command and control legislations." Such laws typically involve three elements:

1. Identification of a type of environmentally harmful activity
2. Imposition of specific conditions or standards on that activity
3. Prohibition of forms of the activity that fail to comply with the imposed conditions or standards.

The most obvious form of regulated activity involves the discharges of pollutants into the environment (*e.g.*, air, water, and groundwater pollution). However, environmental laws may also regulate activities which carry a significant risk of discharging harmful pollutants although they may not cause any harm (*e.g.*, the transportation of hazardous waste, the sale of pesticides, and logging). For actual discharges, environmental laws generally prescribe specific thresholds of allowable pollution. For activities that create a risk of discharge, environmental laws generally establish management practices to reduce that risk. Comprehensive environmental laws impose both environmental-quality and discharge standards and coordinate their use to achieve the desired environmental-quality.

Another type of activity regulated by command-and-control legislation is environmentally harmful trade. Among the most developed regulations are those on trade in wildlife. Once a plant or animal species has been designated as endangered, countries generally prohibit import or export of that species dead or alive or their body parts except in specific circumstances. In 1989, listing of the African elephant as a protected species effectively prohibited most trade in African ivory, which was subsequently banned by Kenya and many other countries. By this time the United States had already banned trade in African ivory, listing the African elephant as a threatened species under its Federal Endangered Species Act (1978). Despite these measures, some countries either failed to prohibit ivory imports (e.g., Japan) or refused to prohibit ivory exports (e.g., Botswana, Namibia, South Africa, and Zimbabwe), and elephants continued to face danger from poachers and smugglers.

2. Laws making Environmental assessment mandatory: Environmental-assessment mandates are another important form of environmental law. Unlike command-and-control regulations, which may directly limit discharges into the environment, mandated environmental assessments protect the environment indirectly from the consequences of the proposed actions. The impact of a proposed human activity on environment is carefully studied. This information potentially improves the decision making of government officials and increases the public's involvement in the creation of environmental policy. The United States National Environmental Policy Act (1969) requires the preparation of an environmental-impact statement for any "major federal action significantly affecting the quality of the human environment." The statement must analyze the environmental impact of the proposed action and consider a range of alternatives, including the so-called "no-action alternative."

3. Laws involving Economic incentives or deterrents: The use of economic instruments to create incentives for environmental protection or deterrents for activities injurious to environment is a popular form of environmental law. Such incentives include pollution taxes, subsidies for clean technologies and practices, and the creation of markets in either environmental protection or pollution. Denmark, The Netherlands, and Sweden, for example, impose taxes on carbon-dioxide emissions, and the EU has debated whether to implement such a tax at the supranational level to combat climate change.

4. Set-aside schemes: A final method of environmental protection is the setting aside of lands and waters in their natural state. In the United States, for example, the vast area of the land owned by the federal government (about one-third of the total land area of the country) is managed with environmental protection in mind. Europe has an extensive network of national parks and preserves on both public and private land, and there are extensive national parks in southern and eastern Africa in which wildlife is protected. There are about 7200 protected areas, parks, sanctuaries, nature reserves or biosphere reserves all round the world. About 655 million hectares of earth's surface, representing about 5% of the total land area of our planet has been set aside for the protection of wild life.

(II) GUIDING PRINCIPLES OF ENVIRONMENTAL LAW

The design and application of much of the modern environmental legislation in force today have been shaped and guided by a set of principles outlined in publications such as Our Common Future (1987), published by the World Commission on Environment and Development, and the Earth Summit's Rio Declaration (1992).

1. The precautionary principle: The precautionary principle states that, if there is a strong suspicion that a certain activity may have environmentally harmful consequences, it is better to control that activity right now rather than to wait for incontrovertible scientific evidence to emerge. Often Environmental legislation has to operate in areas complicated by high levels of scientific uncertainty. As a precaution it is advisable, to control such activity rather than waiting for more evidence to appear. This principle is expressed in the Rio Declaration, which stipulates that, where there are "threats of serious or irreversible damage, lack of full scientific certainty shall not be used as a reason for postponing cost-effective measures to prevent environmental degradation."

2. The prevention principle: Although environmental legislation is drafted often in response to catastrophes, preventing environmental harm is cheaper, easier, and less environmentally dangerous than reacting to environmental harm that has already taken place. The prevention principle is the fundamental notion behind laws regulating the generation, transportation, treatment, storage, and disposal of hazardous waste and laws regulating the use of pesticides. The principle was the foundation of the Basel Convention on the Control of Transboundary Movements of Hazardous Wastes and their Disposal (1989), which sought to minimize the production of hazardous waste and to combat illegal dumping. The prevention principle also was an important element of the European Commission's Third Environmental Action Programme, which was adopted in 1983.

3. The "polluter-pays" principle: Since the early 1970s the "polluter-pays" principle has been a dominant concept in environmental law. Many economists claim that environmental harm is caused by producers who "externalize" the costs of their activities. For example, factories that emit unfiltered exhaust into the atmosphere or discharge untreated chemicals into a river pay little to dispose of their waste. Instead, the cost of waste disposal, in the form of pollution, is borne by the entire community. Accordingly, the purpose of many environmental regulations is to force polluters to bear the real costs of their pollution, though such costs are often difficult to calculate precisely. Such measures encourage producers of pollution to make cleaner products or to use cleaner technologies.

4. The integration principle: Environmental protection requires that due consideration be given to the potential consequences of environmentally fateful decisions. If an industry is being set up which pollutes its surroundings, the cost of the remedial measures should be borne by the industry itself and added to the cost of produce. It should be the consumers of the goods being produced at the industry, which should bear the cost of environmental damage, not the people around the industry that are adversely affected.

5. The public-participation principle: Decisions about environmental protection often formally integrate the views of the public. Generally, government decisions to set up environmental standards for specific types of pollution, to permit significant environmentally damaging activities, or to preserve significant resources are made only after the impending decision has been formally and publicly announced and the public has been given the opportunity to influence the decision through written comments or hearings. In many countries, citizens may challenge government decisions affecting the environment in court or before administrative bodies. These citizen lawsuits have become an important component of environmental decision making at both the national and the international level. Public participation in environmental decision-making has been facilitated in Europe and North America by laws that mandate extensive public access to government information on the environment. Similar measures at the international level are included in the Rio Declaration.

(C) THE INDIAN LAW FOR PROTECTION OF WILD LIFE & ENVIRONMENT

In India, the first codified law, which initiated a series of law for conservation of wild life and prevention of cruelty against animals, dates back to 1873 when the British rulers introduced Madras Wild Elephant Preservation Act 1873 and the All India Elephant Preservation Act 1879 which were closely followed by the Birds protection Act, 1887. As these acts omitted other animals and proved to be insufficient for the protection of wild birds, the British Government passed another act, known as the Wild Birds and Animals Protection Act 1912.

(I) FORTY SECOND AMENDMENT TO THE CONSTITUTION OF INDIA SETS THE JUDICIAL MACHINARY INTO ACTION

Laws concerning wild life and environment were framed for the agencies responsible for ecologically damaging practices and it was considered the duty of the Government to ensure their compliance. The public, which was directly or indirectly affected by the damaged environment, had no role in it earlier. If the authorities of the country failed in their duty there were no other alternatives for the people to look forward to. Such a situation prevailed in a number of countries. Little help could be obtained from the judiciary, as the law did not specifically recognize the right of the people to clean air for breathing, clean water for drinking or a healthy environment to live in!

A similar situation existed in India. It was only after the 42nd Amendment to the Constitution which introduced Article 48-A, that the situation changed a little. This article requires the State to endeavour to protect and improve environment and to safeguard forests and wild life. In addition to it the amendment invoked by Article 51-A made the protection and improvement of the natural environment including forests, lakes, rivers and wild life a fundamental duty of the citizen of India. People could now go to court to seek legal help to guard their rights concerning issues of the environment (18).

The Indian Judiciary on the other hand, also started assisting people in their effort to maintain a healthy environment and wild life by admitting litigations against State's continued neglect of ecological rights of its citizens. In Ratlam Municipal Corporation case, Mr. Justice Krishna Iyer observed that the law is a 'remedial weapon of versatile use', which should be made available to the people in their struggle for environmental justice (19). Justice Ranganath Mishra in a litigation against querying operations in Mussoorie, emphatically stated that 'preservation of environment and keeping ecological balance unaffected is a task which every citizen must undertake' as a part of fundamental duty under the Article 51-A of the Constitution of India. Today, the dockets of every High Court and the Supreme Court of India are full with applications seeking environmental protection. The Supreme Court of India has turned out to be the main hope of the people. In the unique Ganga Pollution case, the judgment delivered so far has resulted in significant development, which the Pollution Control Boards and the Central and State Governments could not do in almost two decades. In the case, M.C. Mehta versus the Union of India, the Supreme Court of India issued directions for the closure of those tanneries at Jajmau, in Kanpur, which had polluted by their discharges the waters of river Ganga and had failed to take minimum steps required for the primary treatment of the industrial effluents (20).

(II) LAW RELATED TO THE PROTECTION OF THE ENVIRONMENT

During late nineteenth century and early twentieth century the pollution of air and water caused by rapid economic expansion and industrialization of the time began to affect public life in India at places where industrial activities were concentrated, like Calcutta, Bombay etc. It was the inconvenience caused to general public which was responsible for their creation. In India there are three important acts which restrict and regulate environmentally harmful activities: These are

(1) The Water (Prevention and Control of Pollution) Act, 1974:

By 1960 the problem of pollution of rivers and streams had assumed considerable importance and urgency. It was, therefore, considered essential to ensure that the domestic and industrial effluents are not allow to be discharged into the water courses without adequate treatment and the first comprehensive act for Prevention and control of water pollution was born. It was felt that a single law regarding measures to deal with water pollution control, both at the central and at State levels should be enacted by the Union Parliament. As legislation in respect to the pollution of rivers and streams is a subject relatable to entry 17 read with entry 6 of List II in the Seventh Schedule to the Constitution, the Union Parliament has no powers to make a law in the State (apart from the provisions of as articles 249 and 250 of the constitution) unless the Legislature of two or more States pass a resolution in pursuance of article 252 of the constitution empowering the Union Parliament to pass the necessary legislation on the subject. Only after the Legislature of the States of Gujarat, Jammu & Kashmir, Kerala, Haryana and Mysore had passed such resolutions that the bill was accepted by the Union Parliament. The bill received the assent of the president of India on March 23, 1974 and soon therefore became an act of Parliament under the short title "The water (Prevention and Control of Pollution) Act, 1974 (21).

The Act provides for the prevention and control of water pollution and for maintaining or restoring wholesomeness of water. With a view to carry out the aforesaid purposes, the Act seeks the establishment of **Pollution Control Boards** at Centre as well as at State levels and confer on and assign to such boards powers and functions relating thereto and for all matters connected therewith. The Pollution Control Boards function in co-ordination with each other and advise the Central and State Governments on matters related with prevention and control of water pollution. They lay down standards for ambient water quality, specify the quality and quantity of effluents their treatment and disposal in various bodies of water. They have been empowered with freedom to enter, inspect works, processes and plants, take samples for analyses and impose bans or penalties for failures to comply with norms laid down by the Board. Under Section 63 of the Water (Prevention and Control of Pollution) Act 1974, the Government in consultation with Pollution Control Boards has laid down rules and procedures for prevention, control and abatement of water pollution (21). In order to promote economic use of water, a precious natural resources, levy and collection of cess has been imposed on persons, industries, corporate bodies and governments under the Water (Prevention and Control of Pollution) Cess act 1977 and Water (Prevention and Control of Pollution) Cess Rules 1978. These provisions are also a source of income for the Boards and provide for the means to carry out their obligations. The Water (Prevention and Control of Pollution) Act of 1974 has been twice amended by the Water (Prevention and Control of Pollution) Amendment Act 1978 and the Water (Prevention and Control of Pollution) Amendment Act, 1988.

(2) The air (Prevention and Control of Pollution) Act, 1981

The Air (Prevention and Control of Pollution) Act, 1981 has important constitutional implications with an international background. The Act draws its inspiration from the proclamation adopted by the United Nations conference on Human Environment held at Stockholm from June 5th to 16th, 1972, of which India was an important signatory (22).The act is specialized legislative measure, meant to tackle one facet of environmental pollution. Its main objectives are:

1. To provide for the prevention, control and abatement of air pollution.
2. To provide for the establishment of Central and State Boards with a view to implement the aforesaid purpose.
3. To provide for conferring on such Boards, the powers and assigning to such Boards function related thereto and such other matters connected therewith.

When the act was taken up for consideration, Central and State Boards for prevention and control of water pollution had already been constituted. As it was felt that there should be an integrated approach for tackling environmental problems relating to pollution, it was decided that scope and responsibilities of these Boards should be enlarged to cover the prevention and control of air pollution as well. These Boards have also been entrusted with the responsibility of organizing through mass media comprehensive programmes to create general awareness regarding the problems concerned with air pollution. They also provide technical assistance, guidance and training to the people engaged in programmes for the prevention, control and abatement of pollution (22).

The Act received the consent of President of India on 29th March 1981 and came into force on 16th May 1981 vide Notification no. G.S.R. 351 (E) dated 15th May 1981. It became an Act of Parliament under the short title "The Air (Prevention and Control of Pollution) Act 1981. The Act has subsequently been amended once in year 1987 by the "Air (Prevention and Control of Pollution) Amendment act, 1987. Under the powers conferred by Section 16 (2: h) of the Act the Board has framed rules for the prevention, control and abatement of air pollution which are: the Air (Prevention and Control of Pollution) Rules, 1982 and the Air (Prevention and Control of Pollution) (Union Territories) Rules 1983.

(3) The Environment (Protection) Act 1986.

United Nations Conference on the Human Environment at Stockholm in June 1972 was an expression of the growing concern over the degenerating state of environment in the second half of the earlier century. The proclamation adopted by the conference reflects the desire of the people of the world to protect and improvement the quality of human environment while rapidly deteriorating conditions added a sense of urgency to whole issue. Environment (Protection) Act 1986, is an outcome of the worldwide desire to provide a legal instrument for the protection and improvement of environment and the prevention of hazards to human beings, other living creatures plants and property. The Act is, therefore, a set of comprehensive legislation regulating activities which are likely to cause or cause damages to the environment and aims at the protection and improvement of the environment and prevention of hazards to human beings, other life forms or property. The concerned act received the assent of the President of India on May 23, 1986 and came into force on November 19, 1986, vide Notification No. G. S. R. 1198, dated November 12, 1986, after which it became an Act of Parliament under the short title 'The environment (Protection) Act 1986' with its jurisdiction extending all over India (23).

Under Section 3(1), the Act provides sweeping powers to the Central Government to take such measures, as it deems necessary or expedient for the purpose of protecting and improving the quality of the environment and preventing, controlling and abating environmental pollution. Section 3(2) of the Act specifies the major concerns with respect of which the Central Government may institute necessary action. Under Section 6(1) and Section 6(2) the Central Government, by notification in the Official Gazette, is empowered to make rules in respect of any of the matters referred to in Section 3 (1-2). The Central Government is empowered to appoint officers or delegate such powers to whomsoever it deems fit, to inspect, supervise or regulate the compliance of the provisions of the act. Sections 15 (1-2), 16(1-2), 17(1-2) of the act contain provisions of punishment for those who fail to comply with or contravene any of the provisions of the Act, the rules made or the directives issue thereunder. Offences under the Act are punishable with imprisonment for a term, which may extend up to five years or a fine, which may extend up to one-lakh rupees or both. In cases where the offence continues an additional fine of rupees five thousand for every day may be imposed. No one, i.e., individuals, companies or government departments are exempt from purview of the act (23).

In exercise of powers conferred by Sections 6 and 25 of the Environment (Protection) Act 1986, the Central Government has made the following rules, which cover most of aspect of protection of environment:

1. The Environment Protection rules, 1986.
2. The Hazardous Waste (Management and Handling) Rules, 1989.
3. The manufacture storage and import of Hazardous Chemical Rules, 1989.
4. The Rules for the manufacture, use, import, export and storage of Hazardous micro-organisms/ genetically engineered organisms or cells, 1989.
5. The Chemical Accidents (Emergency planning, preparedness and response) Rules, 1996.
6. The Bio-Medical Waste (Management and Handling rules) Rules, 1998.
7. The Recycled Plastics Manufacture and Usage Rules, 1999.
8. The Noise Pollution (Regulation and Control) Rules, 2000.
9. The Ozone Depleting substances (Regulations and Control) Rules, 2000.
10. The Municipal Solid Wastes (Management and Handling rules, 2000.
11. The Batteries (Management and Handling) Rules 2001.

(III) THE LAW FOR CONSERVATION OF WILD LIFE & BIODIVERSITY.

The law conserving wildlife has a long history in India. Emperor Ashoka, the Great, has been credited with introducing the earliest codified laws on the matter of preserving wild life and its environment. The first codified law, which initiated a series of law for conservation of wild life and prevention of cruelty against animals, dates back to 1873 when the British Government of India introduced Madras Wild Elephant Preservation Act 1873 and All India Elephant Preservation Act 1879 These were closely followed by the Birds protection Act, 1887 and Wild Birds and Animals Protection Act 1912 which was amended in the year 1935 by the Wild Birds and Animals Protection (Amendment) Act 1935. After the Second World War as the freedom struggle intensified in India the concern for wild life was pushed the background. However, after the independence, the Constituent Assembly placed 'Protection of wild birds and wild animals' at Entry no. 20 in the State list and the State Legislatures were given the powers to frame laws for the same.

1. The Wild Life (Protection) Act, 1972

In our quest for development and economic growth extensive degeneration of natural habitats occurred and plight of wild life received scanty attention in the first one and half decade after independence. It was not until late 1960s that the concern for rapidly diminishing wild life was finally aroused. It was this concern, which manifested itself in the shape of a comprehensive legislation relating wild life protection in 1972. The Act which was known as 'The Wild Life (Protection) Act, 1972, received the consent of the President of India on 9th September 1972 with its jurisdiction extending all over India (24). The Act has subsequently been amended four times, i.e., in 1982, 1986, 1991 and 1993.

The Wild Life (Protection) Act, 1972 contains provision, which, in brief, empower the Central Government to:

1. Create machinery for the purpose of wild life conservation
2. Prohibit killing and hunting of specified animals.
3. Protect specified plants by prohibiting picking, uprooting felling of the plant concerned
4. Constitute sanctuaries, national parks and closed areas for wild life conservation
5. Constitute Central Zoo Authority and recognition of zoos
6. Restrict, regulate or prohibit trade or commerce in wild animals, animal articles and trophies.
7. Prevent and detect offences against wild life and provision of penalties for the same.
8. Frame rules and code of conduct for the protection of wild life and natural resources.

Protection of wild birds and animals including their habitats had been placed as Entry no. 20 by the Constituent assembly as a State subject and State Governments were given powers to regulate, frame rules and enact laws for the same. By the introduction of the Wild Life Protection Act, 1972 the Central Government assumed all powers to regulate, frame rules and device the code of conduct for the purpose (24). The Act has created a versatile and effective mechanism for wild life conservation. which involves:

1. Under Section 3, the Central Government could appoint a Director and Assistant Directors and such other officers, as it may deem necessary for the purpose of Wild life conservation whereas the State Government shall also appoint Chief Wild Life Wardens and Wild life wardens
2. Under Section 6 (1) the state Governments or the in the case of union Territories, the Administrator, are required to constitute a high-powered, **Wild Life Advisory Board**, the duties of which include among others selection and administration of areas declared as sanctuaries, national Parks and wildlife preserves.

The Wild Life Protection Act, 1972, therefore, while taking over the matters concerned with conservation plant and wild life resources established high-powered machinery, both at the Centre and the State levels, to implement the provisions of the Act. In pursuance of powers conferred by the Act the a set of rules have been framed by the Central Government which together regulate hunting, trade in wild life products, capture or maintenance, propagation and conservation of threatened plant and animals.

2. The Biological Diversity Act, 2002

The Biological Diversity Act, 2002 which received the consent of the President of India on February 2003, provides for the conservation of diversity, sustainable use of its components and equitable sharing of the benefits arising out of the use of biological resource, the knowledge and for matter connected therewith or incidental thereto(25).

The act under its section 8(1) seeks to establish a **National Biodiversity Authority** with its head office at Chennai which has been entrusted with the task of regulating access to biological resources, transfer of information and findings of research regarding biological resources of the country. A person who is not a citizen of India, including non-resident Indians, a body corporate, association or organization not incorporated or registered in India shall not be able obtain any biological resources occurring in India. To do so the persons or the party concerned shall have to seek approval of the National Biodiversity Authority which may levy such charges as it deems fit for the purpose. The same provisions apply to the transfer of information, the results of research regarding natural resources. In addition no application for Intellectual property right can be submitted without prior approval of National Biodiversity Authority which may also impose benefit sharing fee or royalty or both on benefits arising out of the commercial utilization of such rights (25).

(IV) LAWS RELATED TO THE CONSERVATIONS OF FORESTS IN INDIA.

The general forest policy and forest management were introduced by the British Rules of India early in the nineteenth century. The stated assumption of such a move was that local communities were incapable of scientific management of forest resources and that trained, centrally organized cadre of officers could manage the forests in a better way.

1. The Indian Forest Act, 1927

The general law relating to forests in British India was contained in the Indian Forests Act, 1878 and it's amending Acts. The Indian Forest Act 1927 is simply an act consolidating the provisions of the original acts. The language has been shortened and ambiguities in the earlier acts have been removed. The act received the consent of Governor-General of India on 21st September 1927 and became a Central Act under the short title "Indian Forest Act 1927". The act was amended and adapted to suite the requirement of administration of the period as the need arose from time to time (26).

The forest policy pursued before independence in India was mainly concerned with extraction forest products for the benefit of British Rulers. Where the interests of local community clashed with those of the rulers the provisions of the Forest Act were invoked to acquire exclusive rights over the forests or assets of the forests by the Government. Therefore, a system of "Reserve Forests or Protected Forests" came into being which were managed by the Government for their own interests and not for the purpose of 'conservation'. The Indian Forest Act, 1927, contains provisions by which the Government may declare forested land as 'reserve forests' or 'protected forests' bar accrual of right over the forest by any person except by succession or under a grant of permit by the Government. Extensive powers were given to the Forest settlement officers or the Collector to curb trespassing, encroachment, felling of trees, grazing by cattle, extraction of firewood, timber, other forest products and land clearance etc. Under Sections 36, 37 the government could by notification in the Official Gazette regulate or prohibit exploitative activities on any forested land or wasteland over which it had no propriety rights and assume their management. Section 52 of the forest act prescribes penalty and punishment for activities and offences committed against the provisions of the Act (26).

2. The Forest (Conservation) Act, 1980

For quiet sometime, the need of a legislation to check extensive deforestation which had been taking place since early 1960s, was being desperately felt as the powers to dereserve a forest land rested with State Government under Section 27(1-2) of the Indian Forest Act, 1927. In order to restrict rather rapid loss of forests, which had begun causing ecological imbalances, the President of India promulgated the Forest (Conservation) Ordinance on October 25th 1980. Subsequently, the concerned bill was introduced in the Parliament which was passed and which received the consent of the President of India on December 27th 1980. It became a Central Act under the short title "The Forest (Conservation) Act, 1980 (27). The act was later amended in the year 1988 by the Forest (Conservation (Amendment) Act 1988.

The Act specifically states that no State Government or any other authority shall order directing any reserved forest or any portion thereof to be 'dereserved' and used for any non-forest purposes, without the prior consent of the Central Government. The Act explicitly directs that no forestland or any portion thereof could be cleared of natural vegetation or trees, which have grown naturally in an area. Sections 3(A-B) provide for penalties which could be imposed for contravention of the provisions of the Act (27).. Section 4 of the Act empowers the Central Government to make rules by notification for carrying out the provisions of the Act in pursuance of which the Forest (Conservation) Rules, 1981 have been framed (Vide G.S.R. 719, dated 20th July 1981, published in the Gazette of India, Extra, Pt.II, Sec.3(i), dated August 1st 1981).

3. The National Forest Policy, 1988

Immediately after our country acquired independence, the erstwhile Ministry of Food and Agriculture, Government of India, framed a national forest policy through its Resolution no. 13/52-F, dated 12th May 1952, which was being followed for the management of State Forests of the Country. Growing pressures of demand for fodder, fuel wood, timber, inadequacy of protection measures, diversion of forestlands to non-forest uses without compensatory afforestation and necessary environmental safe guards etc combined to cause rapid degeneration of our forest wealth. The need to review the situation and evolve for the future a new strategy of forest conservation had become imperative by mid 1980s. In 1988 the policy was revised to include conservation priorities and strategies to ensure environmental stability and ecological balance (28). To achieve the objectives as laid down in the National Forest Policy certain management practices have to be adopted. These essential practices may briefly be summarized as follows:

1. Existing forests and forestlands should be fully protected and their productivity should be improved. Forest and vegetal cover should be increased at a fast pace on hill slopes, in catchment areas of rivers, lakes and reservoirs, on ocean shores and in semi-arid, arid and desert tracts. The desired goal of National forest policy is to have at least one third of the total land area of our country under forest cover.
2. Diversion of good and productive agricultural land to forestry should be discouraged in view of the need for increasing food production.
3. For the conservation of total biological diversity, the network of national parks, sanctuaries, biosphere reserves and other protected areas should be strengthened and extended adequately.
4. Provision of sufficient fodder, fuel and pastures, especially in areas adjoining forests, is important in order to prevent depletion of forests beyond the sustainable limits. Since fuel wood continues to be the predominant source of energy in rural area, the programme of afforestation should be intensified with special emphasis on augmenting fuel wood production to meet the requirement lf the rural people.
5. Minor forest produce provide sustenance to tribal population and to other communities residing in and around the forests. Such produce should be protected, improved and their production enhanced with due regard to generation of employment and income.

To achieve the objectives of the National Forest Policy of our country a multipronged strategy is required. We shall have to conserve and manage existing forests resources effectively and shall have to extend forest cover over barren/marginal/waste/mined lands and watershed areas. We shall have to launch intensive social forestry programme for rural and tribal development. We shall have to develop alternative sources of energy (such as gobar-gas or LPG) and building materials (such as straw boards, ply or plastics) which could replace the use of fuel wood and timber. We shall have to make provision for alternatives to fodder, which could reduce its extraction from forests. We shall have to intensify research related with wild life management of national parks and sanctuaries.

The objectives of the revised Forest policy cannot be achieved without the investment of financial and other resources on a substantial scale. Such investment is indeed fully justified considering the contribution of forests in maintaining the life support systems, essential ecological processes and in preserving diversity of life. Forest should not be looked upon as sources of revenue only. Forests are renewable natural resource. They are a national asset to be protected and enhanced for the well-being of the people of the country.

(V) NATIONAL ENVIRONMENT TRIBUNAL AND APPELLATE AUTHORITY.

When matters concerning human rights are frontally posed, it becomes the duty of the Judiciary to face and resolve them. The Court Craft has to be equipped with adequate scientific and technical inputs, if it has to adjucate complex technical issues. Appropriate steps in this direction were taken in India following the United Nations' conference on Environment and Development in Rio de Janeiro, Brazil in 1992. A bill called the National Environment Tribunal Bill – 1992, was introduced in the Indian Parliament in August 1992. It seeks to establish a high-powered body to decide cases related with Wild Life and Environment in addition to providing compensation to the people for death, injury or damage to property or to the environment. The concerned bill was passed by the parliament and was accorded the consent of the President of India on 17th June 1995 after which it became an act of parliament under the short title 'Nation Environment Tribunal Act, 1995 (No.27 of 1995). The high powered body, the tribunal which this act seeks to establish shall be equipped with Judicial and Technical members with adequate knowledge and experience of legal, administrative, scientific and technical aspects of the problems related with wildlife, environment and problems arising from handling, use or manufacture of hazardous substances. Only two years later, another act concerning matters connected thereof was promulgated under the short title 'National Environment Appellate Authority Act, 1997 The two acts together have now equipped the Courts of our country with judicial as well as technical knowledge to adjucate issues concerned with damaged environment or accidents arising out of handling, manufacture or disposal of hazardous substances (29;30).

The principle of strict civil liability in accident cases arising from the activities, which involve hazardous substances, has been highlighted in many cases by the Supreme Court of India. An enterprise engaged in activities with potential threats to the health and safety of the persons residing in the surrounding areas of a factory owes an absolute duty to the community to ensure that no harm is caused to any one on account of hazardous and inherently dangerous nature of such activities. Cases seeking compensation for damages to human health property and the environment, particularly contamination of sub-surface water are increasing. There is also an increasing trend in number of industrial disasters.

The United Nations Conference on Environment and Development held at Rio de Janeiro in June 1992 in which our country participated also advised the States to develop laws regarding liability and compensation for the victims of pollution and other environmental damages. It was therefore deemed expedient to develop and codify the principles of strict civil liability in respect of all such cases where damage is caused while handling hazardous substances. For an effective and expeditious disposal of cases arising out of industrial accidents and disasters and for providing relief to the persons affected thereof, the Nation Environment Tribunal Act, 1995 and the National Environment Appellate Authority Act, 1997 have been promulgated (29;30).. The two acts contain provisions for:

1. Establishment of high-powered National Environment Tribunal and National environment appellate authority, their composition, the qualification of its members and the manner of transaction of their business.
2. The Jurisdiction and proceedings of the tribunal or appellate authority and manners connected therewith.
3. Penalty and punishment for failures to comply with the orders of the tribunal or the appellate authority.

4. Empower the Central government to make rules by notification to serve the purpose of the act. A number of rules and regulations have been introduced in pursuance of provisions of the act to facilitate speedy disposal of the disputes and provide compensation to the victims of accidents arising thereof.

In pursuance of the powers conferred by Section 22 of the National Environment Appellate Authority Act 1997, the Central Government has made the following rules:

1. The National appellate Authority (Appeal) Rules, 1997.
2. The National Environment Appellate Authority (Financial and Administrative Powers) Rules, 1998.
3. The National Environment Appellate Authority (Salary, Allowances and Conditions of Service of Members) Rules 1998.
4. The National Environment Appellate Authority (Salary, Allowances and Conditions of Service of Chairperson and Vice Chairperson) Rules 1998.

(VI) ENVIRONMENTAL IMPACT ASSESSMENT IN INDIA.

Environmental Impact Assessment studies in India were started in the year 1978 for Environmental clearance on the directives of Planning Commission. However, EIA became mandatory only after 1994 for a number of projects following notification issued by the Ministry of Environment and Forests in exercise of powers conferred to it under Section 3(1) and Section 3(2)(v) of Environment (Protection) Act 1986 read with Rule 5(3)(a) of environment (Protection) Rules 1986. EIA regulations have been amended in 1997 to include mandatory public hearing.

Projects requiring environmental clearance have to obtain no objection certificates (NOC) from respective State and Central Pollution Control Boards and land and water availability approval from the concerned State Government. The ministry of Environment and forests examines both short term and long term impact of the project on environment, natural resources and wildlife. The environmental clearance so given is valid for 5 years. Also if a project once cleared undergoes changes fresh assessment is required. The grant of environmental clearance under Environment (Protection) Act, 1986 could be appealed to National Environmental Appellate Authority. The authority has however the discretion to entertain such appeal even after 30 days but not after 90 days from the date of order/clearance, provided they are satisfied that there is sufficient ground for filing the appeal. The authority shall dispose the petition within 90 days from the date of filing the appeal. The authority while processing or hearing the petition shall be guided by principles of natural justice subject to the provisions of the Act and rules made Central Government.

(D) ARRANGEMENTS AND INSTITUTIONS FOR THE PROTECTION OF THE ENVIRONMENT

The concern for the environment has resulted into the establishment of a number of national and international organizations dedicated to the cause of environmental protection. These arrangements or organizations have developed in response to problems and challenges faced by the world community and are still in the process of evolution dictated by the requirements of the time. Many of these organizations were created and are still being used for purposes not concerned with environment, however, they have played an important role or are still playing an important part in activities concerned with environmental protection. Some of the important instruments or organizations which play an important part in development and implementation of environmental policy are listed below:

(I) THE UNITED NATIONS SYSTEM

United Nations with its headquarters at New York, USA, is an organization which has played and is still playing a very important role in the development of environmental policies and the resolution of many environmental problems. It is a global organization consisting of 185 states which is platform for discussion and resolution of all sort of problems of its member states. The charter for the establishment of this world body was approved and signed by 51 countries representing 80% of the world's population on June 26, 1945, at San Francisco.

United Nations function mainly through a structure of committees, commissions, assemblies, councils and some semi-autonomous bodies. The Secretariat of the United Nations provides administrative services for the General Assembly and its other bodies. It is headed by a Secretary General which is elected by the General Assembly.

United Nations Environment Programme (UNEP) is one of the most important institutions shaping the global environmental policy and its implementation. It was created by the United Nations General Assembly in December 1972, following a decision at the United Nations Conference on the Human Environment at Stockholm (1972) which had recommended the creation of a new U.N. Environment Body. UNEP has been designed to be the environmental conscience of the UN System which plays a catalytic role coordinating the activities of other UN Agencies and prompting them to integrate environmental consideration into their activities. Headquarters of UNEP are located in Nairobi, Kenya.

UNEP's environmental activities involve treaties and guidelines concerning environmental protection. It has been active in the area of International environmental law making and their implementation. More than forty out of nearly a hundred multi-lateral environmental treaties adopted outside the European Community since 1973 were concluded under the auspices of UNEP. Environmental Assessment has been a keystone of UNEP's activity. It administers or co-ordinates activities of Global Environmental Monitoring System (GEMS) and Background Air Pollution Monitoring Network with World Meteorological Organization. With World Health Organization (WHO) it monitors health related information and with Food and Agricultural Organization it monitors food contamination from all sources. UNEP's environmental education programme is aimed at educating the general public. It involves about 10,000 educators in nearly 140 countries of the world. UNEP's important area of activity today has been the building up of a wide spread social and political consensus for sustainable development while it continues to carry out its earlier commitments to the cause of protection of nature, natural resources and the environment.

(II) UNITED NATIONS SPECIALIZED AGENCIES

A large part of the activity of United Nations related with Environment is carried out by specialized agencies established by the General Assembly or by some other sub-division of the United Nations System. Food and Agricultural Organization (FAO), World Health Organization (WHO), United Nations Educational, Scientific and Cultural Organization (UNESCO) etc. are such bodies. These specialized agencies of the United Nations have played a diverse role in development and implementation of global environmental protection policy. They are instrumental in providing information, guidance, necessary means and sponsorships of various projects, conferences and conventions as and when required for the efforts of conservation of nature, natural resources and the environment. These bodies are, however, distinct from semi-autonomous organizations like, United Nations Environment Programme (UNEP) or United Nations Development Programme (UNDP), which are established directly by treaties among member states and are therefore more important organizations devoted to some specific cause such as development or environmental protection.

(III) INTERGOVERNMENTAL PANEL ON CLIMATE CHANGE (IPCC)

The Intergovernmental Panel on Climate Change (IPCC) is a scientific body which was established in 1988 by the World Meteorological Organization (WMO) and the United Nations Environment Programme (UNEP) and was entrusted with the task of evaluating the risk of climate change caused by human activity.

The IPCC does not carry out research, nor does it monitor climate or related phenomena. It bases its assessment mainly on reviews and published scientific literature. IPCC reports are widely cited in almost any debate related to climate change. National and international responses to climate change generally regard the UN climate panel as authoritative. The IPCC first assessment report was completed in 1990, which served as the basis for the United Nations Framework Convention on Climate Change (UNFCCC). A main job of the IPCC, today, is to prepare special reports on topics relevant to the implementation of the UN Framework Convention on Climate Change. The implementation of the UNFCCC led eventually to the Kyoto Protocol.

(IV) POLITICAL ALLIANCES AND ASSOCIATIONS OTHER THAN UN SYSTEM

There are many multi-lateral political alliances and associations of states which deal with environmental matters. No doubt many of these bodies were originally created for entirely different set of problems and environmental matters occupy a relatively small space in their agenda, but at times they participate in transactions of major environmental significance. The important ones among such organization are briefly discussed as follows:

(1) NORTH ATLANTIC TREATY ORGANIZATION (NATO)

North Atlantic Treaty Organization was created in 1949 by Western Allied Forces for the purpose of common defence. However, it also includes economic, cultural and indirectly environmental matters among its functions. NATO's committee on Challenges of Modern Society (CCMS) established in December 1969 has sponsored many studies on pollution control and national and regional impact of air pollution. Restoration of contaminated land, air pollution, estuarine management and environmental impact of military activities have been some of its other activities related with environment.

(2) COUNCIL FOR MUTUAL ECONOMIC ASSISTANCE (CMEA)

The Council for Mutual Economic Assistance created after the Second World War is a Soviet Sponsored Council for mutual economic co-operation and has been primarily a political and economic organization. However, the council has extended its area of activity to cover technical and scientific co-operation and research among its member countries. At its 31st Sesson CMEA adopted an expanded programme of environmental co-operation with the Economic Commission of Europe in areas like transboundary air pollution, waste reduction technology, conservation and rational management of water resources and protection of flora, fauna and natural resources.

(3) EUROPEAN COMMUNITY

European Community is a regional organization which was initiated at the Congress of Europe which met at Hague in 1948. The constituent parts of European Community – the European Coal and Steel Community, European Economic Community and the European Atomic Energy Community – were merged in 1967 to form a single body consisting of the European Parliament and council of ministers the Commission and the European Court of Justice. There are twelve European countries as members of the European Community.

The European Community offers an instructive example of emergence of environment as an important theme for the international policy. The Community was formed primarily for economic purposes but subsequent introduction of environmental matters shows the inter-connectedness of economic and environmental affairs and highlights the growth of environmental awareness among the people of member countries. A comprehensive environmental programme of European Community was approved and launched by its council of Ministers in 1973 which declared that the improvement in the quality of life and protection of natural environment are among the fundamental tasks of the Community. The Second Environmental Action Programme of European Community was started in 1977. The third, fourth and fifth Action Programmes were launched in 1982, 1987 and 1993 respectively. The last of these programmes, the fifth Action programme is expected to continue till 2000 A.D. These programmes have dealt with a variety of issues concerning preservation of nature, natural resources and improvement in the quality of the environment.

(4) ORGANIZATION OF AFRICAN UNITY (OAU)

Organization of African Unity, like other organizations discussed earlier has mainly been concerned with matter other than environmental problems. However, the organization has now become a very important instrument of collective action in the entire African continent for matters related with wild life and environment. In 1968, the OAU sponsored a new treaty for the protection of African wild-life. The organization has also sponsored a College of African Wild life Management at Mweka, Tanzania, which has been doing excellent work of imparting environmental education and training the workers devoted to the cause of wild life conservation. OAU's Lagos Plan of Action (1980–2000) brings the environmental concern directly into the socio-economic planning of many countries of Africa. It was an OAU member state which recommended the draft of World Charter for Nature for adoption by the United Nations' General Assembly.

(5) ORGANIZATION OF AMERICAN STATES (OAS)

The Organization of American States earlier known as Pan American Union, like Organization of African Unity has not shown much concern about environmental affairs. However, it was this body which sponsored in 1940 a convention on protection of nature and wild life in the Western Hemisphere. The organization has also assisted various scientific and technical meetings on environmental matters and has undertaken environmental impact assessment of the proposed river development projects in South American countries in collaboration with UNEP.

(6) ORGANIZATION OF ECONOMIC CO-OPERATION AND DEVELOPMENT (OECD)

Organization of Economic Co-operation and Development with its headquarters at Paris, was initially created for the development and reconstruction of Post World War II, Europe. Originally the countries participating in the Marshal Plan for reconstruction of Europe were signatories to the OECD. Today some other countries, like New Zealand, Australia and Japan have also joined the organization. Although the basic purpose of OECD is to promote economic growth and advancement of international treaties, the organization also provides an important forum for the study, exchange of information relating to environmental problems and co-ordination of policies concerned with environment. The organization has encouraged movement toward a balance between quantitative and qualitative growth and their complementarity and compatibility with environmental policies. In 1970, the Environmental committee of OECD was established at the Ministerial level which deals with the economic implications of environmental problems and the impact of development on the environment. The organization is primarily a consultative body and the majority of work undertaken by the organization is designed to evaluate and recommend policy options together with guidelines and strategies for the implementation.

(7) SOUTH ASIAN ASSOCIATION FOR REGIONAL COOPERATION

The South Asian Association for Regional Cooperation (SAARC) is an economic and political organization of eight countries in Southern Asia. As far as population is concerned, its sphere of influence is the largest of any regional organization. The combined population of its member states is almost 1.5 billion people. It was established on December 8, 1985 by India, Pakistan, Bangladesh,

Sri Lanka, Nepal, Maldives and Bhutan. In April 2007, at the Association's 14th summit, Afghanistan became its eighth member.

(8) SOUTH ASIA CO-OPERATIVE ENVIRONMENT PROGRAMME (SACEP)

South Asia Co-operative Environment Programme consists of nine Asian Countries (seven member countries of SAARC plus Afghanistan and Iran) which was conceived earlier but was given shape at the United Nations' Conference on Environment and Development held at Rio De Janiero in 1992. Our country, India, is an important participant in this organization. The Governing council of SACEP met in September 1992 at Colombo, to discuss the setting up of SACEP trust fund. The meeting also drew up time bound action plan for the next four years dealing with inter-state subjects like Himalayan ecology, river water conservation, oceanography and biotechnology. It was felt that the proposed United Nations' Commission on Sustainable development should be made more powerful so that it could monitor the progress of the programme and the availability of funds. The need of quick development in poor countries with the necessity of preservation of wild life, natural resources and environment was stressed by the governing council.

(V) NON-GOVERNMENTAL ORGANIZATION (NGOS)

Non-governmental organizations have played a very important role in the development of environmental policy in many countries as well as the emergence of global awareness about the state of environment. Non-governmental organizations usually consist of volunteers or group of individuals genuinely interested in the cause of conservation of wild life, natural resources and protection of the environment for which they generate their own resources, contributory funds, grants etc. They are often able to manage things more efficiently and easily than governments which work with authority. They have played a great role in the development of public awareness about environmental matters and much of evolution of national and international environmental co-operation can be attributed to this public awareness – as in a democratic set up it is the general public opinion, which shapes the policies of a country. Some of the more important non-governmental bodies (NGOs) are being discussed here:

(1) INTERNATIONAL COUNCIL OF SCIENTIFIC UNIONS (ICSU)

International Council of Scientific Unions is a federation of federations consisting of about twenty unions which are themselves a composite structure made up of a large number of subsidiary organizations. In order to direct the diverse and highly specialized resources of the Union to the major global problems, ICSU works through inter-union, interdisciplinary committee devoted to particular scientific issue or problem. Among the several types of committee established by ICSU important ones are: Scientific Committee on Oceanic Research (SCOR) formed in 1957, Scientific Committee on Space Research (COSPAR) established in 1958, the Scientific Committee on Antarctic Research (SCAR) and the Scientific Committee on Problems of Environment (SCOPE) established in 1969.

SCOPE is probably one of the most important committee established by ICSU concerned with the problems of environment. Its main objectives are to advance knowledge of the influence of humans on their environment as well as the effect of this environmental change upon the people, their health and their well-fare. It pays more attention to those influences and affects which are either global or are shared by several nations. It serves as a non-governmental, interdisciplinary and international council of scientists which is a source of information, advice and guidelines for the benefit of governments and inter-governmental and non-governmental bodies regarding the matters of environment.

(2) INTERNATIONAL UNION FOR CONSERVATION OF NATURE AND NATURAL RESOURCE (IUCN) OR WORLD CONSERVATION UNION

International Union for Conservation of Nature and Natural Resources was established in 1948, as an International Union for the Protection of Nature (IUPN). The Organization with its wide ranging policies has now assumed world leadership in matters wild life conservation and conservation of natural resources and protection of environment. In 1957, the union renamed itself as International Union for Conservation of Nature and Natural Resources. The Head quarters of the Union are situated in Gland, Switzerland.

IUCN is now globally operated highly influential union of government agencies, national and international non-governmental organizations and loyal supporters who are dedicated to the cause of conservation. The organization is concerned with initiations and promotion of scientifically based action which endeavour to establish links between development and environment, to provide lasting improvement in the quality of life for people all over the world. IUCN works through a General Assembly which is convened at the interval of every three years to determine the policies and broad elements of the organization, a Council elected by the General Assembly which meets at least annually and Commissions consisting of a body of experts and volunteers who make major contribution to the development and execution of IUCN Programmes. Currently there are six commissions, namely: commission on Ecology, on Education and Communication, on Environmental Law, on Environmental Strategy and Planning, on National Parks and Protected Areas and on Species Survival. The IUCN Commissions constitute a global net-work of more than 6,000 scientists and other specialists.

(3) WORLD WILD-LIFE FUND OR WORLD WIDE FUND FOR NATURE (WWF)

World Wild-life Fund which was established in the year 1961, with its headquarters at Glands Switzerland is one of the largest and most important conservation organizations in the world. The WWF logo, a Giant Panda over the letters WWF is perhaps the most widely recognized symbol of Conservation in the world. By 1993, WWF had developed affiliations' with over 28 organizations spread over five continents of the world and had invested more than 330 million dollars in over 10,000 different projects in 130 different countries since its creation in 1961. WWF International provides assistance to scientific research on wild life and supports every day, on the spot conservation programmes and initiatives on four continents – Europe, Asia, Africa and Latin America. Similar efforts are sponsored in U.S. and Canada and Australia by national WWF Organization in these countries. After the Silver Jubilee celebrations in 1986, the World Wild Life Fund was renamed World Wide Fund for Nature.

WWF has enlarged its focus of activity to include issues of sustainable development. It endeavours to integrate wild life conservation with meeting the needs of human society in such ways that the exploitation does not threaten the fragile habitats. Thus, in Nepal, WWF has been planting fuel wood to provide for needs of the people so that natural habitats are not disturbed. It has recognized that to conserve the world's natural resources for future generations we shall have to help alleviate the human crisis and pressure of development which puts wild life and wild land in danger.

In Asia WWF's efforts are especially intensive in countries like Indonesia, Bhutan, Philippines, Thailand, Papua and Nepal. The Trade Record Analysis of Flora and Fauna in Commerce (TRAFFIC) which is a large and influential section of WWF monitoring trade in wild life and wild life products has been very successful in halting the transactions in bear parts, skins and hides of endangered animals and rhino horns. TRAFFIC operates 15 offices all over the world and has been responsible for shutting down smuggling operations involving parrots, spotted cats, sea turtles and tropical timber in Latin America. The fund has also been associated with preservation of Coral reef ecosystems of the world.

(4) NATIONAL AUDUBON SOCIETY

It was as early as 1880 that George Bird Grinnell who grew up in Audubon Park, New York, proposed the creation of an Audubon Society, named in honour of the noted naturalist of the time John James Audubon, dedicated to the cause of protection of wild bird. Following this suggestion nearly 36 Audubon groups were formed by 1905, which were later integrated into the National Committee of Audubon Society. The Society, a strong advocate of protection of wild birds, has been instrumental in the promulgation of important acts aimed for the protection of wild life. Over the years the Society has spread its activity to cover the protection of wild life in general, wild lands, natural habitats and other environmental causes. The Society has over 500 local chapters located in all the 50 states with a membership totaling over 550,000. It maintains education centers, camps, research stations, bird and wild life sanctuaries all over America.

(5) SIERRA CLUB

The Sierra Club was established in 1892 with John Muir as its first president and an elite membership largely drawn from San Francisco bay area. Its objectives included exploration, entertainment and access to mountain regions of the Pacific Coast, publications of authentic information concerning them and enlistment of support and co-operation of the public and government for the preservation of forests and other natural features of Sierra Navada Mountains. Today Sierra Club has expanded its activity to promote the conservation and responsible use of earth's ecosystems and resources of nature, to educate and enlist humanity to protect and restore the quality of natural and human environment. The Organization has more than 50,000 members while its annual budget amounts to about 38 million dollars. The organization is composed of 13 regions, 63 chapters which are subdivided into 400 local groups.

(6) BOMBAY NATURAL HISTORY SOCIETY (BNHS)

Since its inception in 1883, the Society has been an important non-government organization dedicated to the cause of protection of wild life and nature conservation. The Society has been actively involved in collection of scientific information about the flora and fauna all over India. The Society has been instrumental in focusing public and official attention on the need for properly understanding, conserving and developing the rich heritage of India's wild life and natural resources. The prestigious Journal which the Society publishes has been serving well the cause of nature conservation through the publication of articles and new information about the flora and fauna of India.

QUESTIONS

1. Describe the development of environmental legislation in India.
2. Discuss impact of 42nd Amendment to the constitution of India (Introduction of articles 48A 51A) and hesitation of Judiciary decide complex technical issues concerned with environmental litigations.
3. Give salient features of any one of the following Acts:
   a. Water (Prevention and Control of Pollution) Act 1974.
   b. Air (Prevention and control of pollution) Act 1981.
   c. Environment (Protection) Act 198.

4. Water (Prevention and Control of Pollution) Act 1974, Air (Prevention and control of pollution) Act 1981 and Environment (Protection) Act 198 not only provide a comprehensive set of rules for protection of environment but have also equipped the Government with an efficient machinery for the surveillance and implementation of their provisions. Discuss
5. Enumerate different Forest Conservation Acts along with their salient features.
6. Give salient features of National Forest Policy (1988) of India.
7. Give salient features of wild life protection Act 1972. Enumerate the rules made in pursuance of he provision of the act.

REFERENCES CITED

CHAPTER 1: THE ENVIRONMENT

1. **Sytnik KM (Ed), 1985,** *Living in the environment.* Translated from Russian, UNESCO, Naukova, Dumka Publ.
2. **Jacobson Mark Zachary, 2005,** *Fundamentals of Atmospheric Modeling* (2nd Edition ed.). Cambridge University Press.
3. **Annonymous, 1970,** *Meteorological aspects of air pollution.* World Meteorological organization, Tech. Note no. 106, Geneva.
4. **Menon PA, 1989,** *Our Weather.* National Book Trust, New Delhi, India.
5. **Ahrens Donald C, 2006,** *Meteorology Today* (8th Edn). Brooks/Cole Publishing.
6. **Campbell Neil A, Brad Williamson, Robin J Heyden, 2006,** *Biology: Exploring Life.* Boston, Massachusetts: Pearson Prentice Hall.
7. **Frank F (Ed), 1973,** *Water–A comprehensive treatise.* Plenum Press, N.Y.
8. **Holland HB, 1978,** *The Chemistry of atmosphere and oceans.* Wiley-Inter science, N.Y.
9. **Odum EP, 1983,** *Basic ecology.* Saunders College Publ., N.Y.
10. **Penman HL, 1970,** The water cycle. *Sci. American,* **223** (3): 99.
11. **Lutgens Frederick K and Tarbuck Edward J,** *Essentials of Geology,* 7th Ed., Prentice Hall, 2000.
12. **Bailey RA, et al., 1978,** *Chemistry of the environment.* Academic Press, N.Y.
13. **Brady NC, 1974,** *The nature and properties of the soil.* Mc. Milan, N.Y.
14. **Strahler, Alan H and Arthur Strahler, 2003,** *Physical Geography: Science and Systems of the Human Environment.* 2nd Edition
15. **Christopherson RW, 2005,** *Geosystems: An Introduction to Physical Geography.* 5th Edition. Prentice Hall, Upper Saddle River, New Jersey.
16. **Schopf JW, 1978,** The evolution of earliest cells. *Sci. Amer.,* **239**(3):84-102.
17. **Brancazio Peter J, Cameron AGM, 1986,** *The origin and evolution of atmosphere and oceans.* John Wiley & Sons, N.Y.
18. **Cloud P, 1968,** Atmospheric and hydrospheric evolution on primitive earth. *Science,* **160** : 729.

CHAPTER 2: THE ECOSYSTEM AND ECOSYSTEM SERVICES.

1. **Southwick CH, 1972,** *Ecology and quality of our environment.* V on Nostrand Company, N.Y., London
2. **Odum EP, 1983,** *Basic ecology.* Saunders College Publ., N.Y.
3. **Krebs Charles, 1994,** *Ecology* (4th Edn) Harper Collins.
4. **Kormondy E J, 1991,** *Concepts of Ecology.* 3rd Edn. Prentice-Hall of India, Pvt.Ltd.New Delhi.
5. **Barrett GW, 1978,** *Stress ecology.* In *Proc. 2nd Internat. Congr. Ecology,* Jerusalem Abs. Vol I, p.20.
6. **Holling CS, 1973,** Reslience and stability of ecological systems. *Ann. Rev. Ecol. System.,* **5:**1-23.

7. **May RM, 1973,** *Stability and complexity in model ecosystems*. Princeton University press.
8. **Fog GE, 1998.** *The Biology of Polar Habitats.* Oxford University Press, New York.
9. **Brush GS, 1982,** An environmental analysis of forest patterns. *American Scientist*, **70:**18-25.
10. **Richards P, 1973,** The tropical rainforests. *Scientific American*, **229**(6):58-67.
11. **Botkin D and E Keller, 1995,** *Environmental Science: Earth as a Living Planet*. John Wiley & Sons, Inc., Canada.
12. **Chernov YI, 1985,** *The Living Tundra.* Cambridge University Press, Cambridge.
13. **Campbell NA, 1996,** *Biology*, 4th Edition. The Benjamin/Cummings Publishing Company, Inc., Menlo Park, California.
14. **Odum EP, 1993,** *Ecology and Our Endangered Life-Support Systems.* Sinauer Associates, Inc., Sunderland, Massachusetts.
15. **Spurr SH and BV Barnes, 1980,** *Forest Ecology.* John Wiley & Sons, New York
16. **Rodgers CL, and RE Kerstetter, 1974,** *The Ecosphere, Organisms, Habitats, and Disturbances.* Harper & Row, New York.
17. **Brown, L. 1972.** *The Life of the African Plains.* McGraw-Hill Book Company, New York.
18. **Costello DF, 1969,** *The Prairie World.* Thomas Y. Crowell Company, New York.
19. **MacMahon James A, 1985,** *Desert.* Alfred A. Knopf, Inc, New York, 640 p.
20. **Nelson R, 1988,** Dry land management: the desertification problem. Working Paper No. 8, 42 p. World Bank, Washington
21. **Tolba, MK, 1984,** Desertification is stoppable: *Arid Lands Newsletter* No. **21**, p. 2-9.
22. **Farb P, 1970.** *Ecology.* Time, Inc., New York.
23. **Daily GC, 1997,** *Nature's Services: Societal Dependence on Natural Ecosystems.* Island Press, Washington. 392pp.
24. **Pimentel D et al, 1997,** Economic and environmental benefits of Biodiversity. ***Bio-Science***, **47**(11):747-757.
25. **Osborn F, 1948,** *Our Plundered Planet.* Boston. Little, Brown and Company. 217pp.
26. **Vogt W, 1948,** *Road to Survival.* William Sloan: New York. 335pp.
27. **Leopold A, 1949,** *A Sand County Almanac and Sketches from Here and There.* **Oxford University Press, New York. 226pp.**
28. **Sears, P.B. 1956.** The processes of environmental change by man. In: W.L. Thomas, (Ed), *Man's Role in Changing the Face of the Earth* (Vol. 2). University of Chicago Press, Chicago. pp.1193.
29. **Ehrlich PR and A Ehrlich, 1970,** *Population, Resources, Environment: Issues in Human Ecology.* W.H. Freeman, San Francisco. 383pp. (see p.157)
30. **Millennium Ecosystem Assessment (MEA). 2005.** *Ecosystems and Human Well-Being: Synthesis*. Island Press, Washington. 155pp.
31. **Pimentel D, et al, 1995,** Environmental and economic cost of soil erosion and conservation benefits. *Science* **267:**1117-1123.
32. **Pimentel D, et al, 1992,** Conserving biological diversity in agricultural/forestry systems. *Bio Science* **42**:354-362.
33. **USGS 1995,** US Geological survey toxic substances hydrology program, Washington DC: (http://h20.usgs.gov/pulbic/wid/htm/toxic.htm.
34. **Pimentel D, et al, 1997,** Economic and environmental benefits of Biodiversity. ***Bio-Science***, **47**(11):747-757.

35. **Dobereiner J, Baldani VLD, Reis VM, 1995,** Endophytic occurrences of diazotrophic bacteria in non-leguminous crops. pp 3-14. In Frendrick I (Ed) *"Azospirillum VI and related microorganisms: genetics, physiology, ecology"*. Berlin:Springer-Verlag.

36. **Peoples MB, Craswell ET, 1992,** Biological nitrogen fixation: investments, expectations and actual contribution to agriculture. *Plant & Soil* **141**:13-39.

37. **Pimentel D, 1980,** *Handbook of energy utilization in agriculture.* Boca Raton (FL): CRC Press.

38. **Schopf JW** (Ed) **1983,** *Earth's Earliest Biosphere. Its Origin and Evolution.* Princeton, New Jersey, U.S.A.: Princeton Univ. Press.

39. **Van Valen L, 1971,** The history and stability of atmospheric oxygen. *Science,* **171**, 439-443.

40. **Ruddiman William F, 2005,** *Earth's Climate Past and Future.* New York: Princeton University Press.

41. **Chambers S, 1998,** *Short- and Long-Term Effects of Clearing Native Vegetation for Agricultural Purposes.* PhD thesis, Flinders University of South Australia.

42. **Salati E, 1987,** The forest and the hydrological cycle (pp. 273-294). In Dickenson, R. (Ed.), "*The Geophysiology of Amazonia*". New York, John Wiley and Sons.

43. **Laurance WF and Williamson GB, 2001,** Positive Feedbacks among Forest Fragmentation, Drought, and Climate Change in the Amazon. *Conservation Biology*, **15:** 1529-1535.

44. **Wilcove DS, D Rothstein, J Dubow, A Phillips and E Losos, 1998,** Quantifying threats to imperiled species in the United States. *BioScience,* **48:** 607–615.

45. **Levine JM, T Kennedy and S Naeem, 2002,** Neighborhood scale effect of species diversity on biological invasions and their relationship to community patterns. pp. 114–124. In: *Biodiversity and Ecosystem Functioning,* M. Loreau, S. Naeem and P. Inchausti (Eds.), Oxford University Press, Oxford,

46. **Pimentel D, L Lach, R Zuniga, and D Morison, 2000,** Environmental and economic costs of non-indigenous species in the United States. *Bio Science,* **50:** 53–65.

47. **Kennedy TA, S Naeem, KM Howe, JMH Knops, D Tilman, and PB Reich, 2002,** Biodiversity as a barrier to ecological invasion. *Nature,* **417**:636–638.

48. **DeBach P, and Rosen D, 1991,** Biological control by natural enemies. New York: Cambridge University Press.

49. **Oerke EC, Dehne HW, Schonbeck F, Weber A, 1994,** *Crop production and crop protection: estimated losses in major food and cash crops.* Amsterdam: Elsvier.

50. **Pimentel D et al, 1997,** Economic and environmental benefits of Biodiversity. *Bio-Science,* **47**(11):747-757.

51. **Boudreau MA and CC Mundt, 1997:** Ecological approaches to disease control. pp. 33–62. In: *Environmentally Safe Approaches to Crop Disease Control,* Rechcigl NA and JE Rechcigl (Eds), CRC Press, Boca Raton,.

52. **Zhu Y, HR Chen, JH Fan, YY Wang Y Li, JB Chen, JX Fan, SS Yang, LP Hu et al, 2000,** Genetic diversity and disease control in rice. *Nature,* **406:**718–722.

53. **Ngugi HK, SB King, J Holt, and AM Julian, 2001:** Simultaneous temporal progress of sorghum anthracnose and leaf blight in crop mixtures with disparate patterns. *Phytopathology,* **91:** 720–729.

54. **Mundt CC, 2002,** Use of multiline cultivars and cultivar mixtures for disease management. *Ann. Rev. Phytopathology,* **40:**381–410.

55. **Taylor LH, SM Latham and ME, Woolhouse J, 2001**: Risk factors for human disease emergence. *Philosophical Transactions of the Royal Society of London B,* **356:** 983–989.

56. **Buchmann SL and GP Nabhan, 1996,** *The forgotten pollinators.* Island Press, Washington, D.C., 312 pp.

57. **Steiner KE, 1993**, Has *Ixianthes* (Scrophulariaceae) lost its special bee? *Plant Systematics and Evolution,* **185:** 7–16.

58. **Lord JM, 1991**, Pollination and seed dispersal in *Freycinetia baueriana*, a dioecious liane that has lost its bat pollinator. *New Zealand Journal of Botany,* **29:** 83–86.

59. **Johnson SD and WJ Bond, 1994,** Evidence for widespread pollen limitation of fruiting success in Cape wildflowers. *Oecologia,* **109:** 530-534.

60. **Heschel MS and KN Paige, 1995**, Inbreeding depression, environmental stress, and population size variation in scarlet gilia (*Ipomopsis aggregata*). *Conservation Biology,* **9:** 126–33.

61. **Scott-Dupree CD and ML Winston, 1987,** Wild bee pollinator diversity and abundance in orchard and uncultivated habitats in the Okanagan Valley, British Columbia. *Canadian Entomologist,* **119:** 735–745.

62. **Kremen C, NM Williams, and RW Thorp, 2002,** Crop pollination from native bees at risk from agricultural intensification. *Proceedings of the National Academy of Science,* **99:** 16812–16816.

63. **Ricketts TH, GC Daily, PR Ehrlich and CD Michener, 2004,** Economic value of tropical forest production to coffee production. *Proceedings of the National Academy of Sciences USA,* **101:** 12579–12582.

64. **Jordano P, 1987**, Patterns of mutualistic interactions in pollination and seed dispersal: connectance, dependence asymmetries, and co-evolution. *The American Naturalist,* **129:** 657–677.

65. **Jordano P. and EW Schupp, 2000,** Seed disperser effectiveness: The quantity component and patterns of seed rain for *Prunus mahaleb*. *Ecological Monographs,* **70:** 591–615.

66. **Jordano P, 1992**, Fruits and frugivory. In: *Seeds: The Ecology of Regeneration in Plant Communities,* M. Fenner (Ed.), Commonwealth Agricultural Bureau International, Wallingford,

67. **Pacheco LF and JA Simonetti, 2000,** Genetic structure of a mimosoid tree deprived of its seed disperser, the Spider Monkey. *Conservation Biology,* **14:** 1766–1775.

68. **Redford KH and P Feinsinger, 2001,** The half-empty forest: sustainable use and the ecology of interactions. In: *Conservation of Exploited Species* pp. 370–400. J.D. Reynolds, G.M. Mace, K.H. Redford, and J.G. Robinson (Eds), Cambridge University Press, Cambridge,

69. **Terborgh J, 1986,** Community aspects of frugivory in tropical forests. In: *Frugivores and Seed Dispersal,* A. Estrada and T.H. Fleming (Eds), pp. 371–384, Dr. W Junk Publishers, Dordrecht, The Netherlands.

70. **Galetti M and A Aleixo, 1998**, Effects of palm heart harvesting on avian frugivores in the Atlantic rain forest of Brazil. *Journal of Applied Ecology,* **35,** 286–293.

71. **Moegenburg SM, 2002**, Harvest and management of forest fruits by humans: implications for fruit-frugivore interactions. In: *Seed dispersal and frugivory: ecology, evolution and conservation.* D. Levey, W.R. Silva and M. Galetti (Eds). 479–494. Oxon, CABI Publishing.

CHAPTER 3: IMPACT OF HUMAN ACTIVITY ON THE ENVIRONMENT.

1. **Frejka T, 1973,** The prospects of a stationary world population. *Sci. Amer.* **68:**388-397

2. **Frekla T, 1981,** Long term prospects of world population growth. *Population and Development Review,* **7** (3) : 489. Sept. 1981.

3. **Brown LR, 1981,** World population growth, soil erosion and food security. *Science*, **214** : 995.
4. **Vitousek PM, J Lubchenco, HA Mooney, J Melillo, 1997,** Human domination of Earth's ecosystems. *Science* **277:** 494-499.
5. **Whittaker RH, GE Likens, 1975,** The biosphere and man. **In** ***Primary productivity of the biosphere***. Lieth H, R H Whittaker (Eds), Springer-Verlag N.Y.
6. **Olson J et al, 1983,** *Carbon in live vegetation of major ecosystems.* Report for U.S. Department of Energy, TR004, DOE/NRB-0037.
7. **Southwick C.H., 1972,** Ecology and quality of our environment. Von Nostrand Company, N.Y., London.
8. **Holling C S, 1973,** Reslience and stability of ecological systems. Ann. Rev. Ecol. System., 5:1-23.
9. **Krebs Charles, 1994**, Ecology (4th Edn) Harper Collins.
10. **Kormondy EJ, 1991**, Concepts of Ecology. 3rd Edn. Prentice-Hall of India, Pvt.Ltd.New Delhi.
11. **Odum EP, 1983**, Basic ecology. Saunders College Publ., N.Y.
12. **Pimm SL, GJ Russel, JLGittleman, TM Brooks, 1995,** The future of biodiversity, *Science* **269:**347-350.
13. **Kellert RS and EO Wilson, 1993,** The biophilia hypothesis. Washington DC: Island press.
14. **Sasson Albert, 1987,** Conserving plant resources. *The World Scientist*, **16**: 50.
15. **Bolin Bert**, 1970, The Carbon cycle. *Sci. American*, **223** (3): 124.
16. **Bolin Bert, Degens ET, Kempe S, Ketner P, 1979** The global carbon cycle. Chichester: New York: Published on behalf of the Scientific Committee on Problems of the Environment (SCOPE) of the International Council of Scientific Unions (ICSU) by Wiley.
17. **Houghton RA, 2005,** "The contemporary carbon cycle". in William H Schlesinger (Ed). *Biogeochemistry*. Amsterdam: Elsevier Science. pp. 473–513.
18. **Janzen HH, 2004,** "Carbon cycling in earth systems—a soil science perspective". *Agriculture, ecosystems and environment*, **104** (3): 399–417.
19. **IPCC, 2007,** *Climate Change 2007.* Contribution of Working Group I to the Fourth Assessment Report of the Intergovernmental Panel on Climate Change. Intergovernmental Panel on Climate Change, 2007.02.05.
20. **Raven JA et al, 2005,** Ocean acidification due to increasing atmospheric carbon dioxide. Royal Society, London, UK. Policy document 12/05. June 2005. ISBN 0 85403 617 2. www.royalsoc.ac.uk
21. **Caldeira K, Wickett ME, 2003**, "Anthropogenic carbon and ocean pH". *Nature* **425** (6956): 365–365.
22. **Orr James C, Fabry Victoria J, Aumont Olivier, Bopp Laurent, Doney Scott C, Feely Richard A et al 2005,** "Anthropogenic ocean acidification over the twenty-first century and its impact on calcifying organisms". *Nature* **437** (7059): 681-686
23. **Key, RM, Kozyr A, Sabine CL, Lee K, Wanninkhof R, Bullister J, Feely RA, Millero F, Mordy C and Peng T-H, 2004** "A global ocean carbon climatology: Results from GLODAP". *Global Biogeochemical* Cycles 18: GB4031. doi:10.1029/2004GB002247. ISSN 0886-6236.
24. **Cloud P, Gibor A, 1970,** The oxygen cycle. *Sci. American,* **223** (3): 110.
25. **Walker JCG, 1980,** *The oxygen cycle in the natural environment and the biogeochemical cycles*, Springer-Verlag, Berlin,

26. **Ruddiman William F, 2005,** *Earth's Climate Past and Future*. New York: Princeton University Press. ISBN 0-7167-3741-8.

27. **Holland HD, 1984,** *The Chemical Evolution of the Atmosphere and Oceans*, Princeton University Press, Princeton, NJ.

28. **Broecker WS, and Severinghaus JP**, 1992, Diminishing Oxygen, *Nature,* **358**:710-711.

29. **Schlesinger WH, 1991,** *Biogeochemistry: An Analysis of Global Change*. Academic Press, San Diego.

30. **Vitousek PM, Aber J, Howarth RW, Likens GE, Matson PA, Schindler DW, Schlesinger WH, Tilman GD, 1997,** "Human Alteration of the Global Nitrogen Cycle: Causes and Consequences". *Issues in Ecology* 1: 1–17.

31. **Schindler DW and SE Bayley, 1993**, The biosphere as an increasing sink for atmospheric carbon: estimates from increasing nitrogen deposition. *Global Biogeochemical Cycles* **7**:717-734.

32. **Prinn R, D Cunnold, R Rasmussen, P Simmonds, F Alyca, A Crawford, P Fraser, and R Rosen. 1990.** Atmospheric emissions and trends of nitrous oxide deduced from 10 years of ALE-GAGE data. *Journal of Geophysical Research* **95**:18,369-18,385.

33. **Tilman D, 1987**, Secondary succession and the pattern of plant dominance along experimental nitrogen gradients. *Ecological Monographs* **57**(3):189-214.

34. **Nebel Richard J and Richard T Wright, 1998,** *Environmental Science . Prentice Hall*

35. **Fergusson TE, 1982**, *Inorganic chemistry and the earth: chemical resources, their extraction, use and environmental impact*. Pergamon Press, Oxford.

36. **Seinfeld John H, Pandis Spyros N, 1998,** *Atmospheric Chemistry and Physics - From Air Pollution to Climate Change*. John Wiley and Sons, Inc. ISBN 0-471-17816-0

37. **Pimentel D et al, 1997,** Economic and environmental benefits of Biodiversity. Bio-Science, 47(11):747-757.

38. **Pimentel D, et al, 1995,** Environmental and economic cost of soil erosion and conservation benefits. *Science* **267:**1117-1123.

39. **Peoples MB, Craswell ET, 1992,** Biological nitrogen fixation: investments, expectations and actual contribution to agriculture. *Plant & Soil* **141**:13-39.

40. **Ruddiman William F, 2005,** *Earth's Climate Past and Future*. New York: Princeton University Press.

41. **Kennedy TA, S Naeem, KM Howe, JMH Knops, D Tilman, and PB Reich, 2002,** Biodiversity as a barrier to ecological invasion. *Nature,* **417:**636–638.

42. **De Bach P, and Rosen D, 1991,** *Biological control by natural enemies*. New York: Cambridge University Press.

43. **Buchmann SL and GP Nabhan, 1996,** *The forgotten pollinators*. Island Press, Washington, D.C., 312 pp.

44. **Jordano P, 1987**, Patterns of mutualistic interactions in pollination and seed dispersal: connectance, dependence asymmetries, and co-evolution. *The American Naturalist,* **129:** 657–677.

CHAPTER 4: NATURAL RESOURCES

1. **Kesler SE, 1976,** *Our finite mineral resources*. Mac Graw Hill, N.Y.

2. **Dorfman R, 1985,** An economist's view of natural resources and environmental problems. In

The global possible – resource, development and the new century. Ed. Repetto R, Aff EW Press, New Delhi.

3. **Cook E, 1976,** Limit to exploitation of non-renewable resources. *Science,* **191** : 677.
4. **Eckholm EP, 1991,** Global atmospherics. Chapter in *Down to Earth* : *environment and human needs.* Aff. E.W. Press, New Delhi.
5. **Murdoch WW (Ed), 1971,** *Environment, resource, pollution and society.* Sinauer Associates, Stanfard, Connecticut.
6. **Ridker RG, Watson WD, 1980,** *To choose a future.* John Hopkins Univ. Press, Baltimore.
7. **Khoshoo TN, 1990,** Man in nature – past, present and future. In "*Environmental Education and sustainable development*" Desh Bandhu, Harjit Singh, Maitra A.K. (Ed.), Indian Environmental Society, New Delhi.

CHAPTER 5: LAND: A PRECIOUS RESOURCE.

1. **Brady NC, 1974,** *The nature and properties of the soil. Mc. Milan, N.Y.*
2. **Hunt CB, 1972,** *Geology of soils.* W.H. Freeman, San Francisco.
3. **USDA, ERS, World Indices.** World Indices of Agriculture and food Production for various years.
4. **Hrabovazky Janos P, 1991**, Agriclture : the land base. In *The Global Possible : resource, development and the new century. Repetto R.* (Ed.) Aff. E.W. Press, New Delhi.
5. **Valdiya KS, 2004,** *Geology Environment and society* University Press (India) Private Ltd, Hyderbad
6. **Chakrapani GJ, 2005,** Factors controlling variations in river sediment load. *Current Science,* **88**(4):569-573.
7. **Abrol I P, 1994,** Land degradation: A challenge to sustainability. In *"Salinity management for sustainable agriculture"* Eds. Rao D L N, N T Singh, R K Gupta. Central Soil Research Institute, Karnal.
8. **Ghassemi F, A J Jakeman, H A Nix, 1995,** *Salinization of land and water resources.* University of New South Wales Press, Sydney.
9. **Stanley Wood et al 2001** *Report released by International Food Policy Research Institute, 09.02.2001*
10. **Gujja B., H. Shaik, 2003,** Linking rivers: learning from other's mistakes. In "*The Hindu Survey of Environment, 2003*"
11. **Singh KP, Parwana HK 1999** Ground water pollution due to industrial wastewater in Punjab state and strategies for its control. *Indian J Environ Prot,* **19**(4): 241-244. (1999),
12. **Qi Xu 2006** Facing upto "Invisible pollution". *China Environmental Times* Beijing, December 28, 2006. China's soil pollution
13. **Ansari A A, I B Singh, H J Tobschal 1999** Status of anthopogenically induced metal pollution the Kanpur-Unnao industrial region of Ganga plains. *Environ Geol* **38**:25-33 High level of heavy metal in sediments and soil.
14. **Vernon Gill Carter, Tom Dale, 1974,** *Topsoil and Civilization,* University of Oklahoma Press, Norman Oklahoma 292 pp
15. **Gary Gardner, 1996,** "Shrinking Fields: Cropland Loss in a World of Eight Billion", World Watch Paper:131 (July 1996) 56 pp.

16. **Gustave Speth J. 1993** *Towards sustainable food security.* Sir John Crawford Memorial Lecture, October 25, 1993, Published by CGIAR, Washington DC

CHAPTER 6: FRESHWATER.

1. **Penman H.L., 1970,** The water cycle. *Sci. American,* **223** (3) : 99.
2. **L'vovitch M.J., 1979,** *World water resources and their future.* English translation. Ed. R.L. Nace. American Geophysical Union, Washington D.C.
3. **Roggers P.P., 1991,** Fresh-water. In The *Global possible: resource, development and the new century.* Ed. Repetto R., Aff. E.W. Press, New Delhi.
4. **Valdiya KS, 2004,** *Geology, Environment and Society*, University Press (India), Hyderabad.
5. **Bartarya S K, K S Valdiya 1989** Diminishing discharges of mountain springs in a part of Kumaun Himalaya. *Current Science* **58**:417-426
6. **Negi G C S, V Joshi, K Kumar, 1998,** Spring sanctuary development to meet household water demand in mountains: A call for action. In *" Research for mountain development: some initiatives and accomplishment'*. Nainital, Gyanodaya Prakashan, pp 25-46
7. **Upreti B K 1988** Impac of riverbed erosion flood hazards and river shifting in Nepal. In *"Proc. Intern.Symp. Impact of Riverbed erosion Flood Hazard and Problems of Population Desplacement"*, Dhaka, 27 pp.
8. **Kattlemann R, 1987,** Uncertainty in assessing Himalyan water resources. *Mountain Research and Development,* **7**:279-286.
9. **Valdiya KS, 1987,** *Environmental Geology: Indian context.* New Delhi: Tata-McGraw Hill pp 583
10. **Valdiya KS, 1997,** *Developing a paradise in peril.* Kosi-Katarmal: 17th GB Pant Memorial Lecture, GB Pant Institute of Himalyan Environment and Development. 26 pp.
11. **Sloggett G, Dickason C, 1986,** *Ground Water Mining in United States.* USDA, Washington D.C.
12. **Sandra Postel, 1985,** Managing fresh-water supplies. In *State of the world-1985. A Worldwatch Institute Report on progress towards a sustainable society*" Ed. Linda Starke.. Prentice-Hall, New Delhi.
13. **Brown LR, 1987**, Sustaining world agriculture. In *State of the World-1987: A World watch Institute Report on Progress towards a Sustainable Society*" Ed. Linda Starke. W. W, Norton & Co. N.Y..
14. **Kale Vishwas S, 1998,** Monsoon floods in India: A hydrogeomorphic perspective. In "*Flood studies in India*" Kale VS (Ed). Bangalore: Geological Society of India, 229-256.
15. **Kathpalia G., 1985,** Water budgeting and planning. In *Water Pollution and Management,* Ed. Varshney C.K. Wiley Eastern Ltd., New Delhi.

CHAPTER 7: MINERAL RESOURCES

1. **Gibbons D., O'Neill, D., 1990,** "Sustaining our Mineral Development" in *AMIC Mining Review*, July 1990.
2. **Ehrlich, P.R. and A. Ehrlich. 1970.** *Population, Resources, Environment: Issues in Human Ecology***.** W.H. Freeman, San Francisco. 383pp.
3. **Young John E, 1992,** Mining the earth. In "*State of the World*" Brown LR et al (Eds): A Worldwatch Institute Report on Progress Towards a sustainable Society. Horizon India Books, New Delhi.

4. **Down Christopher G, John Stocks, 1977**, *Environmental impact of mining*. New York: John Wiley & Sons.
5. **Rex Bosson and Bension Varon, 1977,** The Mining Industry and the Developing Countries. New York; Oxford University Press.
6. **Cleary David, 1990,** *Anatomy of Amazon Gold Rush.* Iowa City: University of Iowa Press.
7. **Goudie Andrews, 1990**, *The human impact on natural environment.* Cambridge, Mass.:MIT Press
8. **Martyn Kelly, 1988,** *Mining and the freshwater environment.* London: Elsevier Science Publishers.
9. **Rai KL, 2000,** Coal mining environment and related geo-hazards: The Indian Scenario, pp.221-228. In *Contributions to environmental geosciences.* Pathan AM and SS Thigale (Eds). New Delhi: Aravali Books.
10. **Valdiya KS, 2004,** *Geology, Environment and Society*, University Press (India), Hyderabad.
11. **Anonymous 1999**, Indian Mineral Year Book 1998-1999. Nagpur: Indian Bureau of Mines.
12. **Verma OP, 1999**, Emerging trends in mineral exporation and valuation. Roorkee: Indian Geological Congress, 21 pp.

CHAPTER 8: ENERGY RESOURCES

1. **Curran S., 1976,** Society's future energy needs. In *Conservation of resources. Chem. Soc.,* London, Spl. Publ. 27.
2. **Hubbert M.K., 1973,** The energy resources of the Earth. In *Chemistry in the environment.* Readings from Sci. American W.H. Freeman, San Francisco.
3. **Ion D.C., 1980,** *Availability of World Energy resources,* 2nd Ed. Graham and Trotman, London.
4. **Reddy Amulya K.N., 1985,** Energy : Issues and opportunities. In *The Global Possible : resource, development and the new century.* Ed Repetto R., Aff. E.W. Press, New Delhi.
5. **Bhaskaran R., 1995,** Coal: tough challenges for Coal India Ltd. In *The Hindu Survey of Indian Industry 1995,* Madras.
6. **Lipaschutz Ronnie B., 1980,** *Radio-active wastes : politics, technology and risks.* Ballinger, Cambridge, Massachusetts.
7. **Lenssen Nicholas, 1992,** *Confronting nuclear wastes.* In "*State of the World 1992: A Worldwatch Institute Report on progress towards a sustainable society*" Ed. Linda Starke. Horizon India Books Ltd. New Delhi.
8. **Christopher Flavin, 1987,** Reassessing Nuclear Power. In "*State of the World 1992: A Worldwatch Institute Report on progress towards a sustainable society*" Ed. Linda Starke. Prentic-Hall of India Pvt. Ltd. New Delhi.
9. **Ross M., Williums R., 1981,** *Our energy : regaining control.* MacGraw Hill N.Y.
10. **Christopher Flavin, 1992,** Building a bridge to sustainable energy. In "*State of the World 1992: A Worldwatch Institute Report on progress towards a sustainable society*" Ed. Linda Starke. Horizon India Books Ltd. New Delhi.
11. **Darmstadter J., 1971,** *Energy in world economy.* John Hopkins Univ. Press, Baltimore.
12. **Sunderesan B.B., 1991**, Air pollution : The dangerous dimensions. In *The Hindu Survey of Environment-1991,* Madras.
13. **Firor John, J. E. Jacobson, 2003,** *Crowded Greenhouse.* University Press, New Delhi.
14. **Christopher Flavin, Nicholas Lenssen 1990,** Beyond the petroleum age: Designing a solar economy. Worldwatch Paper 100. Washington DC.

CHAPTER 9: FOREST WEALTH.

1. **Whittaker R.H., Likens G.E., 1975,** The biosphere and man. In *Primary productivity of the biosphere.* Eds. Lieth H., Whittaker R.H. Springer-Verlag N.Y.
2. **Reider Parsson, 1974,** *World Forest Resources.* Royal College of Forestry, Stockholm.
3. **FAO-UNEP Studies 1981,** *Forestry resources of Tropical Asia, Africa and America.* F.A.O., Rome.
4. **Eckholm E.P., 1991,** Global atmospherics. Chapter in *Down to Earth : environment and human needs.* Aff. E.W. Press, New Delhi.
5. **Norman Myers, 1980,** *Conversion of tropical moist forests.* Nation Academy of Sciences, Washington D.C.
6. **Skilling F. Robert, Tcheyen N.O., 1979,** *Economic development : prospects of Amazonian Region of Brazil.* John Hopkins School of Advanced studies, Centre for Brazilian studies. Washington D.C. Nov. 1979.
7. **Lawrence S. Hemilton, 1978,** *Panama watershed management project paper.* U.S. Agency for International development, Panama, Nov. 30, 1978.
8. **Fearnside Philip M., 1980,** The effect of cattle pastures on soil fertility in Brazilian Amazon: consequences for beef production sustainability. *Tropical Biology,* **21** (1): 142.
9. **Parson J.J., 1976,** Forest to pastures : development or destruction. In *Revista de Biologia,* **24** : 261.
10. **Tomar M.S., Joshi S.C., 1977,** *Vanishing forests vis-a-vis energy crisis in Madhya Pradesh by 2000 A.D.* State forest resource survey organization, Bhopal, India.
11. **Johnson N.E., Dykstra G.F., 1978,** Maintaining forest production in Kalimantan, Indonesia. In *Proc. 8th World Forestry Congress,* Jakarta, Oct. 16–28, 1978.
12. **Venkataramani G., 1991,** Soil erosion : a man made disaster. In *The Hindu Survey of Environment, 1991,* Madras.
13. **Ehrich C., 1980,** The next flood knocking at the doors : a foreigner's view on the ecological crisis in India. In *Studies on Himalayan Ecology and development strategies.* Eds. Singh T.V., English Book Store, New Delhi.
14. **Meher-Homji V.M., 1991,** Climatic change: the forest link. In *The Hindu Survey of Environment-1991,* Madras.
15. **Likens G.E., et al., 1977,** *Biochemistry of a forested ecosystem.* Springer-Verlag, N.Y.
16. **Boremann F.H., Likens G.E., 1979**, *Pattern and process in a forested ecosystem.* Springer-Verlag, N.Y.
17. **Wank W.T., 1984,** Atmospheric contribution to forest nutrient cycling. *Water Resour. Bull.* **20** : 313.

CHAPTER 10: FOOD RESOURCES

1. **UN FAO-STAT**. Statistics Database. Rome, Available at: http://faostat.fao.org/.
2. **Poehlman JM, D Borthakur, 1969**, *Breeding Asian Field Crops*. Calcutta: Oxford & IBH Publishing Co.
3. **Mazoyer M, Laurence Roudart , 2006**, *A History of World Agriculture - From the Neolithic to the Current Crisis*. Earthscan Publ.

4. **Revelle R., 1985,** Present and future state of living freshwater and marine resources. In the *Global possible-resource, development and the new century.* Ed. Repetto R. Aff. E.W. Press, New Delhi.

5. **Alagarswami K., 1991,** Aqua-culture: ensuring sustainable growth. In *The Hindu Survey of Indian Agriculture – 1991,* Madras.

6. **Wortman S., 1980,** World food and nutrition: The scientific and technological base. *Science,* **209** : 15.

7. **Evans L.T., 1980,** The natural history of crop yield. *American Scientist,* **68** : 388.

8. **Larson W.E., et al, 1983,** The threat of soil erosion to long term crop production. *Science.* **219** : 458.

9. **Scientific American, 1976,** The September-1976 issue devoted to food and agriculture.

10. **Eckholm E.P., 1976**, *Losing grounds : Environmental stress and world food prospects.* W. W. Norton, N.Y.

11. **Roggers P.P., 1991,** Fresh-water. In The *Global possible : resource, development and the new century.* Ed. Repetto R., Aff. E.W. Press, New Delhi.

12. **Sandra Postel, 1985,** Managing fresh-water supplies. In *State of the world-1985. A Worldwatch Institute Report on progress towards a sustainable society*" Ed. Linda Starke.. Prentice-Hall, New Delhi.

13. **Brown L.R., 1987,** Sustaining World Agriculture. In "*State of the World 1987: A World watch Institute Report on Progress towards a Sustainable Society*" Ed. Linda Starke. W. W, Norton & Co. N.Y..

14. **Headly J.C., Kneese A.V., 1969,** Economic implications of pesticide use. *Annals N.Y. Acad. Sci.,* **160 :** 30.

15. **George J.L., Lahmann E., 1966,** Pesticides in the Antarctic. *J. Appl. Ecol.,* **3** : 155 (Suppl.).

16. **Sasson Albert, 1987,** Conserving plant resources. *The World Scientist,* **16** : 50.

17. **Durning A.T., Brought H.B., 1992,** Reforming live stock economy. In *State of the World-1992.* Eds. Brown L.R., et al. Horizon India Books, New Delhi.

18. **Rajan K., 1995,** Dairy: integral part of farming system. In *The Hindu Survey of Indian Agriculture-1995,* Madras.

19. **Hoffman J.S., Wells J.B., 1987,** Changes in green house gases pp. 19–42. In *Conservation Foundation.* Ed. Shands W.E., Hoffman J.S., Washington D.C.

20. **Mitra A.P., 1991,** Global warming : the green house effect. In *The Hindu Survey of Environment-1991,* Madras.

21. **Fearnside Philip M., 1980,** The effect of cattle pastures on soil fertility in Brazilian Amazon: consequences for beef production sustainability. *Tropical Biology,* **21** (1): 142.

22. **Wolf Edward C., 1986,** Managing rangelands. In *State of the world-1986. A Worldwatch Institute Report on progress towards a sustainable society*" Ed. Linda Starke. W.W. Norton, N.Y.

23. **Chary S.N., V. Vyasulu, (Eds.), 2001,** *Environmental Management, an Indian perspective.* McMillan India Ltd.

24. **Radhakrishnan K., 1995,** Marine Fisheries : Strategies to boost landings. In *The Hindu Survey of Indian Agriculture-1995,* Madras.

CHAPTER 11: BIODIVERSITY AND ITS DEGENERAT]ION.

1. **Stork NE, 1988,** Insect Diversity: Facts, fiction and speculations. *Biol. J. Linn. Soc.*, **35** : 213.
2. **Wilson EO, 1988,** The current state of biological diversity. In *Biodiversity*. Eds. Wilson E.O., Peter F.M., National Academy Press, Washingtonn D.C.
3. **McArthur RH, Wilson EO, 1976,** *The Theory of Island Biogeography.* Princeton University Press, Princeton.
4. **Wilson EO, 1985,** Invasion and extinction in the West Indian Ant Fauna: evidence from Dominican Amber. *Science,* **229**:265-67.
5. **Raup DM, Sapkoski JJ, 1984,** Periodicity of Extinctions in the Geologic Past. *Proc. Nat. Acad. Sci.*, U.S.A., **81** : 801.
6. **Raup D.M., 1986,** Biological Extinctions in Earth's History. *Science*, **231** : 1528.
7. **Wilson EO, 1988,** The current state of biological diversity. In *Biodiversity*. Eds. Wilson E.O., Peter F.M., National Academy Press, Washingtonn D.C.
8. **Myers N, 1988,** Threatened biotas: 'Hot-spots' in tropical forests. *The Environmentalists,* **8** : 187.
9. **Myers N, 1990,** Biodiversity Challenge: Expanded Hot-spot Analysis. *The Environmentalist,* **10** : 126.
10. **Kenton R Miller, et al, 1991,** Issue on the preservation of biological diversity. In *Global Possible : Resource, Development and the New Century.* Aff. East-West Press, New Delhi.
11. **Ryan John C, 1992,** Conserving biological diversity. In "*State of the World 1992: A Worldwatch Institute Report on progress towards a sustainable society*" Ed. Linda Starke. Pp.9-26. Prentic-Hall of India Pvt. Ltd. New Delhi.
12. **Prance GT, 1977,** Floristic Inventory of Tropics : Where do we stand? *Ann. Miss. Bot. Gdn,* **64** (4) : 156.
13. **Diamond J.M., 1989,** Overview of recent extinctions. In *Conservation for the Twenty First Century*. Eds. Western D., Pearls M. Oxford University Press, N.Y.
14. **Paine RT, 1966,** Food web complexity and species diversity. *American Naturalist,* **100** : 65.
15. **Terborgh J, 1986,** Key-Stone Species in the Tropical Forests. In *Conservation Biology : the science of scarcity and diversity.* Ed. Soule M.E., Sinauer Associates, Sunderland.
16. **Schneider SH, R Londer, 1984,** *Co-evolution of Climate and life*. Sierra Club Books, San Francisco.
17. **Baker RG, 1983,** Holocene Vegetational History of Western United States. In *Late Quaternary Environment of the United States,* Vol. 2. The Holocene, Eds. H.E.
18. **Bernabo JC, Webb T, 1977,** Changing pattern in the Holocene pollen record of north-eastern North America : a mapped summary. *Quat. Res.*, **8** : 64.

CHAPTER 12: *EX-SITU* CONSERVATION.

1. **Caughley G, ARE Sinclair, 1994,** *Wildlife Ecology and Management*, Blackwell Scientific Publ.
2. **Bolen Eric G, Robinson William, 2002,** *Wildlife Ecology and Management.* 5th Edition, Prentice Hall.
3. **Laurance WF, 1991,** Ecological Co-relates of extinction Proneness in Australion tropical ram forest mammals. *Conservation Biol.* **5** : 79 – 89.

4. **Jenkins M., 1992,** Ex-situ conservation of animals. In "*Global Bio diversity : status of Earth's living resources.* pp. 563 – 575.Ed. Groombridge B., Chapman and Hall London.

5. **Griffith, Brad, Michael Scott, James Carpenter, Christine Reed.** 1989, "Translocation as a species conservation tool: status and strategy.". *Science.* **245** (4917): 477-480.

6. **Stuart S.N., 1991**. Re-introduction to what extent are they needed? In '*Beyond Captive Breeding: Re-introduction of endangered mammals to the wild*' Eds. Gipps J.H.W., pp.27-32 Clarendon Press, Oxford.

7. **Beck B., et al, 1994,** Reintroduction of captive borne animas. In *Creative Conservation : Interactive management of wild and captive animals.* Eds. Olney P.J.S., Mace G.M., Feistner A.T.C., Chapman and Hall, London.

8. **Frankel O.H., Soule M.E., 1981,** Conservation and Evolution. Cambridge University Press, Cambridge.

9. **Olney P.J.S. , P. Ellis 1992,** *International Zoo Yearbook. No.31*, Zoological Society Lonon.

10. **Conway W, 1989,** Can technology aid species preservation. pp 263-268. **In** *Biodiversity.* Ed. Wilson EO, Washington: National Acaemy Press.

11. **Botanic Gardens Conservation International** (BGCI) database, 2000

12. **Wyse Jackson PS, 1999,** Experimentation on a Large Scale-An Analysis of the Holdings and Resources of Botanic Gardens. *BGC News* Vol **3** (3) December 1999. Botanic Gardens Conservation International,UK.

13. **Ashton Peter S, 1989,** Conservation f biological diversity in botanical gardens. pp 269-277. In Biodiversity. Ed. Wilson EO, Washington: National Acaemy Press.

14. **Sasson Albert, 1987,** Conserving plant resources. *The World Scientist,* **16** : 50.

15. **Williums TJ, 1989,** Identifying and protecting the origins of our food plants, pp 240-247. **In** *Biodiversity.* Ed. Wilson EO, Washington: National Acaemy Press.

16. **Seal SU, 1989,** Intensive technology in the care of ex-situ populations of vanishing species. In Biodiversity. Ed. Wilson EO, Washington: National Acaemy Press.

17. **Williums TJ, 1984,** A decade of crop genetic resource, pp 1-17. In *Crop genetic resources: Conservation and evaluation.* Eds Holden JHW and Williums TJ. London :Allen and Unwin.

18. **Zobel B., 1978,** Gene Conservation as viewed by a forest tree breeder. *Forest Ecol. Manag.,* **1** : 339.

CHAPTER 13: *IN-SITU* CONSERVATION.

1. **Mc Neeley J.A.,** *et al*, **1990**, *Conserving World's Biological Diversity.* Gland, Switzerland.

2. **Ryan J.C., 1992,** Conserving Biological Diversity. In "*State of the World 1992: A Worldwatch Institute Report on progress towards a sustainable society*" Ed. Linda Starke. Horizon India Books Ltd. New Delhi.

3. **Tucker, G., P. Bubb, M. de Heer, L. Miles, A. Laurence, J. van Rijsoort, S.B. Bajracharya, R.C. Nepal, R. Serchan and N. Chapagain. 2005.** Guidelines for Biodiversity Assessment and Monitoring for Protected Areas. Cambridge, UK: UNEP-WCMC.

4. **Mittermeier R.A., 1988,** Primate diversity and the tropical forests: Case studies from Brazil and Madagasker and the Importance of Megadiversity countries. In *Biodiversity.* Eds. Wilson E.O., Peter F.M., National Academy Press, Washington D.C.

5. **Mittermeier R.A., Werner T.B., 1990,** Wealth of plant and animal communities in Megadiversity Countries. *Tropicus,* **4** : 4

6. **Myers N., 1988,** Threatened biotas: 'Hot-spots' in tropical forests. *The Environmentalists,* **8** : 187.

7. **Myers N., 1990,** Biodiversity Challenge: Expanded Hot-spot Analysis. *The Environmentalist,* **10** : 126.

8. **Ackery PR, Vanc-Wright RI, 1984,** *Milkweed Butterflies*. London: British Museum of Natural History.

9. **McArthur R.H., Wilson E.O., 1976,** *The Theory of Island Biogeography.* Princeton University Press, Princeton.

10. **Wilson E.O., 1985,** Invasion and extinction in the West Indian Ant fauna: evidence from Dominican Amber, *Science*, **229** : 265 – 67.

11. **Graber J.W., Graber R.R., 1976,** *Environmental Evaluation using birds and their habitats.* Biological Notes No. 297, Illinois Natural History Society, Illinois.

12. **Worboys G, Lockwood M, and De Lacy T, 2001,** Protected Area Management. Principles and Practice. Oxford University Press, South Melbourne, Australia.

13. **Lockwood, M. 2007.** Values and benefits. Managing Protected Areas: A global guide. London: Earthscan.

14. **Keppler C.B., Scott J.M., 1985,** *Conserving the Island Ecosystems. ICBP Tech. Publ. 3* : 255.

15. **Nagata K.M., 1985,** Early Plant Introduction in Hawaii. *Hawaii Journ. History,* **19 : 35.**

16. **Smith C.W., 1985,** Impact of alien plants on Hawaii's native biota. In *Hawaii's Terrestrial Ecosystems : Preservation and Management.* Co-operative National Resource Study Unit, University of Hawaii, Honolulu.

17. **Moulton M.P., Pimm S.L., 1986,** Species Introduction to Hawaii. In *Ecology of Biological invasion of North America and Hawaii.* Eds. Mooney H.A., Drake J.,

18. **Vitousek P.M., 1989,** Diversity and biological invasion of Ocean Islands. In *Biodiversity* Eds. sssWilson E.O., Peter F.M., National Academy Press, Washington D.C.

19. **Margoluis, R., and N. Salafsky. 1998.** Measures of Success: Designing, Managing, and Monitoring Conservation and Development Projects. Island Press, Washington, D.C. Available from Island Press, Washington DC.

20. **Stedman-Edwards, P. 2000**. "A framework for analyzing biodiversity loss" in *The Root Causes of Biodiversity Loss.* A. Woods, P. Stedman-Edwards and J. Mang, eds. London: Earthscan.

21. **Ryan John C, 1992,** Conserving biological diversity. In "*State of the World 1992: A Worldwatch Institute Report on progress towards a sustainable society*" Ed. Linda Starke. Pp.9-26. Prentic-Hall of India Pvt. Ltd. New Delhi.

22. **Brazil Mark, 1991,** Where the Eastern Eagles dare. *New Scientist,* May 4, 1991.

23. **James Rush., 1991,** *The last tree : Reclaiming the Environment in Tropical Asia.* The Asia Society, N.Y.

24. **Jordan III WR, 1989,** Ecological Restoration, pp.311-316. In "*Biodiversity*". Eds. Wilson E.O., Peter F.M., Washington D.C: National Academy Press,

25. **Cairns Jr J, 1989, Increasing diversity by restoring damaged ecosyatem, pp 333-**343. In "*Biodiversity*". Eds. Wilson E.O., Peter F.M., Washington D.C: National Academy Press,

26. **FAO-UNEP Studies 1981,** *Forestry resources of Tropical Asia, Africa and America.* F.A.O., Rome.

27. **Spears John, 1989, Preserving Biological diversity in the tropical forest of asian region, pp.**

28. **Todd John, 1988,** Restoring Diversity: The search for social and economic context. In "*Biodiversity*". Eds. Wilson E.O., Peter F.M., National Academy Press, Washington D.C.

CHAPTER 14: THE POLLUTION OF ENVIRONMENT.

1. **Neely W.B., et al. 1974,** Partition co-efficient to measure bio-concentration potential of organic chemicals in fish. *Environ. Sci. Technol.* **8** : 1113.
2. **Kanazawa J., 1981,** Measurement and bio-concentration factors of pesticide by fresh-water fish and their co-relation with physicochemical properties and acute toxicities. *Pestic. Sci.,* **12** : 417.
3. **Kenaga E.E., Goring C.A.I., 1980,** Relationship between water solubility, soil solution and octanol water partitioning and concentration of chemicals in biota. In *Aquatic toxicology,* Eds. Eaton J.G., Parrish P.R., Hendrick A.C., American Society for testing materials.
4. **Murty A.S., 1986,** *Toxicity of pesticides to fish,* Vol. I. CRC Press Inc. Florida.
5. **Cairn T., Parfitt C.H., 1980,** Persistence and metabolism of TDE in California Clear Lake fish. *Bull. Environ. Contam. Toxicol.,* **24** : 504.
6. **Barker R.J., 1958,** Notes on some ecological effects of DDT spraying on elms. *J. Wild. Mgmt.* **22** : 269.
7. **Hunt E.G., Bischoff A.I., 1960**, Inimical effects on wild-life of periodic DDD application to Clear Lake. *Calif. Fish. Game.* **46** : 4.
8. **Rudd R.L., Herman S.G., 1972,** Ecosystem transfer of pesticide residue in aquatic environment. In *Environmental toxicology of pesticides Part VII.* Eds. Matsumura F., Boush G.M., Misata T., Academic Press, N.Y.
9. **Woodwell G.M., et al., 1967,** DDT Residue in East Coast estuary : a case of biological concentration of persistence insecticide. *Science*, 156 : 821.
10. **Woodwell G.M., Craig P.P., Johnson H.A., 1971**. DDT in biosphere – where does it go. *Science,* **174** : 1101.
11. **Carson Rachel, 1960,** *The Silent Spring.* Houghton, Mifflin, Boston.
12. **Dunlap T.R., 1981,** *DDT, Scientists, Citizens and Public policy.* Princeton Univ. Press, Princeton.

CHAPTER 15: POLLUTION OF EARTH'S SURFACE: WATER AND SOIL, I. SIMPLE ORGANIC WASTES

1. **Youngentob SL, 1999,** Introduction to the sense of smell: understanding odours from the study of human and animal behaviour, pp 23-37. In *Tastes & aromas : the chemical senses in science and industry.* Eds Graham A Bell and Annesley J Watson. Blackwell Science.
2. **Koster Egon Peter, 2002**, The specific characteristics of the sense of smell, pp.27-44, In *Olfaction, taste, and cognition*. Eds. Catherine Rouby and others. New York: Cambridge University Press.
3. **Pelczar MJ, Chan ECS, Krieg NR, 1993,** *Microbiology*, Ch. 35: Epidemiology of Infectious diseases, pp 764-787. Mc Graw-Hill Publishing Co Ltd., New Delhi.
4. **Foehrenbach J. , 1972 ,** Eutrophication. J. Wat. Poll. Contr. Fed. , 44(6): 1150-1159
5. **Rodhe, W. 1969** *Crystallization of eutrophication concepts in North Europe. In: Eutrophication, Causes, Consequences, Correctives.* Washington DC: National Academy of Sciences.
6. **Smith VH, GD Tilman, JC Nekola, 1999,** Eutrophication: impacts of excess nutrient inputs on freshwater, marine, and terrestrial ecosystems". *Environmental Pollution* **100:** 179–196.

7. **Fogg GE, 1969,** The physiology of algal nuisance. *Proc. Roy. Soc. (B), 173: 175-189.*
8. **Kennedy R.H., W. W. Walker, 1990,** Reservoir nutrient dynamics. In "Reservoir limnology: Ecological perspective." Ed. K.W. Thornton et al. *John Wiley & Sons, New York.*
9. **Fitzgerald G. P., 1964,** Biotic relationship within water blooms. In 'Algae and Man' Ed. Jackson D. F., *Plenum Press NY.*
10. **Pelczar MJ, Chan ECS, Krieg NR, 1993,** *Microbiology*, Ch. 27: Microbiology of domestic waters and wastewaters, pp 593-617. Mc Graw-Hill Publishing Co Ltd., New Delhi.
11. **Dugan PR, 1972**, Biochemical ecology of water pollution. New York: Plenum Press
12. **Lathrop, Richard C.; Stephen R. Carpenter, John C. Panuska, Patricia A. Soranno, & Craig A. Stow, 1998,** "Phosphorus loading reductions needed to control blue-green algal blooms in Lake Mendota" (in English). *Canadian J. Fish and Aqu Sci.* **55** (5): 1169–1178.

CHAPTER 16. POLLUTION OF EARTH'S SURFACE:WATER AND SOIL, II. COMPLEX BIODEGRADABLE WASTES

1. **Annonymous, 1979,** The plant protection arsenal. *Environ. Sci. Technol.*, **13** : 1335.
2. **Gruzdyev G.S., et al, 1988,** *Chemical protection of plants*. Mir Publ. Moscow (English translation).
3. **Vas J.G., et al, 1983,** Methods for testing immune effects of toxic chemicals : evaluation of immuno-toxicity of various pesticides in rats. In *Mayamoto J., Keaney P.C., (Eds) : Pesticide chemistry, human welfare and the environment. Vol. 3 : Mode of Action : Metabolism and Toxicology.* Pergamon Press, Oxford.
4. **Street J.C., 1981,** *Pesticide and immune system.* In *Immunologic considerations in Toxicology.* Vol. I. Ed. Sharma R.P., CRC Press, Florida.
5. **IARC, 1983,** *Monographs on evaluation of carcinogenic risks of chemicals to man.* Vol. 30 : Miscelleneous pesticides. International Agency for Research on Cancer-Lyon, France.
6. **Durham W.F., Williums C.H., 1972,** Mutagenic, teratogenic and carcinogenic properties of pesticides. *Ann. Rev. Entomol.* **17** : 123.
7. **Epstein S.S., et al., 1972,** Detection of chemical mutagen by dominant lethal assay in the mouse. *Toxicol. Appl. Pharmacol.*, **23** : 288.
8. **Kamrin Michael A, 1997,** Pesticide profiles –Toxicity, environmental impact and fate. Boca Paton, Florida: CRC Press
9. **Hunt E.G., Bischoff A.I., 1960**, Inimical effects on wild-life of periodic DDD application to Clear Lake. *Calif. Fish. Game.* **46** : 4.
10. **Rudd R.L., Herman S.G., 1972,** Ecosystem transfer of pesticide residue in aquatic environment. In *Environmental toxicology of pesticides Part VII.* Eds. Matsumura F., Boush G.M., Misata T., Academic Press, N.Y.
11. **George J.L., Lahmann E., 1966,** Pesticides in the Antarctic. *J. Appl. Ecol.,* **3** : 155 (Suppl.).
12. **Hocking Martin Blake, 2005,** The handbook of chemical technology and pollution control. 3rd Edn. Academic Press, pp1-799
13. **Ariens E.J., A.M. Simons, J. Offermeier, 1976,** Introduction *to general toxicology.* Academic Press, NY.
14. **Von Miller, et al, 1977,** Increased incidence of neoplasm in rats exposed to low levels of 2-3-7-8 tetrachlorodibenzo-p-dioxin. *Chemosphere,* **6** : 537.

15. **Kociba R.L., et al, 1978,** Results of a two year study on chronic toxicity and oncogenicity of 2-3-7-8 tetrachlorodibenzo-p-dioxin in rats. *Toxic. Apl. Pharmacol.,* **46** : 279.
16. **Young A.I., et al, 1987,** *The toxicology : environmental fate and human risk of herbicide Agent orange and its dioxin.* USAF-DEHL, Technical Reports 78–92, National Technical Information Service, U.S. Department of Commerce, Springfield Va.
17. **Lukens R.J., 1971**, *Chemistry of fungicidal action.* Springer-Verlag N.Y.
18. **Haq I.U., 1963**, Agrosan poisoning in man. *British Medical Journ.*, **1** : 1579.
19. **Edwards CA, 1981,** *Persistent pesticides in the environment.* 4th Edn. Boca Paton, Florida: CRC Press.
20. **Barid, Stuart.** Oil Spills. http://www.iclei.org/efacts/oilspi2.gif

CHAPTER 17. POLLUTION OF EARTH'S SURFACE: LAND AND WATER, III. HEAVY METAL AND TOXIC TRACE ELEMENTS

1. **Young JE, 1992,** Mining the earth. In *State of the World-1992 A Worldwatch Institute Report on progress towards a sustainable society*" Ed. Linda Starke. et al. Horizon India Books, New Delhi.
2. **Howard Michael C., 1991**, *Mining, Politics & development in South Pacific.* Westerview Press, Boulder, Colo.
3. **Tseng W.P., 1977,** Effects and dose response relationship of skin cancer, black foot disease with Arsenic. *Environ. Health Perspective.* **19** : 109.
4. **Kjellstrom T., et al, 1975,** Variations in Cadmium concentration in Swedish wheat and barley, an indication of changes in daily cadmium intake during 19th century. *Arch. Environ. Health,* **30** : 321.
5. **Nordberg G.F., 1978,** Cadmium metabolism and toxicity. *Environ. Physiol. Biochem.*, **2** : 7.
6. **Arumugum V, 1976,** Recovery of chromium from spent chrome liquor by chemical speciation. *Ind. J. Environ. Health.,* **18** : 47.
7. **Chattopadhyaya S.M., et al, 1973,** Characteristics of feasibility of chrome tanning wastes. *Ind. J. Environ. Health,* **15** : 208.
8. **Kothandaraman B., et al, 1976,** Characteristics of wastes from textile mills. Ind. J. Environ. Health, **18** : 99.
9. **Langard S., Norseth T., 1990,** Chromium. In *Handbook on the toxicology of metals.* Eds. Friberg L., Nordberg G.F., Vouk V.B., Elsvier N.Y.
10. **Norseth T., 1981,** The carcinogenecity of Chromium. *Environ. Health Prespect.,* 40 : 121.
11. **Khandekar R.N., et al, 1980**, Lead Chromium, Zinc, Copper and Iron in the atmosphere of Greater Bombay. Atmos. Environ., **14** : 457.
12. **Goyer R.A., 1986,** Toxic effects of metals. In *Casarett and Doull's toxicology : The basic science of poisoning,* 3rd Edn. Eds. Klassen C.D., Amdur M.O., Doull J. Mac Millan N.Y.
13. **Barltop D., Smith A., 1971,** Inter-action of lead with erythrocites. *Experientia,* **27** : 92.
14. **Rutter M., Jones R.R., (Eds.),** 1983, *Lead versus Health: source and effects of low level lead exposures.* John Wiley N.Y.
15. **Needleman H., 1980,** *Low level lead exposure: The clinical implications of current research.* Raven Press, N.Y.
16. **Koller L.D., et al, 1983,** Evaluation of ELISA for detection in-vivo chemical immunomodulation. *J. Toxicol. Environ Health,* **11** : 15.

17. **Moore J.F., et al, 1973,** Lead induced inclusion bodies, solubility, amino-acid content and relationship to residual acidic nuclear proteins. *Lab. Invest.,* **29** : 488.
18. **Gavis J., Fergusan J.F., 1972,** The cycling of mercury through environment. *Water Research,* **6** : 989.
19. **Goldwater L.J., 1972,** *Mercury: a history of quick-silver*. Baltimore, Maryland, USA, York Press.
20. **Bakir F., et al, 1973,** Methyl mercury poisoning in Iraq. *Science,* **181** : 236.
21. **Myron D.R. et al, 1978,** Intake of nickel and vanadium by humans – a survey of selected diet. *Am. J. Cl. Natr.,* **31** : 527–31.
22. **Costa M., 1988,** *Metal carcinogenesis : testing, principles and in-vitro methods.* The Humana Press,
23. **Sunderman F.W. Jr., 1981,** Nickel. In *Disorders of Mineral Metabolism* Vol. I., Eds. Bonner F., Coburn J.W., Academic Press, N.Y.

CHAPTER 18. PROBLEMS CAUSED BY HARMFUL PHYSICAL AGENTS:

1. **Catherine PA and Gary AT, 1983,** *Textbook of Anatomy & Physiology.* St. Louis: Mosby, pages 338-344.
2. **Hirsh IJ, 1952,** *The Measurement of Hearing.* McGraw Hill, New York.
3. **Field JM, 1993,** Effect of personal and situational variables upon noise annoyance in residential areas, Journal of the Acoustical Society of America, **93**: 2753-2763
4. **"*Noise Pollution*"**. World Health Organization. http://www.euro.who.int/Noise.
5. **Kryter Karl D, 1985,** *The Effects of Noise on Man* , Academic Press
6. **Kryter KD, 1994,** *The Handbook of Hearing and the Effects of Noise*. Academic Press, San Diego.
7. **Rosenhall U, Pedersen K, Svanborg A, 1990,** Presbycusis and noise-induced hearing loss. *Ear Hear* **11** (4): 257–63.
8. **Holgers KM, Pettersson B. 2005,** "Noise exposure and subjective hearing symptoms in school children in Sweden". Noise & Health 7 (27): 27–37.
9. **Miller JD, 1974**, Effects of noise on people. *J. Acoust. Soc. Am.,* **56**:729-763.
10. **Hauser MD 1996.** *The Evolution of Communication*. MIT Press, Cambridge, MA.
11. **Edward A Laws, 2000,** *Aquatic Pollution: An Introductory Text*, John Wiley and Sons.
12. **Ristinen RA and Kraushaar JJ, 1998,** *Energy and the Environment*. New York: John Wiley & Sons.
13. **Langford TE, 1990,** *Ecological Effects of Thermal Discharges*. New York: Elsevier Applied Science.
14. **Caufield Catherine, 1989,** *Multiple exposures: Chronicles of the Radiation Age.* Chicago: University of Chicago Press.
15. **Lindee M Susan, 1994,** Suffering Made Real: American Science and the Survivors at Hiroshima. University Of Chicago Press.
16. **NRC, 1989,** Health effects of Exposure to Low Level Ionizing Radiation. Washington National academy Press.
17. **Christopher Flavin, 1987,** Reassessing Nuclear Power. In "*State of the World 1987: A Worldwatch Institute Report on progress towards a sustainable society*" Ed. Linda Starke. W. W. Norton & Co. New York

18. **Lenssen Nicholas, 1992,** *Confronting nuclear wastes*. In "*State of the World 1992: A Worldwatch Institute Report on progress towards a sustainanle society*" Ed. Linda Starke. Horizon India Books Ltd. New Delhi.

CHAPTER 19. AIR POLLUTION.

1. **Sytnik KM (Ed), 1985,** *Living in the environment.* Translated from Russian, UNESCO, Naukova, Dumka Publ.
2. **Amdur M.O., 1986,** Air Pollution. In *Casarett and Doull's toxicology : The basic science of poisoning,* 3rd Edn. Eds. Klassen C.D., Amdur M.O., Doull J. Mac Millan, N.Y.
3. **Ehrlich P.R., Ehrlich A.H., 1970,** *Population, Resource and environment issues in human ecology.* W.H. Freeman, San Francisco.
4. **World Resources Institute. 2007.** Earth Trends: Environmental Information. Available at http://earthtrends.wri.org.com Washington DC: World Resources Institute.
5. **Bidwell R.G.C., Fraser 1972,** Carbon monoxide uptake and metabolism by leaves. *Canad. J. Bot.,* **50** : 1435.
6. **Ingersoll R.B., Inman R.E., Fisher W.R., 1974,** Soil potential as a sink for atmospheric carbon monoxide. *Tellus,* **26** : 151.
7. **US Report, 2008**, World Carbon dioxide levels highest for 650,000 years. The Guardian May 13, 2008, Environmental Correspondent (Available at www.gaurdian.co.uk Retrieved on 04.06.2009).
8. **Hoffman J.S., Wells J.B., 1987,** Changes in green house gases pp. 19–42. In *Conservation Foundation*. Ed. Shands W.E., Hoffman J.S., Washington D.C.
9. **Ramachandran V, 1985,** Trace gases – trends and their potential role in Climatic change. *Journ. Geophysics Res,* **90** : 5547 – 66.
10. **Seth Cagin and Phillip Dray,1993,** *Between Earth and Sky: How CFCs Changed Our World and Endangered the Ozone Layer*. New York: Pantheon Books.
11. **Jonathan PD, Abbott and Mario J. Molina, 1993**, "Status of Stratospheric Ozone Depletion," *Annual Reviews of Energy and Environment*, 18: 1-29.
12. **Chlorofluorocarbons.** The Columbia Encyclopedia, 6th Edition. 2008. Retrieved 09.06.09 from Encyclopedia.com: http://www.encyclopedia.com/doc/1E1-chlorofl.html
13. **Freedman Bill, 1995,** *Environmental Ecology*, 2nd Edn. pp.11- 42. London: Academic press.
14. **Likens, G. E. and F. H. Bormann. 1974.** Acid rain: a serious regional environmental problem. *Science* **184**(4142):1176–1179.
15. **Likens, G. E., R. F. Wright, J. N. Galloway and T. J. Butler. 1979**. Acid rain. *Sci. Amer.* **241**(4):43-51.
16. **Seinfeld,John H, Pandis Spyros N, 1998,** *Atmospheric Chemistry and Physics - From Air Pollution to Climate Change*. John Wiley and Sons, Inc.
17. **Hansen James, Makiko Sato, Pushker Kharecha, Gary Russell, David W. Lea And Mark Siddall, 2007,** Climate change and trace gases. *Phil. Trans. R. Soc. (A),* **365,** 1925–1954
18. **Serruya C., Polinger U.,** 1983, *Lakes of the warm belt*. Cambridge Univ. Press, Cambridge.
19. **Mather I.R., Feddema J., 1986,** Hydrologic consequences of increases in trace gases and carbon dioxide in the atmosphere. Proc. *UNEP and U.S., EPA, International Congress,* Washington D.C.
20. **Caldeira, K.; Wickett, M.E. (2003).** "Anthropogenic carbon and ocean pH". *Nature* 425 (6956): 365–365.

21. **Orr, James C.; Fabry, Victoria J.; Aumont, Olivier; Bopp, Laurent; Doney, Scott C.; Feely, Richard A. et al. (2005).** "Anthropogenic ocean acidification over the twenty-first century and its impact on calcifying organisms". *Nature* 437 (7059): 681–686.

22. **Policy document**, **2005**, Ocean acidification due to increasing atmospheric carbon dioxide. Report of Royal Society, London, 2005. (Available at www.royalsoc.ac.uk)

23. **Ruttiman, J. (2006).** "Sick Seas". *Nature* **442** (7106): 978–980.

24. **Ridgwell, A.; Zondervan, I., Hargreaves, J.C., Bijma, J. and Lenton, T.M. (2007).** "Assessing the potential long-term increase of oceanic fossil fuel CO_2 uptake due to CO_2-calcification feedback". *Biogeosciences* **4**: 481–492.

25. **Tyrrell, T. (2008).** "Calcium carbonate cycling in future oceans and its influence on future climates". *J. Plankton Res.* **30:** 141–156.

26. **Farman JC, BG Gardiner and JD Shanklin, 1985,** "Large losses of total ozone in Antarctica reveal seasonal ClO_x/NO_x interaction". Nature, 315 (6016): 207–210.

27. **Maduro R., Schauerhammer Ralf, 1992,** *Hole in the Ozone scare.* 21st Century Science Associates, Washington D.C.

CHAPTER 20. ENVIRONMENTAL ETHICS AND VALUE EDUCATION

1. **Milbrath L.W., 1981,** Environmental values and beliefs of general public and leaders in United States, England and Germany. In *Environmental Policy formation: Impact of Values, Ideology and Standards.* Lexington Books, Lexington, Massachusetts.

2. **Silver C.S., DeFries R.S., 1991,** "One Earth one Future : Our changing global environment." National Academy of Sciences. Aff. East-West Press, New Delhi.

3. **Cunningham W., M.A. Cunnigham, 2003,** *Principles of Environmental Science: Inquiry and Applications.* 2nd Ed.,Tata McGraw Hill Publication, New D

4. **Ison S., S. Peake, S. Wall, 2002,** *Environmental Issues and Policies.* Prentice Hall.

5. **Ridker R.G., Watson W.D., 1980,** *To choose a future.* John Hopkins Univ. Press, Baltimore.

6. **Scientific American, 1989,** "*Managing Planet Earth,*" Special Issue, (September 1989), New York.

7. **Schrader-Frechete K., 1991**, Environmental Ethics and Global Imperatives. In "*The Global Possible: Resource Development and the new Century*" Ed. Repetto R., Aff. East-West Press, Pvt. Ltd., New Delhi.

8. **Blackstone W.T., 1974,** *Philosophy and Environmental Crisis.* University of Georgia Press, Atlanta, Ga.

9. **Delors Jacques, 1996,** *Learning the treasure within.* UNESCO, Paris

10. **Davis W. H., 1972,** *The land must live.* Environment Action Bulletin, 3:5 (July 15, 1972).

11. **Regan T., P. Singer, (Eds.), 1976**, *Animal rights and human obligations.* Prentice Hall Englewood Cliff. N. J.

12. **Stone C. D. 1974,** Should trees have standing. W. Kaufman, Los Altos.

13. **Leopold Aldo, 1949,** *The Land Ethics. A Sand County Almanac.* Oxford University Press, NY.

14. **Silver C.S., DeFries R.S., 1991,** "One Earth one Future : Our changing global environment." National Academy of Sciences. Aff. East-West Press, New Delhi.

15. **Miller G. Tyler, 2002,** *Living in the Environment.* 12th Edn., Brooks-Cole, Thomas Learning Inc. USA.

16. **Milbreth L W., 1989,** *Envisioning a Sustainable society: Learning our way out.* State University of New York Press, Albany, N.Y.

17. **Frankena W. K., 1979**, Ethics and Environment. In "*Ethics and problems of 21st century*" Ed. Goodpaster K. E.and K. M. .Sayre, University of Notre Dame Press, Notre Dame, Indiana.

18. **Karan Singh, 2004,** Education for Global Society. In "*Silver Jubilee Volume 2004-2005*". Citizenship Development Society, New Delhi. (The article has also been published in the UNESCO Report "Learning The Treasure Within".)

19. **Gangrade K.D., 2004,** Erosion of Values. In "*Silver Jubilee Volume 2004-2005*". Citizenship Development Society, New Delhi.

20. **Sharma J. N., K. M. Gupta, 1996,** *Value Education.* Citizenship Development Society, New Delhi.

CHAPTER 21. THE POPULATION ISSUE

1. **Regina McNamara, 1982,** Demographic Transition Theory. In "*International Encyclopedia of population* Vol.I. Mc Millan Publishing Co. NY.

2. **World Population Datasheets** for various years. Population Reference Bureau, U.S., Washington.

3. **Brown L.R., 1987**, Analyzing the demographic trap. In "*State of the World 1987: A World watch Institute Report on Progress towards a Sustainable Society*" Ed. Linda Starke. W. W, Norton & Co. N.Y.

4. **Meadows D.H., et al, 1972,** *The Limits to Growth.* Universe Books, N.Y.

5. **Dorn H.F., 1962,** World population growth: an international dilemma. *Science,* **135** : 283.

6. **Ehrlich P.R., Ehrlich A.H., 1970,** *Population, Resource and environment issues in human ecology.* W.H. Freeman, San Francisco.

7. **Frekla T., 1981,** Long term prospects of world population growth. *Population and Development Review,* **7** (3) : 489. Sept. 1981.

8. **Hordoy J.E., Daivid Satterthwaite, 1984,** Third World cities and the environment of poverty. *Geoforum,* **15**:3.

9. **Rafael M. Salas, 1986,** *The State of World population,* 1986. UNFPA, New York.

10. **Langer W.L., 1972,** Checks on population growth, 1750–1850. *Sci. American,* **226** (2) : 93.

11. **Scientific American, 1974,** September issue devoted to human populations.

12. **Repetto R., 1991,** Population Resource pressure and Poverty. In "*The Global Possible: Resource Development and the new Century*". Ed. Robert Repetto, Aff. East-West Press, New Delhi.

13. **Inderjit Singh, 1981,** *Small farmers and the landlessness in South Asia.* World Bank, Washington DC.

14. **Radha Singha, 1984,** *Landlessness: A growing Problem,* FAO, Rome.

15. **Brown L.R., Jodi Jacobson, 1987,** Assessing the future of urbanization. In "*State of the World 1987: A Worldwatch Institute Report on Progress towards a Sustainable Society*" Ed. Linda Starke. W. W, Norton & Co. N.Y.

16. **Renaud Bertrand, 1981,** *National Urbanization Policies in Developing Countries.* No.347, World Bank, Washington DC.

17. **Douglas Ian, 1983,** *The urban environment.* Edward Arnold Publishers, Baltimore.

18. **Firor John, J. E. Jacobson, 2003,** *Crowded Greenhouse.* University Press, New Delhi.

19. **Diaz-Briquets Sergio, 1986,** *Conflict in Central America: the demographic dimensions.* Population Trends and Public Policy, No. 10, Population Reference Bureau, Washing DC.

20. **Nagli Choucri, 1983,** *Population and Conflict.* UNFPA, New York.

21. **Wiarda H., I.S. Wiarda, 1985,** *Population, Internal Unrest and U.S. Security.* International Area studies Program, Amherst. Mass.

22. **Anne Geyer, Georgie, 1985,** *Our Disintegrating World: The Menace Of Global Anarchy.* In "Encyclopedia Britannica: Book of the Year." Chicago.

23. **UN. FAO, UNFPA, 1982,** *Expert Consultation Report on Land Resources for Population of the Future.* Rome.

24. **Myers Norman, 1991,** Global Possible: What can be gained? In "*The Global Possible: Resource Development and the new Century*". Ed. Robert Repetto, Aff. East-West Press, New Delhi.

25. **World Population Datasheets** for various years. Population Reference Bureau, U.S., Washington.

22. ACCELERATED URBANIZATION

1. **Mumford Lewis, 1961**, *The City in History.* Harcourt, Brace, Jovanovich, Orland Fla.

2. **Lees Andrew, 1985,** Cities Perceived: Urban Society in European and American Thought. Columbia University Press, NY.

3. **Renaud Bertrand, 1981,** *National Urbanization Policies in Developing Countries.* No.347, World Bank, Washington DC.

4. **World Population Datasheets** for various years. Population Reference Bureau, U.S., Washington.

5. **Brown L.R., Jodi Jacobson, 1987,** Assessing the future of urbanization. In "*State of the World 1987: A Worldwatch Institute Report on Progress towards a Sustainable Society*" Ed. Linda Starke. W. W, Norton & Co. N.Y.

6. **Hamer Andrew, 1982,** *Brazilian industrialization and economic concentration in Sao Paulo: A survey.* Washington DC : World Bank Water supply and Urban Department, 1982

7. **Kandell Jonathan 1985,** Nation in Jeopardy: Mexico city's growth once fostered, turns into economic burden. *The Wall Street Journal*, October 4, 1985.

8. **Barr T.N., 1981,** The world food situation and global grain prospects. *Science*, **214** : 1087.

9. **Douglas Ian, 1983,** *The urban environment.* Edward Arnold Publishers, Baltimore.

10. **Nebel B.J., 1981,** *Environmental Sciences: The way the world works.* Prentice-Hall Inc. Englewood Cliffs.

11. **Sandra Postel, 1984,** Protecting forests. In Lester R.Brown et al (Ed.) '*The state of the world – 1984. A Worldwatch Institute Report on progress towards a sustainable society*" Ed. Linda Starke. New york: W.W.Norton & Co.

12. **Bowonder B., et al, 1985,** *Deforestation and fuel wood use in urban centres.* Hyderabad, India: Centre for energy, environment, technology, and National Remote Sensing Agency.

13. **Shuval Hillel I., et al, 1986,** *Wastewater irrigation in developing countries:* effects and technical solutions, Washington D.C., UNDP & World Bank.

14. **Bastian Robert K., Jay Benforado, 1983,** Waste treatment: Doing what comes naturally. *Technology Review*, February/March 1983.

15. **Shuval Hillel I., 1981,** *Appropriate technology for water supply and sanitation: Night soil composting.* Washington DC., World Bank.

16. **Talashikar S.C., O.P.Vimal, 1984,** From nutrient poor compost to high-grade fertilizer. *Biocycle,* March 1984.

17. **Yeung Yue-Man, 1985,** *Urban agriculture in Asia.* The UN food energy nexus programme. University of Tokyo, September, 1985.

Pimentel D, 1980, *Handbook of energy utilization in agriculture.* Boca Raton, Fla.: CRC Press.

CHAPTER 23: HUMAN HEALTH AND ENVIRONMENT.

1. **IATA 2007**. *Fact sheet: IATA*. Geneva, International Air Transport Association,2007 (http://www.iata.org/pressroom/facts_figures/fact_sheets/iata.htm, accessed 10 May 2007).
2. **Prüss-Üstün, A. et al. (2004),** "Unsafe Water, Sanitation and Hygiene", in M. Ezzati, et al. (Eds.), *Comparative Quantification of Health Risks: Global and Regional Burden of Disease Attributable to Selected Major Risk Factors*, World Health Organisation, Geneva.
3. **Knobler S, Mahmoud A, Lemon S, Pray L, eds. 2006,** *The impact of globalization on infectious disease emergence and control: exploring the consequences and opportunities*. Workshop summary – Forum on Microbial Threats. Washington, DC, The National Academies Press,
4. **Howard-Jones N. 1975,** The scientific background of the International Sanitary Conferences 1851–1938. Geneva, World Health Organization, 1975.
5. **Caldwell L. K., 1991**, *International Environmental Policy: Emergence and Dimensions*. 2nd Edn. Affiliated East West Press Pvt. Ltd. New Delhi.
6. **Anonymous 2006,** *International Health Regulations (2005)*. Geneva, World Health Organization, 2006 (http://www.who.int/csr/ihr/en/, accessed 18 April 2007).
7. **Porter R.1997**, *The greatest benefit to mankind: a medical history of humanity, from antiquity to the present*. London, Harper Collins, 1997
8. **Davey Smith G. 2003**, Behind the Broad Street pump: aetiology, epidemiology and prevention of cholera in mid-19th century Britain [commentary]. *International Journal of Epidemiology,* **31**:920–932.
9. **Cairncross S, 2003**. Water supply and sanitation: some misconceptions [editorial]. *Tropical Medicine and International Health*, **8**:193–195.
10. **Anonymous 2007**, Cholera in the Americas. *Epidemiological Bulletin of the Pan American Health Organization*, 1995, 16(2) (http://www.paho.org/english/sha/epibul_95-98/be952choleraam.htm, accessed 11 April 2007).
11. **Fenner F, Henderson DA, Arita I, Jezek Z, Ladnyi ID. 1988**. *Smallpox and its eradication.* Geneva, World Health Organization, 1988.
12. **Anonymous 2005**. Global smallpox vaccine reserve: report by the Secretariat. Geneva, World Health Organization, 2005 (report to the WHO Executive Board, document EB115/36)
13. **Cohen, A.J. et al. (2004),** "Urban air pollution", in M. Ezzatti, et al. (Eds.), *Comparative Quantification of Health Risks: Global and Regional Burden of Disease due to Selected Major Risk Factors*, World Health Organization, Geneva.
14. **Camp T.R., Meserve R.L., 1974,** *Water and its impurities.* Dowden, Hutchinson and Ross, Pennsylvania.
15. **Lenihen J., Fletcher W.W., 1976,** *Health and the environment.* Academic Press, NY.
16. **Voznaya N.F., 1981**, *Chemistry of water and microbiology*. Mir Pollution Moscow.
17. **Annonymous, 1987,** *Hazardous wastes: uncertainties of existing data*. Report no. GAO/PMED-8711BR, U.S.A.

18. **Cohrrsen J. J., Vincent T. Cavello, 1989,** Risk analysis: guide to principles and methods for analyzing health and environmental risks. Council of Environmental Quality, Washington DC.
19. **Willium W. L., 1976,** *Of acceptable risks.* Kaufman W., Los Altos, Calif.
20. **Ariens E.J., A.M. Simons, J. Offermeier, 1976,** Introduction *to general toxicology*. Academic Press, NY.
21. **Klassen C.D., 1986,** Absorption, distribution and excretion of toxicants. In *"Cassarett and Doull's Toxicology: the basic science of poisoning"*. Eds. Klassen C.D. , M.O.Amdur, J.Doull. McMillan, NY.
22. **Jonathan A. Patz, Diarmid Campbell-Lendrum, 2005**, Tracey Holloway & Jonathan A. Foley 2005, Impact of regional climate change on human health. *Nature*, **438**, 310-317 (17 November 2005)
23. **Haines A, R . Kovats , D . Campbell-Lendrum , C . Corvalan,** 2005? Climate change and human health: impacts, vulnerability, and mitigation . *The Lancet*, **367**(9528): 2101 – 2109.

CHAPTER 24: HUMAN RIGHTS AND THE ENVIRONMENT

1. **Sohn Louis B., 1968,** *United Nations in Actions*. Foundation Press, Mineola, New York.
2. **Gandhi P. R., 1995,** *International Human Rights Documents*, London.
3. **Robertson A.H., J. G. Merrills, 2005,** *Human Rights in the World*. Universal Law publishing co., New Delhi.
4. **Shestack J.J., 1984,** The jurisprudence of human rights. In *"Human rights and international law"*. Ed. Meron T., Oxford.
5. **Kamenka E., A. E-S. Tay, 1978,** *Human Rights*, London.
6. **UN Doc. A/64, 1948**, Universal Declaration of Human Rights; UN General Assembly Resolution 217 A (III), Dec 10, 1948
7. **UN Doc. A/6316, 1966**, International Covenant on Economic, Social, and Cultural Rights; UN General Assembly Resolution 2200A (XX1), 21 UN GAOR, (Supp. No. 16) 52, UN Doc. A/ 6316, 993 U.N.T.S. 3, Dec 16, 1966.
8. **WHO Constitution. 1948**, Constitution of the World Health Organization; Preamble Article 2, April 7, 1948;
9. **Fabra A, 2002**, The Intersection of Human Rights and Environmental Issues: A Review of Institutional Developments at the International Level. Presented at the Joint UNEP–OHCHR Expert Seminar on Human Rights and the Environment, Geneva, Jan 14–16, 2002, Background Paper No. 3.
10. **UN Doc. A/CONF.48/14/Rev 1, 1972**, Declaration on the Human Environment, Principle 1. In Report of the UN Conference on the Human Environment; UN General Assembly Resolution 2997 (XXVIII), New York, 1972.
11. **UN Doc. A/41/53, 1986,** Declaration on the Right to Development, UN General Assembly Resolution 41/128, Supp. No. 53.

25. ACQUIRED IMMUNODEFICIENCY SYNDROME (AIDS)

1. **UNAIDS, WHO (December 2007). "2007** AIDS epidemic update" Retrieved on 2008-03-12. http://data.unaids.org/pub/EPISlides/2007/2007_epiupdate_en.pdf.
2. **Coffin J, Haase A, Levy JA, Montagnier L, Oroszlan S, Teich N, Temin H, Toyoshima K, Varmus H, Vogt P and Weiss RA, 1986** "What to call the AIDS virus?" *Nature* **321** (6065): 10.

3. **Rick Sowadsky, 1999**,"What is HTLV-III?". Retrieved on 2008-04-03.http://www.thebody.com/Forums/AIDS/safesex/Archive/origins/Q8777.html.
4. **Chan D, Kim P (1998).** "HIV entry and its inhibition". Cell 93 (5): 681–4. doi:10.1016/S0092-8674(00)81430-0. PMID 9630213.
5. **Gelderblom, H. R (1997).** "Fine structure of HIV and SIV". in Los Alamos National Laboratory (ed.) (PDF). HIV Sequence Compendium. Los Alamos, New Mexico: Los Alamos National Laboratory. pp. 31–44.
6. **Zheng, Y. H., Lovsin, N. and Peterlin, B. M. (2005).** "Newly identified host factors modulate HIV replication". Immunol. Lett. 97 (2): 225–34. doi:10.1016/j.imlet.2004.11.026. PMID 15752562.
7. **Robertson DL, Hahn BH, Sharp PM. (1995).** "Recombination in AIDS viruses". J Mol Evol. 40 (3): 249–59. doi:10.1007/BF00163230. PMID 7723052.
8. **Greener, R. (2002).** "AIDS and macroeconomic impact". in S, Forsyth (ed.). State of The Art: AIDS and Economics. IAEN. pp. 49–55.
9. **Joint United Nations Programme on HIV/AIDS.** "AIDS epidemic update, 2005". Retrieved on February 2006. http://www.unaids.org/epi/2005/doc/EPIupdate2005_pdf_en/epi-update2005_en.pdf.

26. DISASTER MANAGEMENT, RESETTLEMENT AND REHABILITATION

1. **Haddow, George D.; Jane A. Bullock, 2004,** *Introduction to Emergency Management*. Amsterdam: Butterworth-Heinemann.
2. **Cuny, Fred C, 1983**, *Disasters and Development*. Oxford: Oxford University Press.
3. **Lindell M, Prater C, and Perry R, 2006,** Fundamentals of Emergency Management. Retrieved January 9, 2009 at: http://training.fema.gov/EMIWeb/edu/fem.asp.
4. **Alexander David, 2002,** *Principles of Emergency planning and Management*. Harpenden: Terra Publishing.
5. **Annonymous 2004,** *Disaster management in India*: A status Report. Ministry of Home affairs, National Disaster Management Division, Govt. of India **(August 2004).**
6. **Bradshaw CJ, Sodhi NS, Peh SH, Brook BW, 2007,** Global evidence that deforestation amplifies flood risk and severity in the developing world. *Global Change Biology*, 13: 2379-2395.
7. **Cyclone circulation** Glossary of Meteorology . American Meteorological Society. Retrieved in September 2008
8. **Strahler A.N. Strahler A.H., 1976,** *Geography and man's environment.* John Wiley, N.Y.
9. **Watson, John; Watson, Kathie 1998.** "Volcanoes and Earthquakes". United States Geolocical Survey. http://pubs.usgs.gov/gip/earthq1/volcano.html. Retrieved on May 2009
10. **Natural Hazards** - Landslides. USGS. http://www.usgs.gov/hazards/landslides/. Retrieved in September 2008

27. SUSTAINABLE DEVELOPMENT

1. **World Wide Fund for Nature (2008).** Living Planet Report 2008. Retrieved: March 2009
2. **United Nations General Assembly (1987)** Report of the World Commission on Environment

and Development: Our Common Future. Transmitted to the General Assembly as an Annex to document A/42/427 -

3. **International Institute for Sustainable Development (2009).** "What is Sustainable Development?". http://www.iisd.org/sd/. Retrieved o : Feb,2009.
4. **Kates, R., Parris, T. & Leiserowitz, A. (2005).** "What is Sustainable Development?" Environment 47(3): 8–21. Retrieved on: April 2009.
5. **Marshall, J.D. & Toffel, M.W., 2005,** Framing the illusive concept of sustainability: sustainability hierarchy. *Environmental & Scientific Technology*, **39**(3):673-682.
6. **Redeliff M, 2005, Sustainalle development (1987-2005):** An oxymoron comes of Age. *Sustainable development,* 13(4):212-227
7. **Redclift, M. (2005). "Sustainable Development (1987-2005):** an Oxymoron Comes of Age." *Sustainable Developmenl* 13(4) 212-227
8. **IUCN/IUNEPWWF (1991).** "Caring for the Earth: A Strategy for Sustainable Living." Gland, Switzerland. Retrieved on: March 2009
9. **The Earth Charter Initiative (2000).** "The Earth Charter." Retrieved on: 2009-04-05.
10. **Agenda-21:** http//lhabitat.igc.org/agenda2l/rio-dec.htm
11. **Hasna". AM, 2007,** "Dimensions of sustainability". *Journal of Engineering for Sustainable Development: Energy, Environment. and Health,* 2 (I) 47-57.
12. **Costanza, R. & Patten, B.e (1995).** "Defining and predicting sustainability." *Ecological Economics* 15 (3): 193-196.
13. **Adams WM, SJ Jeanrenand, 2008,** *Transition to Sustainability: Towards a Humane and Diverse World.* IUCN Gland. Switzerland
14. **Sustaillability:** http//len.wikipedia.org/wiki/Sustainabilitv
15. **Wilson E.O., 1988,** The current state of biological diversity. In *Biodiversity.* Eds. Wilson E.O., Peter FM, National Academy Press. Washingtonn D.C.
16. **Diamond J.M., 1989,** Overview of recent extinctions. In *Conservation for Ihe Twenty First Century.* Eds. Western D., Pearls M. Oxford University Press, N.Y.
17. **Kenton R. Miller, et ai, 1991,** Issue on the preservation of biological diversity. In *Global Possible. Resource, Development and the New' Century.* Aff East-West Press, New Delhi.
18. **Jenkins M., 1992,** Ex-situ conservation of animals In *"Global Bio diversity: status of Earth's living resollrces.* pp 563 - 575 cd Groombridge B., Chapman and Hall London.
19. **Olney P.J.S. , P. Ellis 1992,** *International Zoo Yearboak. No.31,* Zoological Society Lonon.
20. **Raven P.H, 1981,** Research in Botanical Gardens. *Bot. Jahrb.,* 102: 52.
21. **See Wilson 1988.**
22. **MC Neeley J.A.,** *et al,* **1990,** *Conserving World's Biological Diversity.* Gland, Switzerland.
23. **Todd .John, 1988,** Restoring Diversity: The search for social and economic context. In *"Biodiversity".* Eds. Wilson E.O., Peter F.M., National Academy Press, Washington D.C.
24. **Brown L.R., Edward C.Wolf, 1987,** Charting a Sustainable Course. In *"Slate of the World* 1987: *A Worldwatch Institllte Report on Progress towards a Sustainable Society"* Ed. Linda Starke. W. W, Norton & CO. N. Y.
25. **Census of India, 2011.** Census Directorate, Home Ministry, Government of india, New Delhi.
26. **Ellis D, 2004.** "What If. The Consequences of Innovation." *Hospital and Health Networks* 78(8) 38-42.
27. **Davey Smith G.S Ebrahim, S Lewis, AL Hansell, L.J Palmer and PR Burton, 2005,** "Genetic Epidemiology and Public Health Hope, type. and Future Prospects." *Lancet* 366: 1484-98.

28. ENVIRONMENT AND LAW: EVOLUTION OF THE LEGAL FRAMEWORK

1. **Caldwell L. K., 1991**, *International Environmental Policy: Emergence and Dimensions*. 2nd Edn. Affiliated East West Press Pvt. Ltd. New Delhi.
2. **Boardman Robert,1981**, *International Organization and Conservation of Nature*. Indiana University Press, Bloomington
3. **Dasmann Raymond F, 1968,** Conservation and Rational Uses of Environment. *Nature and Resources* **4**:2-5, (June 1968)
4. **Nigel Hawkes, 1972,** Human Environment Conference: Search for a modus Vivendi. *Science* 175:736-38. (Feb 1972).
5. **Johnson Brian, 1973,** The Settlement of Stockholm. *Ecologist,* **3**:87-88. (March 1973)
6. **John Baylis, Steve Smith. 2005.** *The Globalization of World Politics* , 3rd Edn.. Oxford University Press Oxford.
7. **John McCormick,** 1995, *The Global Environmental Movement.* John Wiley, London, 1995
8. **Earth Summit:** http://www.un.org/genimfo/bp/enviro.html.
9. **Agenda-21:** http://habitat.igc.org/agenda21/rio-dec.html.
10. Convention on Biodiversity: http://www.cbd.int
11. Kyoto Protocol: http://en.wikipedia.org/wiki/Kyoto Protocol. Retrieved March 2008
12. **Wigley, Tom (1998).** "The Kyoto Protocol: CO2, CH4, and climate implications". *Geophys. Res. Lett.* 25 (13): 2285.
13. **Barker T. and Ekins, P. 2004.** The costs of Kyoto for the US economy. The Energy Journal, volume 25,
14. **De Leo, Giulio A.; Luca Rizzi, Andrea Caizzi and Marino Gatto (2001).** "Carbon emissions: The economic benefits of the Kyoto Protocol". *Nature* 413: 478–479..
15. **Houghton, J.H. 2004.** *Global warming: the complete briefing*. Third edition. Cambridge University Press, Cambridge, UK.
16. **Sibley, Adam et al, 2007,** "World Summit on Sustainable Development (WSSD), Johannesburg, South Africa." In: *Encyclopedia of Earth.* Ed. Cutler J. Cleveland (Washington, D.C.: Environmental Information Coalition, National Council for Science and the Environment).
17. **Najam, A., Poling, J.M., Yamagishi, N., Straub, D.G., Sarno, J., DeRitter, S.M. and Kim, E.M.: 2002,** 'From Rio to Johannesburg: Progress and prospects'. *Environment*, **44**(7):26–38.
18. **Forty second amendment to Indian Constituion:** http://www.constitution.org/cons/india/tamnd42.htm
19. **Ratlam Case 1980,** Municipal Council, Ratlam vs Vardhichand, AIR, 1980 SC, 1622.
20. **Ganga Pollution (Tannaries) Case,** 1988, MC Mehtaa vs Union of India, AIR, 1988,SC, 1037.
21. **The Water (Prevention and Control of Pollution) Act 1981**: http://www.vakilno1.com/bareacts/waterprevcontact/waterprevact.htm
22. **The air (Prevention and Control of Pollution) Act 1974:** http://www.envfor.nic.in/legis/air/air1.html
23. **Environment (Protection) Act 1986:** http://www.envfor.nic.in/legis/env/env1.html
24. **The Wild Life (Protection) Act 1972:** http://www.envfor.nic.in/legis/wildlife/wildlife1.html

25. **The Biological Diversity Act, 2002:** http://www.envfor.nic.in/divisions/biodiv/act/bio_div_act.htm

26. **The Indian Forest Act, 1927**: http://en.wikipedia.org/wiki/Indian_Forest_Act,_1927

27. **The Forest (Conservation) Act, 1980:** http://www.envfor.nic.in/legis/forest/forest2.html

28. **The National Forest Policy, 1988:** http://www.envfor.nic.in/divisions/fp/nfp.pdf

29. **National Environment Tribunal Act 1995:** http://www.envfor.nic.in/legis/others/tribunal.html

30. **The National Environment Appellate Authority Act, 1997:** http://www.envfor.nic.in/legis/others/envapp97.html

GLOSSARY

Activated sludge: Slurry or sediments from bottom of an oxidation pond containing micro-organisms to be used for aerobic decomposition of sewage effluents.

Adaptation: Structural and/or functional characteristics developed in response to environmental factors which help in increasing the growth and reproductive performance of an organism.

Aerosol: Fine solid (smoke) or liquid (mist) particles usually of colloidal dimensions floating in air.

Albedo: The proportion of solar radiations reflected back to the atmosphere.

Albumenuria: Presence of albumen in urine.

Allelopathy: Inhibitory activity of one plant species commonly through exudation of chemicals from their roots on seed germination or growth of other associated species. Also known as allelochemic interaction and antibiosis.

Allergy: Hypersensitivity to certain organisms or chemicals.

Ambient: The outside air.

Anemia : Lack of blood, especially of red blood corpuscles or haemoglobin.

Anesthesia : Insensitibility to pain, general or local loss of feeling.

Anorexia: Want of appetite.

Anoxia: Deficient supply of oxygen to tissues.

Apnea: Difficulty or cessation in breathing.

Aquifer: Under-ground natural storage, sometimes flowing out through artesan wells due to sufficient pressure on the stored water.

Asphyxia: Stoppage of pulse, suspended animation owing to any cause interfering with respiration.

Atmosphere: Gaseous envelop around earth, the air. It has many gases of which N2, O2, CO2 and water vapour are the most important. It exerts a pressure of 14.7 pounds per sq. inch or in terms of column of mercury equivalent to 760 mm.

Atomic wastes: Radioactive material discarded as waste after use. As the radioactivity persists for a long time, these wastes are serious health hazards.

Automobile Exhausts: Gases emitted by automobiles on the burning of petrol and diesel oils. It contains CO, unburnt hydrocarbons, oxides of nitrogen and soot. A serious air pollutant in urban localities.

Background pollution: Level of pollution not from a point source.

Benthos: Organisms living in the bottom of water bodies.

Bio-assay: Use of some organisms to know the biological effect of a substance or factor.

Bio-Chemical Oxygen Demand (BOD): The amount of oxygen required by micro-organisms to decompose organic matter present in water kept at 25°C for five days in an incubator (referred to as BOD5).

Biocides: Collective name for different substances used to kill undesirable organisms such as pesticides, herbicides, bactericides, fungicides etc.

Bio-degradation: Breakdown of organic matter such as the litter or sewage by biological means, *i.e.,* by micro-organisms.

Bio-geo-chemical cycles: Circulation of various chemicals between the living (organisms) and the non-living components of an ecosystem, such as air, soil or water.

Biological magnification: Successive increase in concentration of some chemicals through different food level organisms such as of DDT from water to algae to herbivorous fish to carnivorous fish to fish eating birds and to man.

Biomes: A large unit of climax communities usually representing well defined climatic zones.

Biosphere: All parts of earth where organisms live.

Bronchitis: Inflammation of lining of bronchial tubes.

Calorie: A unit of heat, one calorie is the amount of heat required to raise the temperature of water from 14.5° to 15.5°C. One calorie is equal to 4.2 Joules.

Carcinogen: Agents which induce cancers.

Carcinogenesis: Development and spread of cancer inside the body.

Carcinoma: Disorderly growth of cells which invade adjacent tissues and may spread by lymphatic ducts or blood vessels to other parts of the body.

Cement: A fine powder manufactured from limestone for binding building material. It mainly consists of SiO2, Al2O3, CaO, Fe2O3, MgO, NaO etc. Cement dust is a major particulate pollutant of atmosphere.

CEP: Commission on Environmental Planning (A body of IUCN, Switzerland).

Cesium^{137}: A radio-active isotope released in the atmosphere during atomic explosion. A serious hazard to mankind.

DDT: Dichloro-diphenyl-trichloro-ethane (C_{14} H_9 Cl_{15}), a synthetic organic compound which is a contact poison. Once widely used as insecticide after the World War II, it's use has been banned in a number of countries due to its persistence and accumulation in biological systems including man.

Decibel (dB): A unit of sound intensity.

Demography: Study of populations, particularly human population.

Demographic transition: Change over from slow growth phase, through rapid growth to a phase of stable population growth in which low birth rates equal the death rates.

Decomposers : Micro-organisms, such as bacteria and fungi which break down the complex organic matter into simple compounds.

Dermatitis : Inflammation of skin.

Desertification : The process of desert formation, permanent loss of plant life in a region, loss of capacity to sustain good plant growth.

Detergents : Soap powders and cakes used to wash clothes and utensils etc. They are surface active agents, some of which easily decompose after use and do not harm life but may cause nutrient enrichment of water bodies and eutrophication. However, some of the detergents are non-biodegradable and take a long time to get decomposed and form froth and foam on the surface of water body.

Detritus : The debris of decaying organic matter.

Diarrhoea : Persistent purging or loosening of bowels.

Dichloro-difluoro-methane (Freon-12, a CFC): A colourless, odourless and tasteless gas used in refrigeration. It gradually reaches the upper layers of atmosphere (stratosphere) and damages the ozone layer.

Dieldrin : An insecticide which persists in soil for long time and enters food chain. So its use has been banned in many countries.

Diuretics : Agent causing copious discharge of urine.

Dry Adiabetic Lapoe Rate (DALR): As we move up atmospheric pressure decreases. This causes gases to expand and cool down adiabetically. The decline in temperature of a dry gas as we move up is known as Dry Adiabetic Lapoe Rate. It is a theoretical value and is about **I°C for** every 100 meters of height covered.

Earthwatch: An association under UNEP to monitor pollution of the Earth.

Eco-development: Development taking prior consideration of the holistic impacts of the concerned system.

Ecosystem: The structural and functional entity of biotic communities and their environment. It may be of any size.

Effluent: Liquid discharges or flowing out material usually from industries that often pollute the environment.

Electrostatic precipitator: Device to collect dust from factory chimney smoke to reduce air pollution by imparting electrical charge to dust particles.

Emetic agents: Agents causing vomitting.

Emission Standards: The maximum permissible or safe limit of discharge of pollutants.

Emphysema: An unnatural distention of a body part with air.

Endemism: Restricted area of distribution of a species to an isolated region.

Endrin: Hexachloro-epoxy-octahydro-dimethano-naphthlene — an isomer of Dieldrins. It is persistent chlorinated hydrocarbon, highly poisonous to rodents. It may reach man through food chain.

Environmental Lapse Rate (ELR): Actual decline in temperature per 100 meters of height covered is known as Environmental Lapse Rate. ELR is largely determined by meteorological conditions prevailing in the locality.

EPA: Environmental protection agency of USA. A very effective research and policy planning body for the protection of U.S. Environment, natural resources and pollution control.

Erosion: Surface destruction and removal of soil, mainly by wind and water.

Erythema: Morbid reddening of skin.

Escherichia coli: A rod shaped bacterium found in the intestines of man and other warm blooded animals. It plays an important role in the breakdown of cellulose. Its number per ml determines the safety limits of drinking, bathing and domestic-use water.

Eutrophication: The process of nutrient enrichment of water bodies. This leads to rapid growth of phytoplanktons, algae and macrophytes followed by more organic matter accumulation, and silting up of the bed of water body. Surface run-offs, demestic sewage, detergent bearing waters accelerate the process.

Fall out: The descending radio-active fission products from atmosphere to earth's surface.

FAO: Food and Agricultural Organization of the UNO.

Fault: Deep cracks in earth's surface, factures in rock layers etc. due to geological events.

Fecundity: The rate of offspring production at individual level.

Fibrosis: Morbid growth of fibrous tissues.

Fission: Splitting of atom into lighter fragments and release of energy and neutrons.

Flow meter: Instrument used to measure the quantity of water or effluents flowing out of a system over a unit time.

Flu gas: The mixture of gases emitted from the chimney due to combustion.

Food chain: The sequence of different trophic level organisms starting from primary producers to top carnivores.

Fossil fuels: Fuels derived from fossil plants and animals such as coal, petroleum and natural gas.

Fresh water: Drinking water with less than 0.2% salt content.

Fusion: Combining of nuclear or other things; Combination of atoms such as of hydrogen forming helium and associated with release of enormous amounts of energy.

Gastro-enteritis: Inflammation affecting both stomach and bowels.

Germ-plasm: Any living plant or animal part from which new offsprings may be obtained through tissue culture techniques.

Green house effect: The heating effect through re-radiation. In glass house the sunlight enters, but on re-radiation the heat is trapped inside thus increasing the inside temperature. Similar effect is caused by atmosphere, dust, water vapour, ozone, carbon dioxide etc. which allow the sun light to reach the ground surface but trap the re-radiating heat waves.

Haematuria: Presence of blood in urine.

Half-life: The time taken by radio-active isotopes (or pesticides) to come down to its half strength or decay to its half level from the initial.

Heavy metals: Metals like Fe, Cd, Cu, Pb, Hg, etc. with relative density above 7. Usually toxic in high doses.

Herpes simplex: Infection of either skim or genital organs caused by either of the two strains of Herpes Simplex Virus (HSV). Typical infection of HSV-I is characterized by clusters of small blisters or watery vesicles on the skin or mucous membrane accompanied by itching or burning sensation. HSV-II is associated with genital infections and is characterized by pain or itching sensation following by appearance of clusters of blisters on sex organs which may burst and accompanied by painful burning sensation.

Herpes Zoster: Acute viral infection affecting skin and nerves characterized by group of small blisters appearing along certain nerves accompanied by itching, burning sensation and pain.

Homeostasis: The capacity of ecosystem to resist changes due to disturbances or to return to balanced state.

Humification: Formation of humus from dead organic matter by microbial decomposition.

Hydrology: The study of water storage, distribution and movement on earth's surface.

Hyperglycaemia: Presence of excess of sugar in blood.

Hyperplasia: Over-growth of a part due to excessive multiplication of its cells.

Hypoglycaemia: Deficiency of sugar in blood.

IAP: Index of atmospheric purity.

ICSU: International Council of Scientific Unions : an organization which co-ordinates scientific activities in all sciences.

Inversion: A temporary atmospheric phenomenon in which temperature does not decrease with altitude. It leads to smog formation and heavyness in ground layer.

Ionosphere: The uppermost layer of atmosphere above the stratosphere where ionization takes place. In day time its lower limit is 56 kms and in night 96 kms above earth's surface. It reflects radio-signals.

Isotope: Forms of an element with same number of protons but different number of neutrons. They differ in radio-active behaviour and atomic weights but have the same atomic number and chemical properties.

Itai-itai: A disease of bone atrophy caused by cadmium sulphate in water. It occurred on the banks of Jintsu river in Japan near an abondoned zinc mine.

IUCN: International Union for the Conservation of Nature and Natural Resources with headquarters at Gland, Switzerland. It was founded in 1948 and has published Red Data Book on rare and threatened plant and animal species.

Joule: A unit of energy, 4.2 Joules of which are equal to a calorie.

Kaposi Sarcoma: A rare usually lethal cancer of tissues beneath the surface of skin or mucous membrane characterized by red-purple or blue-brown lesions of skin, mucous membrane or other organs. Common in persons afflicted with AIDS.

Lassitude: Faintness, weariness or weakness.

LC50: Lethal concentration of a chemical that will kill 50% of the target animals.

LD50: Lethal dose of chemical that will kill 50% of the target animals in a natural population.

Leucocythaemia: Large excess of white blood corpuscles in blood associated with changes in lymphatic system and enlargement of spleen.

Leucocytopenea or leucopenia: Deficiency of white blood corpuscles in blood.

Limnology: Study of fresh-water for its various physico-chemical and biological characteristics.

Lithosphere: The solid rocky crust of the Earth.

Littoral: Shallow zone of a water body; also on sea coasts the zone receiving low to high tides.

London smog: An air pollution episode with predominance of reducing material like SO2, smog and soot. It usually occurred in London in morning due to heavy smoke at low temperature and humid atmosphere during fifties and sixties. Pollution control measures have reduced it to a minimum in the city.

Los-Angeles smog: A photochemical smog with predominantly oxidizing materials like ozone and nitrogen oxides. It occurs in clean sky during the noon at high temperature and low humidity.

Lymphocytes: Cells responsible for immunologic reactions to invading foreign cells and other xenobiotics. B-lymphocytes cause development of antibodies, whle T-lymphocytes participate in cell mediated immune response.

Lymphoma: Any of the group of malignant diseases of lymphatic system usually starting in the lymph nodes or in lympoid tissues of other organs.

Minamata tragedy: A disease causing paralysis, loss of hearing and eye-sight, mental disability and rapid death due to methyl mercury pollution. It occurred in Minamata bay in Japan. Mercury reached human body through food chain, from affected waters to plants, then to fishes and finally to man.

Mulch: The organic debris left on soil surface to retain moisture, enrich the soil and avoid weeds.

Mydriatic agent: Agents causing Morbid dilatation of the pupil of eyes.

National Park: A well demarcated and protected area of natural landscape.

Natural Reserve: A well protected landscape to preserve its natural original flora and fauna.

Neuritis: Inflammation of nerves.

Niche: The location and function of an organism in the context of its ecosystem.

Oedema: Pathological accumulation of fluid in tissue spaces, an unhealthy mass of swollen cells.

Oil spills: Large scale oil leakage on land or sea surface affecting the plant and animal life. Millions of tons of oil are spilled in oceans annually which harm sea algae, fishes and birds.

Oropharyngeal candidiasis: An infectious disease caused by yeast like fungus *Candida albicans* which is characterized by white patches of inflamed tissues resembling milk curd. on tongue, in mouth and pharynx.

Over-burden: The deposits of surface soil or other material overlying useful mineral deposits like coal or other ores which have to be removed for open cast mining.

PAN: Peroxyacetylnitrate a secondary pollutant in smog formed by the action of light on hydrocarbons and nitrogen oxides in air.

Paraplasia: Paralysis of lower parts of the body.

PBN: Peroxybenzonylnitrate a secondary pollutant in smog formed like PAN.

Pelagic: The marine communities free from bottom or shore effects.

Permafrost: Permanently frozen zone of the ground.

Pesticides: Chemicals used to kill pests such as weedicides for killing weeds, fungicides for fungi, herbicides for herbs, insecticides for insects etc. Permanent pesticides continue to harm plants and animals for long durations after their application.

***p*H:** Negative logarithm of hydrogen ion concentration ranging from 0 to 14 with 7 as neutral, less than 7 as acidic and more than 7 as alkaline.

Pheromones: A kind of chemicals emitted by animals in very minute quantity which transmits information to other individuals of the same animal species in the vicinity.

Plume: The visible smoke coming out of a chimney.

Pneumonitis: Inflammation of lungs.

Pollution: Man induced change in the natural environmental which is harmful to man, other organisms and damages natural resources.

Polycythaemia: Excess of Red blood corpuscles in blood.

Potable water: Clean water suitable for drinking and cooking.

Prophylaxis: Preventive treatment.

Putrefaction: Microbial anaerobic decomposition of proteins.

PVC: Polyvinyl chloride, a kind of plastic containing carbon, hydrogen and chlorine. It decomposes slowly and on burning produces Hydrochloric acid.

Pyrexia: Fever.

Radio-active pollution: Pollution due to discharge of isotopes emitting radio-activity.

Radio-carbon dating: Determining the age of materials by measuring the proportion of C14 in its total carbon content. The method is useful upto 30,000 yr old specimens.

Recalcitrant: Complex substances not easily degraded by microorganisms.

Remote sensing: Obtaining information about any aspect of ecosystem from far away recording device such as aerial photography, lasers, ultraviolet and infra-red detectors, radars etc.

Resource: Materials useful to man and to other organisms.

Rhizosphere: Zone of soil around the roots and is influenced by the root exudates.

Riparian: Related to rivers and their banks.

Salmonella: A pathogenic bacteria causing gastro-enteritis and other ailments like typhoid (*S.typhi*).

Smog: Visible gaseous and particulate matter concentration near the earth's crust particularly in cities due to SO2, CO2 emission by factories and automobiles under conditions of atmospheric inversion.

Smoke: Ash, soot etc. in air discharged as a result of combustion of organic matter.

Sonic boom: The booming sound and shock waves produced by jet plane when they cross the sound barrier.

Stratosphere: A zone of atmosphere about 30 kms in thickness above the troposphere (see troposphere below). Temperature rises with altitude in this zone which contains ozone shield to protect terrestrial life from biocidal radiations.

Surfactants: The chemicals producing lather in detergents, these are phosphate rich and cause eutrophication.

Synergistic effects: Enhanced effects of two substances — more than the total effect exerted by both the chemicals in isolation.

Tailings: Waste material left after the concentration of ores for extraction of metals. These are usually rich in toxic heavy metals.

Thermal Pollution: Heated discharges from several industries power plants and coolants. These cause harmful effects on aquatic bodies in which these waters are discharged.

Threshold Limit Value (TLV): Maximum permissible concentration of environmental pollutant or toxicant which does not cause any perceptible damage to the plant and animal life.

TOC: Total organic content (in water) expressed as mg of carbon/litre.

Trophic: Related food and feeding habits.

Troposphere: The zone of atmosphere containing clouds and moisture extending upto 15-18 kms in height from the ground surface. Temperature decreases in this zone with altitude.

UNDP: United Nations Development Programme.

UNEP: United Nations Environment Programme — an organization of UNO with its Secretariate at Nairobi (Kenya)

UNESCO: United Naticn's Educational Scientific and Cultural Organization — an organ of UNO located in Paris, France.

UNO: United Nations Organization with headquarters in New York, USA.

Water Pollution: Addition of impurities in water that is harmful or toxic to man and other organisms.

Water-shed: The area drained by a stream or a river.

WHO: World Health Organization — an organ of UNO located in Geneva.

WMO: World Meteorological Organization of UNO

WWF: World Wild life Funds (of IUCN) located at Gland, Switzerland.

ZPG: Zero population growth, *i.e.*, each couple producing only two children so as to stabilize the population at the existing level.

INDEX

A

B

C

D

E

F

G

H

O

P

U

V

W

Z